# Monomers, Oligomers, Polymers, Composites and Nanocomposites Research: Synthesis, Properties and Applications

# MONOMERS, OLIGOMERS, POLYMERS, COMPOSITES AND NANOCOMPOSITES RESEARCH: SYNTHESIS, PROPERTIES AND APPLICATIONS

RICHARD A. PETHRICK,
G.E. ZAIKOV
AND
J. PIELICHOWSKI
EDITORS

**Nova Science Publishers, Inc.**
*New York*

**LIBRARY OF CONGRESS CATALOGING-IN-PUBLICATION DATA**

Monomers, oligomers, polymers, composites and nanocomposites research: synthesis, properties and applications / Richard A. Pethrick , G.E. Zaikov, and J. Pielichowski (editors).
p. cm.
ISBN 978-1-60456-877-6 (hardcover)
1. Polymers. 2. Monomers. 3. Polymeric composites. 4. Nanostructured materials. I. Pethrick, R. A. (Richard Arthur), 1942- II. Zaikov, Gennadii Efremovich. III. Pielichowski, Jan.
QD381.M636 2009
547'.7--dc22

2008042826

*Published by Nova Science Publishers, Inc. ✦ New York*

# CONTENTS

# PREFACE

"In a country that is ruled well, it is a shame to be poor, but in a country that is lead poorly, it is a shame to be wealthy."
Confucius - circa 557-479 B.C.

"The future of the world is in the hands of teachers"
Victor Hugo – 19th century

Knowledge – the hallmark of a flourishing country and mankind. "In the olde world, the wealthiest country was the one who had the wealthiest land, while in today's modern world, the wealthiest country is the one who has the most diverse population. " (Prof. G. Bokle, England) Knowledge is the foundation for mankind's most successful ventures. Today there is a visible shift in science, where people have shifted from chemistry to biology, and from biology to medicine. Everybody wishes to be wealthy and healthy. Jonathan Swift once said, "Everybody wants to live longer but no body wants to be old."

Still, the role of polymer chemistry (pure and applied sciences) is very prominent in the world of science today, but it is heading away from polymers and polymer blends towards composites and nanocomposites. It allows for the creation of new materials with unique properties and new possibilities.

If we measure the world's production of materials by volume and not by weight, then we find that it equals the production of iron, cast, steel, and colored metals together. This volume equals 250-270 million cubic metric of material per year. It is important to note that the rate of production of polymers is overcoming the rate of production of metals by 30-40%. Mankind used to live in the Stone Age, then the Iron Age, then the Bronze Age, and now it has come to be the Age of Polymers (leaning heavily towards composites and nanocomposites.)

There are reviews and essays, according to the opinions of editors, that are helping to further develop polymer science and assist in solving practical applications (new materials with improved properties).

Finally we would like to tell you a little joke with serious conclusion.

One American gentleman visited Poland in time of Polish People's Republic (1980th) and he saw that everywhere it is located two Flags: Polish and Soviet.

American asked one Polish gentleman: “Are you friends with Soviet Union peoples or brothers?”

Polish replied: “We are definitely brothers because we can choose friends!”

So, the contributors of our Handbook are not brothers and not friends of editors of this volume. We selected this people on the base of good scientific results of these scientists.

**Richard A. Pethrick**
University of Strathclyde,
Glasgow, Scotland, UK

**Gennady E. Zaikov**
N.M. Emanuel Institute of Biochemical Physics,
Russian Academy of Sciences
Moscow, Russia

**Jan Pielichowski**
Todeusz Kosciuszko Cracow University of Technology
Cracow, Poland

In: Monomers, Oligomers, Polymers, Composites… ISBN: 978-1-60456-877-6
Editors: R. A. Pethrick, G.E. Zaikov et al. 

*Chapter 1*

# DOING BUSINESS IN CHINA: FROM THEORY TO PRACTICE

***Antonio Ballada***

## FOREWORD

### Introduction

The purpose of this handbook is to try to identify the differences in behavior between Western and Chinese people, with its impact in the business life and to analyze the various steps towards the realization of a typical industrial project in China, to put in evidence what is peculiar and which difficulties are specific in the Chinese environment.

I do not expect reporting anything which cannot be found in the many books published on this very fashionable and trendy subject, the difference is that I will report only facts based on my personal experience.

So let me start confirming that it is right: we are different. According to recent anthropologists' studies, we are 40 thousand years far each other.

The human stream who reached China left the African/European area 40 thousand years ago, which means during the Medium Paleolithic period, 15 thousand years before the paintings in the caves of Lascaux in France or Altamira in Spain.

By the way, we can immediately distinguish a Caucasian and an Asian due to somatic differences and we have to expect that same differences impact the mental processes, the behavior and the body language.

These differences can be occasion of a lot of fun and wonderful relationship or could be the premise of misunderstandings and failure: that is just up to us.

I remember that in Latin the words "guest" and "enemy" share the same root: "hostis" or "hospes" (in English we still have "hostile" and "guest"). Let me recommend that the balance between the "guest" and "enemy" approach has to be carefully watched all the times. Nevertheless my experience confirms that a positive approach is extremely helpful.

So let start dealing with the business issues with positive mindset.

Professor G. Zaikov
Russian Academy of Sciences Moskow

# PART ONE: INTERPERSONAL RELATIONS

## 1. Working with a Chinese Team

A Westerner parachuted in China to take managing responsibilities normally is not prepared to face the cultural shock. Often he would think that to deal with Chinese subordinates or colleagues is only matter of:

- Language (we mean English)
- Education (we mean them to learn our culture)

Unfortunately this would not be fair and for sure is not enough.

As far as languages, I do not mean that we have to learn Mandarin to be able to do business in that language, this is a full-time job; nevertheless I personally found extremely useful the time I spent to realize the structural differences between Mandarin and our Western languages. Mandarin lacks agreement in gender, number and case, lacks declension and conjugation and only cares to establish the sequence of the events more then their exact position in the time continuous. Syntax, on its turn, has to compensate the lack of grammar tools and, as a consequence, obliges to strict rules in the organization of the sentence. The use of ideograms influenced the language, which is still based on syllables: one syllable, one word, one concept, one ideogram. Another interesting effect of the Chinese ideogram writing is that a middle culture Chinese can read what was written centuries in the past: Chinese people live in contact with their history and this for sure has an impact on their "being Chinese" feelings. I could conclude that the study of Mandarin, even if limited to the basics, is a useful and also pleasant way to try to penetrate the mental processes, on which the behavior are normally based, of your Chinese counterpart.

As far as the behavior it is essential to pay attention to the different ways of communication. Semeiologists say that only 30% of men communication flows through the oral language, the rest goes through other means like the body language for example: we have to realize that all those means show differences in use and meaning as compared to western ones.

By the way I anticipated talking not theory but reporting personal experiences and then let me give some of this.

When I was president of a company in Taiwan I realized that anything you ask to a Chinese subordinate the answer is always **"yes"**. Then you start waiting for the follow-up, which will not come. From this behavior originated the Western legend that Chinese are not reliable or are lazy.

The matter is that if the Chinese colleague or subordinate does not understand you, he will never admit it in order not to loose his face. Also worse, if you ask something stupid or impossible, he will never tell you in order not to make you loose your face. I will develop later in this manual the concept of "face" so specific in Chinese culture.

Anyway after some time in Taiwan I learned how to manage.

First of all you never have to take as a given that your counterpart understands English, even if you know that he studied it and so is reported in his CV. This is true even if your counterpart can speak good English. English pronunciation is tricky: when I first arrived to the USA from Italy to work I was able to talk good English but I did not understand in the same way. I had to pretend not to be able to talk to obtain from the American counterpart the use of simple words and a slow talking. For Chinese this situation is also worse and I learned how to politely ask people, before they leave my room, to repeat the conclusions of our meetings and to anticipate to me the means that they were planning to use to perform the task.

If you stay there enough time you will be also able sooner or later to understand the body language and to tell the difference between the various kinds of "yes" some of which simply mean "no way" or "forget it".

### *Anecdote 1: Moving the Headquarter*

After few month from my arrival in Taipei as C.E.O. of XXX, I realized that would have been possible to move the headquarter from Taipei downtown to the same premises where the plant was located in the industrial zone of Kaohsiung, the second town in Taiwan. No technical reason, no Unions interference (unemployment in Taipei was and still is about 4 to 5% would have prevented me from implementing the idea, enjoy nice savings and deserve good bonus. Moving the few key people, even if expensive, still would have been convenient considering the very high cost of the Taipei location. I could not understand why my predecessor didn't think to it but I was too proud to ask. So I called the Vice President in charge of the Human Resources and I asked him to develop the project and to take care of the details.

Of cause the answer was: yes.

After some weeks I asked the Vice President for the status of the project and the answer was: we are working. After some additional weeks I called a meeting to complain for delay in taking action on the project. Fortunately the good personal relationship with the person in charge allowed him to be open with me and this is what I learned:

- In China the family concept is extremely important. The old people are always taken care by the sons' families and they almost always live together with them. In addition to the obvious difficulty of moving the old people from their environment, there was the problem of finding a job in the new location for the various members of the extended family, so typical in China. As a consequence to move the headquarter would have meant to loose all the key people, who, being senior, are normally in charge of large families.
- As far as the other employees, they would have not been damaged in their income, considering the low level of unemployment, but they would have felt betrayed by the Company, showing so poor interest for their experience to renounce to it only for money. This would have had a negative impact on the Company image and on the motivation of everybody. All this comes from a very typical company/employee relationship in China, which will be clear later in this manual when dealing about Confucianism and organization.
- The Chinese shareholder, a local tycoon owning a significant percentage of the company, was informed about my purpose and recommended to discourage the project. In Taiwan for a big company like ours is a must to have the headquarter in the Capital for connections and prestige reasons: moving the headquarter would have damaged, may be, the stock value.

I got enough to give up and so I did.

This example of real life allows various considerations on various aspects of the Chinese mentality, but one thing hit me immediately which is that nobody was ready to object to my request about moving the headquarter: no the Vice President in charge nor the big shareholder. On the contrary, during one of the periodical personal meetings with this gentleman, an old person enjoying great prestige in his Country, I was encouraged, with benevolence, to pursue all my ideas.

Now it is clear to me: I should have understood by myself or I had to get tired and give up. Meanwhile they would have judged me.

As far as *education,* I could experience that many of those confusing differences in behavior clarify a lot if we really try to understand the Chinese culture. And Chinese culture has its roots in Confucianism. At this point, even if it is not a subject to be discussed in a manual, moreover written by a business person, nevertheless I need to try a definition of Confucianism to enlighten my experiences and my interpretations.

I could realize that Confucianism is a high profile philosophy aimed, like other high profile philosophies, to overcome the individual interests in favor of the community and I think that one of the tool Confucianism adopted to pursue this goal is the concept of Organization.

I would say that the Organization plays for a Chinese the same role that Charity or Solidarity play for a Western Christian person and, according to this conclusion, I will use since now the capital initial when mentioning organization with such meaning.

This is the reason why Chinese, and Japanese people as well, are extremely loyal to the organization they belong to, which includes their Companies and their friends.

This is the reason of the strong importance of the personal relationship, the famous *"guan xi"*, for all Chinese.

But I am afraid that for the same reason they do not feel obligated to anybody who is not part of their environment or their *"guan xi"*. I will add some words later on the *"guan xi"* concept.

All above has a big impact in the business life. Consider for example the value that a Westerner and a Chinese recognize to the same contract or any other formal or informal agreement. Another Western legend flourished about that subject, always leading to the conclusion that Chinese are unreliable and not ready to comply with signed agreements.

Actually what counts for a Chinese is the personal relationship: if you enjoy good personal relationship with your counterpart you do not really need a formal contract, in fact an MOU would be enough. If you do not have any personal relationship, no formal contract will be able to protect you. Such behavior, apparently so strange, clarifies if you consider that for a Chinese person any agreement with a Westerner concerns the outside of his Organization, his entourage and at the end his Country. As a consequence any agreement should, and must, be disregarded at the minimum suspect of conflict with the Organization in all its meanings. Let me spend some words on this particular subject.

The concept can be summarized in this scheme, only apparently ironical:

| Personal relationship in place | No personal relationship |
|---|---|
| *you do not need a formal contract* | *no contract will protect you* |

NOTE: you will not know when and if you have a personal relationship or not, particularly if you are a foreigner.

Another personal experience could help to understand the big difference in value of the same contract, or even clause, for a Chinese and a Westerner.

I was negotiating in Beijing a contract to realize a petrochemical plant. As the technology was also involved, the usual secrecy agreement clause had to be drafted.

It took me six months to reach an agreement on the confidentiality clause. I was very worried as, at this rate, considering the complexity of the project, it would have taken years to conclude. Later I understood how important was that clause for the Chinese mentality. The reason is that such clause would have created barriers and constraints in the flow of information in their system of relationship, in their *"guan xi"*, in their Organization, and then had to be carefully considered. Eventually the clause was signed but are we sure about its solidity in case of request from some Authority, may be connected with some of your competitors? Certainly not.

To keep all above into account, I recommend to be very cautious in providing sensitive information and to let them go only step by step accordingly with the progressing of the identification of your interests with those of your Chinese counterpart.

To conclude on this subject about Confucianism, and Organization as its tool, I would list some typical aspects of the Chinese culture which are directly consequence of the importance of the concept of Organization.

Let say that we can expect from Chinese people the good and the bed things of a culture based on organization:

Sense of duty
Reliability
Respect for hierarchy
Respect for old people

But also expect:

Total disregard for whom is perceived as not being part of their world, which means all the Foreigners, the non Chinese
Tendency to bureaucracy
Tendency to gerontocracy
Communication system heavily hierarchical

About this last point, I would comment that also in Europe the communication *top-down or bottom-up* was and still is an issue. In the 90' in Boston a *guru of* business consulting, Mike Hammer, faced that problem analyzing the organization structures in place in that moment, matrix organizations included. From his work the all theory on *Reengineering* was developed and seriously taken into consideration all over the Western business world. I did not find any traces of this studies in the Chinese environment and I deduct that this is due to the fact that a discussion on *horizontal communication* would be against their cultural roots.

***Anecdote 2: Introducing the "Management Committee"***

After some while in my position of CEO, and after having absorbed the first hit from the new culture, I started getting out from my office segregation with the idea to apply some of the rules that a good boss is supposed to apply everywhere in the world. One of this rule is to call regular meetings with your first line people to discuss the most important issues and reach shared decisions. After some of those meetings I realized that there was something wrong. Nobody during the meeting ever gave any suggestion; nobody ever challenged my opinions or decisions. After some efforts with my friends I understood that for a Chinese person to criticize the authority is a bad behavior and, also worse, to show up in public and in presence of the boss with proposals is considered an act of unforgivable arrogance. Was surprising to me to learn that in Chinese schools never and never a student would raise his hand, like in our countries, to offer an explanation or to ask for a question. By the way the "Management Committee" issue was easily resolved as I started asking each person his opinion before revealing mine. After some while things looked like normal. Except one thing which now makes me laugh, and also makes me feel some nostalgia but in those moments made me mad. One of the vice presidents, when unhappy for something or in disagreement with me or with some colleagues, was used to get up and leave the room for a while with some pretext. Later I learned that it was too difficult for him to face any conflict and that I had to fix the issue through person to person meetings.

## 2. Rewarding a Chinese Employee

Just a few words on how to reward a Chinese employee. The compensation in China, more than in Western countries, has the purpose of motivating people and of keeping the good ones, preventing them to go to your competitors as soon as they are trained in your shop.

What I report on this subject is obviously consistent with the insight on their culture, which I developed so far.

The tools available are the same as everywhere else but the meaning for Chinese employees is different. As a consequence is different the relative importance and effectiveness of the same tools and the mix of them which is convenient to apply case by case.

Let me comment the most common compensation means and their Chinese *translation*.

- Salaries and social success has to do with their specific concern about "face". Of course they like it very much but in any case we have to pay attention not to damage the equilibrium or the harmony in the organization. A promotion or a bonus or a salary increase can not be kept confidential in China and if it is not more than justified the reward will charge the incumbent with very high responsibility and will make his colleagues to loose a bit of their face.
- International exposure has to do with their perception of un-justified inferiority to Westerners which inferiority they want to cancel fast. Management courses or training programs are very popular and well accepted as a form of reward. I enjoyed big success offering free English lessons at the end of the working time.

- Training in creativity is necessary to free them from residual mental constraints due to the planned economy environment.

But the most important tool to make your employees happy and loyal to your company is to make them feel part of China and working for China development and benefit, not as part of a Western company aiming only to export the profit, may be competing with other Chinese companies.

Of course if this is the strategy of your company, you cannot change it. If this is the case, then be ready to a big turnover or to react with big benefits and salaries.

## 3. Negotiating with a Chinese Counterpart

Everything I mentioned so far about Chinese culture applies to the relationship with third parties, in particular to the conducting of negotiations.

Focusing on negotiations let me first anticipate a premise: Chinese are smart and tenacious negotiators. This should be surprising considering their recent history, which shows the Chinese locked inside their borders; but in fact this is recent history.

Chinese started trading in Asia very early in the history and their skills where well recognized in the Region. In XIV century they had the most developed navy in the world.

In XII century, in certain sculptures on the walls of Angkor temples in Cambodia, it is easy to recognize Chinese ethnic people, traits and hairdo indicate this on purpose, in the act of trading and shopping with the local Khmer ethnic people in the middle of a Khmer town. (figures 1,2 and 3)

Figure 1. Bayon, external gallery, South side, East wing. Angkor, Cambodia. End of XII century.

Figure 2. Bayon, external gallery, East side, South wing. Angkor, Cambodia. End of XII century.

Figure 3. Bayon, external gallery, South side, East wing. Angkor, Cambodia. End of XII century.

This happens at the end of the XII century and the Chinese presence is not only noticed but its trading role is emphasized. By the way, Khmer people were too busy fighting with Siamese, as you can easily note from the same sculptures in figure 4.

Figure 4. Bayon, external gallery, South side, East wing. Angkor, Cambodia. End of XII century.

Many events took place since then and it is not the purpose of a manual to investigate those, provided that you remember: no matter the way a Chinese appears, his chromosomes are those of an old traders' and travelers' culture.

There is a motto, I do not know if it is Western or Chinese, saying: no matter the way a Chinese appears, if you scratch the surface you will find a farmer. That is true but if you scratch a bit deeper you will find the trader and better you realize this before it is too late.

With this concept in mind let us never undervalue the hard task of negotiating with a Chinese counterpart, also considering that most negotiations take place in China where the Chinese counterpart is inevitably favored.

After this long but necessary premise, let now consider some specific traits of Chinese culture and their impact in conducting a negotiation.

- Patience

Patience is a stereotype about Chinese. Let me confirm that the stereotype corresponds very much to the reality.

We need to be prepared to accept that in the middle of a crucial negotiation the Chinese counterpart disappears. This could happen not only because the counterpart wants to challenge your negotiation strengths but sometimes simply because the passing of time has a different value for the Chinese. Sinologists say that the elapsing of time is linear for Westerners and circular for Chinese. To me it is enough to note that it is normal that in a negotiation the time factor had a different value for the parts involved but that in a Chinese environment to the business differences cultural differences overlap.

If you show yourself nervous you give a weakness signal also in the Western environment but in China this would be a disaster. For us, Western people, you can invoke the "sense of urgency", for the Chinese this concept do not exist at all. If you show sense of urgency this means that you are not high enough in your organization: more urgency means more bosses to answer to.

To our mentality this is difficult to understand but urgency would really reduce your dignity and make you loose a bit of your face.

By the way also in the Western culture "sense of urgency" was not as popular in the past as it is now.

I remember a verse of a famous Italian poem "La divina Commedia" where Dante and Virgilio, having been reproached by Catone, had to accelerate their steps. When eventually they stopped this is the verse which comments the situation:

> "Quando li piedi suoi lasciar la fretta che onestade ad ogni atto dismaga"
> (when their steps gave up the hurry which always reduces dignity)

On the subject of urgency and patience I also remember an old and very Chinese proverb: "Riding the horse you can not feel the smell of the flowers".

***Anecdote 3: Secretaries Chatting***

I was in the Taipei office and I needed to reach Kaoshiong, where the factory is located, to lead the monthly management committee. I had to catch a plane and I was very late. Nevertheless I asked my secretary to call the Kaohsiong hotel just to check if the reservation was ok.

The lady called and then started a long conversation with her counterpart in Kaoshiong hotel.

I was getting nervous and nervous but I resisted. At last the conversation finished and my secretary told me that the reservation was ok.

I was just running out of the office when the lady told me: "Mr. Ballada, in Kaoshiong now it is raining heavily and in addition there is a demonstration in progress, may be you will find difficult to get a cab". This allowed me to call immediately the factory in order to have the company car picking me at the airport, which normally I did not ask being the hotel and the airport close together and both far from the factory.

This saved me a lot of trouble, as a reward for having been patient and having accepted the two ladies to talk enough to consolidate their relationship and to share useful information, as good friends are supposed to do.

- Temper

Chinese people do not like conflicts. They killed each other by the millions just like the Europeans in the past centuries, nevertheless in the personal relationships they follow certain behavioral rules, which we need to adapt or at least to know.

A blunt "no" would be impolite and loosing your temper at a negotiation table would be a barbarism.

The consequences in practical terms are bad. If you loose your temper they will tell you anything you like just to save their faces and your face.

You go home with the impression that you concluded something good and start thinking to the big bonus, instead you will realize soon that you have to start everything from the beginning the next time.

This circumstance adds also some additional explanation of the legend about Chinese disregarding previously reached agreements.

- Face

Face is very important for Chinese in every circumstance.

"Face" is the perfect translation of the Chinese word *mien zi* which, by the way, means exactly face and also is written by an ideogram being the stylization of a face. But this is not enough: it is always difficult to translate a word from a complex culture in the language of another complex culture and this is the case.

For Chinese people Face means dignity (which has to do with the individual values), prestige (which has to do with wealth), and recognition from the others. The importance of Face in the Chinese environment has probably to do with the high importance of the organization for their society and the role that a person plays in it. For a Chinese person Face is like money for us: something you can gain, you can loose, you can trade, you can share, you can inherit or you can leave in heritage.

In any case we never, never have to take advantage of a favorable situation to weaken our counterpart challenging his image or his Face.

This behavior will create for us a tough enemy and will make us to loose the esteem of the others, including those in our side.

I strongly recommend instead to apply this old an nice Chinese proverb: "Give Face and you will receive Face".

***Anecdote 4: Complicated Contracts***

This experience has to do with face and contracts in the same time. The Licensing department of my company needed to ask to a Chinese company a very special favor.

They needed to ask permission to the Chinese company to allow some technicians to be trained in their plants, as only in those plants was in use a certain process technology. Unfortunately in the same Chinese plant certain innovations had been introduced and, to make thinks complicated, such innovation belonged to a different Western company.

The situation was enough complicated and my Licensing department decided to go to visit the Chinese company to explain and negotiate.

Negotiations progressed very well and, so they reported to me, the agreement in principle was reached fast enough being the all matter based on reciprocal good faith.

The two delegations took advantage to touch also other subject about technologies to share and the all mission was accompanied by big dinners and concluded with friendship declarations.

When back from China my company Licensing department sent to the Chinese company a draft of contract and started waiting for the answer.

Time passed, mails were sent to ask for explanation about the delay but no any sign from the Chinese "friends".

After six month of frustration my colleagues asked me for opinion.

Of course I wanted to see the draft contract which was sent after the visit and after having seen it I could guess an explanation for the Chinese behavior. The all matter , considering the specificity of the deal, was about confidentiality and trust. My colleagues prepared a draft not less than one pound of difficult legal English wording which would have costed to the Chinese party months to read and understand.

Moreover I am convinced that the reason why the Chinese did not answer is also another one, having to do with face. The Chinese counterpart would have had difficult time to explain to the bosses that, after a good and happy negotiation, they did not succeed to conquer the Western counterpart reliance, obliging them to send such insulting draft. So I recommended making an exception to the rules, prepare a slim draft agreement and go back to China to apologize and enjoy additional dinners.

***Guan XI***

Anybody dealing with China has to learn soon about this concept.

Chinese manage the business through relationship. They are strongly interconnected and they pay attention to you only after you are properly introduced.

There is a sentence that I heard somewhere depicting perfectly this situation: "Chinese business people do not invest on a project, they prefer to invest on a person".

This is not good business practice of course and probably costed them very much in terms of development, but this is the situation we have to adapt to.

Some analysts say that one of the biggest problems for Chinese future development is the huge bad debt in the Banks books. It easy to conclude that such bad debts has to do with the rules applied in granting the loans. Frequently network, family and loyalty concepts prevail on discounted cash flow considerations and the result are inevitable.

Of course *Guan Xi* has also many good effects.

As a foreigner it is not easy to penetrate the Chinese *Guan Xi*, but staying in China long enough and using the right approach it is possible to get close to it. When this goal is reached the life in China becomes very pleasant and easy.

All friends will do their best to help to cope with big and small problems and to enjoy life; only one warning: friends can ask important favors on their turn and a refusal, with no serious excuse, would be considered very impolite and totally unacceptable.

We have to keep in mind that Confucianism and the concept of Organization, which I mentioned earlier, make everybody to be part of a complex network with rules, not written but highly respected.

Just to give an example: it is considered very impolite to ask for a favor to a higher level person in your *Guan Xi* before having tried with the lower level one, but would be also impolite to ask to the lower level person a favor if only can be done at a higher level in the same *Guan Xi*.

# PART TWO: INDUSTRIAL PROJECT IMPLEMENTATION.

So far I elaborated about people. Now I would like to develop some concepts about the real and typical business issues considering, step by step, the realization of a industrial project.

## 1. Planning Phase

In this section I will try to identify what makes different in China the planning activity and the forecasting which is needed for a conscious and wise planning exercise.

- In China we cannot take advantage of the **many production or export statistics** as we normally do in Europe or in the US.

So it is extremely difficult to evaluate competitor's market share and then to plan a target market share. The same happens with the analysis about planned suppliers and customers which makes in turn difficult to predict the overall competitive position.

- As far as the **market forecasts** the situation is also more confused. China developed at an Average Growth Rate of 8.5%-9% from 1995 to now (including in the average the big Asian Crisis in 1997 and 1998). Is that reasonable to extrapolate such performance? Till how long? I do not try any answer as in the specialized press you can find already all the opinions any also their contrary. Nevertheless I will report one of those opinions as it recalls some of the aspects of the Chinese culture I discussed about in this manual.

Niall Fergusson, professor of International History at Harvard Business School, in one of its recent reports foresees the risk of collapse in the Chinese economy as a consequence of the two networks which represent all over the world basilar key institutions: the credit network and the global information network. The credit network could collapse under the load of the bed debts inherited by the Government owned companies and anyway unable to disregard the *Guan Xi* pressure in the selection of the projects to be financed; the global information network as it opens the world to whom was so far excluded, priming the comparison between what is reported by the official sources and what is internationally accepted. This situation could bring large sectors of the population to an identity crisis with a consequent social protest of unforeseeable effects.

- Another big risk comes from the **big attracting figures** you get extrapolating the Chinese demand.

A typical way or reasoning in the past was for example: one billion Chinese do not have the telephone, so sooner or later they will absorb one billion telephones. This kind of predictions can be wrong for many reasons, in case of the telephones, to remain in the example, was wrong because the Chinese are actually absorbing millions of telephones but portable and not table version.

In this situation would be useful to use the old Michael Porter graph about the dimensions of a business, trying to forecast the future of each dimension: technology, customers, suppliers, etcetera.

- Another issue which has not being taken under enough consideration by the copious literature about "Business in China", is the **change in the trade pattern.** Since few years ago Western countries were used to import from China raw materials and to export finished goods Now it works the other way around. Actually everybody is aware of this new situation so much so that the area between Canton and Hong Kong has been named the "factory of the world". Nevertheless trade patterns take long time to change and to adjust. Logistic infrastructures, commercial relationship had to adapt to the new situation.

Again let me report my personal experience. As a consultant with relationship with China, I am frequently approached by companies or traders looking for raw materials from China to feed their established channels and customers, producing, in Italy or Europe, finished or semi finished products. Very difficult to satisfy those requests as all chemicals, steel and cement stay in China. Everybody is aware of the difficulties now hitting this sector of the industry, I just want to emphasize that the same difficulties are hitting throughout all the supply chain.

Why I emphasize this particular aspect discussing about "doing business in China"? Because when planning a production in China we need to carefully consider that the raw materials availability could be a serious issue. In fact the supply chain to feed a Western project in China from other sources could not exist yet, or it is not yet consolidated, and Chinese raw materials will stay in China with an allocation priority list dictated by the *Guan Xi*.

After having listed the difficulties now would be nice to suggest the tools to overcome such difficulties.

Unfortunately I can not recommend any recipe except to watch the Chinese behavior and try to follow their example. This conclusion looks like banal, nevertheless it is the synthesis of what is repeated in many reports on this subject, which is: if Chinese make money in their environment this should be possible also for Westerners provided that they are able to adapt to the environment. This, in fact, is the actual challenge.

Once again I have to refer to the *Guan Xi*. Chinese business people tend to involve the constituencies of a future project since the planning phase of the same project. They take advantage of their *Guan Xi* and tend to create a new one around the planned project.

I try to translate such behavior in practical recommendations:

- Set a local office as soon as possible and start lobbying.
- Be generous in pre-marketing.
- Create relationship and if convenient also partnership with some key customers or suppliers.

It is obvious that following above recommendations means to reduce the independence of the project but this is exactly what I mean and what I would like to emphasize as a conclusion: it is not wise to invest in China without big connections and the balance between independence and relationship is the name of the game. More: this is true for Chinese business people among themselves.

Before to live the subject of planning and forecasting I would like to report a more general reflection about the Chinese market predictability.

Just because of the fast development, the Chinese Authorities have to balance the following factors:

- the growth of the various business segments;
- the allocation of the limited available financial resources for investment;
- as alternative to local investment, the growth of import, which means need for foreign currency.

Let us make an example. Let imagine that we created in China a company to process plastic materials, PVC for example. Could happen that the companies producing PVC are encouraged to invest in other segments (energy shortage reasons, environment reasons?) or even that the Authorities favor investments in segments different from Plastics: pharmaceuticals for example. The reasons of such decisions would be in any case unpredictable. The outcome for your company would be the need to turn to import to feed your extruders. At this point a foreign currency squeeze could take place. The PVC segment will be penalized and so all private investments in the same segment.

Can you bear this kind of risks?. A normal company can not afford to be caught in the middle of such big Government decisions.

As a consequence it is advisable to progress by steps even if this could not optimize the scale of your assets.

In case it is really necessary to start big (a refinery, for example) than it is better to associate the Chinese Government owned companies in the venture. In fact this is what the major oil company did in order to safely enter in China.

For those who like history, I recall that what I described as a potential risk already happened in the Chinese history. In the sequence of events that I will report a Central Government decision had the greatest consequences not only on a business sector but eventually on all China.

As I already mentioned earlier, in XIV century the Chinese Navy was the strongest in the world. Chinese had invented the compass and anticipated Europe in the use of multiple masts and separate holders in their big boats. This technology allowed them to build the largest boats in the world. Trading flourished and their admirals reached, in those years, the coast of Australia and Africa.

Unfortunately the Ming government in Beijing decided to withdraw support, and indirectly the investments, from the trading activity as it was perceived more important in that moment to convey all the Country resources to some big projects to consolidate the territory. Was considered vital to protect the fields and the population from the recurrent disastrous floods from the big rivers and to protect the Country from the equally recurrent invasions from the northern nomadic populations.

So the Emperor Court was moved from Nanjing to Beijing, big canals were built to connect the Yellow River with the Yang Zi and the Yang Zi with the Pearl River in Canton and also the Great Wall was restored and completed to resist to the pressure from the same northern populations who took over anyway two centuries later.

The consequences were enormous.

- The navy with no support converted into piracy, making unsafe the traveling in the South China See and more and more isolating China from the rest of the world.
- The coastal areas as well became unsafe and people started withdrawing to the internal areas. Somebody said that also the Chinese mistrust for bathing activities and sports has this origin.
- Anyway the biggest of the consequences of the above described Central Government decision is that only few years later the Dutch came with smaller boats and bigger guns and with full support from their Government to colonize the Region and this was the beginning of the end.

## 2. Project Phase

Once completed the planning phase we are ready to start to project the plants.

In this section I describe the specificity in the Project Phase for projects requiring an investment higher than 30 millions US$ and involving a Chinese partner, which is the most common case, particularly for big investments.

- *Negotiations with partners and authorities.* Also in the Western world we need to negotiate with the partners the reciprocal contribution to the venture and both have to negotiate with the Authorities environment specific issues. What is special in China? At least three very important things.

  - First **the alliances**. In the Western world the two partners are normally allied to negotiate from the Authorities the best conditions for the common venture, in China the partner and the Authorities are frequently allied to obtain the maximum contribution from the Western partner.
  - Second **the transparency**. In the Western world we normally know what it is going on between the partners and the Authorities. Needless to say that in China you will never have access to this kind of information. It is true that becoming member of WTO Chinese Authorities accepted epoch-making conditions as far as the transparency in regulations but, just for the newness and importance of such changes, we have to be careful in taking as a given their compliance. Just think that so far the Administrative Authorities had the right to issue some regulations, called "to internal circuit", not to be communicated to everybody but of cause applicable to everybody. This kind of practices have been abrogated by the WTO membership, but few years are really sufficient to modify a consolidated practice, moreover in such a big and diversified territory?
  - Third **the relative importance**. In the Western world the most important negotiation is between you and the partner or the suppliers or the customers. In fact the negotiation between the partners and the Authorities are supposed to take place in the frame of accepted and well known rules. In China the most important negotiation is between you and the authorities. As the rules are not existing, or flexible, or unknown.

- **Regulations.**_What is the problem with regulations? The fact that they change frequently and suddenly and unpredictably. In any case as a "foreigner" you will be the last to know. It happened to me to work for more than one year on a joint venture with one of the two big Companies in the petrochemical business with no result only because of an unforeseeable change of the rules. The project was based on a fifty-fifty approach, officially agreed in an Memorandum of Understanding duly signed by the two top level people of the two Companies involved. Just after one year of negotiations we have been informed that Authorities do not allow any more fifty-fifty deals in the segment we were negotiating, as the same segment became strategic for the Country. In that case my company could not accept the new situation, due to intellectual property reasons, and we had to drop the project. Our Chinese counterpart was unhappy more than us for the lost business opportunity and, for sure, everybody was in good faith. Nevertheless, at the end of the day, my Company lost money and time, not to talk about sensitive information that were made available during the negotiations.

We celebrated the funeral of the big project in a wonderful restaurant in Beijing and I lost one year of my career.

To make the point about regulations, and their changes now taking place, I take as an example the regulations concerning the liberalization in the trade of goods and services.

Up to the approval of the new Trade Law , in 2004, only a certain number of trade companies, Government owned and holding a special license, were allowed to import, export and operate on the international markets. The liberalization took place through various steps. First was allowed also to foreign Companies to conduct trading activities provided that strictly connected to the same Company manufacturing activity. Second everybody was allowed to trade but only in the Special Trade Zones. Lately all over the Country, but limited to Joint Ventures and lastly to everybody. Of cause expect that all above procedure was dressed by exceptions as far as the business segment, plus or minus strategic to the Country, delays in communication and the application of the many "non tariff barriers" invented on purpose to make unprofitable the trade of certain goods.

Moreover we have to keep in mind that any liberalization process in China has to cope with the vastness of the Country, the fast and irregular development of the economy and the resistance to the change of a system used to years of interweaving between business and beaurocracy.

As a consequence of all above, it is logical to recommend caution and insist on the concept that we need to have available all the tools and the appropriate assistance, also legal, before to feel "ready for China".

- **Administrative steps.** In this case makes a big difference if we are dealing with a big project (beyond 30 millions US$ investment) or a smaller project. In the case of a big project we have to follow an established procedure.

MOU signing with the Chinese partner.

Project Proposal (a preliminary study) presentation for approval to the relevant Government Planning Commission.

Negotiation and preparation of the Feasibility Study report.

Negotiation and preparation of the JV contract and the Articles of Association, if this is a JV, or in any case negotiation and preparation of the many contracts for the supply of Utilities, Facilities and Technology, even if it comes from your same company.

Presentation for approval of all the above to the SDPC (State Development Planning Commission) and the other affected Authorities like the Environmental Bureau.

Presentation for approval to the MOFTEC (Ministry Of Foreign Trade and Economic Cooperation).

Application for Enterprise Registration.

Now you can relax and wait for the Business License...

I reported the above boring list with the purpose of emphasizing the importance of being equipped with the necessary tools when projecting a big investment in China: time and money, both patient. May be the above procedure has already changed while I am writing this manual, but anyway something similar would have replaced it.

Last year in 2006 the first petrochemical plants in China, part of the Nanhai project, started up North of Hong Kong.

The project started in 1988 (it is not a typo, I really mean nineteen eighty eight), and Shell opened an office in Beijing to take care of the project.

Generations of managers worked on that project, including myself, and eventually retired before seeing the plants running.

That one is a big project with an investment of four billion US$, but things work in the same way also for smaller projects.

In the case of a medium size fully owned enterprise the project phase can be fast enough but then we have to be prepared to face all unexpected events and to cope with them using our own, and our *Guan Xi,* resources.

Above arguments are not supposed to discourage those who are planning to invest in China. Many already did it and many of them successfully.

So how to face those difficulties if a company is determined to invest in China because it wants to take advantage of a good technology, has already identified a good partner and perceives the need to protect the domestic market in Europe?

It is not impossible, provided that the main rules are followed.

- To make available enough money and time to support the long lasting project phase.
- Create a team sitting in China or familiar enough with Chinese environment. Give to that team enough decision power.
- Build strong relationship with the local partner, or in case you are going "stand alone", with the other constituencies: suppliers, customers, local authorities.
- Retain a local advisor for legal and regulatory matters.

## 3. Construction Phase

In this section I will consider some issues typical of the plant construction phase.

Some of these issues are well known and common to all developing countries. In fact when planning a plant construction in a developing country looks like reasonable to expect the following conditions to apply.

- Manpower cheap and available everywhere.
- Variety of equipment available from foreign JV and now also from local producers.
- Plenty of land available.
- Good and skilled Engineering Companies ready to provide their service cheaper than in developed countries.
- HSE to be less expensive.

Unfortunately the reality is quite different. Actually, considering one by one the above listed expectations, we discover the following.

- **Manpower cheap and available everywhere**. The labor is not necessarily cheap where you need it. In China the differences among the various regions are huge under all standpoints: climatic, cultural, economic, logistic, industrial, commercial and fiscal. There are regions very developed and regions still very poor and other big factors have to be taken into consideration when selecting a site for your initiative. The final decision has to optimize many factors and you can find yourself obliged to select a place where the labor is not cheap as expected.

- **Variety of equipment available.** this is not true for many reasons. First of all some special items or laboratory equipment are not yet so commonly available and, second, in some cases the import of certain materials could be heavily discouraged by tariff or non tariff barriers. As a conclusion about this item, I recommend to **always ask the engineering company in charge to detail the origin of all the items** in their investment estimation.
- **Plenty of land available.** Even if the land is available, normally the site requires significant investments in infrastructures. Local authorities are interested and happy to cooperate to the upgrading of the area, for social or propaganda reasons. Sometimes the local authorities are also available to contribute to the investment but in any case the impact on the schedule of the overall project, could be heavy. In some cases the site preparation involves local residents relocation and, even if the local authorities commit to take care, the schedule can not be kept under control. In practice the only caution to apply is to plan the investment with great care to the "financial charges under construction". This approach requires a different mindset from the Western approach. In Europe, for exempla, we are normally ready to order the so called "long delivery time items" well in advance not to delay the project progress. In China this could be a big risk. To conclude on this item I would say that the PERT chart of a project in China must follow the Chinese concept of time not the Western one.
- **Good and skilled Engineering Companies.** The Engineering Companies are normally controlled by local downstream companies and, at the end, by the Government, so we need to be careful and obtain everybody's agreement before getting to much involved in a project. I was offering to a big company in Beijing a very unique technology and my only competitor was a Japanese company. I was happy and very comfortable to reach an agreement, particularly considering that Chinese do not like to deal with Japanese if they have an alternative. Nevertheless the negotiations were slow and I perceived that there was something wrong. Later I realized that the all matter was about the fact that I put as a condition the engagement of a Western engineering company, while the Chinese were committed to appoint there own. Fortunately I had not contractual obligations with the Western company, which I was proposing only for the fact that I was comfortable about the performance of such well known company. So I could easily resolve the problem accepting the Chinese engineering company with some additional conditions about their expected performance. It is clear that in case I had contractual obligation with the Western company I would have lost the deal. Only an additional remark on the engineering companies subject. In case we are obliged to use a Chinese engineering company, like in the case I reported above, we need to be prepared to see our technology to be copied soon. Again, this could be a problem or not, but in any case we need to know it in order to adopt the right measures.
- **HSE to be less expensive.** May be HSE costs are low for a Chinese producer, not necessarily for a Western investor. Actually the Western investor is expected, by all

the business community and the local authorities, to apply in China the same standard that are applied in the West. This is for many reasons: for "face" reasons but also because the Chinese industrial community wants to learn from the Western initiatives also about this specific side of technology.

## 2. Operations

In this paragraph I consider the issues that normally have to be taken care of when running the plants. Many of these issues are common to the construction phase plus some new others:

- The **supply of the utilities could be a problem**. Even in the case that our plant owns independent utilities supply equipment, still we have to be very careful with the back up requirements. In case of out of service of your utilities plants we do not have to expect the back up from the local infrastructure to be a given. In fact utilities are normally short everywhere in China and in order to be allocated for emergency supply we need to deserve it. Again it is a matter of *Guan Xi.*
- **Logistic costs could be a problem**. Even if the logistic aspect has been considered carefully in the planning phase, still we need to continuously monitor this aspect of the business all the way long. In China the situation is changing very fast in terms of traffic and infrastructures and the balance between them is totally unpredictable. It has been reported that in the Western countries the cost of logistic represents 4% of the total product cost. In China this value reaches 16%. Some analysts predict that, despite the continuous construction of new roads, the traffic could increase more than expected and in the future the logistic costs could probably increase with potential impact on the competitive position of those companies which did not equip on time. The recommendation here is to be very flexible in managing the business and to be continuously updated on the evolution of the area where the plants are located.
- **An additional warning about Quality.** It is not yet an issue except if our production is aimed for export. In this case Quality is more expensive than in Western countries and would be probably necessary to budget important costs to train the operators. An other warning about quality: keep in mind that also quality is an aspect of technology and incorporates important know how. So be careful not to see your quality manager leave your company as soon as he will be able to sell the training received in your factory to some competitor's shop.
- A recurring question about running a company in China is the **management structure**. Western CEO and Chinese Vice-Presidents or Chinese CEO and some Western VP?

In my experience it is very important to have a Western CEO. But not a bureaucrat and not necessarily a technician. We need an aged and experienced person able to gain reputation and to cultivate the relationship, then he will enjoy subordinates loyalty and environment support. The VP's must be local and able to manage the nuances of the day by day business.

The other way around, Chinese CEO and Western technical people, would be a disaster.

The Chinese CEO will soon short circuit with the Chinese subordinates and will isolate the Western employees. In addition the Chinese CEO will be less able to resist to the Chinese environment pressure in case of any conflict of interest between your company and some Chinese constituency.

A personal experience clarifies this point perfectly. I wanted to replace two old extruders with a new one with larger capacity. I immediately anticipated to everybody that the purpose was not to reduce the personnel, actually right in that period the company was recruiting to cope with a debottlenecking in an other section of the factory. So I was surprised when the local Environment Protection Office refused to release the necessary permits. You do not need to be an engineer to understand that when you replace two old machines with a new one with the same total capacity, you do not increase pollution but instead you reduce it. So what was the problem? I must admit that I do not know, even now. When I realized that probably somebody was asking for some favor, I immediately passed the problem to the Manufacturing VP and the Finance and Administration VP asking them to take care together keeping me out of the loop. I had to wait some time but eventually the permits were released. And no evidence in the balance sheet of any cost connected to such issue.

***Anecdote 5: The Japanese Vice president***

The experience I want to report here refers to a Japanese person and took place in Italy and not in China. Nevertheless I think it is very instructive also for our subject here.

Many years ago I was appointed as the Controller of an Italian-Japanese JV producing certain plastic parts. The headquarter was in Milan and the plant in Southern Italy.

The production was based on Japanese technology and the Japanese partner made available the Manufacturing and Technology Vice President while the Italian partner had the CEO position. Looks like very obvious isn't it?

Unfortunately at a certain point the situation in the plant started to be very difficult and the fixed costs got out of control. The two partners began bouncing the responsibility. One blaming the Japanese technology the other the Italian management. At the and my boss, the Italian shareholder of the JV, gave me a flight ticket and asked me to go to the factory and not to come back without an explanation. Fortunately I didn't need much time and I didn't need either to be too smart. The explanation came to me very soon.

The Italian CEO was totally in the hands of the local constituencies: he was obliged to hire unnecessary manpower, obliged to accept extra costs for participation to any kind of local fake social initiatives and so on. The Japanese Vice president was not in a position to perceive all this but he was not able to impose any rule or discipline to the operators and then the plant efficiency and the production quality went fast out of control.

The gentleman was psychologically destroyed and saw in the task I received from the big boss the way out from his situation.

So I learned from him all the details and sent quickly a report to Milan. Eventually I got a career advancement and I was appointed to sell to the Japanese the Italian share of the company.

## 5. CONCLUSIONS

Let me now try to distil some conclusions.

Why to invest in China if it is so difficult and risky?

At this point comes to me very useful an analysis which has been developed by Valeria Gattai of Bocconi University in Milano last year in 2006.

Such analysis refers to the Italian investments in China during the last few years and reports the main motivations which led such investments. Italy is a small country and Italian investments in China are also smaller but, based on my experience, it will not be wrong to extrapolate the results from this sample.

These are the reasons the Italians invested in China:

- 38% to enjoy the large market;
- 23% to take advantage of the cheap manpower;
- 21% to escape the tough competition on Europe;
- 11% to take advantage of an occasion which was offered to the company;
- 4% to by pass the European restrictions in matter of environment protection.

Those who were so patient to read this manual can easily guess how those motivations got to those investors bitter disappointment.

In fact the same analysis reports that only 10% of those investors did not suffered bad surprises either due to the cultural gap, or the communication difficulties, or the invasive bureaucracy, or the legal system, or the corruption and the lack of infrastructures.

Nevertheless the reasons to invest in China are very convincing if you have a good technology and you want to exploit it before it is too late. The same if you want to protect your market share in the West from Chinese competition (tomorrow also Indian), producing at Chinese costs.

In my opinion, in many cases to invest in China it is not an option but a strategic necessity. Nevertheless I recommend to undertake such move, both psychologically and practically, as a defensive need instead then as a aggressive attack. This will avoid you to cherish illusions and in case of failure to reduce the damage.

Said that, I can add that many success cases are also reported.

The overall conclusion is that to invest in China looks like a gamble but it is not if we are culturally equipped and ready to follow the rules. This is also the reason that encouraged me to write this simple manual: to try to describe such rules, which I want to summarize as follows:

- appropriate cultural approach;
- personal relationship, with all constituencies;
- patience and patient money;
- appropriate human resources and management policies;
- in depth risk analysis;
- great flexibility.

I would not consider completed the task I took writing this manual without mentioning also the exceptions which I had the chance to observe to the rules I described so far. I insisted very much on the important heritage of Confucianism in Chinese behavior, and I confirm this, but I want to add a warning. Chinese culture is exposed since some decades to the Western culture to an extent never experienced before in their history. In the next years we can expect the Chinese behavior to slowly mix the Confucian roots with the Western habits, the good ones and the bad ones. This process is already perceived by Chinese themselves and from this they created the derogatory appellative *Xiang Jiao* (banana) for those Chinese who are yellow outside but white inside.

I anticipated at the beginning of this manual to only report personal experiences, so let me confirm that can happen, and actually happened to me, to meet people combining some bad aspects from both cultures. For example the "*mafioso*" behavior coming from the *Guan Xi* concept and the aggressiveness coming from the Western need of straightness and celerity.

So, the overall recommendation is to adopt great flexibility in dealing with everything related to the Chinese world, including this manual.

## APPENDIX: LIVING IN CHINA

In a manual addressed to those who are prepared to go to stay in China can not be missing a paragraph about living in China as an expatriate with the family and children.

Here I am not addressing to the case of a wise traveler by profession or to a journalist, who are equipped by profession with the necessary cultural tools; here I am addressing to those like me, business people, only equipped with the video that the Human Resources guy gave you as a welfare before to leave to China.

Here in some pills what I learned as an expatriate:

- take advantage to learn about the other side of the moon. This will make you a real global business person;
- train yourself in managing new and fast changing business scenarios which is the most important skill required to a business person everywhere not only in China;
- learn to practice the "cultural relativism" from the field instead of from the sociology books;
- in case you have children, take advantage of this opportunity to make them citizens of the world not only of one half of it.

Of cause there is also a typical "wrong approach". Let me give you some examples:

- spending your holidays in traveling back to your Country;
- spending your free time in the local American (or anyway Western) club;
- leaving part of your family in your Country because "schools are better";
- and, last but not least,: looking for Italian food where Italian ingredients do not exist, instead of devoting yourself to discover the real Chinese food, of which the many Chinese restaurants all over the world give a totally wrong idea.

I had the chance to see many colleagues taking this approach and feeling miserable instead of enjoying this outstanding experience.

Let me complete this manual with this last anecdote.

***Anecdote 6: Why to Use the Chopsticks?.***

I mentioned the "cultural relativism". Let me give an example from the field about this very sensitive issue.

When first arriving in China the first things that we have to manage are the chopsticks. Then after having dirtied your tie and your only business suit in the luggage, you start asking yourself: "How come that this people so proud of their civilization did not decide to switch to fork and knife"? You know the comment I got from a Chinese friend about this subject? "How come that you foreigners (Chinese people call all the non-Chinese foreigners) so civilized, show up at the table with all those complicated tools? Why you accept to work to get what you deserve from a lobster, a crab or a good stake instead of letting the job to be done for you in the kitchen, where there is people more skilled, more equipped and paid for that? At the table you should not need more than the chopsticks“.

This is a good lesson about cultural relativism, so I learned fast to use the chopstick. Now, when having breakfast in the morning in the hotel in China, I can eat using only one hand and hold the paper with the other without being obliged to lean the newspaper to the bottle or to the glass like the other Western colleagues.

## Bibliography

I will not add any bibliography as this manual is addressed to people who need very fast and practical information which are available from Internet more than from a bookshelf.

## About the Author

Antonio Ballada was born in Milano Italy on 1944. With a degree in Industrial Chemistry from the Milano University and a Master in Business Administration from Bocconi University, he covered many top positions in the chemical and pharmaceutical business in Italy and the USA.

Since 1996 he works with Greater China, first in Taiwan as CEO of a public company with an affiliate in Hong Kong, later as responsible of the projects in China for a European Multinational.

At present Antonio is running a consultant activity aimed to Medium Small Companies looking for Internationalization of their business. Antonio is based in Milano, Italy.

He is lecturer at Bocconi University and Istituto per gli Studi di Politica Internazionale (ISPI) on various subjects related to the business management in China and is member of the Board of two companies in Greater China: Minerals Co. Taiwan and China Catalyst Ltd. Hong Kong.

Coordinates:
Antonio Ballada
A.B.C.S.
Via Santa Tecla, 4 - 20122 Milano
Tel. : +39 02 861946
Fax : +39 02 87386246
Mobile : +39 335 6950124
E-mail : a.ballada@balladaconsulting.com

In: Monomers, Oligomers, Polymers, Composites...
Editors: R. A. Pethrick, G.E. Zaikov et al.
ISBN: 978-1-60456-877-6


*Chapter 2*

# PREPARATION OF POLY(LACTIC ACID) AND PECTIN COMPOSITE FILMS INTENDED FOR APPLICATION IN ANTIMICROBIAL PACKAGING

***L.S. Liu**[*1], **V.L. Finkenstadt**[2], **C.-K. Liu**[3], **T. Jin**[4], **M.L. Fishman**[1] **and K.B. Hicks**[1]*

[1] Crop Conversion Science and Engineering Research Unit,
[2]Fats, Oils and Animal Co-Products Research Unit,
[4]Food Safety Intervention Technologies Research Unit,
U.S. Department of Agriculture, Agricultural Research Service,
Eastern Regional Research Center,
600 East Mermaid Lane, Wyndmoor, PA 19038
[2] Plant Polymer Research, National Center for
Agricultural Utilization Research, U.S. Department of Agriculture,
Agricultural Research Service, 1815 North University Street,
Peoria, IL 61604

## ABSTRACT

Pectin and poly(lactic acid) (PLA) composite was compounded by extrusion. A model antimicrobial polypeptide, nisin, was loaded into the composite by diffusion. The incorporation of pectin into PLA resulted in a heterogeneous biphasic structure as revealed by scanning electronic microscopy, confocal laser microscopy, and fracture-acoustic emission. The incorporation of pectin also created a rough and cragged surface, which was hydrophilic and facilitated the access and absorption of nisin. The nisin-loaded composite suppressed *L. planturam* growth, as indicated by agar diffusion and liquid phase culture tests. The incorporation of pectin in the amount of ~20% total mass did not alter the Young's modulus of the film from pure PLA. The composite materials were able to retain their tensile strength, flexibility, and toughness to an extent, which

---

* Correspondence to: L.S. Liu (LinShu.Liu@ars.usda.gov).

† Mention of trade names or commercial products in this article is solely for the purpose of providing specific information and does not imply recommendation or endorsement by the U.S. Department of Agriculture.

satisfies the requirements for packaging materials. Results from this research indicate the potential of Pectin/PLA composite for the application of antimicrobial packaging.

**Keywords:** pectin, poly(lactic acid) (PLA), film, nisin, composite, packaging.

## INTRODUCTION

New concepts and new materials for food and non-food packaging result from flourishing manufacture and trade and are promoted by economic globalization. It was reported that the global packaging market was $300 billion in 2004, and 1/3 of it was spent in the U.S. market. Food packaging is a large sector in the packaging industry. Among the $100 billion generated by the U.S. packaging industry, $60 billion was contributed by the food industry; and it will reach a new milestone of $74 billion by 2008.[1,2] Petroleum-derived thermoplastics with great advantages in performance and cost have dominated the packaging market to a major extent for years. Nevertheless, interests have shifted to biobased materials derived from agricultural or forestry resources because of increasing environmental concerns arising from non-biodegradable plastics and an awareness of the limitation of petroleum resources.

Biobased materials include polysaccharides, proteins, lipids, and their polymeric extracts. They also include polymers that can be chemically synthesized from biobased monomers or produced by microorganisms or genetically modified bacteria. PLA is a biodegradable polymer, made from the condensation polymerization of lactic acid. The monomer, which is also the final degradation product, can be derived from the fermentation of carbohydrate feedstocks. PLA, in the form of rigid structures, films, porous scaffolds, and micro/nanospheres, has been used for biomedical applications and disposable plastic products.[3-6] As a packaging material, PLA is attractive because it exhibits a tensile strength comparable to petroleum-derived thermoplastics, degrades under commercial composting conditions, and can be sealed at low temperature. Furthermore, PLA is resistant to oil, is a good water vapor barrier, and has relatively low gas transmittance.[7-9] PLA also has demonstrated an antimicrobial activity, when it is used in a solution of oligomers or in combination with some organic acids or antimicrobial agents.[10-13] Although there are countless publications on PLA-based drug delivery systems, less study has been done on PLA as an antimicrobial carrier for food and non-food packaging. This could be imputed to its drug release mechanism, by which the release of drugs from PLA matrices depends on PLA degradation. Another obstacle is the hydrophobic nature of PLA, to which hydrophilic antimicrobials are less accessible.

Pectin is a film forming agent. Pectin films have shown applications in coating, encapsulating, and thickening for food and pharmaceutical uses. Pectin macromolecules bind with proteins and some organic or inorganic substances via molecular interactions. Pectin can be constructed as matrices to absorb biologically active materials and deliver the pre-absorbed bioactive substances in a controlled manner[14,15]. It is expected that the incorporation of pectin filler with PLA matrix may result in a new complex material, which inherits the advantages of the parent polymers, such as biodegradability, mechanical strength, water resistance, and accessibility to hydrophilic substances. Antimicrobial proteins can be loaded into the complex simply by diffusion-absorption method with higher loading efficiency and biological activity. In this study, we present a new composite material extruded from PLA and pectin. The composite was evaluated for use as a packaging material and for its antimicrobial activity after loading with an antimicrobial polypeptide, nisin.

# MATERIALS AND METHODS

## Materials

Poly(lactic acid) (dl-PLA) was obtained from Dow Cargill (Minneapolis, MN). The average molecular weights were $M_w$ =148,000 and $M_n$ =110,000; the glass transition temperature ($T_g$) was 55-60°C. Nisaplin (containing 2.5% nisin) and pectin sodium salt were purchased from Danisco (Danisco Cultor USA, New Century, KS), the average molecular weight of pectin was $M_w$ =90,000, the degree of esterification was 60% and the water content was 7.8%. Dichloromethane and acetone were from Sigma-Aldrich (Milwaukee, WI). Deionized water (D.I. water) was prepared using a Barnstead E-pure water system (Dubuque, IA).

## Composite Preparation and Physical Characterizations

Compounding was performed using a Werner-Pfleiderer ZSK30 co-rotating twin-screw extruder (Coperion Corporation, Ramsey, NJ). The barrel was comprised of 14 sections, giving a length/diameter ratio of 44:1. The screw configuration was reported earlier.[16] The screw speed was 130 RPM. PLA was fed into barrel section 1 using a gravimetric feeder (Model 3000, AccuRate Inc, Whitewater, WI). After melting the PLA, pectin was fed into barrel section 7 using a loss-in-weight feeder. In all cases, the total feed rate was approximately 75 g/min. The barrel was heated using eight heating zones. The temperature profile was 135°C (zone 1), 190°C (zone 2) and 177°C (zone 3-8). A die plate with 2 holes (4mm diameter) was used. The melt temperature of the exudate at the die was approximately 155°C. Residence time was approximately 2.5 minutes. Die pressure and torque were allowed to stabilize between formulations before sample was collected. Strands were pelletized using a Laboratory (2 inch) Pelletizer (Killion Extruders, Inc, Cedar Grove, NJ).

Thin PLA and Pectin/PLA formulations were prepared with a Brabender single-screw extruder with four temperature zones (150°-170°-170°-150°C). A 3:1 high shear mixing zone screw was employed. Ribbons were extruded using a hangar-type die at 150°C. The thickness of resultant materials was measured by a micrometer (Ames, Waltham, MA).

The appearance of the resultant composites was recorded by a camera, Nikon, Dix, equipped with a 100 mm Nikon macro lens. The resultant composites were characterized for water, PLA, and pectin content by measuring the weight loss after drying and extraction with dichloromethane.[15,17] Briefly, specimens of Pectin/PLA or PLA (~200 mg for each) were weighed, chopped to smaller pieces, placed in a 5.0 ml volumetric flask containing dry-acetone, capped with a pennyhead stopper, and gently shaken at room temperature for 8 h. The acetone was refreshed three times, and then pipetted out; the flask with the contents was vacuum-dried (20 μmHg) at room temperature for 24 h. The weight loss due to the drying process was considered as the water content of the sample. An extraction solution, dichloromethane, 5.0 ml, was added to the flask, which was gently shaken at room temperature for additional 8 h to complete the dissolution of PLA. The extraction solution was removed, the solid phase, pectin particles colored brown, were washed with fresh dichloromethane (3 × 5 ml) and dry ethanol (5 × 5 ml), dried under a dry $N_2$ jet, and weighted immediately. The weight loss due to the dichloromethane extraction reflected the amount of PLA in the composite.

The composites were cut into ASTM D638-99 Type I tensile bars (16.42 × 1.91 cm, *w* × *l*) for mechanical property tests, strips (7.0 × 38.1 mm, *w* x *l*) for dynamic mechanical thermal analysis, discs (1.6 cm in diameter) for nisin loading and antimicrobial activity assay. The bars, strips and discs also were examined microscopically. All sample specimens were stored in a desiccator over desiccant at 4-7°C.

**Scanning Electron Microscopy (SEM)**

Fractured surfaces of PLA and Pectin/PLA samples were examined for morphology and pectin distribution. Fractured surfaces were created either by freeze-fracture using liquid nitrogen or by separating into two parts by a destructive force during tensile testing. Sample fragments were mounted with adhesive to specimen stubs, and the edge was painted with colloidal silver adhesive and sputtered with a thin layer of gold. SEM images were made in a high-vacuum/secondary electron-imaging mode of a Quanta 200 FEG microscope (FEI, Hillsboro, OR). Digital images were collected at 500×, 2500×, and 25000×.

**Confocal Laser Microscopy (CLM)**

For CLM, specimens were glued to a 1 × 3 cm microscope slide and placed on an IRBE optical microscope with a 10x lens integrated with a model TCS-SP laser scanning confocal microscope (Leica Microsystems, Exton, PA). Images were made at 633 nm for confocal reflection and at 425/475 nm (ex./em.) for autofluorescence at two channels.

**Dynamic Mechanical Thermal Analysis (DMA)**

The dynamic mechanical analysis was performed on a Rheometrics RSA II analyzer (Piscataway, NJ). Storage modulus (E') and loss modulus (E") were measured as the function of temperature. The gap between two jaws at the beginning of each test was 23 mm; a nominal strain of 0.1% was used with an applied frequency of 10 rad/s (1.59 Hz). Each sample was equilibrated in the sample chamber under dry nitrogen at -100°C prior to running the test, temperature was increased at the heating rate of 10°C/min; data was collected from -100°C to 200°C and analyzed using Rheometric Scientific Orchestrator software, version 6.5.7.

## Mechanical Test and Acoustic Emission

The mechanical property measurements were performed with a tensile tester, which enabled us to obtain tensile strength, Young's modulus, and toughness of the samples. Tensile strength is defined as the maximum stress to fracture composite specimens. Young's modulus is a physical quantity representing the stiffness of a material. It is determined by measuring the slope of a line tangent to the initial stress-strain curve from the origin to 10 percent strain. Toughness (also called fracture energy) was determined by measuring the energy required to fracture samples, which is the area under the stress-strain curve. Properties were measured at 21°C and 65% RH with a gauge length of 102 mm. An upgraded Instron mechanical property tester, model 1122, and Testworks 4 data acquisition software (MTS Systems Corp., Minneapolis, MN) were used throughout this investigation. The strain rate (cross-head speed) was set at 50 mm/min. The tensile tester was programmed to perform a cyclic test. Samples were loaded into the jaws and the samples were then stretched to 2 % strain at 50 mm/min and then back to 0% strain; once 0% strain was reached the samples were again stretched to 2% strain and then back to 0% strain. A total of 5 cycles were tested and the peak stress was recorded for each cycle.

Acoustic emission (AE) measurements and tensile stress-strain tests were performed simultaneously for the samples previously described. A small piezoelectric transducer was clipped against the samples. This transducer resonates at 150 kHz (Model R15, Physical Acoustics Corp., Princeton, NJ) and is 10 mm in diameter. AE signals emanating from this transducer when the Instron stretched the samples were processed with an upgraded LOCAN-AT acoustic emission analyzer (Physical Acoustics Corp.). The upgraded LOCAN AT, which exceeds the 20 MByte limit of old LOCAN's, is connected to a PC base with enhanced graphing and data acquisition software with all the features and options of the SPARTAN 2000. This AE system has been used in our research center for studying the deformation and fracture mechanisms of bio-composites, fabrics, and leather.

## Nisin Loading and Antimicrobial Activity Test

Nisin was loaded into Pectin/PLA and PLA by soaking samples in a nisin solution. Briefly, 5 specimens were placed in a Petri dish (60 × 15 mm) containing 10.0 ml of nisin solution (1%, w/v; pH 2), and shaken at room temperature at 80 rpm for 18 h. The specimens were removed from the nisin solution, washed 3 times with 10 ml of 1 N NaCl (pH 2), and 3 times with D.I. water by shaking in the solutions for 1 min for each time. The washed sample specimens were dried in a fume hood for 30 min. and stored at 4-7°C in a refrigerator prior to examining for antimicrobial activity.

For the agar diffusion test, each specimen was placed on surface-inoculated MRS agar plate, on which $10^6$ CFU/ml of *L. plantarum* was seeded. The agar plates with the specimens were incubated at 35°C for 48 h. The diameter of the growth inhibition zone was measured with a caliper. The ratio of the diameter of inhibition zone to the diameter of the specimen was used to determine antimicrobial activity. Specimens of PLA, Pectin/PLA with and without nisin loading were tested. Each sample was tested 5 times.

For the liquid culture test, 3 pieces of specimens from either Pectin/PLA composite or PLA (total surface area of ~12.0 $cm^2$ for each) were immersed in a glass tube with 10 ml

MRS broth. The medium was inoculated with 100 μl *L. plantarum* culture and then transferred to a shaker (Innovas 3100, New Brunswick Sci. Inc., Edison, NJ) at room temperature and shaken at 200 rpm. The culture was sampled (1.0 ml) at time points of 0, 2, 16, and 24 h. *L. plantarum* in the culture was serially diluted by sterile phosphate buffer, then pour plated onto MRS agar. Plates were incubated at 35°C for 48 h. A specimen-free inoculated MRS medium served as a control.

All measurements were performed on five samples and data was expressed as the mean ± SD. Significance was determined with the use of a Student's *t*-test.

## Results and Discussion

The physical features of pectin particles were characterized by SEM and CLM and shown in figure 1. The pectin particles were irregular in shape, rough in appearance, and varied in sizes ranging from a few to several hundred micrometers. Images of CLM revealed a strong autofluorescent emission colored as green that outlines the shape and size of pectin particles. The intrinsic fluorescence of pectin was used as a tool for pectin identification in Pectin/PLA composite films through this study.

After composite compounding, PLA and Pectin/PLA specimens were characterized initially for PLA, pectin and water content, thickness, appearance, and surface characteristics. As shown in table 1, the composite contained ~20% pectin and ~7% water. The appearance of PLA and Pectin/PLA composites is shown in figure 2, pectin particles were evenly distributed within the PLA phase. The optical transparency of the composite was inversely reduced with the addition of pectin particles (figure 2A and B). The thin PLA and Pectin/PLA composites displayed negligible change after being bent into circular shape, showing their high flexibility (figure 2C and D) as packaging materials. The surface characteristics of the composites were further identified by CLM and SEM.

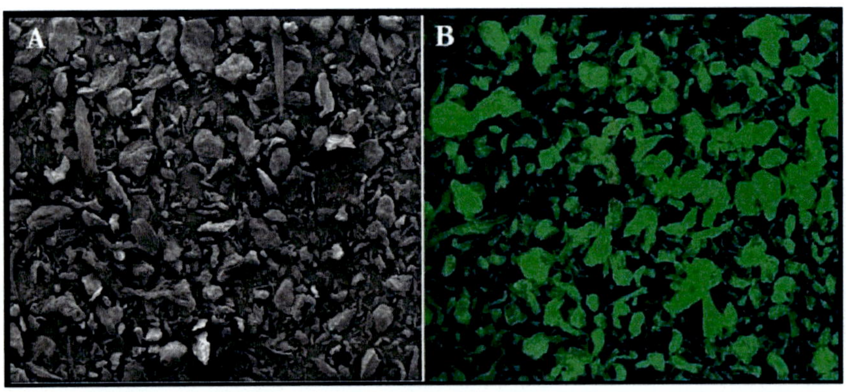


Figure 1. SEM (A) and CLM (B) images of pectin particle prior to extrusion. Field width: 2.6 mm.

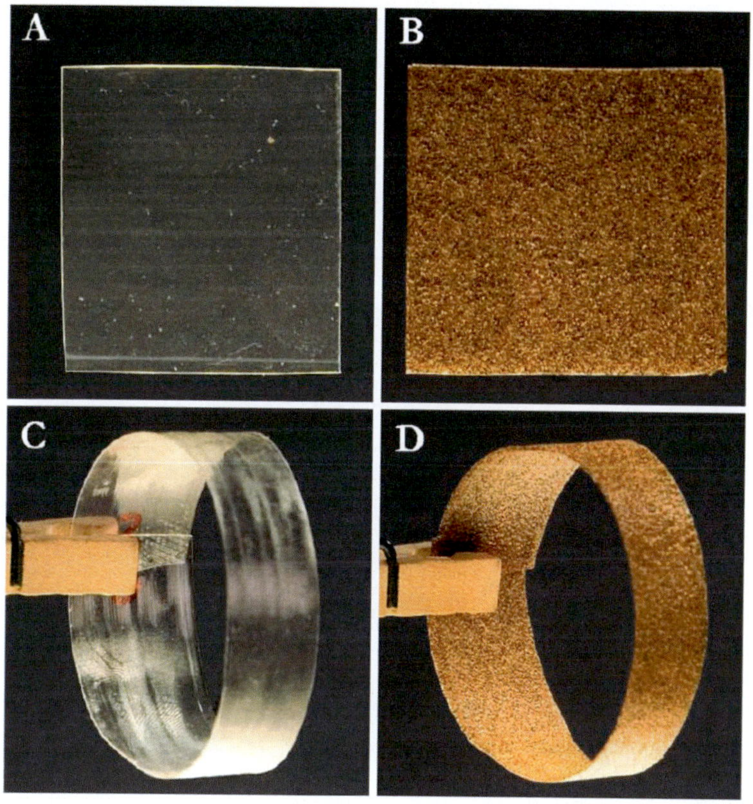


Figure 2. Photographs of PLA samples (A and C) and Pectin/PLA composites (B and D) containing ~19% pectin particles (w/w). A and B: top view; C and D: side view in circular shape. Field width: 6.0 cm.

**Table 1. Components and thickness of PLA and Pectin/PLA composites**

| | Thickness (mm) | PLA content (%) | Pectin content (%) | Water content (%) |
|---|---|---|---|---|
| PLA | 0.54 ± 0.02 | 100 | 0 | < 0.01 |
| Pectin/PLA | 0.55 ± 0.02 | 75.2 ± 3 | 19.1 ± 6.2 | 6.7 ± 1.5 |

Data are expressed as mean ± SD (n = 5).

Confocal fluorescence and confocal reflection microscopy were used in a correlated mode to determine the composite structure. As shown in figure 3A, images of reflection and fluorescence in stereo projection revealed an integrated structure of the two components. The reflection areas colored red contain PLA fibers. Green fluorescence images indicate even distribution of pectin particles, which is consistent with the results shown in figure 2B. Confocal reflection microscopy revealed a continuous, smooth surface for pure PLA sample (figure 3B). Confocal fluorescence indicated a discontinuous morphology for the specimens containing pectin particles. The images of the green pectin areas also revealed a relatively rough morphology, showing a cragged layer of 20-30 μm in thickness laying on the film surfaces (figure 3C). Furthermore, some particles were aggregated to form blocks or penetrated with PLA components (figure 3A). This was confirmed by SEM. Figure 4A shows the image of a vertical section of a composite specimen. The pectin aggregates were located on the surface and extended deeply into the sample. In some areas, the pectin aggregates stretched about 300 μm below the surface, which was 2/3 of the thickness. Higher magnification shows a porous structure of pectin particles embedded in the PLA phase (figure 4B). In comparison with the pectins prior to extrusion (figure 1), the embedded particles showed some changes in appearance, such as more porosity, containing crevices and folds. Since the composites were extruded at high temperature and pressure, the process might cause broken particles and/or adhesion of particles to each other. The melted PLA also could migrate into the pectin particles. As a result, the extruded composites provide a highly porous structure consisting of pectin, which is favorable for the diffusion, adsorption and storage of hydrophilic components. On the other hand, the reduction in the size of the PLA phase (figure 4A) may have an impact on the mechanical properties of resultant composites. All these will be discussed in detail later in this study.

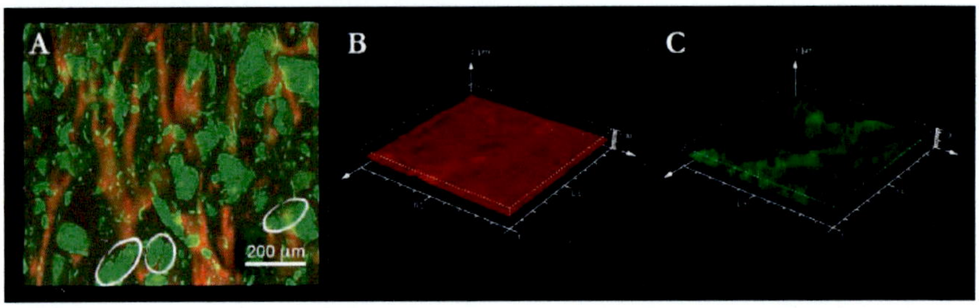


Figure 3. CLM images of (A) Pectin/PLA composites by confocal reflection and confocal fluorescence in two channels, (B) pure PLA samples by confocal reflection, (C) pectin zones by confocal autofluorescence. Areas marked by white circles in 3A indicate pectin aggregation.

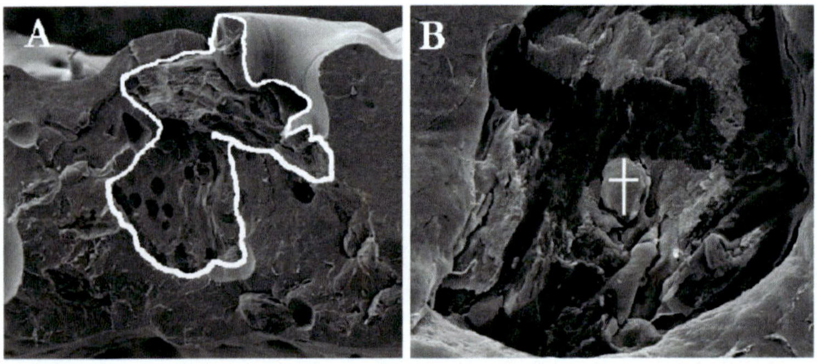


Figure 4. SEM micrographs of the composites indicate that pectin particles were located on the surface and extended into the deep of the specimens (outlined by a white curve), and the penetration of melted PLA into pectin aggregates (+). Field width: A 530 μm, B 56 μm.

As a complement to structural studies, samples were analyzed by dynamic mechanical analysis under a small deformation force. DMA measures the temperature-dependant storage modulus (E') and loss modulus (E"). Comparisons of typical DMA curves of PLA and Pectin/PLA composites are shown in figure 5. There was a sharp decrease in E' beginning at 54°C for both PLA and Pectin/PLA composites, showing a glass transition temperature ($T_g$) at about 57°C for the two films, which was consistent to the $T_g$ of 55-60°C for PLA as provided by the manufacturer. The blend of pectin with PLA did not alter the $T_g$ value of pure PLA. This indicates the good miscibility of pectin with PLA and no chemical interactions between the two phases. Above $T_g$, the E' of PLA samples decreased as the temperature increased, and no response to the force could be recorded at about 120°C, indicating the specimen melted. For composites, a small amount of energy is required to overcome the resistance of pectin macromolecules to thermal movement. The addition of pectin seems to enable the composites to retain certain integrity at a higher temperature. The loss moduli data of PLA are similar to that of Pectin/PLA composites, and show a trend similar to that of the storage moduli.

Samples were then subjected to a destructive analysis for mechanical resistance. Acoustic emission was investigated simultaneously to collect information on structural changes during fracture. At the end of the test, the fractured surfaces of composites were examined by SEM.

Figure 6 shows the fractured surface of composite specimens. A clear and smooth, plate-like image is evidence of the breakdown of both PLA and pectin particles under stress, indicating adhesion between the two phases. However over all fractured surfaces, some pectin pullout also could be observed (data not shown). A decrease in tensile strength of about 19% and fracture energy of about 40% for the composite was recorded (table 2). These decreases are mainly attributed to the reduction of the PLA phase.

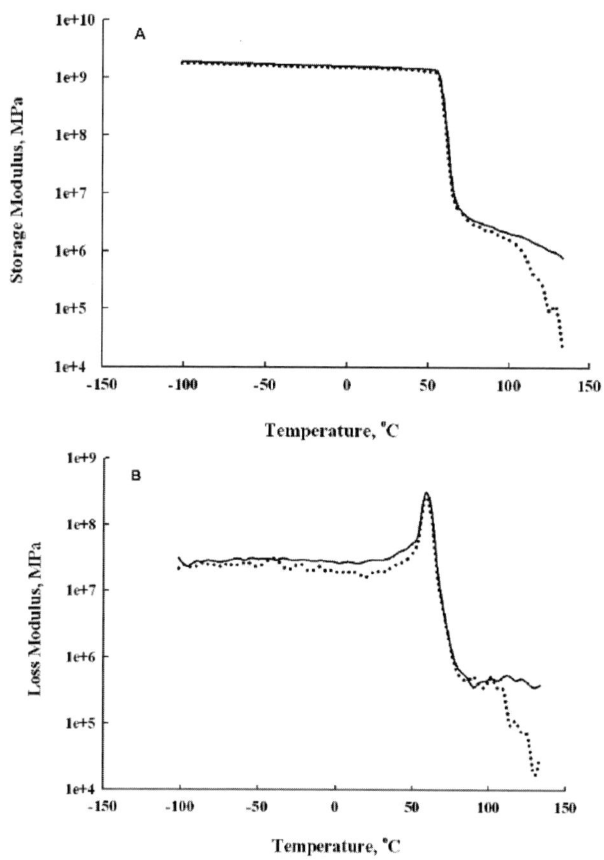


Figure 5. Typical plots of (A) storage modulus and (B) loss modulus as function of temperature. Solid line, Pectin/PLA composites; dotted line, PLA samples.

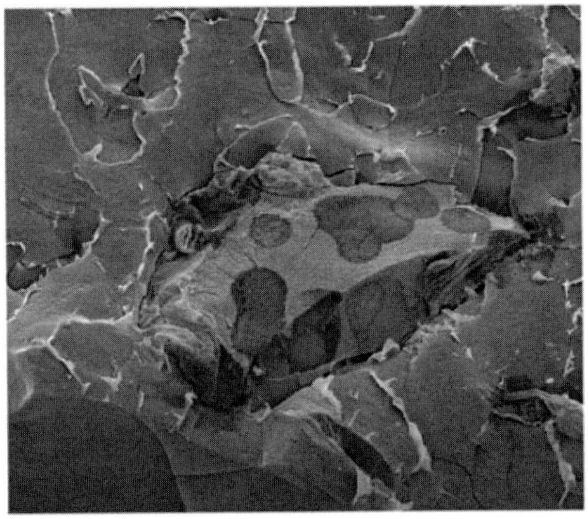

Figure 6. SEM photomicrograph of the Pectin/PLA tensile fracture surface. Field width: 134 μm.

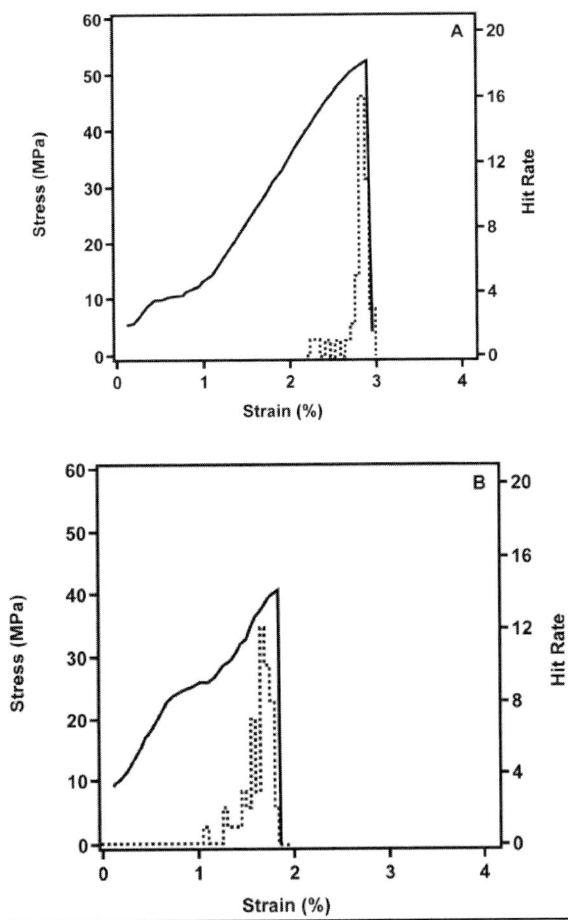


Figure 7. Correlation of strain-stress curve (solid line) with AE hit rate (dotted line). (A) PLA samples, (B) Pectin/PLA composites.

**Table 2. Mechanical Properties of PLA and PLA/Pectin composites**

| Samples | Modulus (MPa) | Tensile Strength (MPa) | Elongation (%) | Fracture Energy ($J/cm^3$) |
|---|---|---|---|---|
| PLA | 2482 ± 99 | 53.4 ± 3.5 | 3.00 ± 0.21 | 0.63 ± 0.11 |
| Pectin/PLA | 2598 ± 100 | 40.2 ± 1.1 | 1.98 ± 0.07 | 0.35 ± 0.02 |

Date are expressed as mean ± SD (n = 5). $P<0.01$.

Figure 7 shows the correlation between the stress-strain curve and strain-AE hit rate pattern. Both PLA and Pectin/PLA composites behaved as linear elastic materials. When the samples were stretched, strain and stress increased simultaneously. For PLA samples, the major AE activities occurred at peak stress when the sample completely destructed, although a few minor AE events were also detected right before destruction, indicating the homogeneous structure of PLA samples. Pectin/PLA composites, unlike pure PLA samples, emitted sound at much earlier stretch. The early occurrence in AE hit correlated to the increase in the slope of the stress-strain curves. This behavior indicated the early defect formation probably due to separation between some pectin particles and PLA at lower stress. Composites fractured at maximum stress, where the largest AE event was recorded. Such behavior is typical for a two-phase composite.

As shown in table 2, the Young's moduli of PLA and Pectin/PLA were similar, ~2500 MPa, indicating that the two materials are similar in stiffness. To fully understand the effects of adding pectin to PLA matrix, we also investigated the mechanical behaviors of composite samples subjected to a cyclic stretch, particularly, the hysteresis that is the energy loss during each cycle of the cycling test. It was calculated by subtracting unloading energy from loading energy. Hysteresis may have a close relationship to resiliency, which governs the dimensional stability of packaging products. Figure 8 shows the stress-strain curves as a function of the number of stretch cycles. We observed that the pure PLA samples (figure 8A) had a higher stress at loading compared to the Pectin/PLA composite samples (figure 8B). These stress-strain curves reveal the mechanical behavior differences between these two samples. Particularly, in the first cycle, the loop (hysteresis) for the composite sample is significantly bigger than pure PLA samples. Figure 9 shows the relationship between hysteresis and stretch cycle. At the first cycle, it was evident that the hysteresis for the composite samples was greater than that of the pure PLA samples. Presumably, adding pectin to PLA led to a decrease in elasticity of samples, therefore increasing the hysteresis (energy loss) in cyclic tests. However, after the first cycle, there appeared to be little difference between the pure PLA and Pectin/PLA composite samples. Figure 10 demonstrates the stress as a function of stretch cycles. For the pure PLA sample, figure 10A shows very little change in peak stress, whereas figure 10B clearly demonstrates that the peak stress steadily decreases as number of stretch cycles increase. This behavior implies that the addition of pectin weakened the composites and caused more permanent deformation at the first stress, therefore less force is needed to further stretch the sample. On the other hand, pure PLA samples have a homogenous structure, and have a higher peak stress than the Pectin/PLA composites. This higher stress indicates that PLA samples are structurally more resistant to a deformation than the Pectin/PLA composite. Because of its high resistance to deformation, the peak stress for PLA samples remains constant through various stretching cycles.

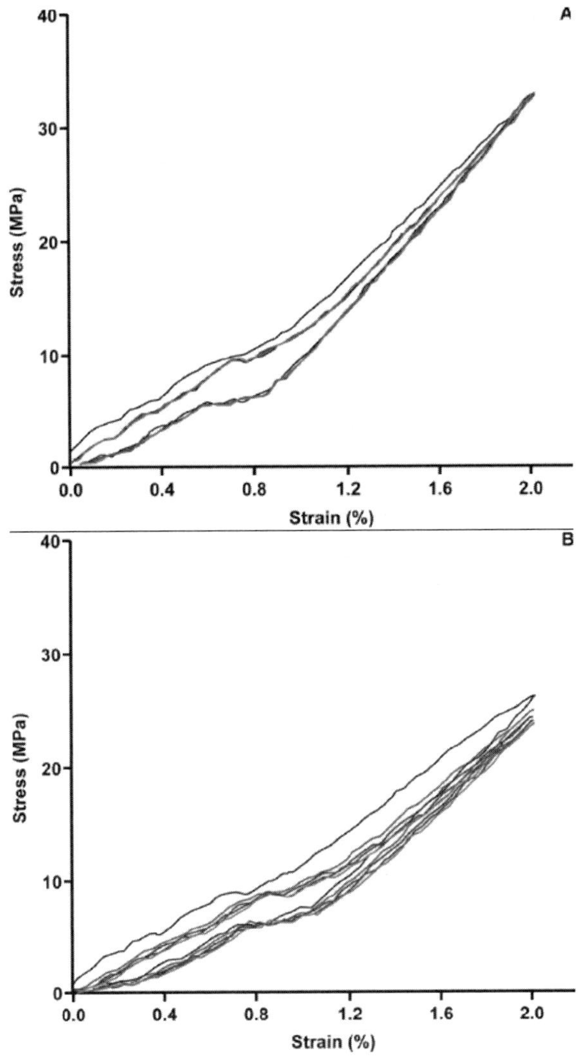


Figure 8. Stress-strain curves observed for cyclic tensile tests; (A) PLA samples and (B) Pectin/PLA composites. Cycling #1-#5 in sequence colored as black, purple, red, brown, and green. (Color figure can be only viewed in the online issue).

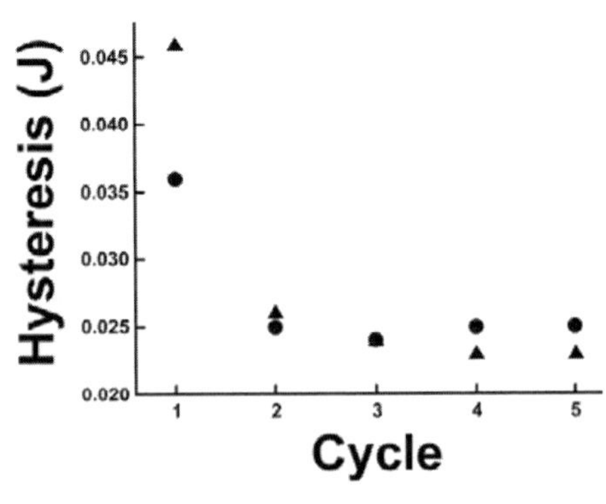


Figure 9. Energy loss (hysteresis) as a function of cycle. PLA samples ●, Pectin/PLA composites ▲.

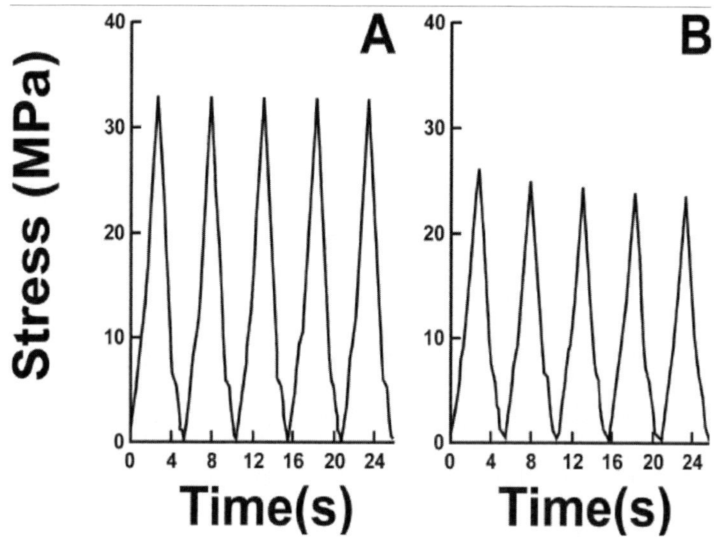


Figure 10. Stress curves as a function of time; (A) PLA samples and (B) Pectin/PLA composites.

The inclusion of pectin reduced the tensile strength and elongation to break of PLA. Nevertheless, the Pectin/PLA composites are still strong enough to serve as a packaging material, if one compared their tensile strength with other polymeric packaging materials, such as biodegradable blends from soybean flour protein and carboxymethylated corncob, 29 MPa,[17] or non-biodegradable polyvinyl chloride, 35 MPa, and polystyrene, 55 MPa.[19]

Samples were tested for antimicrobial activity by two methods: one is the agar diffusion method and the other, liquid phase culture. Figure 11 shows the images taken from the agar diffusion test. Without the pre-treatment of nisin solution, PLA (sample #1) and Pectin/PLA (sample #2) samples showed no antimicrobial activity. With the pre-treatment of nisin solution, PLA samples (sample #3) also showed no antimicrobial activity. Probably, it could be attributed to the hydrophobic nature of PLA surfaces that limits the nisin binding, while facilitating the loss of bound nisin during the washing process. In contrast, the Pectin/PLA composite samples that were pre-treated with nisin solution (sample #4) demonstrated a significant antimicrobial activity against *L. planturam*. No bacterial growth could be detected on the agar that was covered by sample #4. Furthermore, the diameter of inhibition zone around sample #4 was 2.41 ± 0.05 cm, whereas the ratio of the diameter of the zones of inhibition to the diameter of specimen was 1.5. This indicates that nisin was released from the composite film into the agar layer and inhibited the microbial cells growth on the agar.

To confirm this result, the samples were tested by incubation with a liquid medium containing same bacterial under standard condition. As shown in figure 12, there were no differences in microbial counts between sample #3 and the control at time points of 0, 2, 16, and 24 h. However, sample #4 exhibited a strong activity against *L. plantarum*. At 2 h incubation time, sample #4 had already reduced the cells from 5.1 logs to 2.5 logs. No colony in sample #4 was detected at a $10^{-1}$ dilution level (<10 cfu/ml) at 16 h and 24 h, whereas sample #3 had 9 logs and 9.2 logs of the cells, respectively.

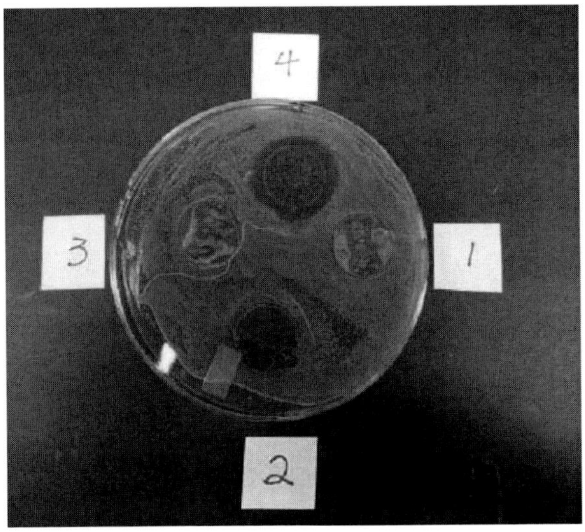


Figure 11 Antimicrobial effect on *L. Plantarum* growth by the agar diffusion method. Samples #1 and #3, PLA samples; #2 and #4, Pectin/PLA composites; sample #1 and #2, specimens without nisin solution pre-treatment; #3 and #4, with nisin solution pretreatment.

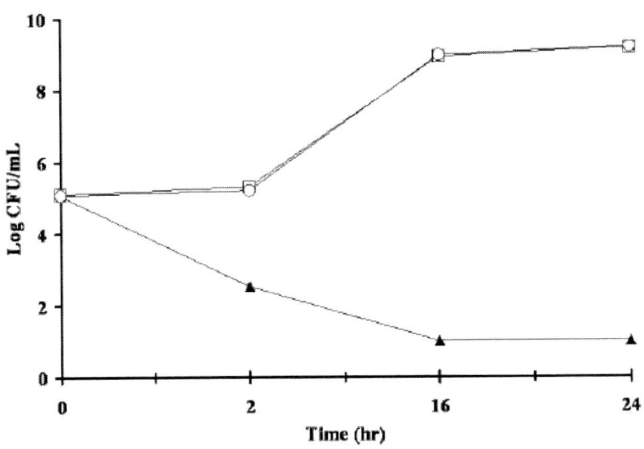


Figure 12 Antimicrobial effect on *L. Plantarum* growth by the liquid culture method. ▲ Pectin/PLA composites with nisin solution pre-treatment, ○ PLA samples with nisin solution pre-treatment, and □ control.

## CONCLUSIONS

By the incorporation of pectin particles into PLA matrix, we have prepared a composite that can absorb and store hydrophilic antimicrobial compounds, such as nisin. The resultant composite was able to inhibit bacterial growth in aqueous or gel phases by releasing the absorbed nisin. The incorporated pectin particles were located on the surface and extended deep into the materials, facilitating access and absorption of nisin into the composites. Although the mechanical properties of the composite were somewhat poorer than the films made from PLA alone, they were sufficiently good to produce a viable packaging material. Further research is needed to optimize the ratio of pectin to PLA. The goal is to balance absorption of antimicrobial reagents and retention of mechanical properties. Furthermore, the diffusion and release kinetics of nisin into and out of the films needs to be evaluated.

## ACKNOWLEDGMENT

We thank Dr. Peter H. Cooke, Mr. Nichols Latona, Ms. Guoping Bao, Mr. Brian Jasberg, Mr. Rick Haig, Ms. Kathy Hornback, and Mr. Gary Grose for technical and scientific assistance.

## REFERENCES

[1] Food Technology, www.itf.org, 03.06
[2] Cutter, C. N. Meat Science. 2006 74 131-142.
[3] Weber, C. J.; Huagaard, V.; Festersen, R.; Bertelsen, G. Food Additives and Contaminants. 2002, 19, 172.
[4] Langer, R.; Peppas, N. AIChE Journal. 2003, 49, 2990.
[5] Langer, R. Nature. 1998, 392, 5.
[6] Garlotta, D.; Doane, W. M.; Shogren, R. L.; Lawton, J. W.; Willett, J. L. J. Applied Polymer Science. 2003, 88, 1775.
[7] Suyatma, N.E.; Copinet, A.; Tighzert, L.; Coma, V. J. Polymer and the Environment. 2004, 12, 1.
[8] Petersen, K.; Nielsen, P.V.; Bertelsen, G.; Lawther, M.; Olsen, M.B.; Nilsson, N.H.; Mortensen, G. Trends in Food Science and Technology. 1999, 10, 52.
[9] Södergaard, A. Proceedings of the Food Biopack Conference, Copenhagen, 2000, 14.
[10] Ariyapitipun, T.; Mustapha, A.; Clarke, A.D. J. Food Prot. 1999, 63(1), 131.
[11] Alanson, A. J. Food Safety. 2000, 20, 13.
[12] Mustapha, A.; Ariyapitipun, T.; Clarke, A.D. J. Food Science. 2002, 67(1), 262.
[13] Krishnamurthy, K.; Demirci, A.; Puri, V.; Cutter, C.N. Transaction of the American Society of Agricultural Engineers. 2004, 47, 1141
[14] Liu, L.S.; Cooke, P. H.; Coffin, D. R.; Fishman, M. L.; Hicks, K. B. J. Applied Polymer Science. 2004, 92, 1893.
[15] Liu, L.S.; Won, Y. J.; Cooke, P. H.; Coffin, D. R.; Fishman, M. L.; Hicks, K. B.; Ma, P. X. Biomaterials. 2004, 25, 3201.
[16] Finkenstadt, V.; Liu, L.S.; Willett, J. L. J. Polym. Env. in press.
[17] Liu, L.S.; Fishman, M. L.; Hicks, K. B.; Liu, C. K. J. Agr. Food Chem. 2005, 53, 9017.
[18] Schilling, C. H., Tomasik, P., Karpovich, D. S., Hart, B., Garcha, J., Boettcher, P. T. J. Polym. Environ. 2005, 13(1), 57.
[19] Callister, W. D. Materials Science and Engineering, 5th Ed. John Wiley and Sons: New York, 2000; p 800.

In: Monomers, Oligomers, Polymers, Composites… ISBN: 978-1-60456-877-6
Editors: R. A. Pethrick, G.E. Zaikov et al. 

*Chapter 3*

# PECTIN COMPOSITE FILMS

***LinShu Liu*, Marshall L. Fishman and Kevin B. Hicks***

Eastern Regional Research Center, ARS, U.S. Department of Agriculture
600 East Mermaid Lane, Wyndmoor, PA 19038

## ABSTRACT

Blends of pectin with starch or proteins, or with synthetic hydrocolloids were investigated to characterize their ability to form strong self-supporting films. Microscopic analysis indicated that the blends were biphasic structure, except for pectin/proteins composites. Pure pectin films exhibited no thermal transitions, whereas the inclusion of plasticizers introduced a glass transition temperature to the pectin blends, as revealed by dynamic thermal mechanical analysis. The variation in composition, ratio and plasticizers determined the thermal transition property and mechanical properties of the blends, such as storage modulus, loss modulus, tensile strength and maximal elongation. Treating the pectin/proteins composite films with glutaraldehyde/methanol induced chemical cross-linking with the proteins and reduced the interstitial spaces among the macromolecules and, consequently, improved their mechanical properties and water resistance. Furthermore, antimicrobial reagents can be loaded onto the composite films, the resultant films can be used for food packaging or wrapping. The article summarizes the research results on pectin films conducted at the Eastern Regional Research Center, Agricultural Research Service, USDA, exploring the new utilities of pectin.

## 1. INTRODUCTION

Pectin is a cell wall polysaccharide. It consists of one third of the cell wall dry substances of higher plants, and occupies a much less proportion in lower plants such as grasses.

---

* Correspondence to LS Liu. Tel. +1-215-233-6486; Fax: +1-215-233-6406. Email address: LinShu.Liu@ars.usda.gov;
Mention of brand or firm name does not constitute an endorsement by the U.S. Department of Agriculture above others of similar nature not mentioned.

Commercial pectin is mainly derived from citrus peels and apple pomace, also from sugar beet pulp and sunflower heads. All those are the products, coproducts, or byproducts of agricultural processing. Thus, the production of pectin and the research and development on pectin derived non-food products address two important issues in agribusiness: the use of surplus and the development of value-added materials.

The majority of pectin structure consists of homopolymeric partially methylated poly-α-(1⟶4)-D-galacturonic acid residues ("smooth" areas), but there are substantial "hairy" areas of alternating α -(1⟶2)-L-rhamnosyl-α -(1⟶4)-D-galacturonosyl sections containing branch-points with mostly neutral side chains (1 - 20 residues) of mainly L-arabinose and D-galactose (rhamnogalacturonan I). Pectins may also contain rhamnogalacturonan II with sidechains containing other residues such as D-xylose, L-fucose, D-glucuronic acid, D-apiose, 3-deoxy-D-*manno*-2-octulosonic acid and 3-deoxy-D-*lyxo*-2-heptulosonic acid attached to poly-α-(1⟶4)-D-galacturonic acid regions. The types and amount of substructural entities in pectin preparations depend on their source and extraction methodology. Commercial extraction causes extensive degradation of the neutral sugar-containing side chains. Pectin molecule does not adopt a straight conformation in solution, but is extended and curved with a large amount of flexibility. The carboxylate groups tend to expand the structure of pectin. Methylation of these carboxylic acid groups forms their methyl esters, which are much more hydrophobic and have a different effect on the structuring of surrounding water. Thus, the properties of pectin depend on the degree of esterification (D.E.). High D.E. pectin (> 40% esterified) tends to gel by the formation of hydrogen-bonding and hydrophobic interactions at low solution pH (pH ~3.0, to reduce electrostatic repulsions) or in the presence of sugars (> 70% esterified); while, low D.E. pectin (< 40% esterified) gels by calcium di-cation bridging between adjacent two-fold helical chains forming so-called "egg-box" junction zone structures so long as a minimum of 14-20 residues can cooperate [1-4].

Closely related to the gelling properties, pectin is a well established film forming agent. In cell wall, pectin associates with cellulose, hemicellulose, lignin, proteins and metal ions to form a physical barrier, which conducts mass transportation, signal transduction and protect the cells from environmental invasion. In isolated form, pectin readily reassociates or aggregates to form networks, and interacts with proteins, other polysaccharides and synthetic hydrocolloids via hydrogen bonding, ionic or hydrophobic interactions. This character has led to applications of pectin in encapsulation, coating, packaging and wrapping for food and pharmaceutical products.

In this review article, we summarized the recent work in our laboratory on the fabrication and characterization of pectin composite films. Materials used for blending with pectin include high-amylose starch [5-7], fish skin gelatin (FSG), soybean flour protein (SFP) [8, 9], poly(vinyl alcohol) (PVA) [10,11], and poly(ethylene oxide) [12]. The resultant pectin composite films have demonstrated to possess diverse physical, biological, and chemical properties, which can be tailored to satisfy various applications. Starch, gelatin and oilseed proteins have demonstrated a good film-forming property and have a long history of safe use in the food and food packaging industries, and so the synthetic hydrocolloids of PVA and PEO [12-14].

## 2. PECTIN/STARCH FILMS

Pectin/starch composite films were prepared by mixing solutions of pectin and glycerine with gelatinized starch solutions, casting them on a LEXAN plate using a Micro film applicator (Paul N. Gardner Co., Pompano Beach, FL), and allowing the films to air-dry overnight. After air-drying, the samples were vacuum-dried for 30 min at room temperature. Films were removed from the coating plates with a razor blade. For comparison purpose, other pectin composite films discussed in this article were also prepared by the same method. The resultant films were tested for structural and mechanical properties and other properties. Small deformation analysis, using a Rheometrics RSA I1 solids analyzer (Piscataway, NJ), was performed as a complement of microscopic examination.

In general, the pectin/starch films showed a two-phase structure. The pectin formed a homogeneous smooth matrix phase, within which the starch particles distributed and embedded. The size and distribution of the filler phase highly depended on the degree of starch gelatinization. The more the starch granules gelatinized (i.e., treated with microwave oven for a longer time), the more homogeneous films could be obtained.

The presence of the starch had an effect on the mechanical properties of composite films. The film containing 35% starch was noticeably more brittle than films containing less starch. As the amount of starch present increased from 0% to 35% by weight, both the storage modulus (E') and loss modulus (E") of pectin/starch composite films gradually decrease. At 35% starch both moduli were one-third lower than for the sample containing no starch. These differences were consistent over the temperatures ranging from 25 °C to 210 °C. In comparison with some commercially available thermoplastics, all samples tested had values of E' and E" at room temperature that were approximately an order of magnitude higher than those for polyethylene. While the high film moduli were encouraging because they open up many potential uses, the films were too brittle for use in many applications. Therefore, plasticizers were added to the system to obtain films which were more flexible and less susceptible to brittle failure. Four plasticizers were used: urea, glycerine, poly(ethylene glycol) 300, and poly(ethylene glycol) 1000. Glycerine was distinguished from the others. The glycerol breaks intermolecular pectin-pectin hydrogen bonds through preferential solvation (i.e. substitution of pectin-pectin hydrogen bonds with pectin-glycerol hydrogen bonds), and thereby facilitates chain slippage when pectin films are stressed. The level of glycerine present in the films had a noticeable effect on the tenacity and elongation to break of the films. Both elongation and tenacity roughly doubled as the glycerine content was raised from 9% to 19%. No further increase was seen at 27% glycerine. Significantly higher plasticizer levels may be necessary to obtain large increases in elongation to break.

It was also noted that films containing only pectin were slightly tacky. This tackiness seemed to be reduced or eliminated in the samples that contained starch. Increasing the glycerine content increased the tack of the films; however, this was overcome at higher starch levels. It appears that the starch is a useful additive to control or eliminate tack in these films.

In conclusion, plasticized pectin/starch blends can be made into strong, fairly flexible films with tensile strengths on the order of 3 X $10^8$ dyn/cm$^2$, approximating those of commercial plastic films, and elongations of 1-3 %. The room temperature storage and loss moduli of the films were 1.5-6 x $10^{10}$ and 1.3-5 x $10^9$ dyn/cm$^2$, respectively, depending on composition. This is equal to or higher than what is found in many commercial films.

## 3. Pectin/Proteins Filmns

Pectin/protein composite films were prepared from pectin with several proteins, such as bovine serum albumin (BSA), chicken egg albumin (CEA), type B bovine skin gelatin (BSG), type A porcine skin gelatin (PSG), fish skin gelatin (FSG), and type I soybean flour protein (SFP). The pectin and protein composite films possess diverse physical, biological, and chemical properties, which can be tailored to satisfy various applications.

The inclusion of protein promoted molecular interactions, resulting in a homogeneous structure, as revealed by scanning electron microscopy, confocal laser scanning microscopy, and fracture-acoustic emission analysis [8, 9]. The treatment of pectin/protein composite films with glutaraldehyde/methanol resulted in a well organized microstructure (figure 1). The glutaraldehyde treatment introduced protein interactions via bridging of their $-NH_2$ groups. The resultant protein packs (with the size of 100-300 nm scale) were arranged into fibers; the fibers were then parallel to each other to form a tightly packed, non-woven structure. The micrographs also showed there were gaps between adjacent packs with the size of <5 nm (figure 1, indicated by arrow), forming closer connections, and crevices ranging from 50 to 100 nm between parallel fibers (figure 1, indicated by triangle), forming relatively loose connections. Since under the experimental conditions, the glutaraldehyde-mediated reactions only occurred between the active aldehyde and primary amines of the proteins and could not be referred to the hydroxyl groups of polysaccharides [15], the "free" pectin macromolecules, supposedly, were packaged within or penetrated through the protein packs [16]. Although the glutaraldehyde does not cause pectin-pectin or pectin-protein cross-linking, the methanol, as a dehydrate reagent, could reduce the free spaces between macromolecules and, consequently, enhance the polysaccharide chain-chain interactions, such as hydrogen bonding and hydrophobic interactions.

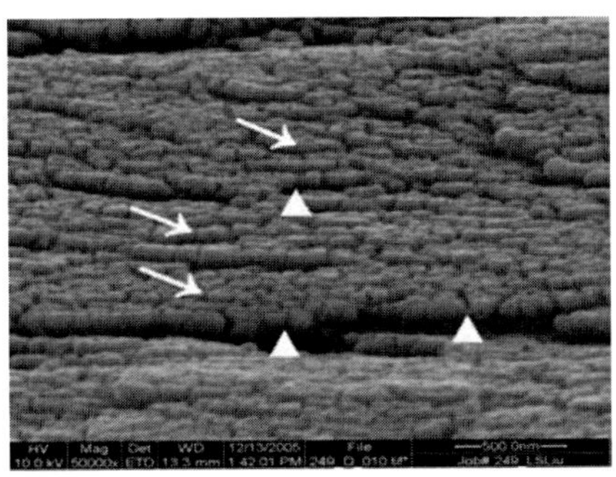

Figure 1. SEM photograph of frozen-fractured pectin/FSG film pretreated with 0.1% glutaraldehyde in methanol, showing gaps (arrow) between adjacent packs with the size of <5 nm, and crevices (triangle) with the size ranging from 50 to 100 nm between parallel fibers.

To verify this hypothesis, the pectin/protein films were immersed in solutions at three different pH, their swelling behavior and the amount of released pectin and proteins were measured. It was found that the chemical treatment suppressed protein release, but not pectin release. The dissolution of pectin/proteins composite films was pH-dependent. The dissolution rate increased in the sequence of pH 4.0 < pH 7.2 < pH 8.5. The pectin-FSG and pectin-SFP films did not dissolve at any pH tested, but did display a pH-dependent swelling behavior. Both composite films swelled least at pH 4.0. As the solution pH increased, the composite films were more swelled. Measurements at the same pH revealed that pectin films containing SFP swelled into a larger size than pectin films containing an equal amount of FSG. The effect of pH on the dissolution of the paired biopolymers could be referred to the electrostatic interactions between the two biopolymers. These electrostatic interactions are responsible for the mechanism of the formation-deformation of the electrostatic complex. Accordingly, the values of p$K_a$ or p$I$, respectively, are around 4 for pectin, 4.8-5.2 for FSG, and 4.5-5.1 for SFP; electrostatic complexes between pectin and proteins could be formed at pH 4 solution. This may stabilize the films and suppress their swelling. For the composite films without chemical treatment, about 50% of pectin or proteins were released into the dissolution buffer at pH 4.0 in 48 h, whereas the values increased as the solution pH increased and reached the highest decomposition around 80-90% at pH 8.5. The chemical treatment did not influence the release of pectin from composite films, but did suppress the release of protein. This was more pronounced for the films containing FSG than SFP; less than 20% of the incorporated FSP released into the three dissolution buffers, whereas the SFP release seemed to be more pH-dependent even after chemical treatment. This could be attributed to more primary amine being available in FSG than in SFP, which generated more cross-linked bonds with FSG films than with SFP films.

The effects of protein blending and the following chemical treatment on other water resistant properties of resultant pectin composite films, such as water adsorption and water vapor transmission were summarized on table 1. The protein-free pectin films showed the highest values in both water adsorption and water vapor permeability. The chemical treatment showed little impact on water adsorption of pectin films but dramatically reduced the water vapor permeability of the films. As discussed in the above section, film dehydration in anhydrous methanol may enhance the level of chain-chain packing and reduce interstitial spaces among the pectin molecules. However, this is a reversible process. After conditioning in a more humid environment for a longer period, that is, at 95% relative humidity for 2 weeks as in this study, the dehydrated films could gradually be re-hydrated; thus, their water adsorption capability could be partially recovered. For pectin-FSG composite films, both water adsorption and water vapor permeability were suppressed. The smaller values of water adsorption and WVTR were obtained at higher FSG content. The chemical cross-linking further suppressed water adsorption and water vapor transmission; these values were even lower than those obtained for transglutaminase-modified protein and pectin-protein films [16, 17]. The blends of SFP with pectin and the subsequent chemical cross-linking showed an impact on water adsorption and penetration, which had the same trend as the inclusion of FSG but smaller. It is consistent with the results from the dissolution studies.

The structural characteristics have an impact on the mechanical properties of the blend. Table 2 shows the stiffness, strength, and flexibility of some blends. The inclusion of FSG or SFP remarkably enhanced both the tensile strength and the elongation of pectin films. The chemical treatment further strengthened the films. The chemical treatment also has a tendency to produce a stiffer composite as indicated by the increase in tensile modulus and the decrease in elongation. Adequate tensile strength is very important in manufacturing polymeric films, where the material is often subjected to a force during mechanical stretching. In a variety of end uses, products must be capable of resisting considerable stress without fracture. The pectin-protein composite films exhibit tensile strengths as high as 24-59 MPa (table 2). By comparison with the tensile strength of 29 MPa for biodegradable blends from SFP and carboxymethylated corncob [17], 35 MPa for non-biodegradable polyvinyl chloride, and 55

MPa for polystyrene [18], the composite materials presented in the current study appear to be promising candidates for biodegradable wrapping and packaging materials.

**Table 1. Water Resistant Property of Pectin/Protein Films**

| material[a] | water adsorption[b] (%) | WVTR[c] ($g \times m^{-2} \times day^{-1}$) |
|---|---|---|
| pectin ⌧,§,⊥,∧,◆ | 47 ± 6 | 226 ± 11 |
| pectin-gt[d]* ⌧ | 42 ± 7 | 178 ± 23 |
| FSG ▲ | 28 ± 4 | 98 ± 11 |
| FSG-gt ▲ | 6 ± 2 | 112 ± 15 |
| pectin-FSG (0.1)[e,#,⊥,§] | 32 ± 4 | 147 ± 7 |
| pectin-FSG (0.1)-gt §,† | 12 ± 2 | 103 ± 12 |
| pectin-SFP (0.1) ⊥,‡ | 37 ± 5 | 196 ± 16 |
| pectin-SFP (0.1)-gt ⊥,†,‡ | 17 ± 6 | 158 ± 17 |
| pectin-FSG (0.2) ◆,§ | 26 ± 4 | 121 ± 10 |
| pectin-FSG (0.2)-gt § | 8 ± 3 | 114 ± 9 |
| pectin-SFP (0.2) ∧,· | 29 ± 6 | 184 ± 15 |
| pectin-SFP (0.2)-gt · | 11 ± 2 | 161 ± 18 |

[a] The following symbols indicate statistical significance ($p < 0.01$): ⌧, §, ⊥, †, ◆, ‡, ·, ▲, ∧. [b] Measured by weight gain after conditioning at 95% relative humidity at room temperature. [c] Determined by ISO 2528 (1995E). [d] -gt* indicates the treatment of films with 0.1% glutaraldehyde/methanol solution. [e] Data in parentheses indicate the weight percent of protein in composite.

**Table 2. Mechanical Properties of Pectin/Proteins Composite Films[a]**

| naterial | tensile modulus (MPa) | tensile strength (MPa) | elongation (max) (%) |
|---|---|---|---|
| pectin •,#,▲,§ | 1082 ± 168 | 17.0 ± 3.4 | 2.5 ± 0.6 |
| pectin-gt*[b] • | 4139 ± 766 | 59.2 ± 11.1 | 1.7 ± 0.5 |
| FSG ‡ | 1906 ± 34 | 71.8 ± 0.9 | 6.4 ± 2.4 |
| FSG-gt ‡ | 3132 ± 319 | 99.2 ± 6.1 | 3.6 ± 0.6 |
| pectin-FSG (0.1)[c,#,∧] | 1825 ± 43 | 43.5 ± 7.6 | 3.0 ± 1.5 |
| pectin-FSG (0.1)-gt †,∧ | 3306 ± 86 | 54.2 ± 6.9 | 2.1 ± 0.4 |
| pectin-SFP (0.1) §,⌧ | 1213 ± 286 | 24.0 ± 3.0 | 2.9 ± 0.9 |
| pectin-SFP (0.1)-gt †,⌧ | 2158 ± 113 | 33.0 ± 1.7 | 1.7 ± 0.4 |
| pectin-FSG (0.2) ▲ | 2178 ± 224 | 59.1 ± 12.4 | 3.2 ± 1.1 |
| pectin-FSG (0.2)-gt | 3016 ± 58 | 58.9 ± 1.8 | 2.6 ± 0.3 |

[a] The following symbols indicate statistical significance ($p < 0.01$): #, †, ‡, ▲, •, ∧, §, ⌧. [b] -gt* indicates the treatment of film with 0.1% glutaraldehyde/methanol.

[c] Data in parentheses indicate the weight percent of protein in composite.

During a tensile test, composite deformation and fracture are accompanied by a rapid movement, relocation, or breaking of structural elements such as fillers, fibers, matrices, and their interfacial areas. As a result, sound waves are produced that can be detected by an acoustic transducer and converted into electronic signals named as "hit" using an acoustic emission (AE) analyzer. Figure 2 showed the relationship between the total elastic energy released by an acoustic event in response to the maximal stress that specimens were subjected to. There is a clear correlation between the tensile strength and AE energy released at fracture. However, some spreading is also observed. For instance, the AE energy released for composite films with 10% proteins was detected lower than those of protein-free pectin films [comparing pectin film (●) with the composite films of 10% SFP (○) or FSG (□) shown in figure 2]. The energy loss could be attributed to some structural factors, such as signal attenuation due to scattering or absorption losses during sound wave propagation from the AE source to the transducer and internal energy dissipation by the friction and toughening mechanisms. The preliminary results suggested that proteins might function as a lubricant in the blend structure. Observations indicated that the composite films, after blending with an adequate amount of protein, might be able to abate wave propagation. Moreover, data also showed AE energy increases with protein content.

We also studied the correlation between the stress-strain curve and AE hit rate pattern by referring one to another in order to get better understanding on the fracture mechanisms and how the structure changes effect the property of a sample. For films consisting of a single component, there were no acoustic events before the peak stress; the AE activities occurred exclusively at the peak stress when the specimens completely fractured. This behavior is due to their homogeneous structure, in which the single component specimens were able to transfer the stress evenly. In contrast, the composite films emitted sound at an earlier strain due to the micro-structural movement of individual components, which was correlated to the increase in the initial slope of the stress-strain curves (data not shown). Observation also revealed that a sudden increase in

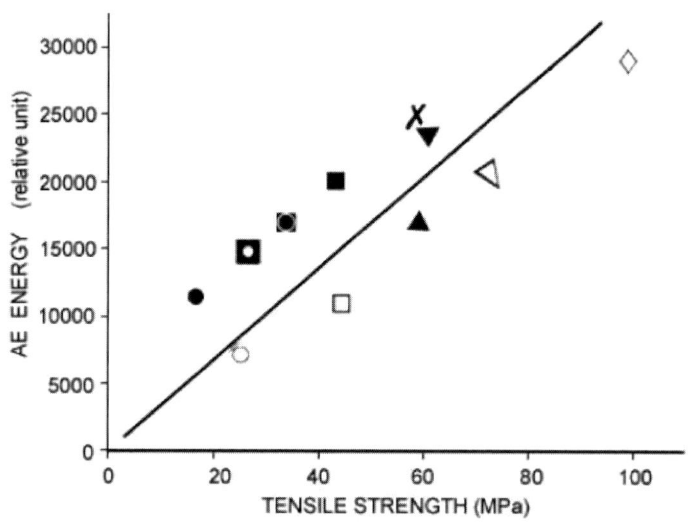


Figure 2. Acoustic energy versus tensile strength of various pectin/protein films: pectin (closed circle), pectin-gt (X), FSG (open triangle), SFG-gt (diamond), pectin-FSG (0.1) (open square); pectin-SFP (0.1) (open circle), pectin-SFP (0.1)-gt (closed circle in a square); pectin-FSG (0.2) (up closed triangle); pectin-FSG (0.2)-gt (closed down triangle); pectin-SFP (0.2) (open circle in square); pectin-SFP (0.2)-gt (closed square).

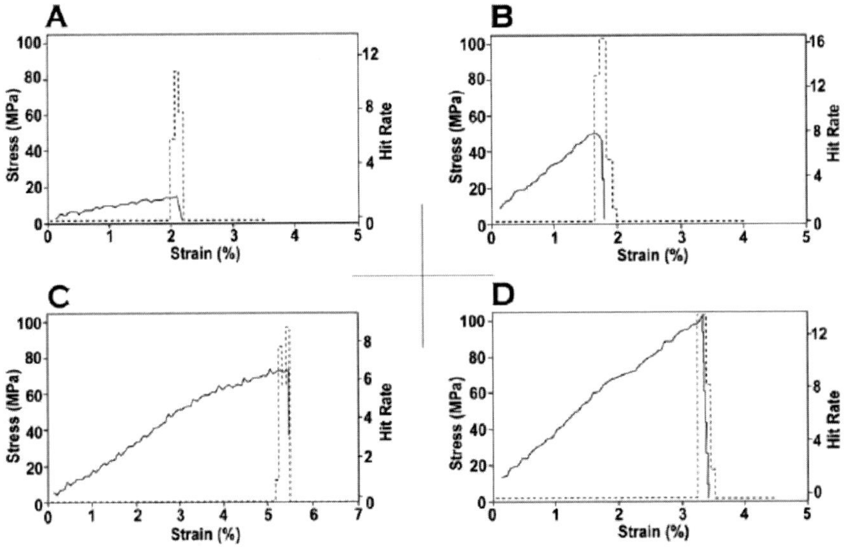


Figure 3. Relationship of strain-stress curve (solid line) with the acoustic emission hits (dotted line): pectin (A) and FSG (C) films; glutaraldehyde/methanol-treated pectin (B) and FSG (D) films.

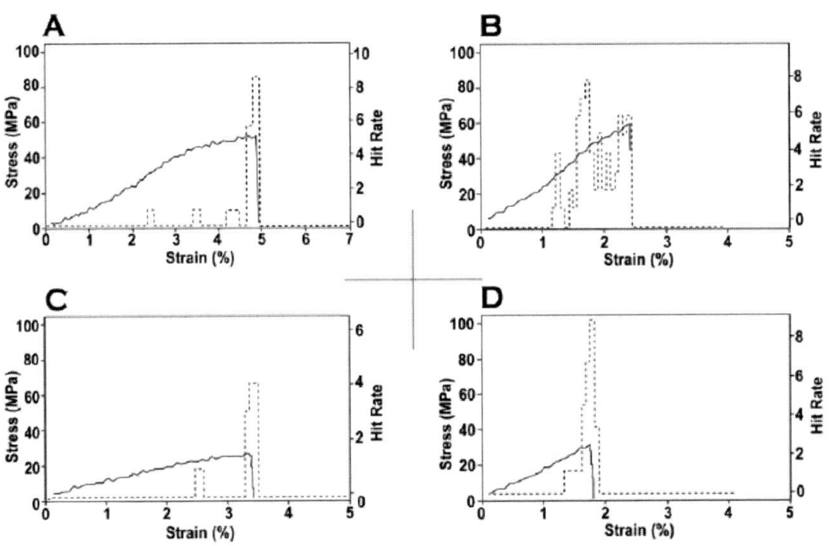


Figure 4. Relationship of strain-stress curve (solid line) with the acoustic emission hits (dotted line): pectin-FSG (A) and pectin-SFP (C) films; glutaraldehyde/methanol-treated pectin-FSG (B) and pectin-SFP (D) films.

AE hits occurred at the peak stress. Furthermore, AE hits are more frequent and are more evident for cross-linked films than for those not cross-linked. The cross-linked composite films produced a wide band of acoustic waves at a much earlier strain (figure 4 B and D). This behavior may be ascribable to structural defects such as crevices found between fibers in cross-linked blend films (figure 1). If the direction of those crevices and the direction of film elongation coincide, AE events would be further enhanced.

In conclusions, the results presented in this study indicate that inclusion of proteins into pectin films improved both mechanical strength and flexibility. The treatments of resultant composite films with glutaraldehyde/methanol further enhanced film strength and reduced water vapor permeability, while retaining the flexibility of the original pectin films to some degree. However, it appears that only chemical cross-linking can suppress the films' solubility in water, because methanol treatment is a dehydration process that reduces only the interstitial spaces between macromolecular chains and is reversible. The results suggest the potential of pectin and protein composite films in the applications of wrapping or packaging materials compared to other commercial films, which requires moderate mechanical strength and low water vapor transmission.

## 4. PECTIN/PVA FILMS

PVA is constructed from nonrenewable resources and is not as environmentally friendly as pectin. It would be advantageous to replace PVA with pectin, because PVA films are widely used in industries. Examples include water-soluble pouches for packaging detergents and insecticides, flushable liners and laundry bags for contaminated linens, coating for paper and a temporary protective coating for metal finishes. Since the utility of a material as a coating is closely related to its film forming properties, pectin might replace PVA in these applications.

It is of interest to note that pectin/PVA films for the most part appear transparent to the naked eye. Optically dense phases appeared to be interconnected through lucent phases. The existence of two distinct phases in pectin/PVA films was confirmed by phase-contrast optical microscope. Pure pectin film was uneven and consisted of small ridges and crevices, which were oriented parallel to the film plane (data not shown). Pure PVA film appears smooth except for a few small linear ridges (data not shown). For composite films, the fracture plane becomes increasingly smoother with increasingly lower weight ratios of pectin to PVA. This may indicate the coating of pectin with PVA at higher PVA content. Pectin/PVA films form compatible biphasic composites over a wide range of compositions.

These films undergo brittle to ductile transitions with increasing concentrations of PVA. Addition of glycerol to pectin/PVA films significantly increased the ductility of these films when relatively brittle. The solubility of pectin/PVA films were temperature and composition dependent. At 30, 50 and 70°C respectively, films containing 30% or less PVA were more soluble than pure PVA films measured at the same temperature. At 70°C all compositions of films containing pectin/PVA are more soluble than pure PVA films. In general, addition of PVA to pectin films resulted in films with more PVA-like properties, and addition of pectin to PVA films gave more pectin-like properties to PVA films. Increasing the amount of PVA in the blends reduced the storage and loss modulus of the films above the glass transition temperature. The values of glass transition temperature observed decreased as the amount of PVA in the blend increased. Addition of glycerol depressed the glass transition temperature of PVA and merged it into the glass transition temperature of the pectin/glycerol blend. Changes in the molecular weight and degree of ester hydrolysis of PVA exerted a rather small effect on the blends.

Because PVA has good tensile strength and high elongation to break, the inclusion of PVA to pectin blends had good impact on the toughness of the resultant composites. The water solubility of the two polymers facilitates blending and processing.

## 5. PECTIN/PEO FILMS

Poly(ethylene oxide), PEO, is a food hydrocolloids. Pectin and PEO composite films were prepared for food packaging purpose. Microscopic analysis revealed the composite films with well-mixed integrated structures made of evenly distributed synthetic hydrocolloids in the biopolymers (figure 5).

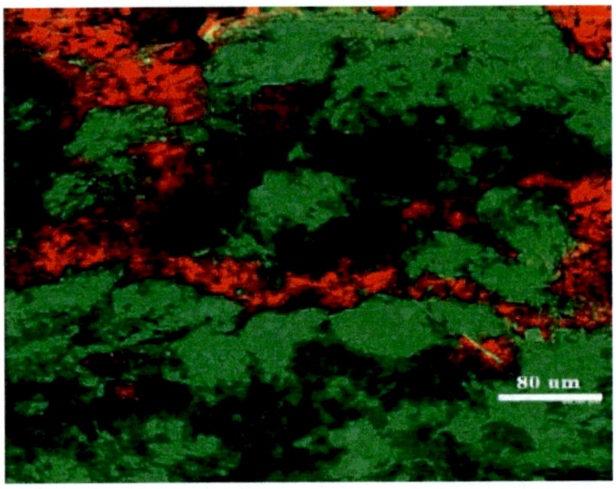


Figure 5. Confocal laser scanning microimages of pectin/PEO composite film. The organization of the biopolymers were resolved by confocal fluorescence (excitation 484 nm, emission 520-580 nm), the PEO was defined by confocal reflection (633 nm). The micrograph was collected in stereo projection in extended focus images of 20-30 micrometer-thick slabs of the film. Field width, 470 μm.

In a variety of end uses, packaging and wrapping materials are often subjected to a force during tensile strain. The materials must be able to resist considerable stress without failing to a fracture at a designed stress. Furthermore, as edible food wrapping materials, the materials may be taken with foods together either for convenient purpose or to enhance or alter the food texture. In these cases, their mechanical properties directly relate to the mouth feel, which is an important measurement of food quality. Table 3 shows the mechanical properties of selected samples of compression and blown films. In general, the composite films have mechanical properties, such as Young's modulus and tensile strength that are similar to cast films from most natural hydrocolloids that are consumed in our ordinary life. The changes in PEO content dramatically enhanced the mechanical properties of the composite films. The results of the stress-strain cyclic tests are shown in figure 6. For the PEO-free samples, the loop created by the first cycle is bigger than those created by following cycles. Among the following cycles, the difference in loop size is not significant. For samples containing PEO, the size of loop decreased gradually as cycled, then, became constant in the last two cycles. At the end of the cyclic test, the PEO containing sample expressed a much higher stress than the PEO-free sample. Although the PEO free samples are more resistant to mechanical force and less permanent deformation occurred than with the PEO-containing sample; the inclusion of PEO strengthens the composite films.

**Table 3. Mechanical properties of pectin/PEO composite films**

| Samples | Tensile strength (MPa) | Elongation at break (%) | Young's modulus (MPa) | Toughness (J/cm$^3$) |
|---|---|---|---|---|
| I, (10:90) | 2.35 ± 0.09 | 6.14 ± 0.6 | 77.2 ± 4.49 | 0.16 ± 0.08 |
| II (30:70) | 8.8 ± 0.3 | 13.1 ± 2.0 | 353.4 ± 26.9 | 0.90 ± 0.12 |
| III (50:50) | 3.1 ± 0.8 | 7.3 ± 1.5 | 125.2 ± 25.3 | 0.20 ± 0.07 |
| IV (70:30) | 2.3 ± 0.3 | 2.8 ± 0.3 | 151.2 ± 20.8 | 0.03 ± 0.01 |

Ratio in parenthesis: pectin/PEO, w/w.

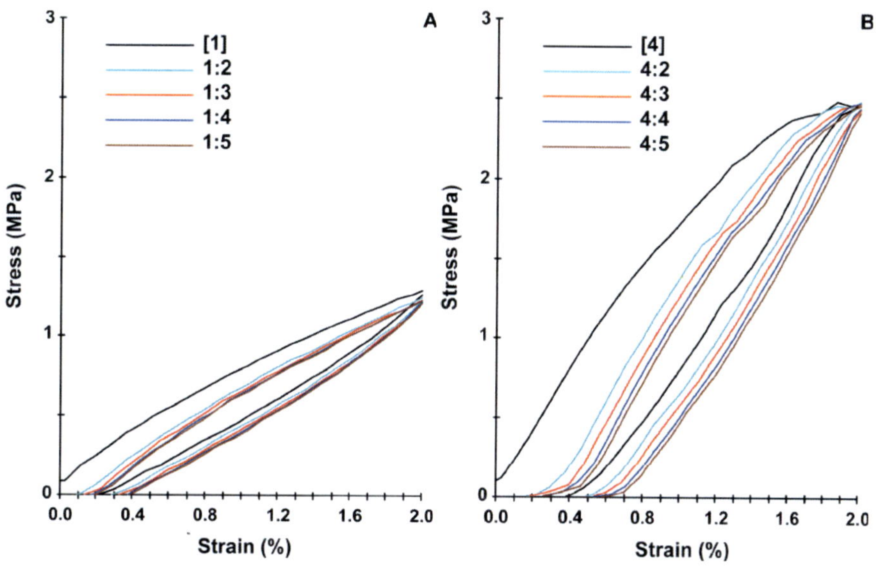


Figure 6. Stress-strain curves obtained from cyclic tensile tests for samples of pectin without PEO (A) and with PEO (B). For the PEO free films, the loop created in the first cycle is larger than following cycles. For the PEO-containing films, the size of loops gradually decreased as cycled, then became constant in the last two cycles.

The addition of PEO has an influence on film destruction caused by an external destructive force. Without PEO, the external force created a clear-cut fracture surface (data not shown), indicating the good adhesion between the two biopolymers. With the inclusion of PEO, the deformation created a fibrous surface (data not shown). This can be seen clearly from SEM and fluorescent microscopy. As shown in figure 7 A and B, fibers were pulled out, extended, and then, broken, but still embedded in the matrix phase. We examine the fibers with confocal reflection and confocal fluorescence in two channels. It confirms that the main component of the fibers is PEO; however, the biopolymers were either inserted or encapsulated within the fibers (figure 7C).

Figure 7. Microscopic images of the fractural surfaces of pectin/PEO composite films obtained by scanning electron microscope (left), laser microscope (middle), and confocal laser scanning microscope in confocal fluorescence and confocal reflection two channels (right). Field width: 520 μm (left) and 480 μm (middle and right).

In the spectrum of food packaging and wrapping materials, antimicrobial film is a new member. Besides providing a physical barrier, the films function by prohibition, protection and suppression of microbial migration to or growth in the packages by creating antimicrobial surfaces or releasing antimicrobial substances. The use of antimicrobial materials in food packaging improves food safety and is more convenient to the consumers; therefore, the market for antimicrobial food packaging is projected to grow rapidly [19, 20]. In the present study, we incorporated an antimicrobial polypeptide, nisin, into the film formulation. The antimicrobial activity of resultant films is shown in figure 8. Without nisin incorporation, films were not active at all (diamond). In contract, the nisin incorporated composite films inhibited bacterial growth.

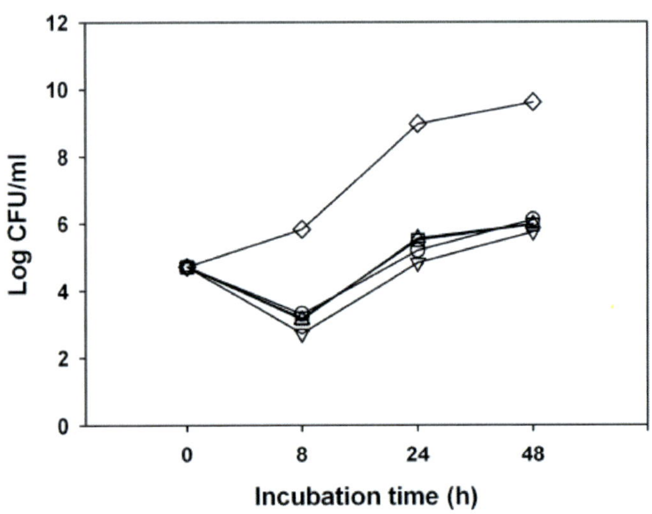


Figure 8. Growth of *Listeria monocytogenes* in BHI broth at 24 °C. Control (diamond), Nisaplin® prior to loading (square), Nisaplin® post loading (circle), Nisaplin® in composite film I (up triangle), and Nisaplin® in composite film II (down triangle).

## 6. SUMMARY

Pectin is used as food ingredients, although pectin is not a energy contributor to the body and only digested by colonic microflora. Pectin is water soluble and readily to associate with other polysaccharides, proteins, polyelectrolytes, and synthetic hydrocolloids to form networks. New uses for pectin derived products are needed to better utilize abundant crop processing residues. Edible and/or antimicrobial packaging materials were developed from pectin and starch, pectin and proteins, or pectin and food-grade synthetic hydrocolloids. The composite films possess the mechanical properties similar to those of petroleum-derived thermoplastics that are currently used for food packaging, and can absorb antimicrobial agents and control their release, serving as an active barrier for the inhibition of bacterial migration and growth. The edible films provide a barrier to moisture, lipids, and aromas, and can be used to alter or retain the food texture, based on the ratio of pectin and other film components. These new packaging materials are green biobased products that can replace petroleum-based products.

## REFRENCES

[1] Voragen, A.G.J., Plinik, W. Pectins. In *Food Polysaccharides and Their Applications*. Stephen, M.A., Ed. New York, NY: Dekker, 1995; pp 287-323.

[2] Yamada, Y. In *Pectins and Pectinase*. Visser, J., Voragen, A.G.J., Eds. Amsterdam: Elsevier, 1996; pp 173-190.

[3] Glicksman, M. Food Hydrocolloids; CRC Press: Boca Raton, FL, 1984;Vol. 111.

[4] Morris, G.A., Foster, T.J., Harding, S.E. The effect of the degree of esterification on the hydrodynamic properties of citrus pectin. *Food Hydrocolloids.* 14:227-325(2004).

[5] Fishman, M.L., Coffin, D.R., Unruh, J.J., Ly, T. Pectin/starch/glycerol films: blends or composites? *J. Macromol. Sci.* A33 (5):639-654 (1966).

[6] Coffin, D.R., Fishman, M.L. Viscoelastic properties of pectin/starch blends. *J. Agric. Food Chem.* 41:1192-1197(1993).

[7] Coffin, D.R.; Fishman, M.L. Physical and mechanical properties of highly plasticized pectin/starch films. *J. Appl. Polym. Sci.* 54:1311-1320(1994).

[8] Liu, LS., Liu, C.K., Fishman, M.L., Hicks, K.B. Composite films from pectin and fish skin gelatin or soybean flour protein. *J. Agr. Food Chem.* 55(6):2349-2355(2007)

[9] Liu, LS.; Tunick, M.; Fishman, M.L.; Hicks, K.B.; Cooke, P.H.; Coffin, D.R. Pectin based networks for non-food application. In *Advances in Biopolymers: Molecules, Clusters, Networks and Interactions*; ACS Symposium Series 935; Fishman, M.L., Wicker, L., Qi, P.X., Eds.; American Chemical Society: Washington, DC, 2006; pp 272-283.

[10] Fishman, M.L.; Coffin, D.R. Mechanical, microstructural and solubility properties of pectin/poly(vinyl alcohol) blends. *Carbohydr Polym.* 35:195-203(1998).

[11] Coffin, D.R., Fishman, M.L., Ly, T.V. Thermomechanical properties of blends of pectin and poly(vinyl alcohol) . *J. Appl. Polym. Sci.* 61:71-79(1996).

[12] Liu, LS., Jin, T., Liu, C.K., Hicks, K.B., Mohanty, A.K., Bhardwaj, R., Misra, M. Incorporation of bacteriocin in edible pectin films for antimicrobial packaging.

Presented at the 233rd National meeting of the American Chemical Society, Chicago, IL, March 25-29, 2007: AGRO 202.

[13] Cha, D.S., Chinnan, M.S. Biopolymer-based antimicrobial packaging: a review. *Crit. Rev. Food Sci. Nutr.* 44:223-237(2004).

[14] Cutter, C.N. Opportunities for bio-based packaging technologies to improve the quality and safety of fresh and further processed muscle foods. *Meat Sci.* 74:131-142(2006).

[15] Liu, L.S., Thompson, A.Y., Heidaran, M.A., Poser, J.W., Spiro, R.C. An osteoconductive collagen/hyaluronate matrix for bone regeneration. *Biomaterials* 20, 1097-1108(1999).

[16] Pierro, P.Di., Mariniello, L., Giosafatto, C.V.L., Masi, P., Porta, R. Solubility and permeability properties of edible pectin-soy flour films obtained in the absence or presence of transglutaminase. *Food Biotechnol.* 19, 37-49(2005).

[17] Schilling, C.H., Tomasik, P., Karpovich, D.S., Hart, B., Garcha, J., Boettcher, P.T. Preliminary study on converting agricultural waste into biodegradable plastics. Part II: corncobs. *J. Polym. Environ.* 13(1), 57-63(2005).

[18] Callister, W.D. *Materials Science and Engineering*, 5th ed.; Wiley: New York, 2000; pp 800-801.

[19] Cutter, C.N. Opportunities for Bio-Based Packaging Technologies to Improve the Quality and Safety of Fresh and Further Processed Muscle Foods. *Meat Science* 74, 131-142(2006).

[20] Ozdemir, M., Floros, J.H. Active Food Packaging Technologies Critical Reviews. *Food and Nutrition* 44,185-193(2004).

In: Monomers, Oligomers, Polymers, Composites… ISBN: 978-1-60456-877-6
Editors: R. A. Pethrick, G.E. Zaikov et al. 

*Chapter 4*

# Features of Mechanism of Free Radical Initiation in Polymers under Exposure to Nitrogen Oxides

***E. Ya. Davydov, I. S. Gaponova, T. V. Pokholok, G. B. Pariyskii and G. E. Zaikov***

Emanuel Institute of Biochemical Physics, Russian Academy of Sciences.
4 Kosygin Street. 119334 Moscow, Russia
Fax: (495) 939-71-03, E-mail: *pgb@sky.chph.ras.ru*

## Abstract

The mechanism of interactions of NO, $NO_2$, $NO_3$ as well as dimers of $NO_2$ with functional groups of organic molecules and polymers are considered. It is demonstrated that nitrogen oxides are active initiators of radical reactions as a result of which various molecular products of nitration and stable nitrogen-containing radicals are formed. Features of the initiation mechanism determining the composition of molecular and radical products of conversions of the compounds on exposure to nitrogen oxides have been revealed.

**Keywords:** nitrogen oxides, polymers, macroradicals, EPR spectra, mechanism of reactions

## I. Introduction

Nitrogen oxides play the important role in various chemical processes proceeding in an atmosphere, and affect on an environment[1-4]. These compounds ejected to atmosphere by the factories and transport in huge amounts creates thus serious problems for fauna and flora as well as synthetic polymeric materials. In this connection, the researches of the mechanism of nitrogen oxide reactions with various organic compounds including polymers are important for definition of stability of these materials in polluted atmosphere. On the other hand, these researches can be used in synthetic chemistry[5-7], in particular for development of methods of the polymer modification, for example, for preparation of spin-labelled macromolecules[8,9]. The generation of spin labels occurs thus in consecutive reactions including formation and conversion of specific intermediate molecular products and active free radicals. It is necessary to note essential advantages of such way of obtaining spin labels not requiring application of complex synthetic methods based on reactions of stable nitroxyl radicals with functional groups of macroradicals[10]. If the polymers are capable of reacting with nitrogen oxides, the formation of stable radicals in them takes place spontaneously or by thermolysis of molecular products of nitration[11,12].

Most important for reactions with various organic compounds and polymers are nitrogen oxides of the three types: NO, $NO_2$, $NO_3$. All of them represent free radicals with different reactivity[13]. In the present review the features of the mechanism of reactions of these oxides and also $NO_2$ dimeric forms with a number of polymers and low-molecular compounds are considered. The especial attention is given to the analysis of structure of stable nitrogen-containing radicals and kinetic features of their formation. On the basis of results of the analysis, the conclusions on the mechanism of primary reactions of initiation and intermediate stages of complex radical processes under action of nitrogen oxides are given. The principles of use of nitrogen oxides for grafting spin labels to various polymers are considered.

## II. Interaction of Nitrogen Oxides with Products of Photolysis and Radiolysis of Polymers

From three nitrogen oxides examined, the radical NO is least reactive. It is not capable of abstracting hydrogen atoms even from least strong tertiary or allyl C–H bonds; the strength of H–NO bond[14] makes only 205 $kJ\cdot mol^{-1}$. Nitrogen oxide cannot join to isolated double C=C bonds of alkenes[15]. For NO, the recombination with free radicals with formation of effective spin traps (nitroso compounds) is characteristic process. The structure of stable nitrogen-containing radicals formed from nitroso compounds in the subsequent reactions gives information on the mechanism of proceeding radical process in the given reacting system.

During photolysis of polymethylmethacrylate (PMMA) in atmosphere of NO by unfiltered light of a mercury lamp at 298 K, the formation of acylalkylaminoxyl radicals such as $R_1N(O^{\bullet})C(=O)OR_2$ was observed with typical parameters of anisotropic triplet EPR spectrum in a solid phase[16]: $A_{II}^{N} = 2.1 \pm 0.1$ mT and $g_{II}$=2.0027 ± 0.0005. If photolysis of the same samples to carry out at 383 K, dialkylaminoxyl radicals $RN(O^{\bullet})R$ in addition to acylalkylaminoxyl radicals are formed with parameters of EPR spectrum: $A_{II}^{N} = 3.2 \pm 0.1$ mT and $g_{II}$=2.0026 ± 0.0005. Under action of filtered UV light (260-400 nm) at room temperature there is a third type of stable radicals namely iminoxyls $(R_1R_2)C{=}NO^{\bullet}$, the triplet EPR spectrum of which in benzene solution of PMMA is characterized by the parameters: $a^{N}$ = 2.8 mT and $g$ = 2.005. The occurrence of acylalkylaminoxyl radicals is evidence of eliminating of methoxycarbonyl radicals in a course of the polymer photolysis:

$$\sim(CH_3)C(COOCH_3)CH_2(CH_3)C(COOCH_3)\sim \xrightarrow{h\nu}$$
$$\sim(CH_3)C(COOCH_3)CH_2C^{\bullet}(CH_3)\sim (R^{\bullet}) + {}^{\bullet}COOCH_3 \qquad (1)$$

The subsequent reaction with participation of NO gives acylalkylaminoxyl radicals:

$$^{\bullet}COOCH_3 + NO \rightarrow O{=}N{-}COOCH_3 \xrightarrow{+R^{\bullet}} RN(O^{\bullet})COOCH_3 \qquad (2)$$

Dialkylaminoxyl radicals are formed by decomposition of macroradicals $R^{\bullet}$:

$$R^{\bullet} \xrightarrow{kT} \sim(CH_3)C(COOCH_3)CH_2{=}C(CH_3)CH_2\sim + {}^{\bullet}C(CH_3)(COOCH_3)\sim$$
$$\xrightarrow{+NO, R^{\bullet}}$$
$$RN(O^{\bullet})C(CH_3)(COOCH_3)\sim \qquad (3)$$

Iminoxyl radicals are appeared as a result of generation of macroradicals $\sim(CH_3)C(COOCH_3)C^{\bullet}H(CH_3)C(COOCH_3)\sim$ ($R_1^{\bullet}$) which in NO atmosphere are converted into nitroso compounds and further into oximes:

$$R_1^{\bullet} + NO \rightarrow R_1NO \rightarrow \sim(CH_3)C(COOCH_3)C({=}NOH)(CH_3)C(COOCH_3)\sim(R_1R_2NOH) \qquad (4)$$

As a result of the mobile hydrogen atom abstraction from oximes by, for example, methoxycarbonyl radical, iminoxyls are formed:

$$R_1R_2NOH + {}^{\bullet}COOCH_3 \rightarrow HCOOCH_3 + R_1R_2NO^{\bullet} \qquad (5)$$

A limiting stage of reaction (2) is nitrogen oxide diffusion in a polymeric matrix. The rate of reaction (3) should much more depend on mobility of macromolecular reagents. Therefore, the distinction in composition of radicals in PMMA photolysed at room temperature and 383 K is observed. At room temperature, acylalkylaminoxyl radicals are formed due to accepting low-molecular methoxycarbonyl radicals $^{\bullet}COOCH_3$ by nitroso compounds. At 383K, when molecular mobility essentially grows, dialkylaminoxyl radicals $R_1N(O^{\bullet})R_2$ are formed because the meeting two macromolecular particles is provided. The results obtained demonstrate an opportunity of NO use for an elucidation of the polymer photolysis mechanism. With the help of this reactant, it was possible to establish a nature and mechanism of formation of intermediate short-lived radicals in photochemical process using EPR spectra of stable aminoxy radicals.

The application of nitrogen oxide enables to prepare spin-labelled macromolecules in chemically inert and insoluble polymers, for example, in polyperfluoroalkanes. As was shown in the work[16], the radiolysis of oriented films of polytetrafluoroethylene (PTFE) and copolymer of tetrafluoroethylene with hexafluoropropylene initiates reactions with formation of iminoxyl macroradical according to the following scheme:

$$PTFE \xrightarrow{\gamma} \sim CF_2C^{\bullet}FCF_2 \sim + NO \rightarrow \sim CF_2CF(NO)CF_2 \sim \xrightarrow{\gamma}$$
$$\sim CF_2C^{\bullet}(NO)CF_2 \sim \rightarrow \sim CF_2C(=NO^{\bullet})CF_2 \sim \quad (6)$$

$$\sim CF_2CF(CF_3)CF_2 \sim \xrightarrow{\gamma, NO} \sim CF_2CF(CF_2NO)CF_2 \sim \xrightarrow{\gamma}$$
$$\sim CF_2CF(C^{\bullet}FNO)CF_2 \sim \rightarrow \sim CF_2CF(CF{=}NO^{\bullet})CF_2 \sim \quad (7)$$

The EPR spectrum of iminoxyl radicals in oriented PTFE films represents a triplet ($A_{II}^{N}$=4.1 ± 0.1mT) of septets ($A_{II}^{F}$= 0.5 ±0.1 mT) with $g_{II}$=2.0029 ±0.0005. The unpaired electron in fluoroiminoxyl radicals interacts with a nitrogen nucleus and four in pairs magnetically equivalent fluorine nucleuses with hyperfine splittings of 1.0 ± 0.1 and 0.5 ± 0.1 mT.

Aminoxyl radicals in polyperfluoroalkanes are not formed in these conditions. However, if to carry out a preliminary γ–irradiation on air, middle and end peroxide macroradicals appear with conversion into fluoroaminoxyl macroradicals $\sim CF_2CF(NO^{\bullet})CF_2 \sim$ by the subsequent exposure to NO. Their EPR spectra in oriented films are quintet of triplets with the parameters: $A_{II}^{N}$=0.46 mT, $A_{II}^{F}$=1.11 mT, $g_{II}$=2.006; $A_{\perp}^{N}$=1.12 mT, $A_{\perp}^{F}$=1.61 mT and $g_{\perp}$=2,0071. The following mechanism of formation of aminoxyl radicals in these conditions is offered[17]:

$$\sim CF_2\,CF_2\,C^{\bullet}F\,CF_2 \sim \rightarrow \sim C^{\bullet}F_2 + CF_2{=}CFCF_2{\sim} \quad (8)$$

The end alkyl radical is oxidized into end peroxide radical:

$$\sim C^{\bullet}F_2 + O_2 \rightarrow \sim CF_2OO^{\bullet} \quad (9)$$

In NO atmosphere, end peroxide radical is converted as follows:

$$\sim CF_2CF_2OO^{\bullet} + NO \rightarrow \sim CF_2CF_2O^{\bullet} + NO_2 \quad (10)$$

$$\sim CF_2CF_2O^{\bullet} + NO{\sim} \leftrightarrow CF_2CF_2ONO \quad (11)$$

$$\sim CF_2CF_2O^{\bullet} \rightarrow \sim C^{\bullet}F_2 + COF_2 \quad (12)$$

$$\sim C^{\bullet}F_2 + NO \rightarrow \sim CF_2NO \quad (13)$$

$$\sim CF_2NO + CF_2{=}CFCF_2{\sim} \xrightarrow{NO} {\sim}CF_2N(O^{\bullet})CF_2CF(NO)CF_2{\sim} \quad (14)$$

In copolymer of tetrafluoroethylene with hexafluoropropylene, the radicals ${\sim}CF_2N(O^{\bullet})CF_3$ are formed in the same conditions[17].

Thus, spin-labelled macromolecules of fluoroalkyl polymers can be prepared using postradiational free radical reactions in NO atmosphere. The advantage of this method of introduction of spin labels lies in the fact that the radical centre can be located both in the end and in middle of macrochains, that allows basically to obtain the optimum information on molecular dynamics.

## III. Radical Reactions Initiated by Nitrogen Trioxide

The radicals $NO_3$ play an essential role in chemical processes proceeding in the top layers of atmosphere[18]. These radicals are formed in reaction of nitrogen dioxide with ozone:

$$NO_2 + O_3 \rightarrow NO_3 + O_2 \quad (15)$$

Under action of daylight, nitrogen trioxide is consumed with releasing atomic oxygen:

$$NO_3 \xrightarrow{h\nu} NO_2 + O \quad (16)$$

Its disappearance occurs at night in reaction with nitrogen dioxide too:

$$NO_3 + NO_2 \rightarrow N_2O_5 \quad (17)$$

The $NO^3$ radicals are characterized by high reactivity in reactions with various organic compounds[18-22]. Typical reactions of these radicals are abstraction of hydrogen atoms from C–H and addition to double bonds. Along with these reactions, the radicals $NO_3$ are decomposed in thermal and photochemical processes. The thermal decomposition of nitrogen trioxide generated by pulse radiolysis of concentrated water solutions of a nitric acid takes place with high rate ($k_{208K} = 8{\cdot}10^{3}$, $s^{-1}$)[23]:

$$NO_3 \xrightarrow{kT} NO_2 + O \quad (18)$$

The radicals $NO_3$ have three intensive bands of absorption in visible region of an optical spectrum with $\lambda_{max}$ = 600, 640 и 675 nm, and in UV region at 340–360 nm[18,19,23-27]. The action of light on $NO_3$ results in dissociation of them by two mechanisms including formation of $NO_2$ and atomic oxygen similarly to reaction (18) or nitrogen oxide and molecular oxygen[1]:

$$NO_3 \rightarrow NO + O_2 \quad (19)$$

The efficiency of the $NO_3$ conversion by one or either mechanisms is determined by spectral composition of light. The approximate border of wave lengths for photodissociation by reactions (18) or (19) lays at 570 nm. Above 570 nm, $NO_3$ decomposes on NO and $O_2$ with the very high (~ 1) quantum yield; the basic products of the photolysis are $NO_2$ + O below 570 nm [1].

One of the most widely widespread ways of the $NO_3$ generation is the photolysis of Ce (IV) nitrates in particular ceric ammonium nitrate (CAN) $(NH_4)_2Ce(NO_3)_6$. The absorption spectrum of CAN has a wide and intensive band with a maximum at 305 nm ($\varepsilon$ = 5890 l·mol$^{-1}$·см$^{-1}$) which is conditioned by an electron transfer to $Ce^{4+}$from nitrate anion[26]. Under action of light in the given spectral region there is a photoreduction of CAN [26,27]:

$$Ce^{4+} NO_3^- \rightarrow Ce^{3+} + NO_3 \quad (20)$$

Thus, the CAN photolysis gives different active radical particles: NO, $NO_2$, $NO_3$ and atomic oxygen. Use of light of various spectral composition allows thus to generate these particles in different ratio.

Atomic oxygen being very active reactant [14] interacts with C—H bonds of organic compounds. Macroradicals formed by action of atoms O on polymers can be converted in the presence of NO into stable aminoxyl radicals. The possibility using these processes for the purposes of chemical modification of polymers is considered by the example of polyvinylpyrrolidone (PVP) in the works [28,29] In PVP with CAN (0,05—0,2 mol kg$^{-1}$) in the course of photolysis by light with $\lambda$ > 280 nm, the formation of alkyl macroradicals is registered by EPR as a result of the reaction:

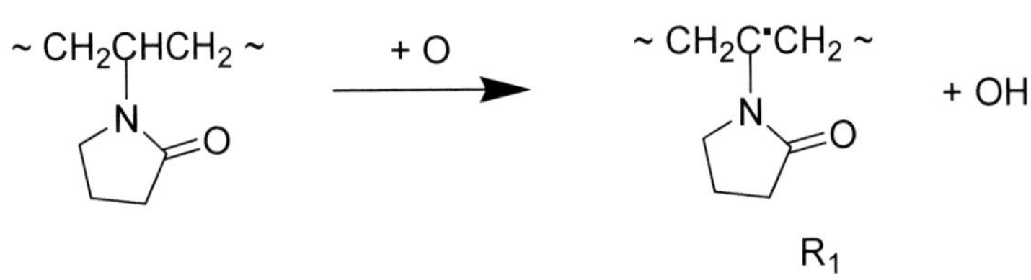


(21)

The radicals $R_1$ are stabilized only at low temperatures (77 K). Photolysis of samples at 298 K by the same light results in the production of stable dialkylaminoxyl radicals characterized by anisotropic EPR triplet spectrum with parameters of $A_{II}^{N}$ = 3.18 mT and $g_{II}$ = 2.0024. The cross-linkage of PVP macromolecules takes place in these conditions with formation of a gel-fraction as a result recombination of macroradicals $R_1$ with nitrogen oxide:

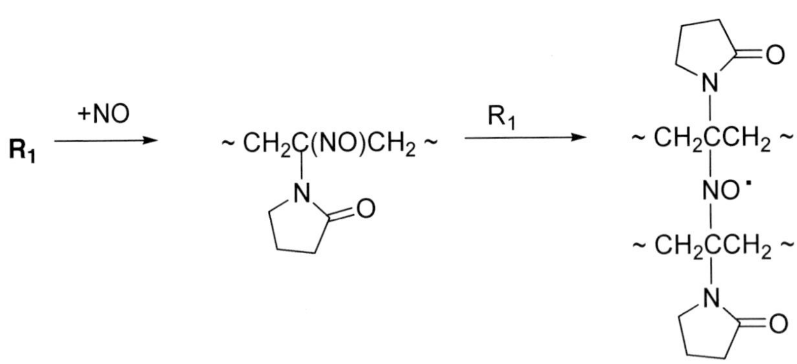


(22)

As shown in the work [28], the yield of a gel-fraction directly correlates with the amounts of stable radicals which in the given system represent cross-linkages of macromolecules.

Such simple method of cross-linking PVP can be applied for obtaining hydrogels used as specific sorbents [30].

# IV. Free Radical and Ion - Radical Reactions under Action of Nitrogen Dioxide and Its Dimers

Nitrogen dioxide effectively reacts with various low- and high-molecular organic compounds [5-7, 31]. However it must be emphasized that $NO_2$ is a free radical of moderate reactivity, and the ONO—H bond strength [14] makes up 320 kJ mol$^{-1}$. Therefore radicals $NO_2$ are capable of initiating free radical reactions by abstraction of hydrogen atoms from least strong, for example, allyl C—H bonds or addition to double C=C bonds [32,33]:

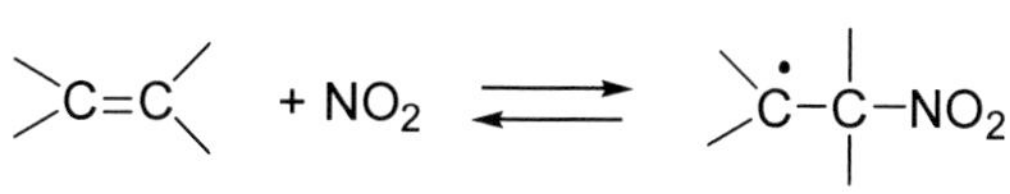

(23)

This process causes further radical conversions of olefins with formation of dinitro compounds and nitro nitrites:

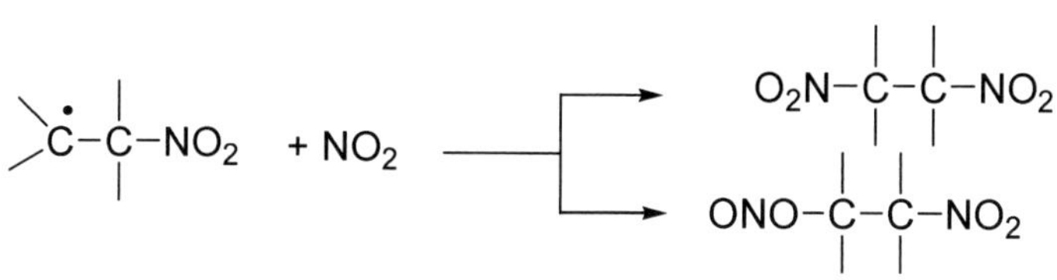

(24,25)

From Jellinek's data [31], butyl rubber destroys under the action of $NO_2$:

$$\sim C(CH_3) = CH\sim \quad + NO_2 \rightleftharpoons \sim\dot{C}(CH_3) - CH(NO_2) \sim \longrightarrow \text{destruction} \qquad (26)$$

As a consequence of primary reactions of nitrogen dioxide radicals with the isolated double bonds, stable aminoxyl radicals can be generated. Such transformations are characteristic for rubbers.

## IV.1. Preparation of Spin-Labelled Rubbers

The possibility of obtaining spin-labelled rubbers by interaction of their solutions in the inert solvents with a mixture of nitrogen dioxide and oxygen has been demonstrated in the work [34]. Such rubbers can be prepared simply and rapidly by reactions of block polymeric samples with gaseous $NO_2$ [35]. The experiments were carried on 1,4-*cis*-polyisoprene (PI) and copolymer of ethylene, propylene and dicyclopentadiene. The samples had the form of cylinders of 1.5 cm height and 0.4 cm in diameter. On exposure these polymers to $NO_2$ ($10^{-5}$–$2.3 \cdot 10^{-3}$ mol·l$^{-1}$) at 293 K, identical EPR spectra were registered. The spectra represent an anisotropic triplet with parameters which are typical for dialkylaminoxyl radicals with $A_{II}^{N}$ = 3.1 mT and $g_{II}$ = 2.0028 ± 0.0005. The spectra with such parameters testify that the correlation time of rotational mobility $\tau_c$ at the given temperature exceeds $10^{-9}$ s. At increasing temperature up to 373 K, the isotropic triplet signal with $a^N = 1.53 \pm 0.03$ mT and $g = 2.0057 \pm 0.0005$ was observed, that is caused by essential decreasing correlation time ($5{\cdot}10^{-11} < \tau_c < 10^{-9}$ s). The change of $\tau_c$ with temperature is described by the relation $\tau_c = \tau_0 \exp(E/RT)$, where $\log \tau_0 = -14.2$, and $E$ is the activation energy of rotational diffusion (34.7 kJ·mol $^{-1}$)

The study of kinetics of the aminoxyl radical accumulation has shown that concentrations of the radicals pass through a maximum determined by concentrations of $NO_2$ in a gas phase. So the concentration of radicals peaks in 40 times time more slowly when [$NO_2$] falls from $2.3{\cdot}10^{-3}$ up to $10^{-5}$ mol·l$^{-1}$. It is suggested that such kinetic features are due to disappearance of aminoxyl radicals in reactions of their oxidation by $NO_2$ dimers. The presence of oxygen in a gas mixture results in the long induction period for kinetics of the radical formation, however their maximum concentration in this case are approximately in 3 times higher than on exposure of rubbers to pure $NO_2$. The scheme of the aminoxyl radical formation includes four basic stages: generation of macroradicals in reaction of $NO_2$ with rubbers; synthesis of macromolecular, spin trapping of macroradicals by nitroso compounds, and destruction of aminoxyl radicals:

$$\sim CH_2C(CH_3) = CH - CH_2 \sim + NO_2 \rightarrow \sim CH_2C^{\bullet}(CH_3) - CH(NO_2) - CH_2 \sim (R_1^{\bullet})$$
$$\sim CH_2C(CH_3) = CH - CH_2 \sim + NO_2 \rightarrow \sim CH_2C^{\bullet}(CH_3) - CH(ONO) - CH_2 \sim (R_2^{\bullet})$$
$$\sim CH_2C(CH_3) = CH - CH_2 \sim + NO_2 \rightarrow HNO_2 + \sim CH_2C(CH_3) = CH - C^{\bullet}H \sim (R_3^{\bullet}) \quad (27)$$

$$R_1^{\bullet}(R_2^{\bullet}, R_3^{\bullet}) + NO_2 \rightarrow \text{продукты} \quad (28)$$
$$2HNO_2 \rightleftharpoons NO + NO_2 + H_2O \quad (29)$$
$$R_1^{\bullet}(R_2^{\bullet}, R_3^{\bullet}) + NO \rightarrow R_1NO(R_2NO, R_3NO) \quad (30)$$
$$RNO + R_1^{\bullet}(R_2^{\bullet}, R_3^{\bullet}) \rightarrow (R)_2N{-}O^{\bullet} \quad (31)$$
$$(R)_2N{-}O^{\bullet} + R^{\bullet} \rightarrow (R)_2N{-}OR \quad (32)$$
$$(R)_2N{-}O^{\bullet} + N_2O_4 \rightarrow [(R)_2N^{+}ONO_3^{-}] + NO \quad (33)$$

As can be seen from this scheme, the accumulation of aminoxyl radicals should be accompanied by cross-linkage of macromolecules. The presence of oxygen inhibits the aminoxyl accumulation as a result of conversions of primary nitroalkyl and allyl macroradicals into peroxide radicals.

In the course of interaction of solid polymers with $NO_2$ one can expect a non-uniform distribution of formed stable nitrogen-containing radicals in sample owing to diffusion restrictions. A possibility of an examination of the nitration reaction front thus is created in polymeric materials by measurement of spatial distribution of aminoxyl radicals. Such studies can be performed using the method of EPR tomography (or the EPR imaging technique) described in the works [36,37]. During the reaction of $NO_2$ with double bonds of rubbers, the spatial distribution of aminoxyls characterizes kinetics of the reaction front progress and thus structural - physical properties of a concrete sample. The EPR tomograms in a non-uniform magnetic field (figures 1a and b) were obtained for cylindrical samples PI of 0.4 cm in diameter and height of 1.5 cm. The accumulation of aminoxyl macroradicals up to maximum concentrations in enough thick layers (~1 mm) of samples shows that a spatial grid of cross-linked PI hindered diffusion of $NO_2$ has not time to be generated. The introduction of $O_2$ in gaseous mixture narrows front of reaction and reduces rate of the aminoxyl radical formation. This effect is connected with the additional channel of an intensive consumption of intermediate macroradicals $R_1^{\bullet}$ ($R_2^{\bullet}$, $R_3^{\bullet}$) and decreasing their equilibrium concentration. The results obtained by EPR tomography have shown that on exposure of rubbers to $NO_2$ at enough high concentrations ($10^{-4}$ – $2{,}3 \cdot 10^{-3}$ mol·l$^{-1}$) the chemical and structural modification occurs in a superficial layer and does not take place in deeper layers.

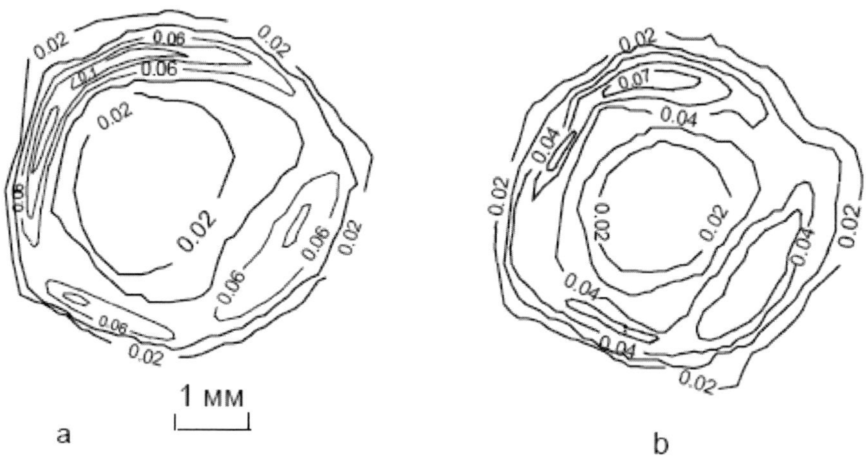


Figure 1. Bulk distribution of dialkylaminoxyl radicals produced upon nitration of PI by $NO_2$ during 2.5 (a) and 740 hours (b).

## V. On the Mechanism of Initiation of Radical Reactions by Nitrogen Dioxide Dimers

As noted above, nitrogen dioxide can initiate free radical reaction in compounds containing least strong C—H bonds or double C=C bonds. However, the effective formation of stable nitrogen-containing radicals was observed also in aromatic polyamidoimides, polycaproamide, polyvinylpyrrolidone (PVP)[8] and also aromatic polyamide (AP) [12]. These facts allow considering other probable mechanisms of the radical processes initiation. The fact is that the basic radical products of interaction of nitrogen oxide with polymers containing amide groups are iminoxyl and acylalkylaminoxyl radicals which are produced from oximes and acylnitroso compounds [8,12]. The occurrence of these predecessors of stable radicals is in turn connected with the nitrogen oxide formation.

In this connection, it is necessary to suppose a participation of $NO_2$ dimeric forms in radical initiation. The main dimers of $NO_2$ are planar nitrogen tetroxide $O_2N\text{-}NO_2$ (PD) and nitrosyl nitrate $ONONO_2$ (NN). Ab initio calculations [33] show that these dimers are formed with the most probability in $NO_2$ atmosphere; the form of nitrosyl peroxynitrite ONOONO is too unstable to be considered as efficient participant of reactions, however it can play a role of intermediate compound at oxidation of nitrogen oxide by oxygen [38]. As NN has strong oxidative properties [39], the generation of radicals can take place by an electron transfer from donor functional groups with the formation of intermediate radical cations [9,40]:

$$RH + ONONO_2 \rightarrow [R^{\bullet}H^{+}(NO\cdots ONO_2^{-})] \rightarrow R^{\bullet} + NO + H^{+} + ONO_2^{-} \quad (36)$$

The recombination of radicals with nitric oxide gives nitroso compounds that undergo isomerisation into oximes [41] to produce iminoxyl radicals in the reaction with $NO_2$:

$$>C{=}NOH + NO_2 \rightarrow \; C{=}NO^{\bullet} + HNO_2 \quad (37)$$

The tertiary nitroso compounds are effective spin traps and a source of stable aminoxyl radicals:

$$RN{=}O + R_1^{\bullet} \rightarrow R(R_1)N\text{-}O^{\bullet} \quad (38)$$

Thus the mechanism involving reactions (36-38) formally could explain an appearance of stable radicals in the polymers not containing specific chemical bonds reacting with $NO_2$ mono radicals. However, there are certain obstacles connected with energetic properties of $NO_2$ dimers [33] for realizing such mechanism; the energy of syn- and anti forms of NN exceeds that of PD respectively 29.8 and 18.4 кJ·mol$^{-1}$; that is the equilibrium

$$O_2N\text{-}NO_2 \rightleftharpoons 2NO_2 \rightleftharpoons ONONO_2 \quad (39)$$

should be shifted to PD in gas phase.

The diamagnetic (PD) is capable of generating nitrogen-containing radicals in specific reaction with system of the connected double bonds of $p-$quinines (Q) [42]. On exposure of BQ to nitrogen dioxide, the formation of radicals I of oxyaminoxyl type ($R_{oxy}$) [43] takes place by the following scheme:

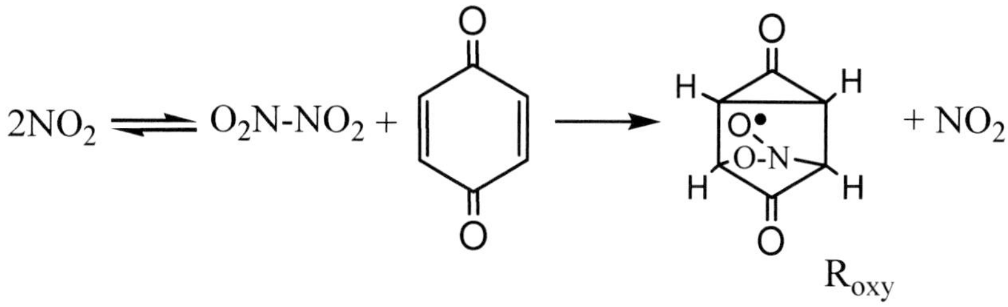


(40)

The triplet EPR spectrum of radicals $R_{oxy}$ in Q (figure 2a) has parameters: $a^N = 2.82$ mT and $g = 2.0053$. The scheme (40) is confirmed by kinetic data according to which the rate of the radical accumulation is proportional to a square of $NO_2$ concentration in gas phase.

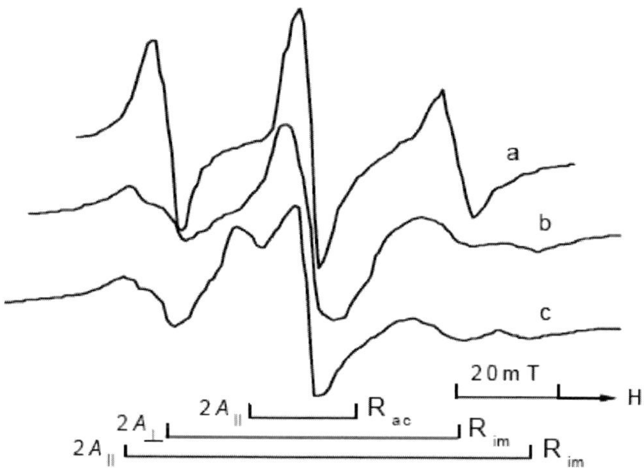


Figure 2. EPR spectra of Q after exposure to $NO_2$ (**a**); Q + AP (b) and Q+ PVP (c) after preliminary exposure to $NO_2$ and subsequent pumping-out the composites.

Thus, both dimer forms of $NO_2$ can be active. On this basis it is possible to suggest that the shift of equilibrium (39) to the formation of NN in PVP and AP is caused by specific donor-acceptor interaction of PD with amide groups which induce the conversion into NN and ion - radical process by scheme (36). As the indicator of conversion of PD into NN, the dependence of a yield of radicals $R_{oxy}$ on the contents of AP and PVP was used in composites: Q + AP and Q + PVP. To increase a surface of interaction with nitrogen dioxide, silica gel with particles of 100-160 μ in diameter was added to the composites. Samples for measurement of EPR spectra contained constant quantities of Q (100 mg), $SiO_2$ (100 mg) and variable quantities of PVP (10-30 mg) or AP (10-50 mg). In addition to $R_{oxy}$, iminoxyl radicals $R_{im}$ [8] occur in composites of BQ with AP on exposure to nitrogen dioxide. Under the same conditions, the sum of radicals $R_{im}$ and acylalkylaminoxyl radicals $R_{ac}$, [8] along with $R_{oxy}$, was registered in composites of BQ with PVP. Signals of radicals $R_{im}$ and $R_{ac}$ are masked by an intense signal of radicals $R_{oxy}$ in the EPR spectrum. However on can separate spectra of radicals $R_{im}$ and $R_{ac}$ using the fact that radicals $R_{oxy}$ exist only in $NO_2$ atmosphere. In view of rather low thermal stability, radicals $R_{oxy}$ quickly disappear at room temperature within several minutes after pumping out nitrogen dioxide from the samples. Remaining spectra of stable radicals $R_{im}$ in AP and the sum of $R_{im}$+$R_{ac}$ in PVP are shown respectively in figures 2b and 2c. They represent anisotropic triplets with $A_{\|}^{N}$ = 4.1 mT, $g_{\|}$ = 2.0024 and $A_{\perp}^{N}$ = 2.6 mT, $g_{\perp}$ = 2.005 ($R_{im}$) [12] and with $A_{\|}^{N}$ = 1.94 mT, $g_{\|}$ = 2.003 ($R_{ac}$) [8]. Using this procedure, the maximum concentrations of radicals $R_{oxy}$, $R_{im}$ and $R_{ac}$ were separately determined in composites with the various contents of AP and PVP after exposure to $NO_2$ within 24 hours. The results obtained are shown in figures 3a and 3b. As is seen from the figures, the concentration of radicals $R_{oxy}$ accumulated monotonously falls as the relative contents of AP and PVP is increased, while concentrations of radicals $R_{im}$ and $R_{im}$ + $R_{ac}$ vary within 10 - 20 % of the average value, that is within the accuracy of integration of EPR spectra. This fact is indicative of obvious dependence of the radical $R_{oxy}$ yield on the contents of polymers with amide groups in composites, suggesting that PD is converted under the influence of amide groups into NN that generates stable radicals $R_{im}$ and $R_{ac}$ in the polymeric phases. It is significant that an appreciable decrease of the yield of radicals $R_{oxy}$ was not observed in control experiments when polymers of other chemical structure, for example, acetyl cellulose were used in composites. Therefore one can conclude that amide groups play special role in the process PD → NN.

The scheme of $R_{im}$ formation in AP can be presented as follows:

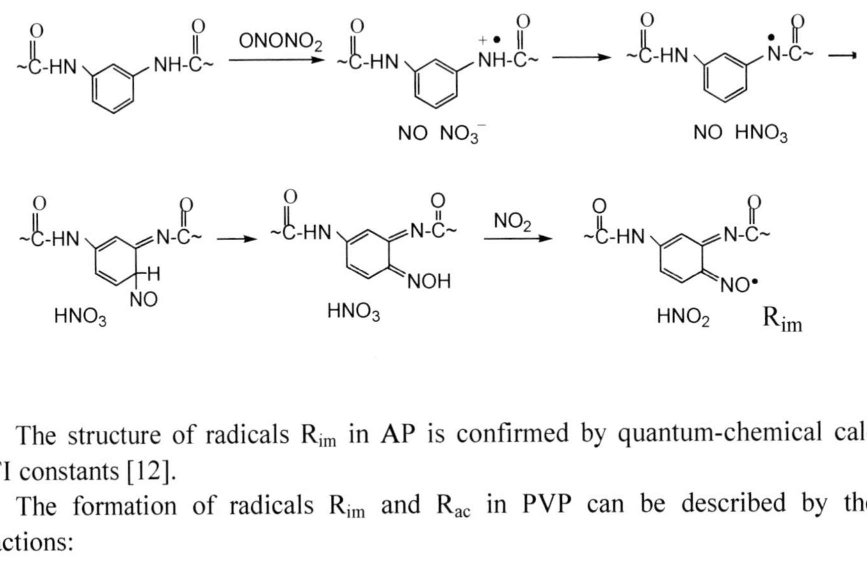


(41)

The structure of radicals $R_{im}$ in AP is confirmed by quantum-chemical calculations of HFI constants [12].

The formation of radicals $R_{im}$ and $R_{ac}$ in PVP can be described by the following reactions:

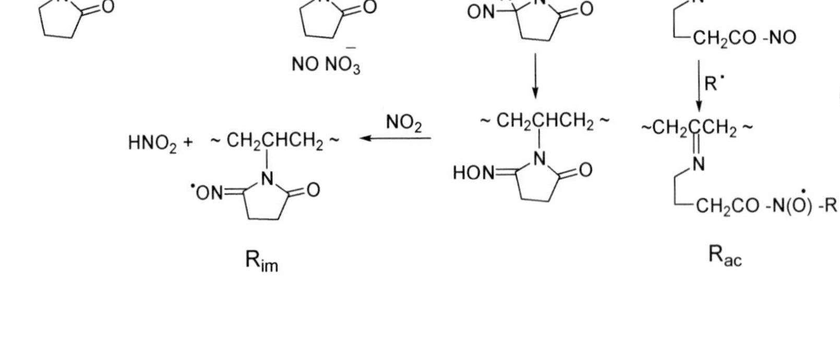


(42)

where $R^{\bullet}$ appears as a result of the radical cation decomposition:

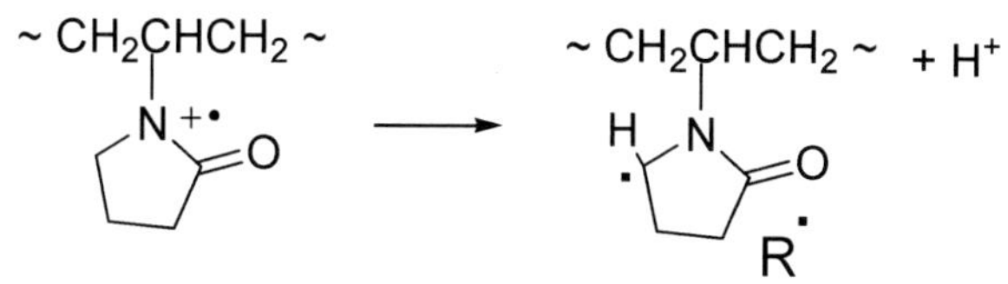

(43)

The decrease of relative yield of radicals $R_{im}$ on addition of polymers with amide groups to composites (figure 3) is apparent from the formal kinetic scheme:

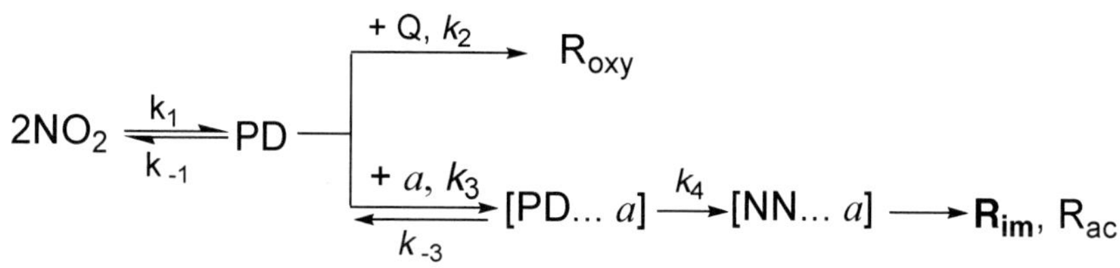

(44)

where *a* is an amide group. Taking into consideration stationary state for concentrations of PD, NN, [PD···*a*], [NN···*a*] and invariance of Q contents in composites, the following equations for rates of accumulation of radicals $R_{oxy}$, $R_{im}$ and $R_{ac}$ can be obtained:

$$\frac{d[\mathrm{R_{oxy}}]}{dt} = \frac{k_1 k_2 (k_{-3} + k_4)[\mathrm{NO_2}]^2}{(k_{-3} + k_4)(k_{-1} + k_2 + k_3[a]) - k_{-3}k_3[a]} \tag{45}$$

$$\frac{d[\mathrm{R_{im}}],[\mathrm{R_{ac}}]}{dt} = \frac{k_1 k_3 k_4 [a][\mathrm{NO_2}]^2}{(k_{-3} + k_4)(k_{-1} + k_2 + k_3[a]) - k_{-3}k_3[a]}, \tag{46}$$

where $[NO_2]$ is the concentration of nitrogen dioxide in gas phase, [*a*] is the surface concentration of amide groups. These equations can be simplified if concentrations of amide groups in composites are comparatively high, and the conversion of PD into NN occurs enough effectively, that is $k_3[a] >> k_{-1} + k_2$. Then

$$\frac{d[\mathrm{R_{oxy}}]}{dt} = \frac{k_1 k_2 (k_{-3} + k_4)[\mathrm{NO_2}]^2}{k_3 k_4 [a]} \tag{47}$$

$$\frac{d[\mathrm{R_{in}}],[\mathrm{R_{ac}}]}{dt} = k_1\,[\mathrm{NO_2}]^2 \tag{48}$$

Thus the rate of accumulation of radicals $R_{im}$ and $R_{ac}$ is determined by $[NO_2]$, and concentrations of these radicals accumulated on exposure to nitrogen dioxide do not depend appreciably on AP and PVP contents (figure 3 a, b (curve 2)). In contrast, the yield of $R_{oxy}$ decreases as polyamides are added to composites and [*a*] is increased. These plots in character are representative of competitive pathways for PD interactions with Q and amide groups. Note that the yield of $R_{oxy}$ is not changed in the $NO_2$ atmosphere in composites of Q with other polymers, for instance, acetyl cellulose at any ratio of the components.

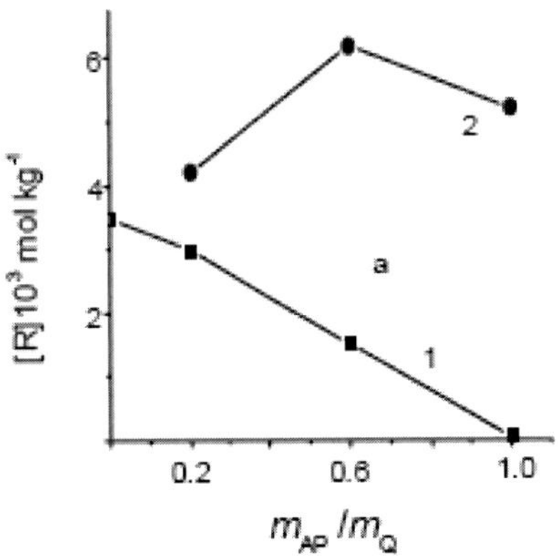


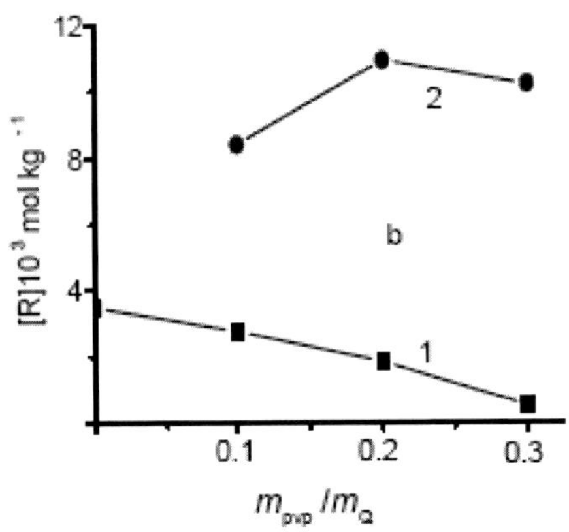


Figure 3. Dependence of concentrations of radicals $R_{oxy}$ (1), $R_{im}$ (2) in Q + AP (**a**) and $R_{oxy}$ (1), $R_{im}$ + $R_{ac}$ (2) in Q + PVP (b) after exposure to $NO_2$ on weight ratio of Q, AP and PVP.

## VI. Ab Initio Calculations of Energies for Conversions of Nitrogen Dioxide Dimmers

For validating the mechanism proposed of the conversion of PD into NN, the calculations of energy changes in process of nitrogen dioxide interaction with simplest amide (formamide) have been carried out within the framework of density functional theory by the Gaussian 98 program [44]. The B3LYP restricted method for closed and open shells was used. The intention of the calculations is to correlate energy consumptions for PD → NN with those for other stages of the radical generation process. Energies of the following states according to scheme (44) were calculated:

$2NO_2 + NH_2COH$ (49)
$O_2N\text{-}NO_2 + NH_2COH$ (50)
$ONONO_2 + NH_2COH$ (51)
$[O_2N\text{-}NO_2\cdots NH_2COH]$ (52)
$[ONONO_2\cdots NH_2COH]$ (53)
$N^{\bullet}HCOH + NO + HNO_3$ (54)
$NH_2CO^{\bullet} + NO + HNO_3$ (55)

The geometry optimization of all structures was performed applying the basis set 6-31G (d, p). The given process includes intermediate molecular complexes of PD and NN with formamide (52, 53). The changes of minimum energies are shown in figure 4. One can see that the formation of PD from $NO_2$ is energetically advantageous process [33], whereas NN is generated from $NO_2$ in an endothermic reaction. The complexation of PD with formamide is accompanied by release of energy: $\Delta E$ = 28 kJ·mol$^{-1}$. However, PD in complex (52) is not capable of reacting with formamide and can only leave the reacting cage. At the same time, PD in the complex can be converted approximately with the same energy consumption into NN (53), which further reacts by the electron transfer reactions (54, 55) giving radicals, nitric oxide, nitric acid and significant release of energy (44-57 kJ·mol$^{-1}$). Such sequence of transformations seems to be more efficient in comparison with a direct interaction of NN and formamide by state (51), as the energy of dimers in complexes (52) and (53) is lower than that of initial state (49).

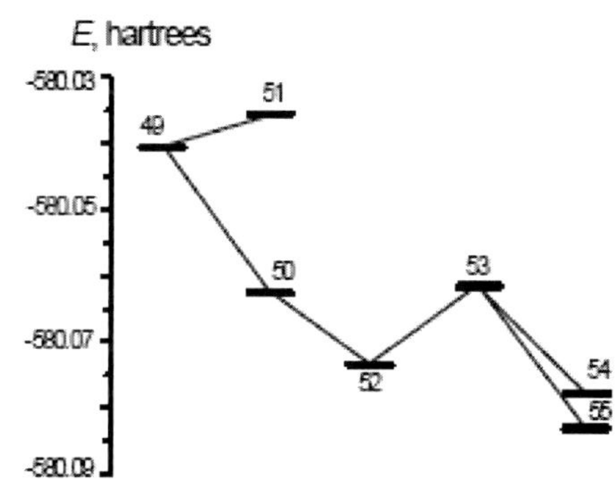


Figure 4. Changes of minimum energies calculated for reactions of $NO_2$ with formamide.

## VII. Detecting Radical Cations in Reactions of Electron Transfer to NN

Registration of radical cations by the EPR method in the presence of nitrogen dioxide could provide direct experimental evidence that the initiation proceeds through scheme (36).

However, because of high reactivity and fast decomposition [45], these particles are difficult to detect by this method. Nevertheless, the formation of radical cations can be revealed indirectly in the act of their decomposition with emission of a proton. Pyridine is known to be capable of accepting protons to yield pyridinium cations. Hence, if protons are formed in during decomposition of radical cations by reactions (36), they can be detected easily from IR spectra typical of pyridinium cations. Note that pyridine can be nitrated only under quite severe conditions. For example, *N*-nitropyridinium nitrate was obtained only when pyridine was treated with a NO2 –ozone mixture in an inert solvent [46]. This is evident from the IR spectrum (figure 5) of 1:1 mixture of pyridine and *N*-methylpyrrolidone (low-molecular analog of PVP) after treating it by nitrogen dioxide. In the spectrum there are two intense bands at 2400-2600 and 2200 $cm^{-1}$ of pyridinium cations [47]. The scheme of reactions in the given system involves the following consecutive stages:

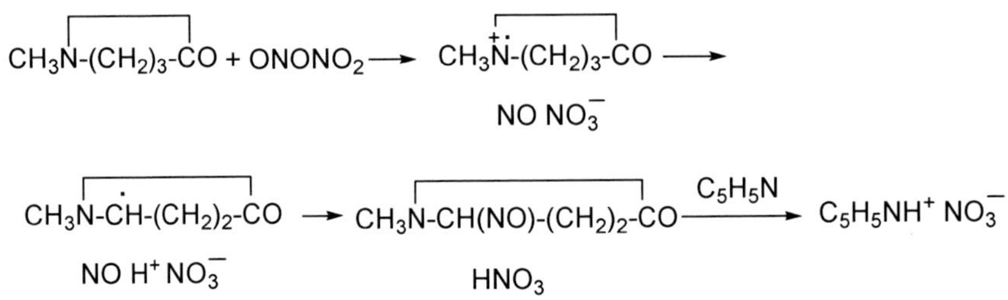


(56)

Thus pyridine fixes occurrence of radical cations as a result of an electron transfer to NN, and in doing so the ion - radical mechanism (36) is experimentally confirmed. It is possible to believe that such oxidative mechanism of generating radicals is characteristic for compounds with the ionization potential providing an electron transfer from donor groups of molecules to NN. This regularity can be followed from examples of thermal and photochemical nitration of aromatic compounds under the action of nitrogen dioxide. According UV spectroscopy data, NN can form charge-transfer complexes with methylbenzenes [48,49]. For the complexes, a bathochromic shift in the corresponding absorption bands was observed with an increase in number of methyl substituents in the benzene ring. Given this, this shift correlates with a decrease in the ionization potential from 8.44 to 7.85 eV on passage from *p*-xylene to hexamethylbenzene.

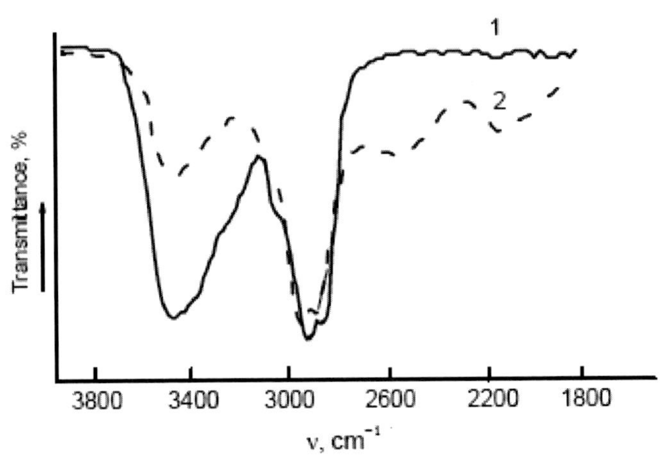


Figure 5. IR spectra of 1:1 *N*-methylpyrrolidone and pyridine mixture (1) and after exposure of the mixture to $NO_2$ (2).

## CONCLUSION

Nitrogen oxides are the effective initiators of radical reactions in a number polymers with formation the various molecular nitration products and stable nitrogen-containing radicals. Nitrogen oxide does not react directly, but it recombines with free radicals formed in polymers by UV photolysis or $\gamma$ — radiolysis. The nitro compounds formed in the subsequent reactions are converted into stable radicals. By this way, spin labels can be inserted even to chemically inert polyperfluoroalkanes. However, the chemical structure of macromolecules can essentially change in enough severe conditions of generating radicals. In this connection, nitrogen trioxide obtained by photolysis of the Ce (IV) nitrates is promising in application for spin labels synthesis. Under the action of visible and near UV light on these additives, the radicals and nitrogen oxide are formed simultaneously with transformation finally into spin labels. The light of such spectral composition does not cause undesirable side effects on macromolecules. Nitrogen dioxide is capable of interacting with least strong C—H and double C=C bonds initiating thus radical reactions in the given system. The dimeric forms of nitrogen dioxide actively react by mechanism depending on the chemical structure of those. The dimers in the form of nitrosyl nitrate represent oxidizing agent initiating ion-radical reactions with formation of stable nitrogen-containing radicals. Amide groups can induce a transition of energetically steadier planar dimers of $NO_2$ into nitrosyl nitrate. This specific ion-radical mechanism determines high activity relative to $NO_2$ even such stable polymers as aromatic polyamides.

## REFERENCES


[1] R. A. Graham, and H. S. Johnston, *J. Phys. Chem.*, 1978, 82 (3), p. 254.

[2] H. S. Johnston and R. A. Graham, *Canad. J. Chem.*, 1974, 52 (8), p. 1415.

[3] C. Stroud, S. Madronich, E. Atlas, B. Ridley, F. Flocke, A. W. Tallot, A. Fried, B. Wert, R. Shetter, B. Lefler, M. Coffey and B. Heik, *Atmospheric Environment,* 2003, 37 (24), p. 3351.

[4] D. Q. Tong, D. Kang, V. P.Aneja and J. D. Ray, *Ibid.*, 2005, 39 (2), p. 315.

[5] A. I Titov, *Tetrahedron*, 1963, 19, p. 557.

[6] A. V. Topchiev, *Nitration of hydrocarbons and other organic compounds,* Moscow, Academy of Sciences of the USSR, 1956.

[7] S. S. Novikov, G. A. Shveyhgeymer, V. V. Sevastyanova and V. A. Shlyapochnikov, *In Chemistry of aliphatic alicyclic nitro compounds.* 1974, Moscow, Khimiya.

[8] G. B. Pariiskii, I. S. Gaponova and E. Ya. Davydov, *Russ. Chem. Rev.*, 2000, 69 (11), p. 985.

[9] G. B. Pariiskii, I. S. Gaponova, E. Ya Davydov and T. V Pokholok, In *Aging of polymers, polymer blends and polymer composites.* Ed. G. E. Zaikov, A. L. Buchachenko, V. B Ivanov, New York, Nova Science Publishers, 2002.

[10] A. M. Vasserman and A. L. Kovarsky, Spin labels and spin probes in physical chemistry of polymers, Moscow, Nauka, 1986.

[11] I. S. Gaponova, E. Ya. Davydov, G. B. Pariiskii and V. P. Pustoshny, *Vysokomol. Soed.*, ser. A, 2001, 43, (1), p. 98.

[12] T. V. Pokholok, I. S. Gaponova, E. Ya. Davydov and G. B. Pariiskii, *Polym. Degrad. Stability*, 2006, 91 (10), p. 2423.

[13] F. T. Bonner and G. Stedman, *In Methods in nitric oxide research*, Ed. M. Feelish and J. S. Stamler, Chichester, Wiley, 1996.

[14] B. Rånby and J. F. Rabek, *Photodegradation, photo-oxidation and photostabilization of polymers,* London, Wiley, 1975.

[15] J. S. B. Park and J. C. Walton, *J. Chem. Soc., Perkin Trans. 2*, 1997, p. 2579.

[16] I. S. Gaponova, G. B. Pariiskii and D. Ya. Toptygin, *Vysokomolek. Soed.*, 1988, 30 (2), p. 262.

[17] I. S. Gaponova, G. B. Pariiskii and D. Ya. Toptygin, *Khimich. Fizika*, 1997, 16 (10), p. 49.

[18] P. Neta and R. E. Huie, *J. Phys. Chem.,* 1986, 90 (19), p. 4644.

[19] B. Vencatachalapathy and R. Ramamurthy, *J. Photochem. Photobiology A: Chem.,* 1996, 93, p. 1.

[20] S. M. Japar and H. H. Niki, J. *Phys. Chem.,* 1975, 79 (16), p. 1629.

[21] R. Atkinson, C. N. Plum, W. P. L. Carter, A. M. Winer and J. N. Pitts, *J. Phys. Chem.,* 88 (6), p. 1210.

[22] O. Itho, S. Akiho and M. Iino, *J. Org. Chem.,* 1989, 54 (10), p. 2436.

[23] A. K. Pikaev, G. K. Sibirskaya, E. M. Shyrshov, P. Ya Glazunov and V. I. Spicyn, *Dokl. Akad. Nauk SSSR,* 1974, 215 (3), p. 645.

[24] E. Hayon and E. Saito, *J. Chem Phys.,* 1965, 43 (12), p. 4314.

[25] L. Dagliotti and E. Hayon, *J. Phys. Chem.,* 1967, 71 (12), p. 3802.

[26] P. W. Glass and T. W. Martin, *J. Am. Chem. Soc.,* 92 (17), p. 5084.

[27] P. H. Wine, R. L. Mauldin and R. P. Thorn, *J. Phys. Chem.,* 1988, 92 (5), p. 1156.

[28] E. Ya. Davydov, E. N. Afanas'eva, I. S. Gaponova and G. B. Pariiskii, *Org. Biomol. Chem.,* 2004, 2 (9), p. 1339.

[29] E. Ya. Davydov, I. S. Gaponova and Pariiskii, *Vysokomolek. Soed.,* ser. A, 2003, 45, (4), p. 581.

[30] H. J. Naghash, A. Massah and A. Erfan, *Eur. Polym. J.,* 2002, 38, (1), p. 147.

[31] H. H. G. Jellinek, *Aspects of degradation and stabilization of polymers. Chapter 9.* New York, Elsevier, 1978.

[32] D. H. Giamalva, G. B. Kenion and W. A. Pryor, *J. Am. Chem. Soc.,* 1987, v. 109. № 23, p. 7059.

[33] P. Golding, J. L. Powell and J. H. Ridd, *J. Chem. Soc., Perkin Trans. 2*, 1996, p. 813.

[34] M. Gyor, A. Rockenbauer and F. Tüdos, *Tetrahedron Lett.*, 1986, 27 (39), p. 4795.

[35] T. V. Pokholok and G. B. Pariiskii, *Vysokomolek. Soed., ser. A*, 1977, 39 (7), p. 1152.

[36] A. I. Smirnov, E. N. Degtyarev, O. E. Yakimchenko and Ya. S. Lebedev, *Pribory Tekhn. Eksperim*, 1991, 1,p. 195.

[37] E. N. Degtyarev, T. V. Polholok, G. B. Pariiskii O. E. Yakimchenko, *Zh. Fiz. Khim.,* 1994, 68 (3), p. 461.

[38] M. L. McKee, *J. Am. Chem. Soc.,* 1995, 117 (8), p. 1629.

[39] E. H. White, *Ibid.,* 1955, 77, (20), p. 6008.

[40] E. Ya. Davydov, I. S. Gaponova, G. B. Pariiskii and T. V. Pokholok, *Polm. Sci., ser.A*, 2006, 48 (4), p. 375.

[41] H. Feuer, *The chemistry of the nitro and nitroso groups*, New York, Wiley, 1969.

[42] E. Ya. Davydov, I. S. Gaponova and G. B. Pariiskii, *J. Chem. Soc., Perkin Trans. 2,* 2002, p. 1359.
[43] I. Gabr and M. C. R. Symons, *J. Chem. Soc., Faraday Trans.*, 1996, 92 (10), p. 1767.
[44] M. J. Frisch, G. W. Trucks, H. B. Schlegel, G. E. Scuseria, M. A. Robb, J. R. Cheeseman et al.: Gaussian 98. Pitsburgh, PA, Gaussian Inc., 1998.
[45] D. Greatorex and J. Kemp, *J. Chem. Soc., Faraday Trans.*, 1972, 68 (1), p. 121.
[46] H. Suzuki, M. Iwaya and T. Mori, *Tetrahedron Lett.*, 1997, 38 (32), p. 5647.
[47] L. J. Bellamy, *The infra-red spectra of complex molecules.* London, Methuen, 1957.
[48] E. Bosch and J. K. Kochi, *J. Org. Chem.*, 1994, 59 (12), p. 3314.
[49] T. M. Bockman, Z. J. Karpinski, S. Sankararaman and J. K. Kochi, *J. Am. Chem. Soc.*, 1992, 114, (6), p. 1970.

In: Monomers, Oligomers, Polymers, Composites... ISBN: 978-1-60456-877-6
Editors: R. A. Pethrick, G.E. Zaikov et al. 

*Chapter 5*

# A NOVEL TECHNIQUE FOR MEASUREMENT OF ELECTROSPUN NANOFIBER

***M. Ziabari, V. Mottaghitalab and A. K. Haghi****
University of Guilan, P. O. Box 3756, Rasht, Iran

## ABSTRACT

In electrospinning, fiber diameter is an important structural characteristic because it directly affects the properties of the produced webs. In this chapter, we have developed an image analysis based method called direct tracking for measuring electrospun fiber diameter. In order to evaluate its accuracy, samples with known characteristics have been generated using a simulation scheme known as μ-randomness. To verify the applicability of the method, some real webs obtained from electrospinning of PVA have been used. Due to the need of binary input images, micrographs of the real webs obtained from scanning electron microscopy were segmented using local thresholding. The method was compared with the distance transform method. Results obtained by direct tracking were significantly better than distance transform, indicating that the method could be used successfully for measuring electrospun fiber diameter.

**Keywords**: Electrospinning, Fiber diameter, Image analysis, Direct tracking, Distance transform

## 1. INTRODUCTION

Conventional fiber spinning (like melt, dry and wet spinning) produce fibers with diameter in the range of micrometer. In recent years, electrospinning has gained much attention as a useful method to prepare fibers in nanometer diameter range. These ultra-fine fibers are classified as nanofibers. The unique combination of high specific surface area, extremely small pore size, flexibility ad superior directional strength makes nanofibers a

* Haghi@Guilan.ac.ir

preferred material form for many applications. Proposed uses of nanofibers include wound dressing, drug delivery, tissue scaffolds, protective clothing, filtration, reinforcement and micro-electronics.

In the electrospinning process, a polymer solution held by its surface tension at the end of a capillary tube is subjected to an electric field. Charge is induced on the liquid surface by an electric field. Mutual charge repulsion causes a force directly opposite to the surface tension. As the intensity of the electric field is increased, the hemispherical surface of the solution at the tip of the capillary tube elongates to form a conical shape known as the Taylor cone. When the electric field reaches a critical value at which the repulsive electric force overcomes the surface tension force, a charged jet of the solution is ejected from the tip of the cone. Since this jet is charged, its trajectory can be controlled by an electric field. As the jet travels in air, the solvent evaporates, leaving behind a charged polymer fiber which lays itself randomly on a collecting metal screen. Thus, continuous fibers are laid to form a nonwoven fabric. Figure 1 illustrates the electrospinning setup [1-[6].

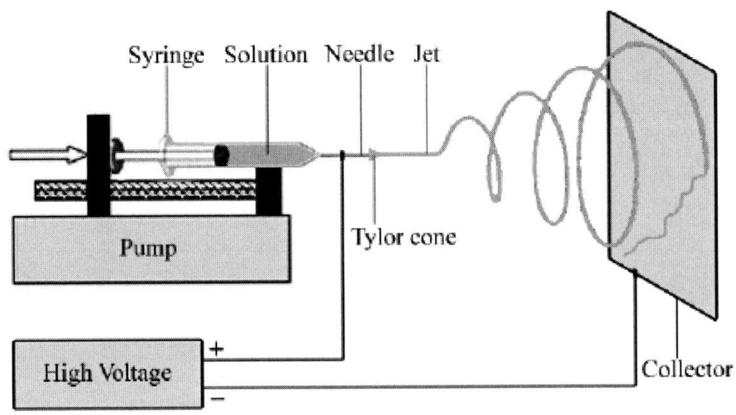


Figure 1. Electrospinning setup.

The properties of electrospun nonwoven webs depend on the nature of the component fiber as well as its structural characteristics such as fiber orientation [7]-[12] fiber diameter[13], pore size [15], uniformity and other structural features [16]. Analyzing the electrospun webs yield results and information, which will help researchers in improving the quality and predicting the overall performance of the electrospun webs. Some of the reasons for characterization may be process control, process development and product or quality control. Fiber diameter is the most important structural characteristics in electrospun nonwoven webs. Depending on the process and material variables the diameter of the fibers produced by electrospinning varies. Almost all researches who have done characterization, have reported the effects of processing variables on electrospun fiber diameter. There is no standard technique to measure the fiber diameter and analyze its distribution. This explains the need to study the fiber diameter of electrospun webs. Recently, image analysis has been used to identify fibers and measure structural characteristics in nonwovens. However, the accuracy and the limitations of these techniques have not been verified.

The objective of our research is to use image analysis for measuring electrospun fiber diameter.

# 2. METHODOLOGY

## 2.1. Simulation of Electrospun Web

In order to reliably evaluate the accuracy of the developed methods, samples with known characteristics are required. Since this end cannot be met with experiment, a simulation algorithm has been employed for generating nonwovens with known characteristics. The use of simulation is not a new idea. It was used by Abdel-Ghani and Davis [17]] and Pourdeyhimi et. al. [7] for simulation of nonwovens with both continuous and discontinuous fibers by the use of idealized straight lines. The most important component of simulation is the way in which lines or curves are generated. Abdel-Ghani and Davis [17] presented three methods for generating a random network of lines.

1. Surface randomness known as S-randomness
2. Mean free path known as μ-randomness
3. Internal randomness known as I-randomness

It is assumed that the lines are infinitely long (continuous filament) so that, at least in the image plane, all lines intersect the boundaries.

The aim is to obtain unbiased arrays which are spatially homogeneous for infinitely long lines. Lately it was discovered by Pourdeyhimi et. al. [7] that the best way to simulate nonwovens of continuous fibers is through the second method. Under this scheme, a line with a specified thickness is defined by the perpendicular distance d from a fixed reference point O located in the center of the image and the angular position of the perpendicular α. Distance d is limited to the diagonal of the image [7], [17]. Figure 2 demonstrates this procedure.

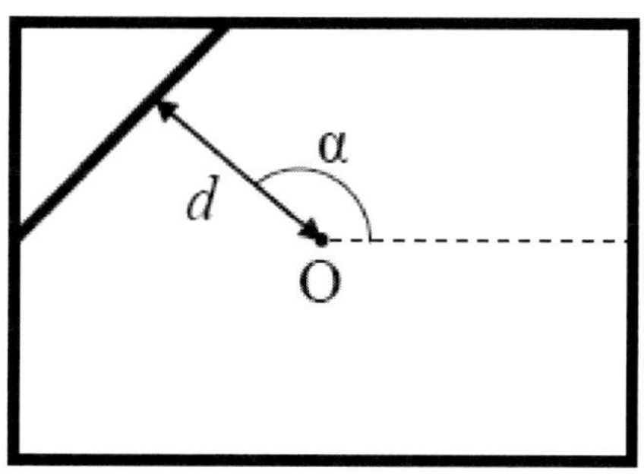


Figure 2. Procedure for μ-randomness.

Seeveral variables are allowed to be controlled during the simulation.

1. Web density: can be controlled using the line density which is the number of lines to be generated in the image.
2. Angular density: useful for generating fibrous structures with specific orientation distribution. The orientation may be sampled from either a normal or a uniform random distribution.

3. Distance from the reference point: varies between 0 and the diagonal of the image, restricted by the boundary of the image and is sampled from a uniform random distribution.
4. Line thickness (fiber diameter): is sampled from a normal distribution. The mean diameter and its standard deviation are needed.
5. Image size: can also be chosen as required.

### 2.2. Fiber Diameter Measurement

Electrospinning process produces very fine fibers and this is one of the few methods from which fibers of sub-micron size can be produced. So it becomes immensely important to understand the behavior of fiber diameter and fiber diameter distribution in the electrospun web as impacted by the processing parameters. The first step in determining fiber diameter is to produce a high quality image of the web at a suitable magnification using electron microscopy techniques, called micrograph. The methods for measuring electrospun fiber diameter are described in following sections.

#### *2.2.1. Manual Method*

The conventional method of measuring the fiber diameter of electrospun webs is to analyze the micrograph manually. The manual analysis usually consists of the following steps, determining the length of a pixel of the image (setting the scale), identifying the edges of the fibers in the image and counting the number of pixels between two edges of the fiber (the measurements are made perpendicular to the direction of fiber-axis), converting the number of pixels to nm using the scale and recording the result. Typically 100 measurements are carried out. Figure 3 illustrates this process.

However, this process is tedious and time-consuming especially for large number of samples. Furthermore, it cannot be used as on-line method for quality control since an operator is needed for performing the measurements. Thus, developing automated techniques which eliminate the use of operator and has the capability of being employed as on-line quality control is of great importance.

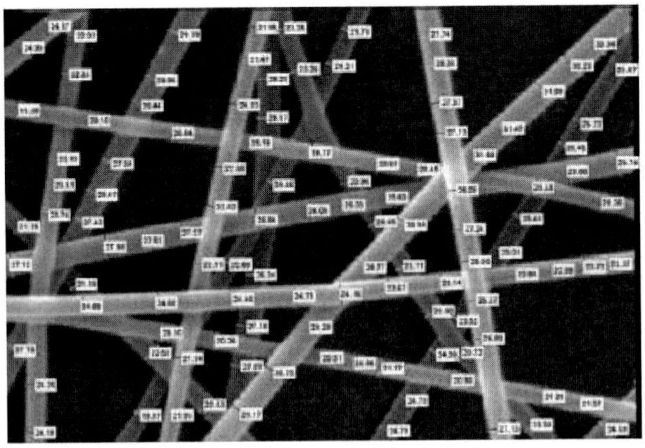

Figure 3. Manual method.

#### 2.2.2. Distance Transform

The distance transform of a binary image is the distance from every pixel to the nearest nonzero-valued pixel. The center of an object in the distance transform image will have the highest value and lie exactly over the object's skeleton. The skeleton of the object can be obtained by the process of skeletonization or thinning. The algorithm removes pixels on the boundaries of objects but does not allow objects to break apart. This reduces a thick object to its corresponding object with one pixel width. Skeletonization or thinning often produces short spurs which can be cleaned up automatically with a pruning procedure [18].

The algorithm for determining fiber diameter uses a binary input image and creates its skeleton and distance transformed image. The skeleton acts as a guide for tracking the distance transformed image by recording the intensities to compute the diameter at all points along the skeleton. This method was proposed by Pourdeyhimi et. al. [13]. Figure 4 shows a simple simulated image, its skeleton overlaid on its distance transform and the histogram of fiber diameter obtained by this method.

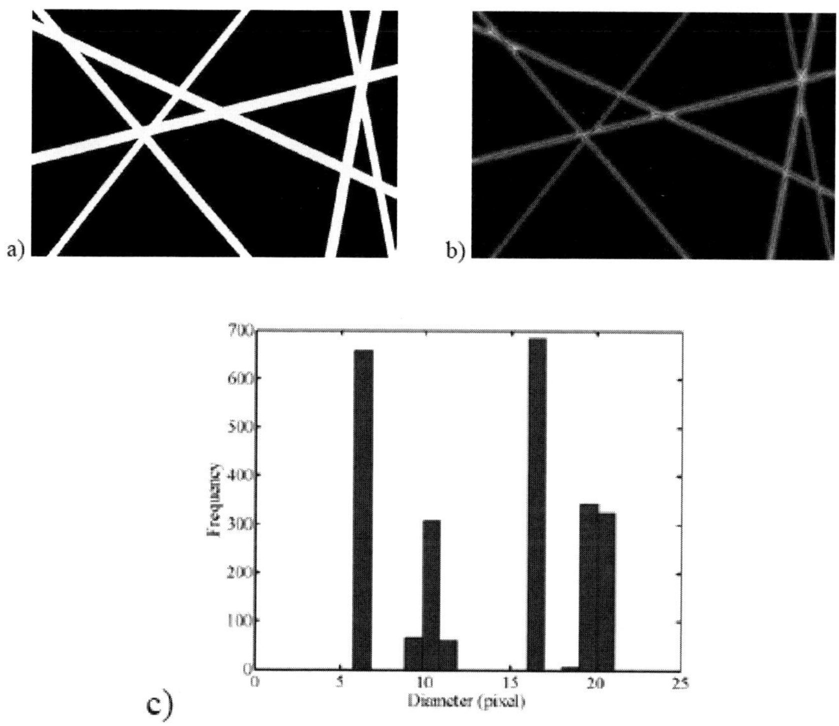


Figure 4. a) A simple simulated image, b) Its skeleton overlaid on its distance transform, c) Histogram of fiber diameter distribution obtained by distance transform.

#### 2.2.3. Direct Tracking

In order to measure the electrospun fiber diameter, we developed an image analysis based method called Direct Tracking. This method in which a binary image is used as an input determines fiber diameter on the basis of two scans; first a horizontal and then a vertical scan. In the first scan, the algorithm searches for the first white pixel adjacent to a black. Pixels are counted until reaching the first black. The second scan is then started from the mid point of horizontal scan and pixels are counted until the first black is encountered. If the black pixel isn't found, the direction changes. Having the number of horizontal and vertical scans, the number of pixels in perpendicular direction which is the fiber diameter could be measured from a geometrical relationship. The process is illustrated in Figure 5.

In electrospun webs, fibers cross each other at intersection points and this brings about some untrue measurements in these areas. To circumvent this problem, black regions are labeled and it is identified which couple of regions consist of fiber. Then, in order to enhance the processing speed, the image is cropped to the size of selected regions. Afterwards, fiber diameter is measured according to the explained algorithm. Finally, the data in pixels are converted to nm and the histogram of fiber diameter distribution is plotted. Figure 6 shows a labeled simple simulated image and the histogram of fiber diameter obtained by this method.

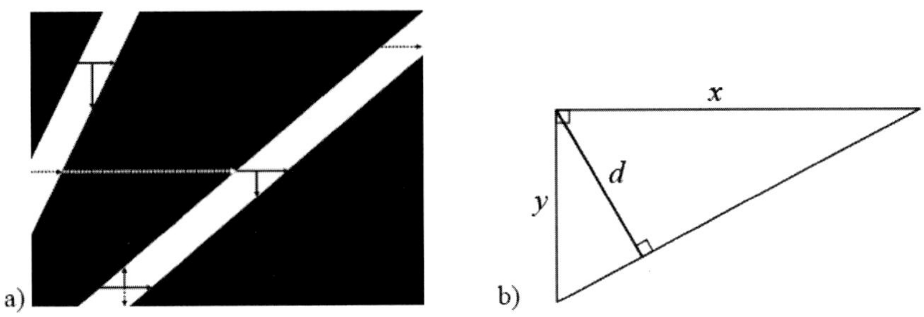


Figure 5. a) Direct tracking, b) Fiber diameter from the number of vertical and horizontal pixels.

Figure 6. a) A simple simulated image which is labeled, b) Histogram of fiber diameter distribution obtained by direct tracking.

## 2.3. Real Webs Treatment

Both of distance transform and direct tracking algorithm for measuring fiber diameter require binary image as input. Hence, the micrographs first have to be converted to black and white. This is done by thresholding (known also as segmentation) which produces binary image from a grayscale (intensity) image. This is critical step because the segmentation affects the result. In the simplest thresholding technique, called global thresholding, the image is partitioned using a single constant threshold. One simple way to choose a threshold is by trial and error. Then each pixel is labeled as object or background depending on whether the gray level of that pixel is greater or less than the value of threshold respectively.

The main problem here is global thresholding can fail in the presence of non-uniform illumination or local gray level unevenness. An alternative to circumvent this problem is to use local thresholding instead. In this approach, the original image is divided to subimages and different thresholds are used for segmentation. Another variation of this approach which has been used in this study consists of estimating the background illumination using morphological opening operation, subtracting the obtained background from the original image and applying a global thresholding to produce the binary version of the image. In order to automatically compute the appropriate threshold, Ostu's method could be employed. This method chooses the threshold to minimize interaclass variance of the black and white pixels. Prior to the segmentation, an intensity adjustment operation and a two dimensional median filter were applied in order to enhance the contrast of the image and remove noise [8]-[21]. As it is shown in Figure 7, global thresholding resulted in some broken fiber segments. This problem was solved using local thresholding.

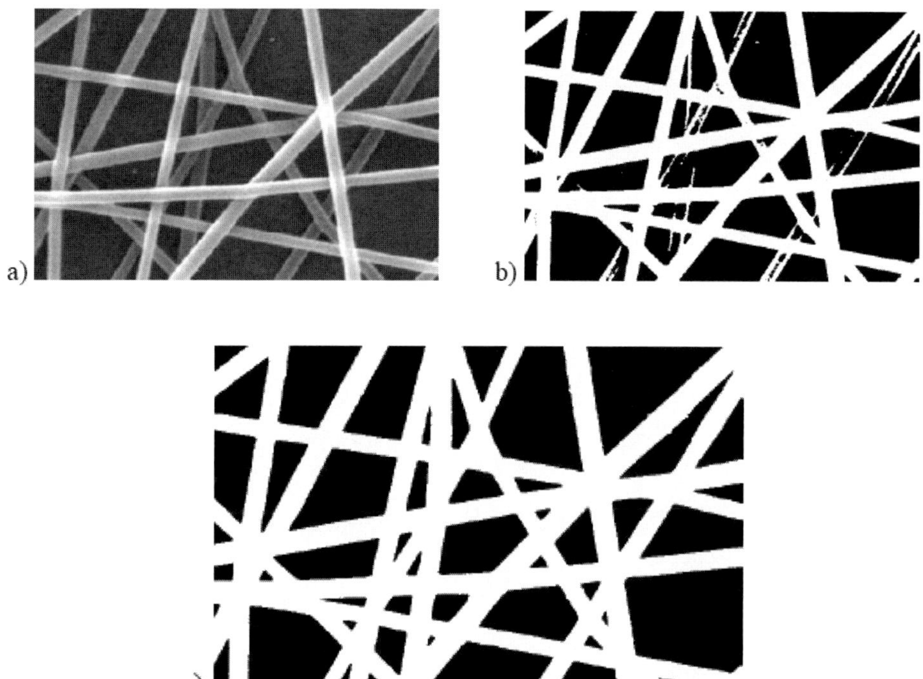


Figure 7. a) A real web, b) Global thresholding, c) Local thresholding.

## 3. EXPERIMENTAL

Electrospun nonwoven webs used as real webs in image analysis obtained from electrospinning of PVA with average molecular weight of 72000g/mol, purchased from MERCK Company, at different processing parameters. The micrographs of the webs were obtained using Philips (XL-30) environmental Scanning Electron Microscope (SEM) under magnification of 10000X after being gold coated. Figure 8 shows the micrographs of the electrospun webs used as real webs.

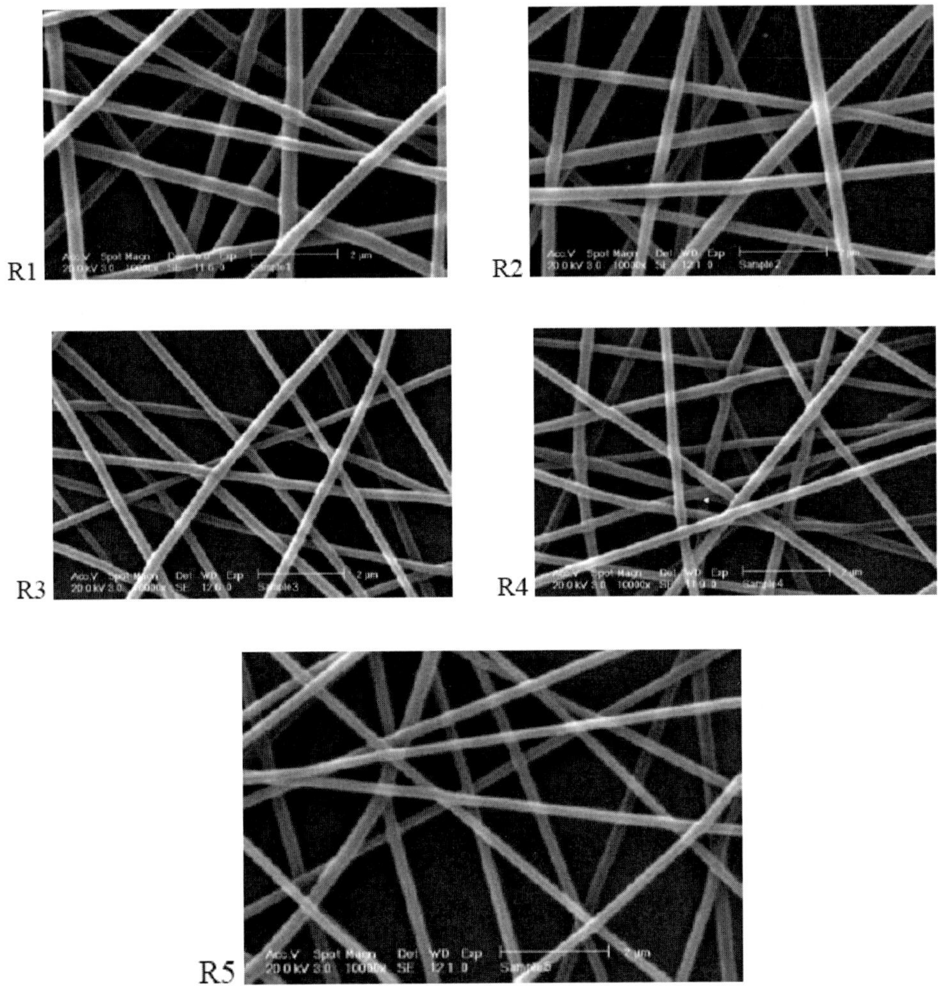


Figure 8. Micrographs of the electrospun webs.

## 4. Results and Discussion

Two sets each composed of five simulated images generated by μ-randomness procedure were used as samples with known characteristics to demonstrate the validity of the techniques. The first set had random orientation with increasing constant diameters; the second was also randomly oriented but with varying diameter sampled from normal distributions with a mean of 15 pixels and standard deviations ranging from 2 to 10 pixels. Table 1 and Table 2 show the structural features of these simulated images which are shown in Figure 9 and Figure 10. Moreover, the applicability of the techniques was tested using five real webs obtained from electrospinning of PVA.

**Table 1. Structural characteristics of first set images**

| Image No. | Angular Range | Line Density | Line Thickness |
|---|---|---|---|
| C1 | 0-360 | 30 | 5 |
| C2 | 0-360 | 30 | 10 |
| C3 | 0-360 | 30 | 15 |
| C4 | 0-360 | 30 | 20 |
| C5 | 0-360 | 30 | 25 |

**Table 2. Structural characteristics of second set images**

| Image No. | Angular Range | Line Density | Line Thickness | |
|---|---|---|---|---|
| | | | Mean | Std |
| V1 | 0-360 | 30 | 15 | 2 |
| V2 | 0-360 | 30 | 15 | 4 |
| V3 | 0-360 | 30 | 15 | 6 |
| V4 | 0-360 | 30 | 15 | 8 |
| V5 | 0-360 | 30 | 15 | 10 |

Mean and standard deviation of the simulated images in the first and second set are shown in Table 3 and Table 4 respectively. Table 5 shows the results for real webs in term of pixel and nm. show histograms of fiber diameter distribution for simulated images in the first and second set respectively. Histograms for real webs are given in figure 13..

In the first set, for simulated images with the line thickness of 5 and 10 pixels, distance transform presents mean and standard deviation of fiber diameter closer to the simulation. For the line thickness of 15, the standard deviation of diameter obtained from direct tracking method is closer to that of the simulation. However, in this case distance transform measured the average diameter more accurately. For the simulated webs with line thickness more than 15 in the first set, direct tracking method resulted in better estimation of the mean and standard deviation of fiber diameter. This is due to the fact that as the lines get thicker, there is higher possibility of branching during the skeletonization (or thinning) and these branches remain even after pruning. Although these branches are small, their orientation is typically normal to the fiber axis; thus causing widening the distribution.

**Table 3. Mean and standard deviation for series 1**

| | | C1 | C2 | C3 | C4 | C5 |
|---|---|---|---|---|---|---|
| Simulation | mean | 5 | 10 | 15 | 20 | 25 |
| | std | 0 | 0 | 0 | 0 | 0 |
| Distance Transform | mean | 5.486 | 10.450 | 16.573 | 23.016 | 30.063 |
| | std | 1.089 | 2.300 | 5.137 | 6.913 | 10.205 |
| Direct Tracking | mean | 5.625 | 11.313 | 17.589 | 22.864 | 29.469 |
| | std | 1.113 | 2.370 | 4.492 | 5.655 | 7.241 |

**Table 4. Mean and standard deviation for series 2**

| | | V1 | V2 | V3 | V4 | V5 |
|---|---|---|---|---|---|---|
| Simulation | mean | 15.247 | 15.350 | 15.243 | 15.367 | 16.628 |
| | std | 1.998 | 4.466 | 5.766 | 8.129 | 9.799 |
| Distance Transform | mean | 16.517 | 16.593 | 17.135 | 17.865 | 19.394 |
| | std | 5.350 | 6.165 | 7.597 | 9.553 | 11.961 |
| Direct Tracking | mean | 16.075 | 15.803 | 16.252 | 16.770 | 18.756 |
| | std | 2.606 | 5.007 | 6.129 | 9.319 | 10.251 |

**Table 5. Mean and standard deviation for real webs**

| | | | R1 | R2 | R3 | R4 | R5 |
|---|---|---|---|---|---|---|---|
| Manual | mean | pixel | 24.358 | 24.633 | 18.583 | 18.827 | 17.437 |
| | | nm | 318.67 | 322.27 | 243.11 | 246.31 | 228.12 |
| | std | pixel | 3.193 | 3.179 | 2.163 | 1.984 | 2.230 |
| | | nm | 41.77 | 41.59 | 28.30 | 25.96 | 29.18 |
| Distance Transform | mean | pixel | 27.250 | 27.870 | 20.028 | 23.079 | 20.345 |
| | | nm | 356.49 | 364.61 | 262.01 | 301.94 | 266.17 |
| | std | pixel | 8.125 | 7.462 | 4.906 | 7.005 | 6.207 |
| | | nm | 106.30 | 97.62 | 64.18 | 91.64 | 81.21 |
| Direct Tracking | mean | pixel | 27.195 | 27.606 | 20.638 | 21.913 | 20.145 |
| | | nm | 355.78 | 361.15 | 269.99 | 286.68 | 263.55 |
| | std | pixel | 4.123 | 5.409 | 4.148 | 4.214 | 3.800 |
| | | nm | 53.94 | 70.77 | 54.27 | 55.14 | 49.72 |

Furthermore, distance transform method fails in measuring the diameter of fiber in intersections. The intersections cause to overestimate fiber diameter. Since in direct tracking method, image is divided into parts where single fibers exist, the effect of intersections which causes in inaccurate measurement of fiber diameter is eliminated. Therefore, there will be a better estimate of fiber diameter.

In the second set, independent of the line thickness in the simulation, for all simulated webs, direct tracking resulted in better measurement of mean and standard deviation of fiber diameter. For the real webs, mean and standard deviation of fiber diameter for direct tracking were closer to those of manual method which concurs with the trends observed for the simulated images. Figure 10. Moreover, the applicability of the techniques was tested using five real webs obtained from electrospinning of PVA.

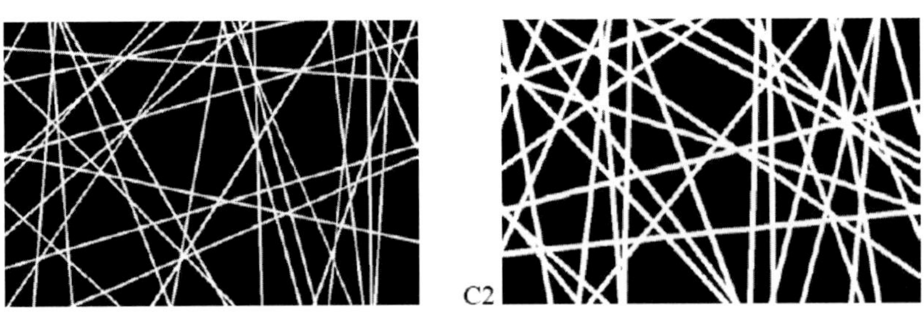

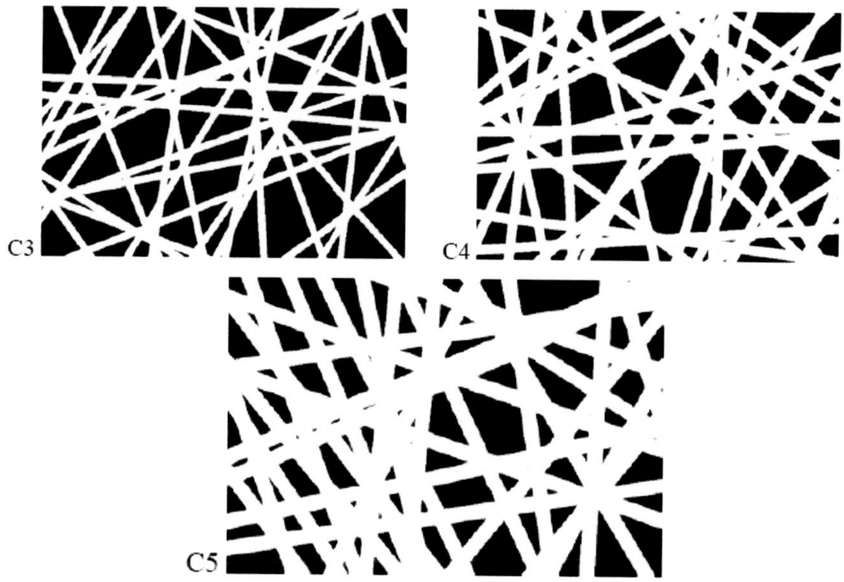


Figure 9. Simulated images with constant diameter.

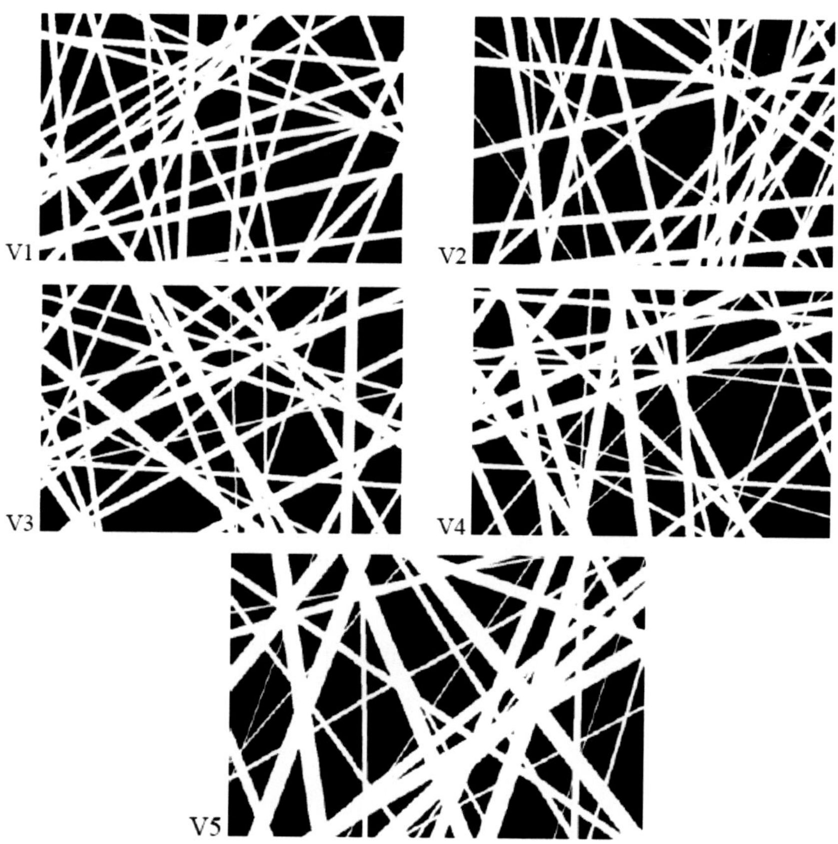


Figure 10. Simulated images with varying diameter.

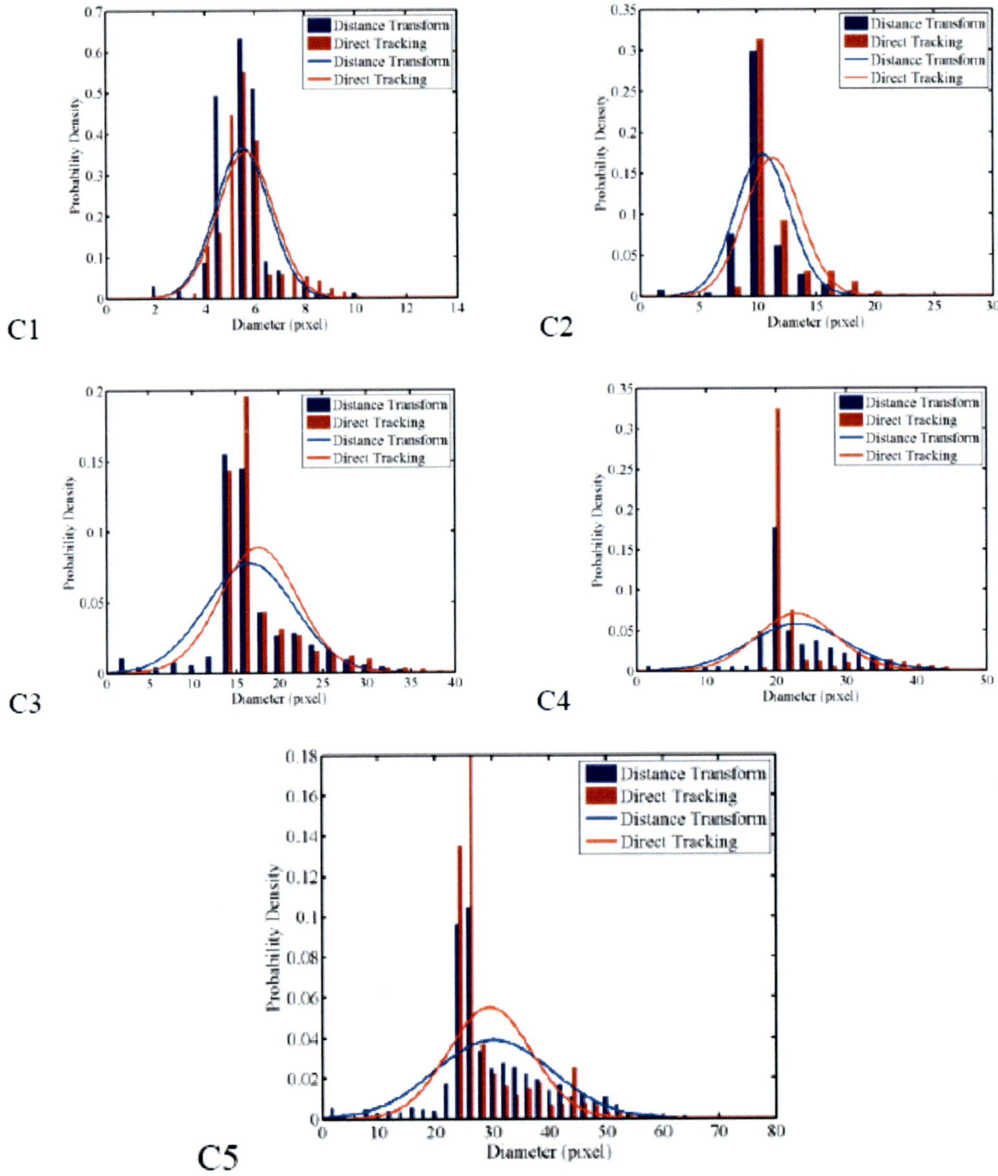


Figure 11. Histograms for simulated images with constant diameter.

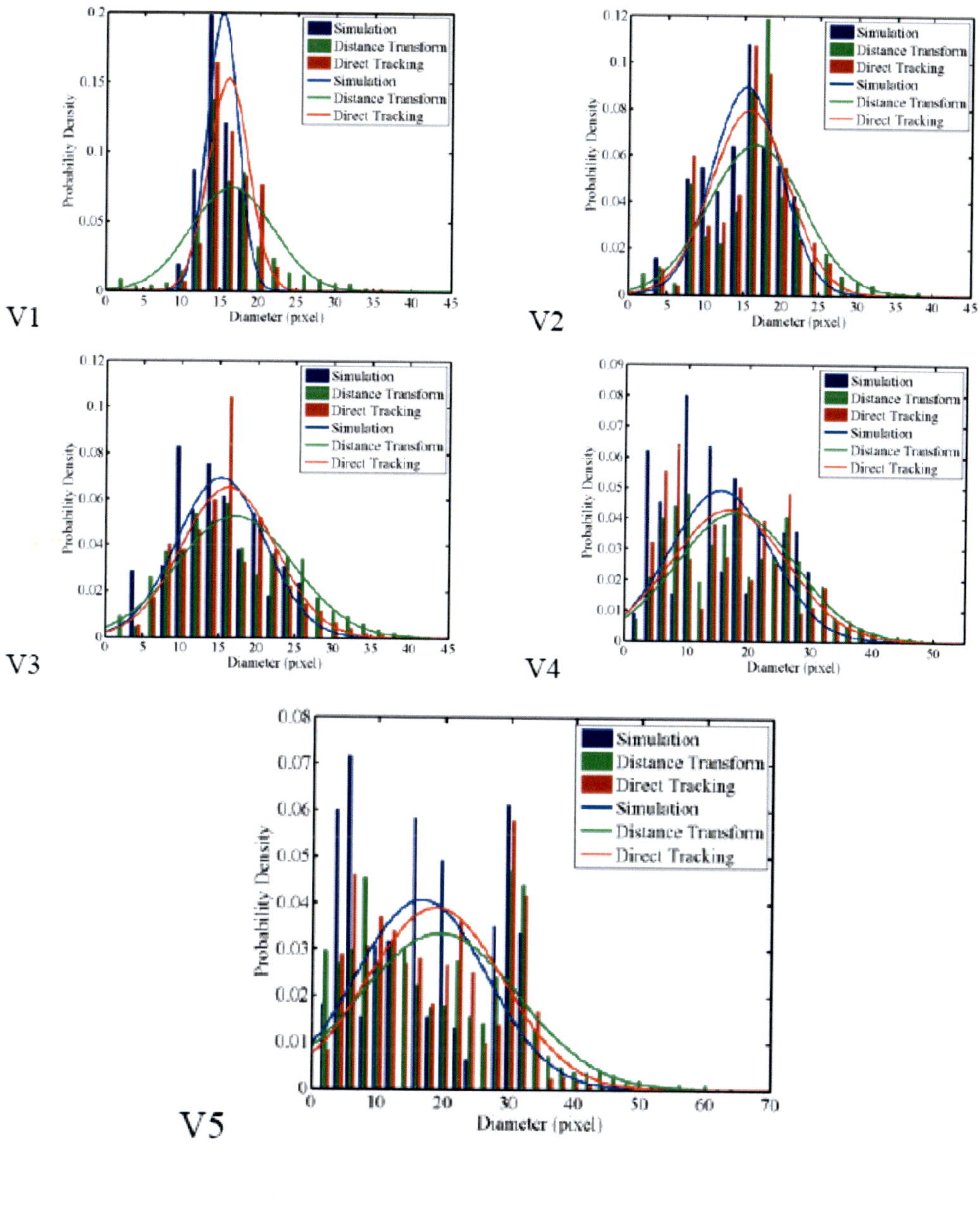


Figure 12. Histograms for simulated images with varying diameter.

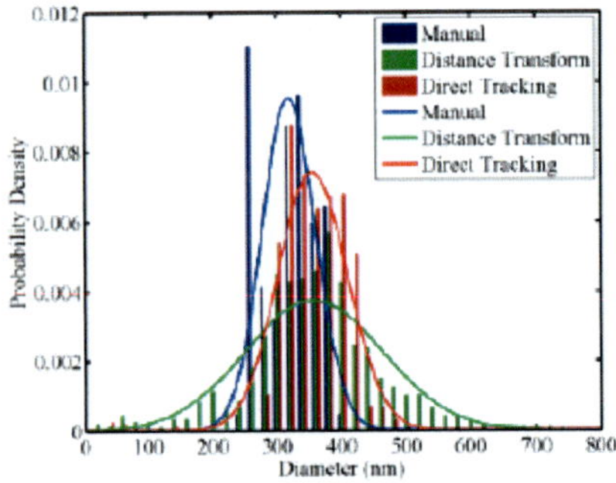


R1

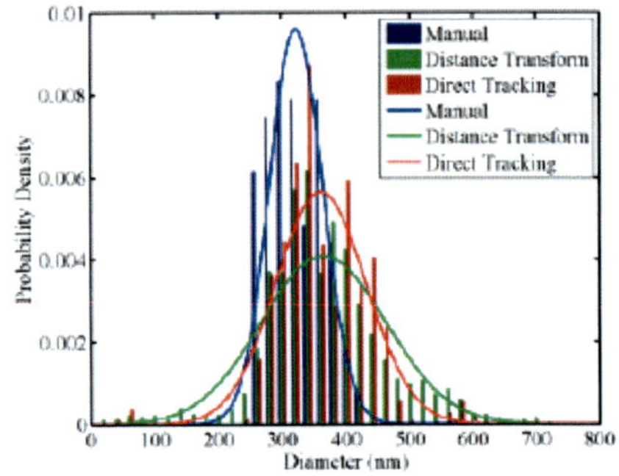


R2

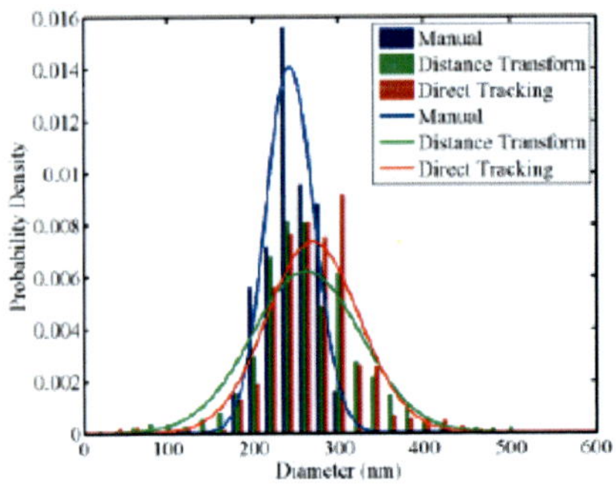


R3

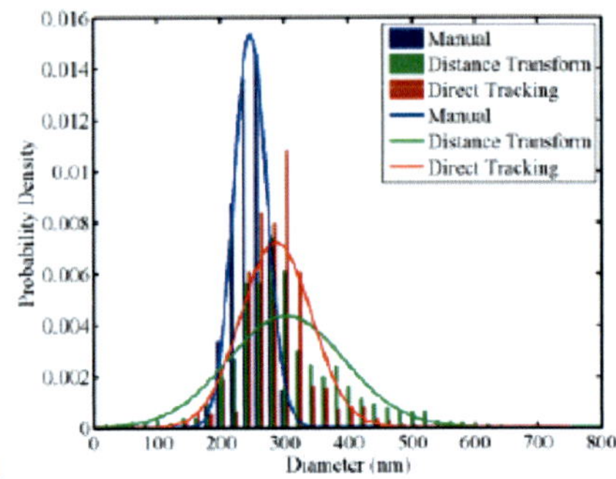


R4

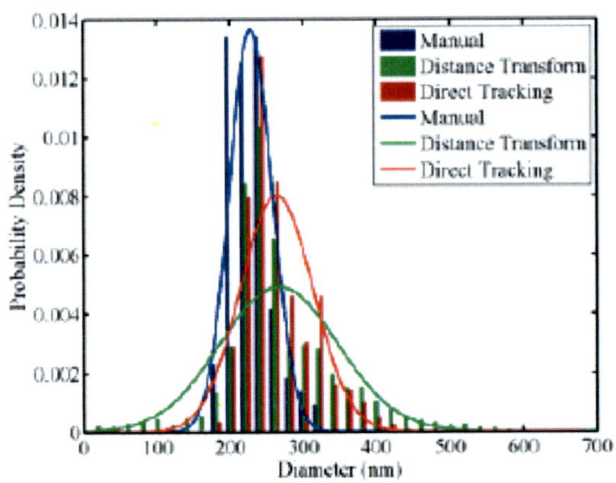


R5

Figure 13. Histograms for real webs.

## 5. Conclusion

Fiber diameter is the most important structural characteristics in electrospun nonwoven webs. The typical way of measuring electrospun fiber diameter is through manual method which is a tedious, time consuming and an operator-based method and cannot be used as an automated technique for quality control. We investigated the use of image analysis for determining fiber diameter and developed an automated method called direct tracking. Since this is a new technique its accuracy needs to be evaluated using samples with known characteristics. To that end, μ-randomness procedure was used in order to simulate the electrospun nonwoven webs. Based on this scheme, two sets of simulated images, each containing 5 webs, were generated. The first set had random orientation with increasing constant diameter. For the second set, the diameter values were sampled from normal distributions with a mean of 15 and standard deviation ranging from 2 to 10 pixels. We compared our method with the distance transform method. For all the simulated webs with varying diameter and for those with constant diameter more than 15, direct tracking method resulted in the mean and standard deviation closer to the simulation. However, for the simulated webs with smaller constant diameter, distance transform measured the mean and standard deviation of fiber diameter more accurately. The results suggest that the direct tracking method is an accurate, direct measurement technique, because it extracts the fiber diameter for the samples by tracking fixed segment of the fiber and eliminates the effect of intersections.

We have demonstrated the general applicability of the method using real webs. 5 real electrospun nonwoven webs obtained by electrospinning of PVA were used. Since the methods needed binary images as input, the images first had to be segmented. A local thresholding method together with Ostu's method was employed in order to automatically compute the appropriate threshold. The results obtained for real webs confirm the trends suggested by simulated images. The results show that the use of image analysis in order to determine the fiber diameter in electrospun nonwoven webs has been successful.

## REFERENCES


[1] A. K. Haghi, M. Akbari, *Physica Status Solidi.* (a) 204 (2007) 1830-1834.

[2] J. Doshi, D. H. Reneker, *Journal of Electrostatics.* 35 (1995) 151-160.

[3] D. H. Reneker, I. Chun, *Nonotechnology.* 7 (1996) 216-223.

[4] H. Fong, D. H. Reneker, Electrospinning and the Formation of Nanofibers, In: D. R. Salem, Structure Formation in polymeric Fibers, Hanser, Cincinnati, 2001, pp. 225-246.

[5] Th. Subbiah, G. S. Bhat, R. W. Tock, S. Parameswaran, S. S. Ramkumar, *Journal of Applied Polymer Science.* 96 (2005) 557-569.

[6] W. Zhang, Z. Huang, E. Yan, Ch. Wang, Y. Xin, Q. Zhao, Y. Tong, *Materials Science and Engineering.* A 443 (2007) 292–295.

[7] B. Pourdeyhimi, R. Ramanathan, R. Dent, *Textile Research Journal.* 66 (1996) 713-722.

[8] B. Pourdeyhimi, R. Ramanathan, R. Dent, *Textile Research Journal.* 66 (1996) 747-753.

[9] B. Pourdeyhimi, R. Dent, H. Davis, *Textile Research Journal.* 67 (1997) 143-151.

[10] B. Pourdeyhimi, R. Dent, *Textile Research Journal.* 67 (1997) 181-190.

[11] B. Pourdeyhimi, R. Dent, A. Jerbi, S. Tanaka, A. Deshpande, *Textile Research Journal.* 69 (1999) 185-92.

[12] B. Pourdeyhimi, H.S. Kim, *Textile Research Journal.* 72 (2002) 803-809.

[13] B. Pourdeyhimi, R. Dent, *Textile Research Journal.* 69 (1999) 233-236.

[14] B. Xu, Y. L. Ting, *Textile Research Journal.* 65 (1995) 41-48.
[15] A. H. Aydilek, S. H. Oguz, T. B. Edil, *Journal of Computing in Civil engineering.* (2002) 280-290.
[16] R. Chhabra, International Nonwoven Journal. (2003) 43-50.
[17] M. S. Abdel-Ghani, G. A. Davis, *Chemical Engineering Science.* 40 (1985) 117-129.
[18] R. C. Gonzalez, R. E. Woods, *Digital Image Processing,* 2nd Edition, Prentice Hall, 2001.
[19] B. Jähne, Digital Image Processing, 5th Revised and Extended Edition, Springer, 2002.
[20] M. Petrou, P. Bosdogianni, Image Processing the Fundamentals, John Wiley and Sons, 1999.
[21] J. Serra, Image Analysis and Mathematical Morphology, Academic Press, London, 1982.

In: Monomers, Oligomers, Polymers, Composites…
Editors: R. A. Pethrick, G.E. Zaikov et al.
ISBN: 978-1-60456-877-6


*Chapter 6*

# A Study on the Effects of Recycled Glass, Silica Fume and Rice Husk Ash on the Interfacial and Mechanical Properties of Cementitious Composite

***A. Sadrmomtazi and A.K. Haghi****
University of Guilan, P. O. Box 3756, Rasht, Iran

## Abstract

The carpet waste fibers as well as waste glass generated each year create a serious environmental problem and accumulated in the landfills which can be converted into useful products. The use of these waste materials in a cement-based composite can be a promising direction for waste reduction and resources conservation. In this study, several tests carried out to investigate the performance of ordinary Portland cement using recycled glass and fibers as a fraction of aggregates used in a cement-based composite. These tests included compressive strength, flexural strength, flexural toughness and water absorption. The results of this study revealed that carpet waste fiber and waste glass could be reused as substitutes for conventional materials in cement-based composites.

## 1. Introduction

As the world population grows, so do the amount and type of waste being generated. Many of the wastes produced today will remain in the environment for years to come. The creation of nondecaying waste materials, combined with a growing consumer population, has resulted in a waste disposal crisis. So one of the solutions to improve the environment is recycle and reuse these waste materials in construction.

A great amount of fibrous textile waste is discarded into landfills each year all over the world. More than half of this waste is from carpets, with the main constituents being plastic

* Corresponding author: e-mail: Haghi@Guilan.ac.ir

and polymeric fiber, which decays at a very slow rate and which is difficult to handle in landfills. The development of low-cost technologies to convert textile waste into useful products could reduce disposal pressures on landfills and result in cost savings for carpet industries. Accordingly, many carpet and textile manufacturers, fiber and chemical suppliers, recycling companies, and academic institutions are actively pursuing various methods to recycle and convert fibrous waste into various useful products [ ].

One promising reuse of these wastes lies in concrete reinforcement and construction applications. Extensive studies have been reported in the literature that indicate the use of natural and synthetic fibers in concrete can enhance the tensile strength and flexural toughness characteristics of concrete,

Waste carpet fibers have been used in cement-based composites concrete since the late 1960s [1]. Natural and other synthetic fibers are added to cement as secondary reinforcement to control plastic shrinkage [2]. The effect of polypropylene fibers on the properties of cement-based composites varies depending on the type, length, and volume fraction of fiber, the mixture design, and the nature of materials used.

Although many researchers studied the reuses of polypropylene fibers in cement-based composite but admixture of waste glass in fiber-reinforced cement-based composite have never been the issue.

The main objective of this study is to evaluate the performance of a cement-based composite under the effect of using waste carpet fiber (polypropylene) and waste glass.

## 2. Experimental Approach

### 2. 1. Material Specifications

The fibers included in this in this study (table 1) were 4.5 denier monofilament fibers obtained from recycled raw materials that were cut to 6.35 mm in length.

**Table 1. Properties of Polypropylene fibers reused in this study**

| Property | Polypropylene |
|---|---|
| Unit weight [g/cm3] | 0.9 - 0.91 |
| Reaction with water | Hydrophobic |
| Tensile strength [ksi] | 4.5 - 6.0 |
| Elongation at break [%] | 100 – 600 |
| Melting point [°C] | 175 |
| Thermal conductivity [W/m/K] | 0.12 |

The chemical composition of the reused glass was analyzed using an X-ray microprobe analyzer and listed in table 2 together with that of silica fume and rice husk ash for comparison.

In accordance to ASTM C618, the glass satisfies the basic chemical requirements for a pozzolanic material. However, it dose not meet the optional requirement for the alkali content because of high percentage of $Na_2O$.

**Table 2. Chemical composition of materials**

| | Content (%) | | |
|---|---|---|---|
| Oxide | Glass C | Silica fume | Rice husk ash |
| $SiO_2$ | 72.5 | 91.1 | 92.1 |
| $Al_2O_3$ | 1.06 | 1.55 | 0.41 |
| $Fe_2O_3$ | 0.36 | 2 | 0.21 |
| CaO | 8 | 2.24 | 0.41 |
| MgO | 4.18 | 0.6 | 0.45 |
| $Na_2O$ | 13.1 | - | 0.08 |
| $K_2O$ | 0.26 | - | 2.31 |
| CL | 0.05 | - | - |
| $SO_3$ | 0.18 | 0.45 | - |
| L.O.I | - | 2.1 | - |

To satisfy the physical requirements for fineness, the glass has to be grounded to pass a 45μm sieve. This was accomplished by crushing and grinding of glass in the laboratory, and by sieving the ground glass to the desired particle size. To study the particle size effect, two different ground glasses were used, namely:

- Type I: ground glass having particles passing a #80 sieve (180μm);
- Type II: ground glass having particles passing a #200 sieve (75μm).

The particle size distribution for two types of ground glass, silica fume, rice husk ash and ordinary Portland cement were analyzed by laser particle size set and have shown in figure 1. As it can be seen in figure 1 silica fume has the finest particle size. According to ASTM C618, fine ground glasses under 45μm qualify as a pozzolan due to the fine particle size. Moreover glass type I and II respectively have 42% and 70% fine particles smaller than 45μm that causes pozzolanic behaviour and doesn't occur any excessive expansion due to alkali–silica reaction (ASR).

Also standard river sand and Portland cement type1 were used for preparing the specimens.

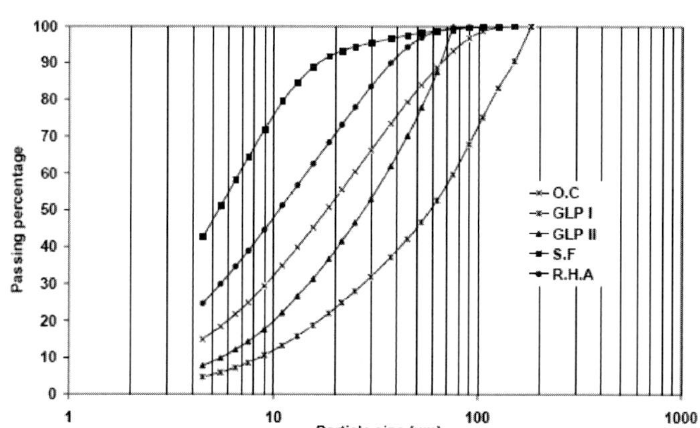


Figure 1. Particle size distribution of ground waste glass type I, II, silica fume, rice husk ash and ordinary cement.

**Table 3. Mixture proportions**

| batch No | sand/c | w/c | Content (by weight) | | | | | PP fibers | batch No | sand/c | w/c | Content (by weight) | | | | | PP fibers |
|---|---|---|---|---|---|---|---|---|---|---|---|---|---|---|---|---|---|
| | | | O.C | GI | GII | SF | RH | (by volume) | | | | O.C | GI | GII | SF | RH | (by volume) |
| 1 | 2.25 | 0.47 | 100 | - | - | - | - | 0 | 11 | 2.25 | 0.6 | 100 | - | - | - | - | 1 |
| 2 | 2.25 | 0.47 | 90 | 10 | - | - | - | 0 | 12 | 2.25 | 0.6 | 90 | 10 | - | - | - | 1 |
| 3 | 2.25 | 0.47 | 90 | - | 10 | - | - | 0 | 13 | 2.25 | 0.6 | 90 | - | 10 | - | - | 1 |
| 4 | 2.25 | 0.47 | 90 | - | - | 10 | - | 0 | 14 | 2.25 | 0.6 | 90 | - | - | 10 | - | 1 |
| 5 | 2.25 | 0.47 | 90 | - | - | - | 10 | 0 | 15 | 2.25 | 0.6 | 90 | - | - | - | 10 | 1 |
| 6 | 2.25 | 0.6 | 100 | - | - | - | - | 0.5 | 16 | 2.25 | 0.6 | 100 | - | - | - | - | 1.5 |
| 7 | 2.25 | 0.6 | 90 | 10 | - | - | - | 0.5 | 17 | 2.25 | 0.6 | 90 | 10 | - | - | - | 1.5 |
| 8 | 2.25 | 0.6 | 90 | - | 10 | - | - | 0.5 | 18 | 2.25 | 0.6 | 90 | - | 10 | - | - | 1.5 |
| 9 | 2.25 | 0.6 | 90 | - | - | 10 | - | 0.5 | 19 | 2.25 | 0.6 | 90 | - | - | 10 | - | 1.5 |
| 10 | 2.25 | 0.6 | 90 | - | - | - | 10 | 0.5 | 20 | 2.25 | 0.6 | 90 | - | - | - | 10 | 1.5 |

### 2.2. Test Program

For the present study, twenty batches were prepared. Control mixes was designed containing natural sand at a ratio of 2.25:1 to the cement matrix. Concerning %10 of cement weight replaced by the ground waste glass type I and II(GI,GII ) , silica fume(SF) and rice husk ash (RH). Meanwhile, modified batches were designed with the same admixtures but reinforced with 0.5%, 1% and 1.5% (by volume) of polypropylene fibers (PP). In the control batches, water to cementitious ratio of 0.47 was used whereas in modified mixes (with different amount of PP fibers) water to cementitious ratio was 0.6. The mortar mixture proportions are represented in table 3.

The strength criteria of mortar specimens and impacts of polypropylene fibers on characteristics of these specimens were evaluated at the age of 7, 28 and 60 days.

In our laboratory, the test program conducted as follows:

1. The fibers were placed in the mixer.
2. Three-quarters of the water was added to the fibers while the mixer was running at 60 rpm; mixing continues for 1 minute.
3. The cement was gradually added while the mixer was still running. After adding the cement, the mixer is allowed to run for two minute to allow the cement to mix with the water.
4. The sand and remaining water were added, and the mixer was allowed to run for another two minutes.

After mixing, the samples were casted into the forms $50\times50\times50mm$ for compressive strength and $50\times50\times200mm$ for flexural strength. All the moulds were coated with mineral oil to facilitate demoulding. The samples were placed in two layers. Each layer was tamped at least 25 times using a hard rubber mallet. The sample surfaces were finished using a metal spatula. After 24 hours, the specimens were demoulded and cured in water at $20^{\circ}C$.

## 3. Results and Desiccations

### 3.1. Compressive Strength

The variations of compressive strength with age are presented in figure 2. The compressive strength of the composite in comparison with control specimens containing 0% fiber is represented in figure 2.a. A significant decrease in compressive strength can be observed. In view of the above, the lowest decline belongs to the composite containing 10% silica fume (SF) and the highest are related to composites containing 10% glass I (GI) and rice husk ash (RH). The compressive strength of SF specimens is close to target specimens due to the finest particle size and highest pozzolanic behaviour. This is however confirmed according to ASTM C618. Meanwhile, glass type 1(GI) due to coarse size of aggregates exhibit low pozzolanic behavior.

In addition, however the pozzolanic behaviour of rice husk ash in early age is low but by ageing the specimens, this effect will rise significantly.

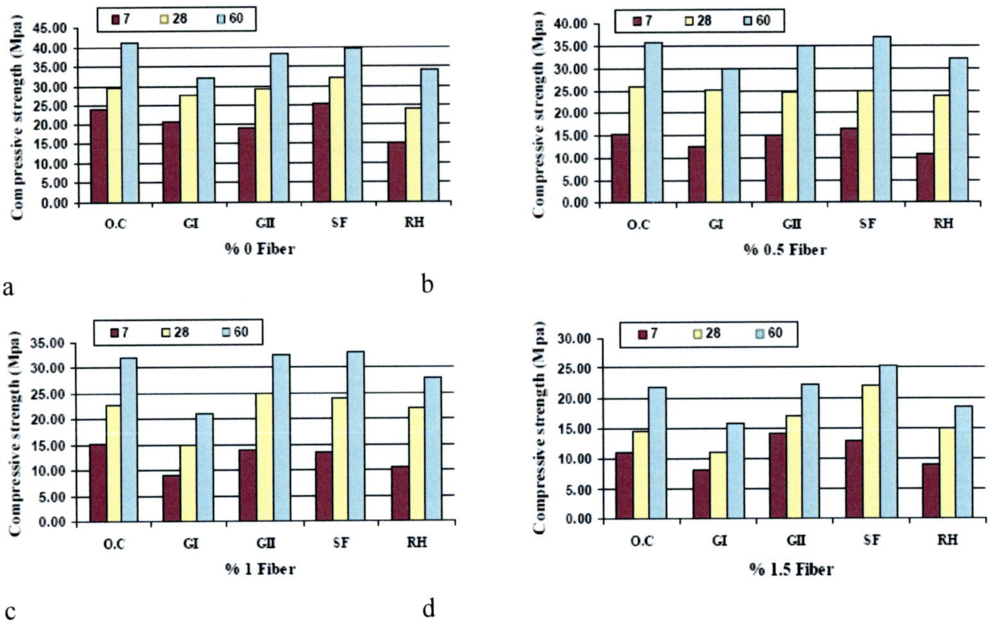


Figure 2. Compressive strength of mortars in different ages.

The variation of compressive strength with age shows a continuous increase by decreasing the amount of polypropylene fibers in matrix (figures 2.b, c and d.). The specimens contain 1.5% PP fiber by volume exhibit a considerable reduction around 50% in compressive strength comparing to target specimen. Although the variation of strength in mortars contained 0.5% PP by volume is not significant. Despite of compressive strength decrement in reinforced matrix contained PP fiber, the specimens which contains silica fume (SF) and PP fiber have shown greater compressive strength than the others. Moreover GII approximately indicates the same results to target specimens.

In view of the above, the replacement SF and GII (in which 70% of fine particles are smaller than 45μm )is directly proportional to their respective pozzolanic activity index values (according to ASTM C618 and C989 and table 2).

### 3.2. Flexural Strength

The effect of fiber volume fraction, and the age of the mortar matrix on the flexural strength are represented in figure 3.

In all cases, by increasing the amount of fiber in matrix and ageing of specimens, the flexural strength tends to increase.

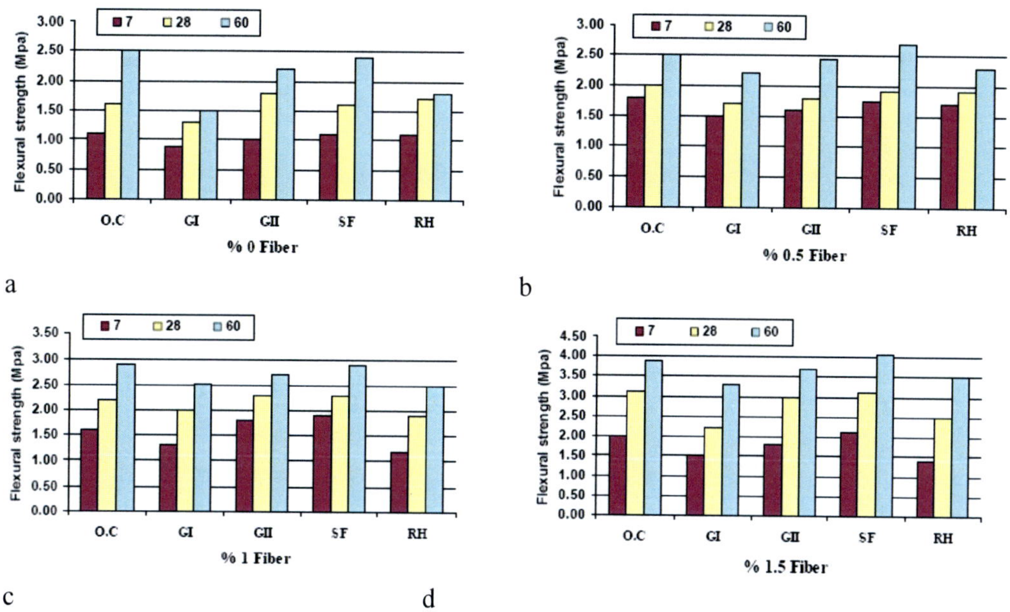


Figure 3. Flexural strength of mortars in different ages.

In view of the above, the highest flexural strength between control specimens belong to O.C, 10% SF and 10% GII replacement in cement which is relevant to pozzolanic activity which mentioned before.

The flexural strength of specimens contained 0.5% PP fiber doesn't show considerable increment. Nevertheless, in mixes contained 1% and 1.5% PP fibers, represent a relative increment in flexural strength.

It is also observed that the lowest strength is related to GI in all mixes due to course particle size and low pozzolanic behaviour.

### 3.3. Flexural Toughness

Toughness is defined as the ability of materials to sustain load after initial cracking and is measured as the total strain experienced at failure. Upon cracking, fibers are able to bridge the initial crack and hold the crack together until the fibers either pull-out from the matrix (in early age) or fracture that say flexural toughness.

Figure 4 represents typical load-deflection responses of plain matrix and mixtures contained 0.5%, 1%, and 1.5% fibers.

For plain mortars, the behaviour was in a brittle manner. When the strain energy was high enough to cause the crack to self-propagate, fracture occurred almost while the peak load was reached (this is due to the tremendous amount of energy being released). For fiber reinforced mixes, the fiber bridging effect helped to control the rate of energy release. Thus, fibers still can carry load even after the peak. With the effect of fibers bridging across the crack surface, fiber was able to maintain the load carrying ability even after the mortar had been cracked. According to ASTM C1018 [11], toughness or energy absorption defined as the area under the load-deflection curve from crack point to 1/150 of span.

Results indicate that, by increasing the amount of fiber in matrix flexural toughness rise significantly.

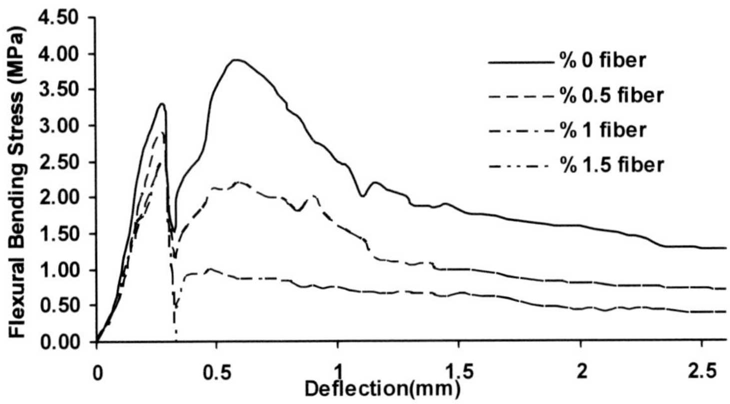


Figure 4. Load-Deflection Response of plain and polypropylene fiber reinforced mortar.

### 3.4. Water Absorption

Water absorption of specimens is measured and evaluated in table 4 according to ASTM C642.

In general, the incorporation of SF improves the water absorption properties of the material because of the reduction of permeable voids. The same behavior was reported by Aguilar et al. [5] when they studied steel and nylon fiber reinforced Portland cement matrices with additions of SF and GGBS (granulated blast furnace).

**Table 4. Water absorption of all mixes**

| batch No | Water absorption | batch No | Water absorption | batch No | Water absorption | batch No | Water absorption |
|---|---|---|---|---|---|---|---|
| 1 | 5.83 | 6 | 7.10 | 11 | 8.50 | 16 | 10.80 |
| 2 | 5.30 | 7 | 7.42 | 12 | 8.80 | 17 | 10.52 |
| 3 | 5.16 | 8 | 6.59 | 13 | 8.12 | 18 | 10.22 |
| 4 | 4.36 | 9 | 5.42 | 14 | 6.37 | 19 | 8.15 |
| 5 | 4.77 | 10 | 5.30 | 15 | 6.79 | 20 | 8.42 |

A reduction in water absorption is expected when a pozzolanic material is added to the matrix. SF addition had the greatest effect, followed by RH and GII addition but GI doesn't show a significant effect on water absorption. This is due to particle size and then, pozzolanic activity of materials.

This explains that by increasing the fiber percentage in the matrix of mortars, water absorption will increase, as well.

## 4. Conclusion

This research study proves the feasibility of the use of glass as aggregates replacement and PP fibers for composite reinforcement. Based on the experimental results of this investigation the following conclusion can be drawn:

- Application of fibers in matrix causes the noticeable reduction in compressive strength and negligible rise in flexural strength.
- Mixture of cement based composite with GII and SF containing different percentage of fibers shown close properties to target specimens. So results show the great possibility of usage of ground glass and silica fume in mortars as a replacement in cement.
- Increasing the fibers in matrix causes low compressive strength, but it increases the flexural toughness substantially. In fact the ability for absorbing the energy rises up.
- Water absorption will decrease by addition of pozzolanic materials in to the matrix because of fine particle size and covering the permeable voids.

## References

[1] Bentur A, Mindess S. Fiber reinforced cementitious on durability of concrete. Barking: Elsevier, 1990.

[2] Balaguru PN, Shah SP. Fiber reinforced cement composites. New York: McGraw-Hill, Inc, 1992:367p.

[3] Duckett, K., "Olefin Fiber" and "Nylon Fibers", Project: Textile Science 526 - Nonwovens Science and Technology II, (http://trcs.he.utk.edu/textile/nonwovens/), Fall 1999.

[4] Callister, W.D., Material Science and Engineering - An Introduction, 4th ed., John Wiley & Sons, New York, 1997.

[5] Aguilar MTP, Gomes AM, Cetlin PR, Friche GHS. Proc. II International Conference on High Performance Concrete, ramado, Brazil, 1999. SP58

[6] S. Kumar, M. B. Polk, and Y. Wang, Fundamental studies on the utilization of carpet waste, presented at the SMART (Secondary Materials & Recycled Textiles, An International Association) 1994 Mid-Year Conference July, 1994, Atlanta, GA.

[7] Y. Wang, S. Kumar, and M.B. Polk, Fundamental studies on the utilization of carpet waste, presented at The Fiber Society Spring Technical Conference, May, 1994, Annapolis, MD.

[8] Y. Wang, B.S. Cho, and A. H. Zureick, Fiber Reinforced Concrete Using Recycled Carpet Industrial Waste and Its Potential Use in Highway Construction, in Proceedings of the Symposium on Recovery & Effective Reuse of Discarded Materials & By-Products for Construction of Highway Facilities, U.S. Department of ransportation, October, 1993, Denver, CO.

[9] Y. Wang, A. Zureick, B.S. Cho, D. Scott, "Properties of Fiber Reinforced Concrete Using Recycled Fibers from Carpet Industrial Waste", Journal of Materials Science, in print, 1994.

[10] R. K. Datta, M. B. Polk, and S. Kumar, Reactive Extrusion of Polypropylene and Nylon, Polymer-Plastics Technology and Engineering, in print, 1994.

[11] ASTM C 1018. Test method for flexural toughness and first crack strength of fiber reinforced concrete using beam with third-point loading. Annual Book of ASTM Standards, 04.02, 1988:491]496

In: Monomers, Oligomers, Polymers, Composites... ISBN: 978-1-60456-877-6
Editors: R. A. Pethrick, G.E. Zaikov et al. 

*Chapter 7*

# The Synthesis and Properties of Unsaturated Halogen -Containing Poly (Arylene Ether Ketone)S

***A.M. Kharayev, A.K. Mikitaev, G.E. Zaikov* and R.Ch. Bazheva***
Kabardino-Balkar State University,
360004, Nalchik, Chernishevskaya st. 173, KBR, Russia.
*Institute of biochemical Physics
Russian Academy of Sciences
119334 Moscow, Kosygin str., 4, Russia

## Abstract

Using dihydroxyl-containing olygoketones with different condensation degrees and unsaturated halogen - containing dichloroanhydrides, aromatic poly(arylene ether ketone)s (PAEK) were synthesized by means of acceptor-catalytic polycondensation. The correlation between the structure of olygoketones and properties of poly(arylene ether ketone)s were investigated. It was shown that the PAEK was characterized by a high physical and chemical properties.

**Keywords:** poly(arylene ether ketone), copolymerization, glass transition, dielectric properties, viscosity

## Introduction

Polyethers (PEK) and polyetheretherketones (PEEK), possessing a complex of valuable properties, take an important place among thermal and heat resistant polymers; due to this some of the mentioned polymers are used in production[1-3]. Under the synthesis of PEEK the methods and conditions of synthesis play a significant role. Two approaches are noted here: the synthesis of PEEK on the basis of 4,4'-dichlorobenzophenone in high boiling organic solvents[4] and synthesis of PEEK of block structure on the basis of 4,4'-dichlorobenzophenone by means of acceptor-catalytic polycondensation[5].

There is a great number of information in literature about the synthesis of polyetherketones on the basis of 4,4'-dichlorobenzophenone and widely known bisphenols, including 4,4'-dioxyphenyl - 2,2-propane and phenolphthalein. A great shortage of these methods is the duration of the synthesis (about 30 hours) and the high temperature (about $300^0$ C and more). Besides, even in such stringent term synthesis other halogenated diphenylketones do not actively react[6-14].

In this connection, the investigation of a possibility of carrying out of a polycondensation process of getting poly(arylene ether ketone)s in greatly more soft conditions with the usage of 4,4'-dichlorobenzophenone. As 4,4'-dichlorobenzophenone in the medium of dimethylsulphoxide 150-$160^0$ C doesn't give with 4,4'-dioxyphenyl - 2,2-propane, and phenolphthalein polymers with high molecular weight, we used the getting block-copolyetheresterketones.

As one of monomers 1,1-bis(4-chlorophenyl)-2,2,2-trichloroethane (DDT) is chosen. A chose of this monomer is explained so that the prohibition of 1,1-bis(4-chlorophenyl)-2,2,2-trichloroethane a global scale as an insecticide faced the humanity a necessity of finding of an another application for it. As till the present day the mentioned matter is in great quantity in the storehouses and arouses a serious ecological threat to the environment, so each method of its utilization demands proper attention. The getting of different monomers from 1,1-bis(4-chlorophenyl)-2,2,2-trichloroethane for their further usage as monomers in the process of synthesis of fire-resistant polyesters is highly perspective in this regard.

With this aim 2,2-bis(4-chlorophenyl)-1,1-dichloroethylene had been got from 1,1,1-bis(4-chlorophenyl)-2,2,2-trichloroethane, and then it was used for synthesis of 4,4'-dichlorobenzophenone.

As acid component dichloroanhydride 1,1-dichloro-2,2di-(4-carboxyphenyl)ethylene was used. The leading in the structure of polymers of >C=$CCl_2$ group promotes an increase of refractoriness in conditions of preservation of high importance of mechanical properties.

## Experimental Procedure

In the present paper, poly(arylene ether ketone)s were synthesized in two steps on account of low activity of 4,4'-dichlorobenzophenone in polycondensation process and also having a desire to hold the process of synthesis under softer conditions. PAEKs were synthesized by means of acceptor-catalytic polycondensation from olygoketones (OK) on the basis of 4,4'-dioxyphenyl - 2,2-propane, containing two hydroxyl end groups and dichloroanhydride 1,1-dichloro-2,2di-(4-carboxyphenyl)ethylene. Olygoketones with different condensation degrees were synthesized by means of interaction of disodium salt of 4,4'-dioxydiphenyl-2,2-propane with 4,4'- dichlorobenzophenone in anhydrous dimethylsulphoxide (DMSO) according to the known method [14], where n=1,5,10 and 20 *(scheme 1).*

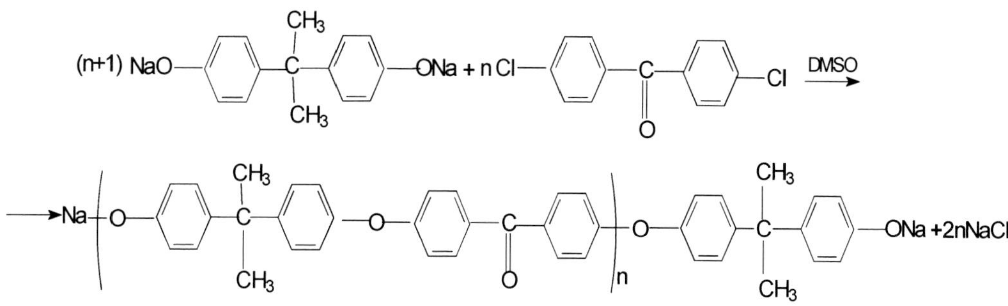


Scheme 1.

4,4'- dichlorobenzophenone was obtained from 1,1-bis(4-chlorophenyl)-2,2,2-trichloroethane (DDT) in two steps: in the first stage 2,2-bis(4-chlorophenyl)-1,1-dichloroethylene was synthesized from DDT by means of dehydrochloration; in the second stage, this product was reacted with oxide of six-valent chromium in ice acetic acid and 4,4'-dichlorobenzophenone with a melting temperature of 146°C was obtained. The given process was described by Yanota et al[15].

Some properties of olygoketones are given in table 1.

**Table 1. The properties of olygoketones**

| Olygoketones | Condensation degree | Softening temperature(C) | Molecular weight (in carbon units) | OH-group content (%) | |
|---|---|---|---|---|---|
| | | | | Calculated | Measured |
| OK-1D | 1 | 129-132 | 647,78 | 5,36 | 5,30 |
| OK-5D | 5 | 147-152 | 2260,72 | 1,50 | 1,35 |
| OK-10D | 10 | 160-165 | 4293,17 | 0,79 | 0,75 |
| OK-20D | 20 | 167-175 | 8358,43 | 0,41 | 0,40 |

Dichloroanhydride 1,1-dichloro-2,2di-(4-carboxyphenyl)ethylene was obtained according to the *scheme 2*:

Poly(arylene ether ketone)s was obtained by the technique of acceptor-catalytic polycondensation in 1,2-dichloroethane for 1 hour at 20°C. Triethylamine was used as an acceptor-catalyser, and isopropanol - as a precipitator. The synthesized PAEK possesses the following structure (*Scheme 3):*

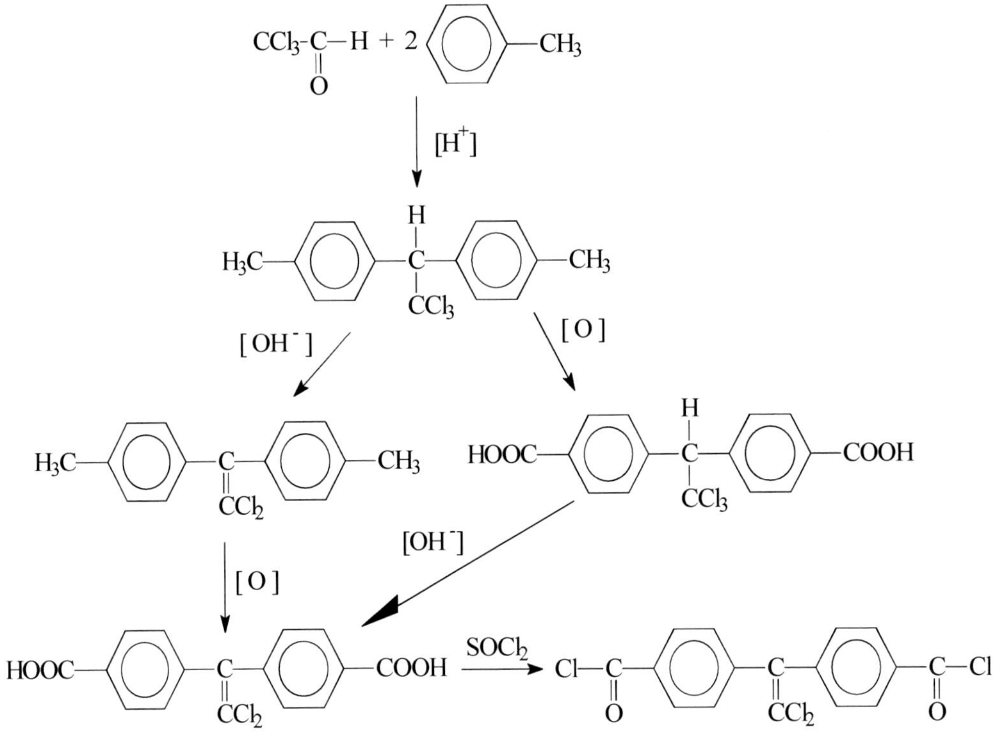


Scheme 2.

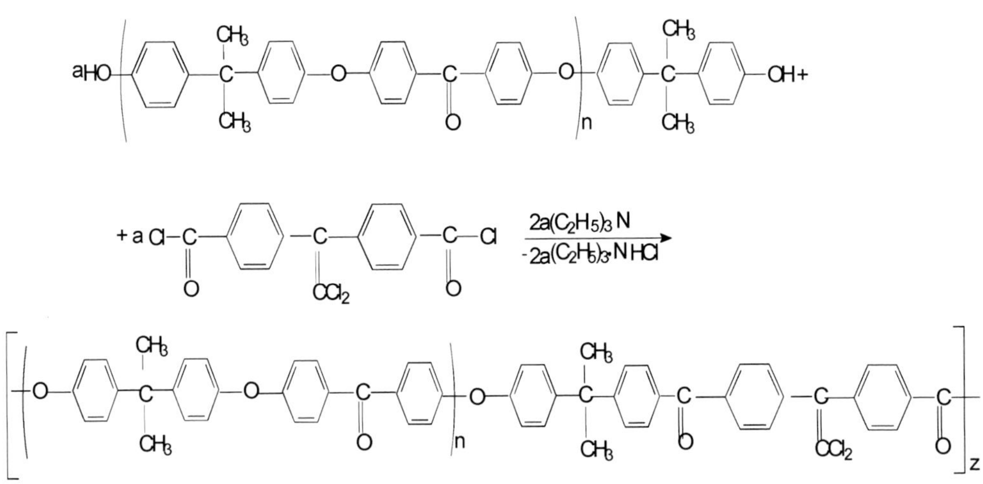


Scheme 3.

## TEST METHODS

The experiments were conducted at a room temperature. The viscosity measurements were made according to GOST 10028-81 (Russian State Standards) using an Ubbelohde viscometer with a diameter of 0,34 mm. The viscosity was measured in 1,2-dichloroethane and the density of the solution was 0,5g/dl.

Poly(arylene ether ketone)s were investigated aiming at definition the possibilities of their use as construction materials. Mechanical properties of film PAEK specimens (100mm x l0mm x 0,lmm) were tested (GOST 17-316-71) on a MRS-500 with a constant strain rate of 40 mm/min at 20°C. The film specimens were obtained by pouring the polymeric solution into 1,2-dichloroethane.

The poly(arylene ether ketone)s structures were confirmed from the data of elemental analysis and infrared spectroscopy. Infrared spectra of olygomers and polymers were taken from peels and films with the use of IK-Fuye FTJR-850 spectrophotometer (Shimadzu, Japan);'windows from KBr, stratum 0,05 mm.

The molecular masses of poly(arylene ether ketone)s were measured by sedimentation in an ultracentrifuge "MOM" (type 317 B) through the balance approach technique.

Thermo-gravimetric analysis of poly(arylene ether ketone)s was performed on the derivatograph "MOM" under a temperature increase rate at 5 degree/min in the atmosphere (MOM is the abbreviation of the Hungarian company that manufactured the derivatograph).

The investigation of the polydispersity of the block copolymers was conducted by the turbidimetric titration method on an FEC-56M device. The principle of titration is that the diluted polymeric solution will become turbid if a precipitator is added and will have a different optical density from the original solution. The density of the solution was 0,01 g/ml. 1,2-dichloroethane was used as a solvent, and isopropyl alcohol - as precipitator.

Investigations of thermomechanical properties of the film specimens were carried out on a device thermo-mechanical analyzer. The temperature was increased at the rate of 4 degree/min and the load was kept to give a constant stress value of 0,05 MPa. The thickness and width of the test specimens were 0,1 mm and 8 mm, respectively, while the distance between the clamps was 80 mm. A thermal camera with a diameter of 30 mm and length of 150 mm was employed. The temperature gradient of the thermal camera was 2 degree along the length and 0,1 degree along the diameter. The deformation of the specimen was measured using a strain gauge.

The tests of the dielectric properties of poly(arylene ether ketone)s were carried out at a set with a Kumetre VM-560 Tesla machine at a frequency of $10^6$ Hz. The values of dielectric permeability and tan δ of the dielectric loss angle were found for all specimens over the temperature range 20-300 °C.

Investigation of chemical resistance of poly(arylene ether ketone)s was carried out on the film specimens in the shape of a disk with a diameter of 5 sm. The specimens were tested in aggressive environment at 20°C for 16 days (384 hours). The specimens were measured every 12 hours.

The thermal -resistance was tested on the film specimens (strips) fixed vertically in a cylindrical camera, and laminar stream of nitrogen and oxygen mixture of the given correlation was put through it. The investigation was carried out under various structures of the gas mixture unless the optimal structure, which provided burning of the specimen was found. Thermal-resistance was evaluated by the percentage of oxide, contained in the specimen (GOST 21207-75).

## RESULTS AND DISCUSSION

The obtaining of poly(arylene ether ketone)s of the expected structure was confirmed by the data of infrared spectroscopy. The presence of the absorption bands which correspond to ether bonds, isopropylidene groups, diarylketone groups, $Ar_2C=C$, and the absence of the absorption bonds of corresponding hydroxyl groups points to the fact that polycondensation of olygoketones and dichloroanhydride 1,1-dichloro-2,2di-(4-carboxyphenyl) ethylene was thoroughly completed (figure 1).

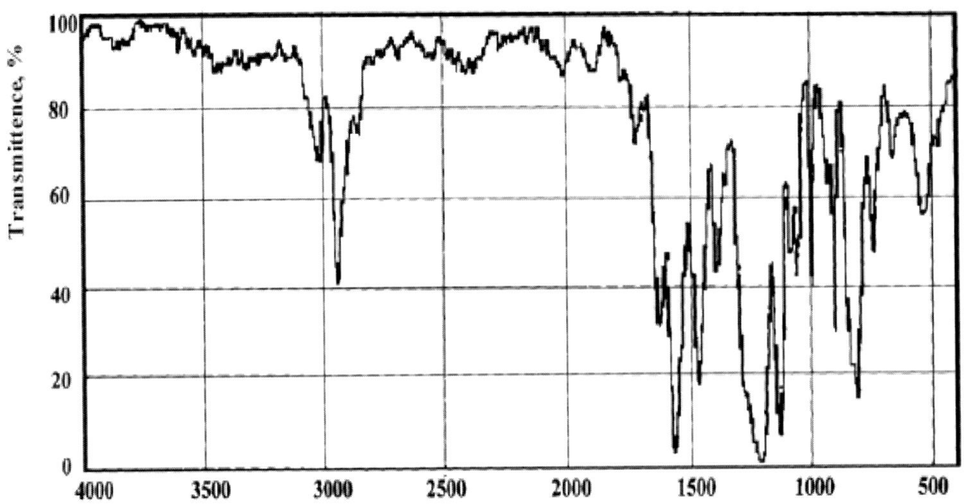


Figure 1. IR spectrum PAEK obtained from OK-20D.

The obtained PAEK is characterized by a high viscosity (*table 2*).

The results of turbidimetric titration also confirmed the structure of the above mentioned polymers. This fact was documented by the presence of only one maximum on the differential curves which means that the polymer reactants were statistically mixed (figure 2).

**Table 2. Some properties of PAEK**

| PAEK obtained from | Intrinsic viscosity ($m^3$/kg) | Molecular weight x $10^3$ | $T_g$ ($^0$C) | $\sigma_{break}$ MPa | ε, % | Oxygen Index, % | Mass loss temp. ($^0$C) | | | |
|---|---|---|---|---|---|---|---|---|---|---|
| | | | | | | | $T_{2\%}$ | $T_{10\%}$ | $T_{50\%}$ | $T_{2*\%}$ |
| OK-1D | 0,078 | 80 | 200 | 72,5 | 14,5 | 35,5 | 367 | 430 | 560 | 392 |
| OK-5D | 0,075 | 70 | 190 | 74,5 | 13,6 | 32,0 | 372 | 438 | 557 | 390 |
| OK-10D | 0,049 | 47 | 182 | 78,7 | 11,2 | 31,0 | 376 | 464 | 567 | 390 |
| OK-20D | 0,040 | 40 | 175 | 82,7 | 8,10 | 30,5 | 388 | 512 | 583 | 400 |

$T_2$*%- the PAEK loss in mass after thermal treatment at 300 °C for 30 minutes.

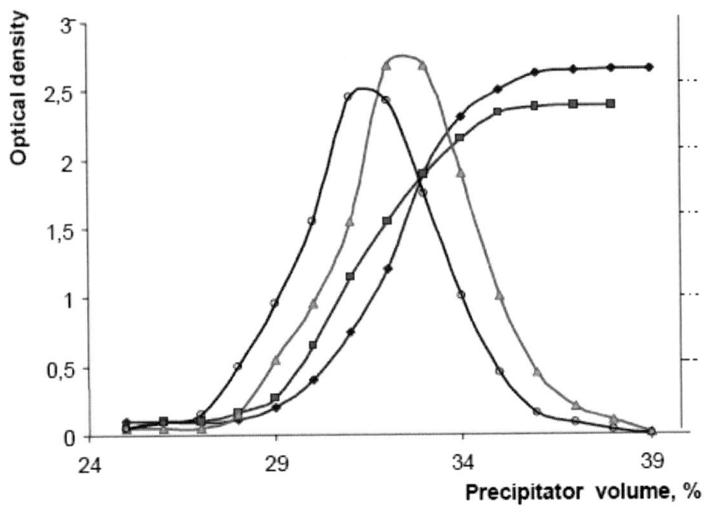


Figure 2. The turbidimetric titration curves of PAEK obtained from OK-1D (♦, Δ) and PAEK obtained from OK-20D (■,○); integral curves (♦,■), differential curves (Δ,○).

All the poly(arylene ether ketone)s, obtained were well dissolved by 1,2-dichloroethane, chloroform, tetrahydrofuran, etc.

The molecular weights of poly(arylene ether ketone)s were found to be in the interval 40000-80000. The highest molecular weight values correspond to poly(arylene ether ketone)s on the basis of short olygoketones, which confirms their high activity.

The thermomechanical tests showed that the values of the glass transition and of fluidity temperature of these polymers depended on the presence of a great number of simple flexible ethereal bonds in their macro chains. The portion of simple ethereal bonds in the chain becomes higher with the increase of length (condensation degree) of olygoketones, which provides decreased $T_g$ and $T_f$ of these polymers. PAEK on the basis of OK-20D, where the portion of ether bonds is the highest, and the portion of dichloroanhydride 1,1-dichloro-2,2di-(4-carboxyphenyl)ethylene remnants in the chain is the lowest, possesses the structure, closest of to that of olygoketones, and the latter has a low glass transition temperature. In the case of PAEK, based on OK-ID, higher $T_g$ defined by the most "severe" component - dichloroanhydride 1,1-dichloro-2,2di-(4-carboxyphenyl)ethylene.

The comparative analysis of the poly(arylene ether ketone)s thermal properties shows that with the increase of the initial olygomers length, a significant increase of thermal-resistance is observed. This kind of changes of thermal-resistance can be explained by the fact that, on the one hand, a saturation of polymeric chain by thermal-resistant stable ether bonds is observed, and on the other hand, the portion of non-stable ester bonds, that are brought in by the remnants of dichloroanhydride 1,1-dichloro-2,2di-(4-carboxyphenyl)ethylene, decreases abruptly. Moreover, with the increase of polycondensation degree of the initial olygoketones the packing density of polyetherketones increases. It is obvious, that these three factors promote to the growth of thermal-resistance of these poly(arylene ether ketone)s.

The data obtained from thermo-gravimetric analysis have shown that in case of thermal oxidizing destruction of PAEK, processes of intensive structuring are observed simultaneously with decomposition reactions, and this peculiarity is connected with the presence of group >C=$CCl_2$ in the structure of the given polymers, which is able to connect with the disclose of double bonds, what causes the destruction process lowering .

Under the investigation of a number of other properties, including infrared spectroscopy of specimens, passed thermal treatment, the structuring of aromatic unsaturated poly(arylene ether ketone)s according to the place of a double bond is confirmed by the *scheme 4*:

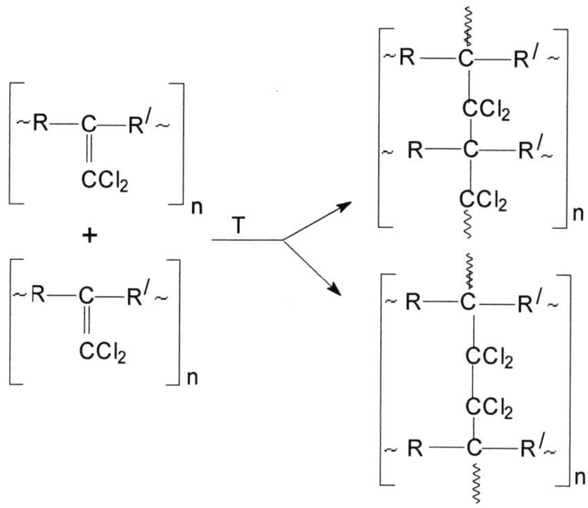


Scheme 4.

The investigation of thermal-resistance of PAEK specimens, kept preliminarily for 30 minutes at 300°C, has shown a significant increase of temperatures of loss of 2% of mass (table 2), which is strongly marked for poly(arylene ether ketone)s based on OK-1D and OK-5D.

For poly(arylene ether ketone)s with a less quantity of double bonds in macro chains thermal treatment has less effect. This phenomenon confirms, that some properties of poly(arylene ether ketone)s may be increased by an optimal choice of thermal treatment.

By their strength properties the given unsaturated poly(arylene ether ketone)s can be compared with the similar PAEKs on the basis of dichloroanhydrides of isophthalic and terephthalic acids. However, it must be noticed that the presence of a double bond in the structure of the given poly(arylene ether ketone)s gives an opportunity to increase the resources of these polymers. Thus, thermal treatment of PAEK on the basis of OK-ID under the pointed above conditions allowed to increase the value of the breaking strength up to 90 MPa, preserving the relative lengthening at the 10% level.

As it was expected, the given PAEK, as well as halogen -containing polyethers, possesses a high thermal-resistance. The obtained data show that poly(arylene ether ketone)s, which are more saturated by the chlorine atoms, are characterized by a higher level of their oxygen indices. All the obtained poly(arylene ether ketone)s are related to the self-damping polyethers.

The investigation of dielectric permeability (ε') and tan of the dielectric loss (tan δ) has shown that these indices are not high and relatively stable in a glassy condition. The insignificant fall of ε' from 3,5 to 3,3 under transition from PAEK on the basis of OK-ID to the ones on the basis of OK-20D, may be connected both with the lowering of the macro chain capacity for dipolar orientation, and the decrease of contents of the polar group $C = CCl_2$ in the macro chain.

The further increase of temperature facilitates the dipolar polarization by the decrease of the environment viscosity. The curve and dipoles orientation is accompanied by a loss of energy and tan δ increases. However, when a certain temperature is reached, the viscosity of the environment decreases and the loss of energy reduces too. The curves (*see figure 3*) show that with the increase of length of initial olygoketones, maximums on the curves, which are related to α-process, displace to the side of lower temperatures, and it is corresponded with the thermo-mechanical analysis data.

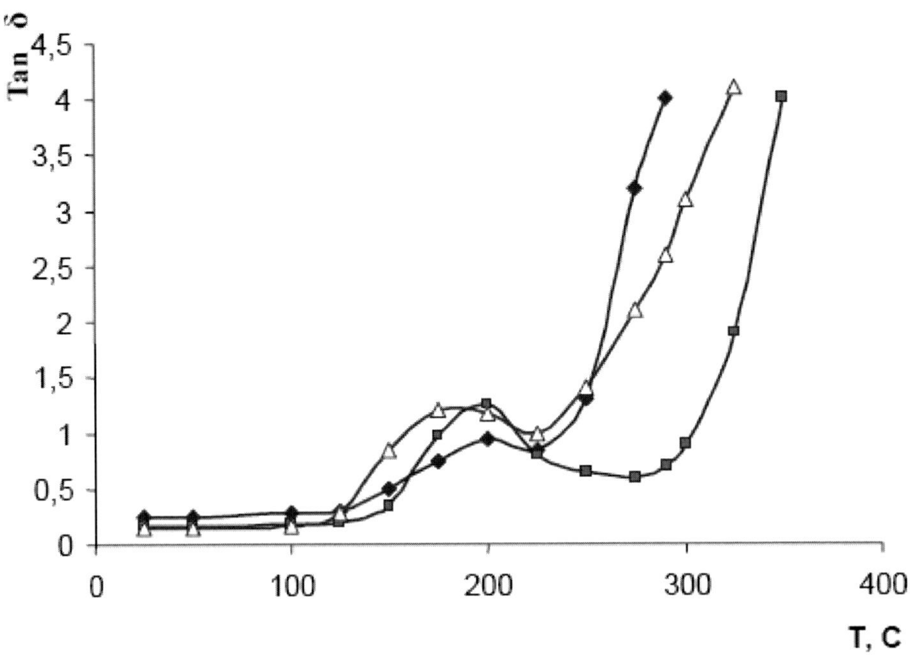


Figure 3. The relationship between the tan δ and temperature for PAEK obtained from OK-1D (♦), OK-10D (■) and OK-20D (Δ).

Due to the fact that constructional polymeric materials are often used in aggressive environment, the obtained poly(arylene ether ketone)s were tested for resistance to different acids and alkali. The tests showed (see table 3), that PAEK demonstrate a good resistance to the diluted mineral acids and to the concentrated hydrochloric acids. However, the poly(arylene ether ketone)s are easily dissolved both in the diluted and in the concentrated alkali, which may be connected with the presence of the chemically, non-stable complex-ethereal bonds in macro chains of the poly(arylene ether ketone)s.

**Table 3. Chemical stability of PAEK**

| PAEK obtained from | Exposure time (h) | Weight variations (%) | | | | |
|---|---|---|---|---|---|---|
| | | Diluted acid ($H_2SO_4$) | | Diluted alkali (NaOH) | | Conc. acid (HCl) |
| | | 10% | 30% | 10% | 30% | |
| OK-1D | 24 | 0,38 | 0,34 | 0,93 | 0,39 | 0,74 |
| | 48 | 1,04 | 0,98 | 2,59 | -0,55 | 1,64 |
| | 96 | 1,71 | 1,82 | 2,71 | -1,84 | 2,85 |
| | 384 | 1,87 | 1,29 | -0,12 | -9,68 | 2,91 |
| OK-10D | 24 | 0,33 | 0,29 | 0,90 | 0,32 | 0,65 |
| | 48 | 0,51 | 0,54 | 1,82 | -0,62 | 1,62 |
| | 96 | 1,22 | 1,27 | 2,24 | -1,74 | 2,18 |
| | 384 | 1,25 | 1,26 | 1,03 | -7,19 | 2,29 |
| OK-20D | 24 | 0,14 | 0,19 | 0,87 | 0,22 | 0,14 |
| | 48 | 0,39 | 0,44 | 1,69 | -0,52 | 1,52 |
| | 96 | 0,74 | 0,99 | 1,97 | -0,91 | 1,91 |
| | 384 | 0,85 | 1,14 | 1,65 | -5,92 | 2,07 |

## CONCLUSION

Poly(arylene ether ketone)s on the basis of olygoketones with different condensation degrees and dichloroanhydrides 1,l-dichloro-2,2-di(4-carboxylphenyl) ethylene were obtained. It was shown that poly(arylene ether ketone)s were characterized by high indices of stability. They are not flammable, they show high resistance to the diluted mineral acids. It was also shown, that a number of common features of poly(arylene ether ketone)s may be improved by thermal treatment due to the presence of unsaturated bond in the macro chain. The dependence of change of poly(arylene ether ketone)s properties on their condensation degree of the initial olygoketones was confirmed.

The used research methods are precise and reliable.

## REFERENCES

[1] Shaov A.H., Charayev A.M., Mikitaev A.K., Kardanov A.Z. Khasbulatova Z.S. Aromatic polyethers and polyetherethers, *Plastich. Massy.*, 11, 14 (1990).

[2] Charayev A.M., Shaov A.H., Mikitaev A.K., Matvelashvili G.S., Khasbulatova Z.S. Polymeric composite materials on a basis polyetherethers, *Plastich. Massy.* 3, 3 (1992).

[3] Charayev A.M., Shaov A.Kh., Shustov G.B., Mashukov N.I. Blok copolyetherketons, Third Russian-Chinese Simpo-sium " *Advanced materials and processes* ", Kaluga, Russia, 245, (1995).

[4] Sharapov V.V., Shaposhnikova V.V., Salaskin S.N. Influence of conditions of polycondensation on synthesis polyarylenetherketones, *Vysokomolek. Soed.*, 45, 113, (2003).

[5] Ozden S., Charayev A.M., Shaov A.H. The synthesis of polyetheretherketones and investigations of their properties. *J. Mater. Sci.*- 34, 2741, (1999).

[6] Ozden S., Shaov A.H., Charayev A.M., Bidanikov A.Y. Compositions Based in Aromatic Block Copolyester and p-Bytoxyphenyl Cyclohexyl Phosphinic Acid.- *Polym. and polym. Comp.*-1998.-Vol. 6.- № 2.- P.103-107.

[7] Ozden S., Charayev A.M., Shaov A.H., Shustov G.B. Synthesis and Assessment of the Properties of Polyetherketones (PEK) Based on Olygoketone-phenolphtaleines (OKPP)-Polyester Blok Copolymers.- *J. Appl. Polym. Sci.*. 68. 1013 (1998).

[8] Charayev A.M., Shaov A.H., Shustov G.B. Mikitaev A.K. Synthesis and Properties of Polyetheretherketones –*Chemistry and chemical technology*.- 41, 78 (1998).

[9] Osden S., Shaov A.H., Charaev A.M., Gurdaliyev X.X.. The effects of p-butoxy-phenylcyclohexylphosphinic acid on the properties of PC based on bisphenol A.- *J. Appl. Polym. Sci.* 80, 2113 (2001).

[10] Ozden S., Shaov A.H., Charayev A.M., Mikitaev A.K., Bedanokov A.Y. Aromatic block copolyesters stabilised with metallic salts of phosphinic acid.- *Polymers and Polymer Composites.* 9, 213 (2001).

[11] Ozden S., Charaev A.M., Shaov A.H. High impact thermally stable block copolyethers.- *J. Mater. Sci.* 36, 4479 (2001).

[12] A.Kh. Shaov, N.N. Amerkhanova, A.M. Kharayev. Phosphorous Organic Compounds as the stabilizers of the Aromatic Blockpolyethers.–Aging of polymers, polymer blends and polymer composites. – *Nova Sci. Publ.* – New York. – 2002.- Vol. 2.- pp. 161-166.

[13] S. Ozden , A. M. Charayev, A. H. Shaov. Synthesis of block copolyetherether ketones and investigations of their properties.- *J. Appl. Polym.* Sci. 85, 485 (2002).

[14] Patent №1736128 (USSR)

[15] Yanota H., Alov E.M., Moskviechev YU.F., Mironov G.S. *J. Org. Himii,* 21, 365 (1985).

In: Monomers, Oligomers, Polymers, Composites… ISBN: 978-1-60456-877-6
Editors: R. A. Pethrick, G.E. Zaikov et al. 

*Chapter 8*

# The Ethanol Influence on Acrylic Acid Polymerization Kinetics and Mechanism in Inverse Emulsions Stabilized by Lecithin

***S.A. Apoyan, R.S. Harutynnyan, J.D. Grigoryan and N.M. Beylerian***
Yerevan State University; E-mail: norbey@ysu.am

## Abstract

The functions which describe the dependence of polymerization rate ($R_{pol}$) of acrylic acid (AAc) initiated with potassium persulfate (PP) in inverse emulsions stabilized by lecithin (Le) in presence of ethanol (Et) on $[Et]_0$ and $[Le]_0$ are bell shape curves.

$R_{pol} \sim [PP]_o^n [AAc]^m$. With increase of $[Et]_0$ n increases from 0.5 to 1. m=1 when $0 < [AAc] \leq 1M$. With increase of $[AAc]_o$ more than 1M m decreases striving 0. Both phenomenons are explained. It is established that the polymerization initiation rate ($R_{in}$) does not depend on both $[Et]_o$ and $[Le]_o$. They have considerable influence on colloid-chemical properties of the emulsion assisting to $R_{pol}$ increase.It is concluded that the coexistence of Et +Le displays simultanously positive and negative effects which condition the appearance of maximums on curves $R_{pol} = f([Et]_o)$ and $R_{pol} = f([Le]_o)$ functions.

## Introduction

The influence of some aliphatic alcohols on kinetics of styrene and chloroprene polymerization kinetics is studied [1-3]. In particular it has been established that alcohols are chain transfer agents, so they are being used to regulate the mean molecular masses (MMM) of obtained polymers. Apart it is shown that they influence on colloid-chemical properties of polymerization systems. So by this means they act on the over-all polymerization kinetics.

In [4] it is shown that in the case of AAe polymerization initiated with PP in inverse emulsions (IE) stabilized by Le in a more extent unsoluble polyAAe is being obtained.

Taking into consideration data presented [1-4] it is assumed to use alcohols to obtain watersoluble polyAAe in IEs.

The aim of the present research-work is to study ethanol's influence on AAc's $R_{pol}$ initiated with PP in IEs stabilized by Le, as well on obtained polyAAe MMM.

It is worthy to note that Et is not toxic and is available.

## EXPERIMENTAL

$R_{pol}$ is determined making use dilatometry. The volume ratio water/oil (toluene)=1:2=const. Water was bidistilled.

PP is recrystallized fivefolds from water solution. The purity was 99,8% (iodometric determination). Le was 10% ethanol solution ("standard" grade). The prepared Le initial solution was kept in oxygen-free medium in refrigerator. MMM is determined using viscositnetric method [5]. The following formulae is used: $[\eta]=6{,}6\cdot10^{-4}\cdot M^{0,5}$ at 243k and for 0,2% HCl solution of AAe.

The surface and interface tensions are determined by Rebinder's method[6] (determination of the maximal pressure in bubles).

The emulsion's stability has been determined at room temperature measuring the moving on of the separation boundary between two phases[7].

## RESULTS AND DISSCUSSION

The curves depicted on fig.1 present the AAc $R_{pol}$ and polyAAc MMM`S dependences on $[Et]_0$ in Absence of Le (1) and in its presence (2).

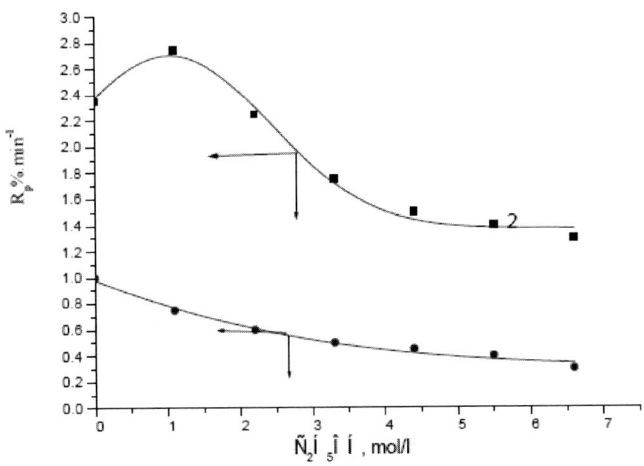


Figure 1. Dependence of AAc polymerization rate ( $R_{pol}$ in %· $min^{-1}$) and polyAAc MMM on $[El]_o$ (M) in absence (1) and in presence (2,3) of L. $[L]_o$=1% (2) and 0.5% (2,3). $[AAc]_o$ =1M, $[PP]_o = 2\cdot 10^{-3}$ .

The curves depicted on fig.1 present the AAc $R_{pol}$ and polyAAc MMM's dependences on $[Et]_o$ in absence and in presence of Le. In Le absence with increase in $[Et]_o$ $R_{pol}$ decreases. But in presence of Le (1% in toluene) $R_{pol}$ has maximum value at $[Et]_o$ =1M.

Same regularities are established for MMM=f($[Et]_o$). At $[Le]_o \leq 0,5\%$ and in presence of Et the obtained polyAAc is water soluble. MMM decreases with $[Et]_o$ increase. But increasing $[Le]_o$ the obtained polyAAe solubility decreases.

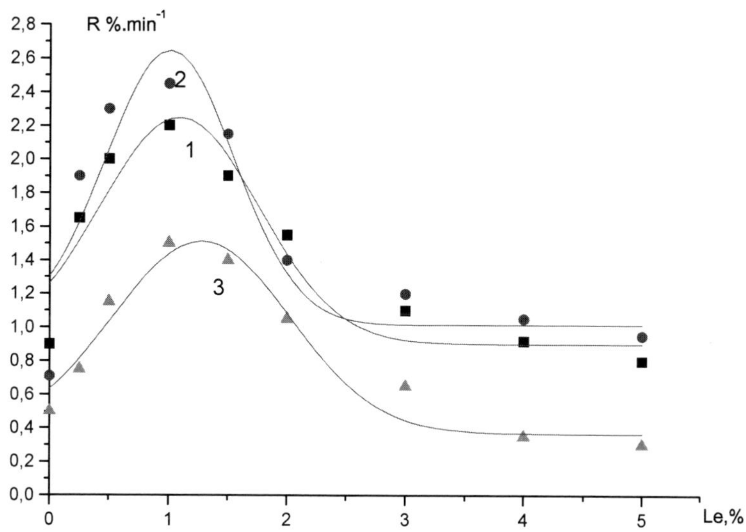


Figure 2. Dependence of $R_{pol}$ on $[Le]_o$ in presence of different amounts of Et. $[Et]_o$ : 0(1); 1,1 M (2) and 6,6(3) at $[AAc]_o$ = 1M, $[PP]_o = 2 \cdot 10^{-3}$.

The observed maximum corresponds to $[Le]_o$=1%. It depends on a few extent on $[Et]_o$. Further kinetic studies are carried out for this condition.

The study of $R_{pol}$ dependence on $[PP]_o$ show that $R_{pol} \sim [PP]_o^{\ n}$, where n=f($[Et]_o$) (see table 1).

**Table 1. $R_{pol}$'s dependence on $[PP]_o$ at $[Le]_o$=1% and 318K $[AAc]_o$=1M.**

| $[Et]_o$ / $10^3[PP]_o$,M | 0 | 1.1 | 6.6 | |
|---|---|---|---|---|
| 0.5 | 1.0 | 1.03 | 0.42 | $R_{pol}$ %·min$^{-1}$ |
| 1 | 1.35 | 1.58 | 0.8 | |
| 1.5 | 1.65 | 2.43 | 1.32 | |
| 2.0 | 2.20 | 2.65 | 1.52 | |
| n | 0.5 | 0.73 | 1.0 | |

These data show that in presence of Le with increase in $[Et]_o$ the mechanism of chain termination is being changed. In absence of Et the chain termination occurs by quadratic mechanism, while at $[Et]_o$ =6,6M it becomes linear.

The following step is the study of $R_{pol}$ dependence on $[AAe]_o$ in presence of different amounts of Et.

The obtained data are summarized in table2.

**Table 2. $[Le]_o$=1%, $[PP]_o=2\cdot10^{-3}$M, $V_w/V_{tol}$ =1:1, T=318K**

| $[Et]_o$ / [AAc]M | 0 | 1.1 | 6.6 | |
|---|---|---|---|---|
| 0.5 | 1.3 | 1.5 | 0.5 | $R_{pol}$ $\%\cdot min^{-1}$ |
| 0.75 | 1.8 | 2.0 | 1.0 | |
| 1.0 | 2.2 | 2.65 | 1.52 | |
| 1.5 | 2.3 | 2.82 | 1.9 | |
| 2.0 | 2.5 | 2.95 | 1.95 | |

From data presented in table 2 it follows that $R_{pol}\sim [AAc]^m$. It is easy to show that when $[AAc]_o\leq$ 1M m=1. But when $[AAc]_o$>1M m tends to 0.

To explain this regularity at first the $R_{in}$ dependence on $[Et]_o$ end $[Le]_o$ has been studied. Inhibitory method has been used to determine $R_{in}$. The inhibitor was TEMPO.

Obtained kinetic data unambiguously show that $R_{in}$ does not depend on $[AAc]_o$ , [Et] and $[Le]_o$. These results are in full accordance with data obtained earlier [8-10]. So the primary radicals are being formed in the water phase as result of persulfate dianion monomolecular decomposition.

Therefore , Et and Le may influence on chain propagation , termination and transfer reactions. But the following circumstance must be considered too.

As was mentioned the polymerization is being carried out in IE. Principally the influence of Et and Le on the emulsion colloid-chemical properties may not be excluded. Data given in table 3 confirm the expressed assumption.

**Table 3.**

| $[C_2H_5OH]$ M | $T_{ss}$ min | σ dyne/cm | | [AAc] % | |
|---|---|---|---|---|---|
| | | $[Le]_o$=0% | $[Le]_o$=1% | In water | In toluene |
| 0 | 6 | 34.4 | 11.5 | 70 | 30 |
| 1.1 | 8 | 22.9 | 9.5 | 66 | 34 |
| 6.6 | 14 | 5.7 | 5.0 | 50 | 50 |

$[AAc]_o$=1M , $VH_2O$: V $C_6H_5CH_3$ =1:1, T=303K.

$t_{ss}$ – is the emulsion semidecay time, σ – the surface tension. It is obvious that Et increases the emulsion stability. On the other hand they both diminish the surface tension which results in decrease of micelles size , which in its turn must result in enhance of micelles(the polymerization loci) number on which depends $R_{pol}$ . This is the positive influence of the Et + Le mixture on $R_{pol}$.

But at the same time that mixture displays also a negative influence on $R_{pol}$ , as well on MMM. It is due to the fact that they both efficiently transfer the chain , in the case of Le- with degradation. This may be the reason why at high concentration of Et , when Le solubility in the water phase may be increased , the chains are being terminated by linear mechanism.

What which conceurns the peculiar dependency of $R_{pol}$ on$[AAc]_o$ , one cane say the following.

The main polymerization loci is the water phase. It is well know that all carbonic acids , AAc too, form dimers :2AAc↔dimer. It is obvious that the dimer is less polar than the monomeric acid. So it is more probable that the dimer molecules cross from the water phase into the "oil".

If $[AAc]_o$ is the AAc's initial concentration , so a part of AAc molecules will be dimerized resulting in decrease of AAe's real concentration ([AAc]real)(see table 3). As was mentioned both forms of AAc are in equilibrium. So ; $[AAc]_{real} = [AAc]_o - [dimer]$. In the propagation reaction takes part monomerie molecules of AAc , which concentration in the reaction zone (in the water phase) is equal to $[AAc]_{real} < [AAc]_o$. It is easy to note that increasing $[AAe]_o$ $[AAc]_{real}$ may be increased only in a very few extent. Probably for this reason increasing $[AAc]_o$ the reaction order with respect to acrylic acid tends to zero.

## References


[1] Storoge Gu. F., Yurjenko A.U., Coll.J., 1963.v.25, No,p.77(in Russion)

[2] Harutunian R.S., Beylerian N.M., Trans. of Yerevan State University, 1982, N2, p. 79(in Russian)

[3] Harutyunyan R.S., Dissertation (for PhD degree), Yerevan, 1984 (in Russia)

[4] Apoyan S.H., Grigorian J.D., Harutyunyan R.S., Beylerian N.M., Physical chemistry of Polymers, Coll. of scientific papers, 'Tver state university (in Russian), 2004, issue10, c.79

[5] Rafikov S. R., Pavlova S.A., Tverdekhlebova I.I., Methods to determine molecular weights and polydispersity of highmolecular compounds, Moscow, Ed.ofURSS Academy of Sciences, 1963, p.301.

[6] Ayvazian V.B., Textbook for practical works in chemistry of surface phenomenon and adsorption, Moscow, Ed. of High School, 1973, p.19.

[7] Poustovalov N.N., Poushkarev V.V., Berezok V.G, Coll.J.,1974 v.36,N 1, p. 171

[8] Chaltikian H.H., Khachatrian A.Gu., Beylerian N.M., Kinetika I kataliz , 1971, v,12,N4, p. 1049(in Russia)

[9] Beylerian N.M., Khachatrian A.Gu., Chaltikyan H.H., *Armenian chem.. J.,* 1970,v,23,N7, p.575(in Russia)

[10] Beylerian N.M., *Acta Polymerica,* 1982,v,33,N5, p.339.

In: Monomers, Oligomers, Polymers, Composites… ISBN: 978-1-60456-877-6
Editors: R. A. Pethrick, G.E. Zaikov et al. 

*Chapter 9*

# SMOOTHED PARTICLE HYDRODYNAMICS (SPH) ALGORITHM FOR NUMERICAL FLUID-STRUCTURE INTERACTION STUDIES IN POROUS MEDIA – NEW TRENDS AND ACHIEVEMENTS

*N. Amanifard and A. K. Haghi*[*]
University of Guilan, P. O. Box 3756, Rasht, Iran

## ABSTRACT

In this chapter, SPH (Smoothed Particle Hydrodynamics )has been introduced to investigate pore-scale flow phenomena in porous media . Using this approach, fluid velocity and pressure distributions, discharge velocity, and fluid particle paths can be computed, as well as other information that would be difficult or impossible to observe experimentally

## 1. INTRODUCTION

Numerical methods applied to porous media flow include finite difference methods [1], finite element methods [2-6], and boundary integral methods [7,8]. It can be difficult, however, to apply such methods to flows involving immiscible fluids or mobile solid boundaries. Lagrangian particle techniques, such as smoothed particle hydrodynamics (SPH) and lattice Boltzmann [9], provide an alternative approach which is more easily applied to such problems.

SPH has been used to investigate pore-scale flow phenomena in porous media [10-13]. The SPH method firstly, developed for astrophysical applications [14, 15], and it is a fully Lagrangian and mesh-less. Using this approach, fluid velocity and pressure distributions, discharge velocity, and fluid particle paths can be computed, as well as other information that

[*] Corresponding author e-mail: Haghi@Guilan.ac.ir

would be difficult or impossible to observe experimentally. SPH has certain advantages over other fluid dynamical methods, which may encounter difficulty with deformable boundaries, multiphase flows, free surfaces, and the extension to three dimensions. For example, if the soil grains which form a porous structure are allowed to move in response to fluid flow, other techniques may require continual remeshing of the flow domain, leading to increased numerical diffusion and algorithmic complexity. In addition, SPH uses an interpolation method which simplifies the inclusion of chemical and thermal effects.

While SPH is versatile, errors can sometimes be larger than those obtained using grid-based methods tailored for specific problems. Moreover, SPH can be computationally expensive for certain applications. For example, individual time-steps for comparable numbers of nodes with SPH and finite element methods take similar computational effort. However, many more steps are typically required by SPH to obtain a solution to steady-state problems. For time-dependent problems at low Reynolds number, the computational effort is similar to a finite element method employing explicit time integration.

The computational expense of SPH is in part a consequence of its versatility.

More specialized numerical techniques employ computational nodes which have relatively small regions over which they interact. For Eulerian methods, these regions remain fixed throughout the computation (provided nodes are not added or deleted). In addition, the weights associated with nodes are often time-independent.

The adaptive smoothed particle hydrodynamics (ASPH) is introduced by Owen et al., (1998) and Fulbright and Benz,( 1995), and has been developed for microchannel studies as a novel attempt by Liu and Liu (2005).

As a mesh-free particle Lagrangian method, SPH has special advantages, and has been extended to different areas with various applications (Liu et al., 2002; 2003a;b). It can naturally determine the location of non-homogeneities, free surfaces, and moving interfaces. With SPH, the nodes (particles) are free to move, producing variations in the nearest neighbors of each particle. Thus, the weights associated with particle interactions are time-dependent and must be recalculated at each time step. The word "particle" does not mean a physical mass instead it refers to a region in space. Field variable are associated to these particles and at any other point in space are found by averaging, or smoothing, the particle values over the region of interest. This is fulfilled by an interpolation or weight function which is often called the interpolation kernel. It should be pointed out that SPH method was then successfully applied to the study of various fluid dynamics problems, such as free-surface incompressible flows [19], and viscous flows [20,21]. Since the early 1990s, SPH was applied also to the simulation of elasticity and fragmentation in solids. SPH has been also used to simulate the interaction between different fluids [18, 22], different solids [23] and between fluids and structures [24] in presence of explosions.

The objective of this chapter is to perform a fundamental base on numerical observations by the SPH algorithm in microchannels and porous media problems, and the authors believe that it can be used with some further studies for a wide range of Knudson numbers from continuous approach to the particle-molecular flow zones.

## 2. Basic Concepts

The SPH method is based on the interpolation theory. The method allows any function to be expressed in terms of its values at a set of disordered points representing particle points using a kernel function. The kernel function refers to a weighting function and specifies the contribution of a typical field variable, $A(r)$ at a certain position, $r$ in space. The kernel estimate of $A(r)$ is defined as [28]:

$$\langle A_h(r) \rangle = \int_{Space} A(r')\, W(r - r', h)\, dr' \tag{1}$$

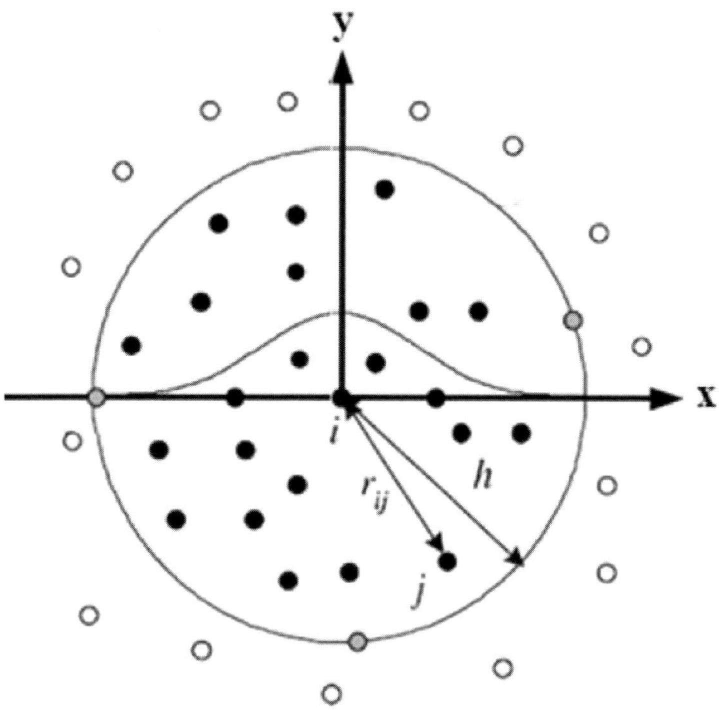


Figure 1. SPH Characteristics in a two-dimensionl space. Th smoothing length $h$ which the gives the boundaries of domain of influence is shown.

Where the smoothing length $h$ represents the effective width of the kernel and $W$ is a weighting function with the following properties[28]:

$$\int_V W(r - r', h)\, dr' = 1 \quad \lim_{h \to 0} W(r - r', h) = \delta(r - r') \tag{2}$$

If $A(r')$ is known only at a discrete set of $N$ point $r_1, r_2, \ldots, r_N$ then $A(r')$ can be approximated as follows [28]:

$$A(r') = \sum_{j=1}^{N} \delta(r' - r_j) A(r_j)(dV)_j \tag{3}$$

Where $(dV)_j$ is the differential volume element around the point $r_j$. Combining Eq. (1) and Eq(3) yields[28]:

$$\langle A_h(r) \rangle = \sum_{j=1}^{N} \int \delta(r' - r_j) A(r_j)(dV)_j W(r - r', h) dr' \tag{4}$$

After integration, and replacing the differential volume element $(dV)_j$ by $m_j / \rho_j$ one gets [28]:

$$\langle A_h(r) \rangle = \sum_{j=1}^{N} \frac{m_j}{\rho_j} A_j W(r - r', h) \tag{5}$$

Where the summation index j denotes a particle label and particle j carries a mass $m_j$ at position $r_j$, a density $\rho_j$ and a velocity $v_j$. The value of $A$ at $j-\text{th}$ particle is shown by $A_j$. The summation is over particles which lie within a circle of radius $2h$ centered at $r$.

### 2.1. Kernel Function

The kernel function is the most important ingredient of the SPH method. Various forms of kernels with different compact support were proposed by many researches. Recently studies [21-23] indicate that stability of SPH algorithm depends strongly upon the second derivative of the kernel.

Using different kernel in SPH method is similar to using different schemes in finite difference methods. One of the most popular kernels is based on spline functions [23]:

$$W(r,h) = \frac{\sigma}{h^{\nu}} \times \begin{cases} 1 - \frac{3}{2}s^2 + \frac{3}{4}s^3 & 0 \le s < 1 \\ \frac{1}{4}(2-s)^3 & 1 \le s < 2 \\ 0 & 2 \le s \end{cases} \qquad s = \frac{|\mathbf{r}|}{h} \tag{6}$$

Where $\nu$ is number of dimensions and $\sigma$ is normalization constant with the values: $\frac{2}{3}, \frac{10}{7\pi}, \frac{1}{\pi}$ in one, two and three dimensions respectively. This kernel has compact support which is equal to $2h$, it means that interactions are exactly zero for $r > 2h$. The second derivatives of this kernel is continues and the dominant error term in the integral interpolant is $O(h^2)$. Higher order splines can be used, but they interact at further distances and thus require more computational time.

### 2.2. Gradient and Divergence

The gradient and divergence operators need to be formulated in a SPH algorithm if simulation of the Navier-Stokes equations is to be attempted. In this work, the following commonly used forms are employed for gradient of a scalar $A$ and divergence of a vector $u$ [29]:

$$\frac{1}{\rho_i}\nabla_i A = \sum_j m_j \left( \frac{A_i}{\rho_i^2} + \frac{A_j}{\rho_j^2} \right) \nabla_i W_{ij} \tag{7}$$

$$\frac{1}{\rho_i}\nabla_i . \mathbf{u}_i = \sum_j m_j \left( \frac{\mathbf{u}_i}{\rho_i^2} + \frac{\mathbf{u}_j}{\rho_j^2} \right) \nabla_i W_{ij} \tag{8}$$

Where $\nabla_i W_{ij}$ is gradient of kernel function $W(|r_i - r_j|, h)$ with respect to $r_i$, the position of particle $i$. This choice of discretization operators ensure that an exact projection algorithm is produced.

### 2.3. Laplacian Formulation

A simple way to formulate the Laplacian operator is to envisage it as dot product of the divergence and gradient operators. This approach proved to be problematic as the resulting second derivative of the kernel is very sensitive to particle disorder and can easily lead to pressure instability and decoupling in the computation due to the co-location of the velocity and pressure. In this paper, the following alternative approach is adopted [30]:

$$\nabla . \left( \frac{1}{\rho} \nabla A \right)_i = \sum_j m_j \frac{8}{(\rho_i + \rho_j)^2} \frac{A_{ij} r_{ij} . \nabla_i W_{ij}}{|r_{ij}|^2 + \eta^2} \tag{9}$$

Where $A_{ij} = A_i - A_j$, $r_{ij} = r_i - r_j$ and $\eta$ is a small number introduced to avoid a zero denominator during computations and is set to $0.1h$.

# 3. GOVERNING EQUATIONS

In this chapter a new algorithm for elastic deformation modeling of solid particles is presented which doesn't utilize any artificial viscosity and artificial stress terms. The proposed algorithm for solid particles modeling is completely compatible with fluid particles and it Permits to easily follow the motion of fluid–solid interface in time without any specific treatment.

## 3.1. Governing Differential Equations for Fluid Particles

The governing equations for transient compressible fluid flow include the conservation of mass and momentum equations. In a Lagrangian framework these can be written as:

$$\frac{1}{\rho}\frac{D\rho}{Dt}+\nabla.\mathbf{V}=0 \tag{10}$$

$$\frac{D\mathbf{V}}{Dt}= g+\frac{1}{\rho}\nabla.\overset{\Rightarrow}{\tau}-\frac{1}{\rho}\nabla P \tag{11}$$

Where $t$ is time, $g$ is the gravitational acceleration, $P$ is pressure, $V$ is the velocity vector , $\overset{\Rightarrow}{\tau}$ is viscous stress tensor and D/Dt refers to the material derivative. The density $\rho$ has been intentionally kept in the equations to be able to enforce the incompressibility of the fluid. The momentum equations include three driving force terms, i.e., body force, forces due to divergence of stress tensor and the pressure gradient. These must be handled along with the incompressibility constraint. In a SPH formulation the above system of governing equations must be solved for each particle at each time-step. The sequence with which the force terms are incorporated can be different from one algorithm to another.

## 3.2. Viscous Terms in Fluid Zone

It is known that In the Newtonian fluids, $\overset{\Rightarrow}{\tau}$ in equation (11) must be defined as below:

$$\overset{\Rightarrow}{\tau}=2\mu\,\mathbf{D} \tag{12}$$

$$\mathbf{D}=\tfrac{1}{2}\left(\nabla\mathbf{v}+\nabla\mathbf{v}^{t}\right) \tag{13}$$

$$\mathbf{v} = u\,\vec{i} + v\,\vec{j} \tag{14}$$

$\overset{\Rightarrow}{\tau}$ in the momentum equation appears as a divergence form which can be written in SPH formulation as below (the details of this formulation was presented in my last paper)

$$\left(\frac{1}{\rho}\nabla.\overset{\Rightarrow}{\tau}\right)_i = \sum_j m_j\left(\frac{\overset{\Rightarrow}{\tau}_i}{\rho_i^2} + \frac{\overset{\Rightarrow}{\tau}_j}{\rho_j^2}\right).\nabla W\left(\vec{\mathbf{r}}_i - \vec{\mathbf{r}}_j, h\right) \tag{15}$$

### 3.3. Governing Differential Equations for Elastic Solid Zone

The acceleration equations (Momentum equation) for Elastic Medium can be written as below [26]:

$$\frac{\mathrm{d}\mathbf{V}}{\mathrm{d}t} = g + \frac{1}{\rho}\nabla.\overset{\Rightarrow}{\sigma} \tag{16}$$

In above equations $d/dt$ denotes a derivative following the motion, $\sigma$ is the stress tensor and can be written as [26]:

$$\sigma^{ij} = -P\delta^{ij} + S^{ij} \tag{17}$$

By Combining Eq. (16) and Eq. (17) the final form of governing equations for Elastic Medium Motion includes the continuity and the acceleration equations can be written as below

$$\frac{1}{\rho}\frac{D\rho}{Dt} + \nabla.\mathbf{V} = 0 \tag{18}$$

$$\frac{\mathrm{d}\mathbf{V}}{\mathrm{d}t} = g + \frac{1}{\rho}\nabla.\overset{\Rightarrow}{S} - \frac{1}{\rho}\nabla P \tag{19}$$

Where $g$ is the gravitational acceleration, $P$ is pressure, $\overset{\Rightarrow}{S}$ is the deviatoric stress tensor and its rate of change is given by:

$$\frac{\mathrm{d}S^{ij}}{\mathrm{d}t} = 2\mu\left(\dot{\varepsilon}^{ij} - \frac{1}{3}\delta^{ij}\dot{\varepsilon}^{ij}\right) + S^{ik}\omega^{jk} + \omega^{ik}S^{kj} \tag{20}$$

Where:

$$\dot{\varepsilon}^{ij} = \frac{1}{2}\left(\frac{\partial \mathbf{v}^i}{\partial x^j} + \frac{\partial \mathbf{v}^j}{\partial x^i}\right) \tag{21}$$

And the rotation tensor $\omega^{ij}$ is:

$$\omega^{ij} = \frac{1}{2}\left(\frac{\partial \mathbf{v}^i}{\partial x^j} - \frac{\partial \mathbf{v}^j}{\partial x^i}\right) \tag{22}$$

## 4. Solution Algorithm

In this chapter, a fully explicit three-step algorithm is used for both fluid and elastic solid particles which will be explained in details.

### 4.1. Solution Algorithm for Fluid Particles

In the first step of this algorithm, the momentum equation is solved in the presence of the body forces neglecting all other forces. So, an intermediate velocity is computed as

$$u^* = u_{t-\Delta t} + g_x \Delta t \tag{23}$$

$$v^* = v_{t-\Delta t} + g_y \Delta t \tag{24}$$

As said before, in the second step of fluid flow simulation the temporal velocity is employed to calculate the divergence of viscous stress tensor. Note that the divergence of viscous stress tensor is a vector $\vec{T}$ given by:

$$\left(\frac{1}{\rho}\nabla.\overset{\Rightarrow}{\tau}\right)_i = \vec{T} = T_x \vec{i} + T_y \vec{j} \tag{25}$$

At the end of the second step, the velocity of particle is updated according to

$$u^{**} = u^* + T_x \Delta t = u_{t-\Delta t} + g_x \Delta t + T_x \Delta t \tag{26}$$

$$v^{**} = v^* + T_y \Delta t = v_{t-\Delta t} + g_y \Delta t + T_y \Delta t \tag{27}$$

At this stage particle moved according to temporal velocities ($u^{**}$ ,$v^{**}$) and temporal position of particle is:

$$x^{*} = x_{t-\Delta t} + u^{**}\Delta t \tag{28}$$

$$y^{*} = y_{t-\Delta t} + v^{**}\Delta t \tag{29}$$

Thus far no constraint has been imposed to satisfy the incompressibility of the fluid and it is expected that the density of some particles change during this updating. In fact, with the help of the continuity equation one can calculate the density variations of each particle as

$$\frac{D\rho_i}{Dt} = \sum_j m_j (\mathbf{v}_i - \mathbf{v}_j) \nabla_i W(r_{ij}, h) \tag{32}$$

Where $\rho_i$ and $\mathbf{v}_i$ are the density and velocity of particle $i$. When two particles approach each other, their relative velocity and therefore the gradient of kernel function become negative, so $D\rho_i / Dt$ will be positive and $\rho_i$ will increase. Consequently, this will produce a repulsive force between the approaching particles. In a similar fashion, if two particles are repulsed from each other, an attractive force will be produced to stop this. This interaction based on the relative velocity of particles and the resulting coupling between the pressure and density will enforce incompressibility in the solution procedure.

The velocity field $\hat{\mathbf{v}} = (\hat{u}, \hat{v})$ which is needed to restore the density of particles to their original value is now calculated. To do this, in the third step of the algorithm, the momentum equation with the pressure gradient term as a source term is combined with the continuity equation (10) as

$$\frac{1}{\rho_0}\frac{\rho_0 - \rho^{*}}{\Delta t} + \nabla.(\hat{\mathbf{v}}) = 0 \tag{33}$$

$$\hat{\mathbf{v}} = -\left(\frac{1}{\rho^{*}}\nabla P\right)\Delta t \tag{34}$$

to obtain the following pressure Poisson equation

$$\nabla.\left(\frac{1}{\rho^{*}}\nabla P\right) = \frac{\rho_0 - \rho^{*}}{\rho_0\,\Delta t^2} \tag{35}$$

Equation (35) can be discretized according to equation (9) to obtain pressure of each particle as

$$P_i = \left( \frac{\rho_0 - \rho_*}{\rho_0 \Delta t^2} + \sum_j \frac{8m_j}{(\rho_i + \rho_j)^2} \frac{P_j \vec{r}_{ij}.\nabla_i W_{ij}}{\left|\vec{r}_{ij}\right|^2 + \eta^2} \right) \Bigg/ \left( \sum_j \frac{8m_j}{(\rho_i + \rho_j)^2} \frac{\vec{r}_{ij}.\nabla_i W_{ij}}{\left|\vec{r}_{ij}\right|^2 + \eta^2} \right) \tag{36}$$

using this value for the pressure of each particle one can calculate $\hat{\mathbf{v}}$ according to equation (34) and (7) as

$$\hat{\mathbf{v}}_i = \sum_j m_j \left( \frac{P_i}{\rho_{i\,*}^2} + \frac{P_j}{\rho_j^2} \right) \nabla_i W_{ij} \tag{37}$$

Finally, overall velocity of each particle at the end of time step will be obtained as

$$u_{t+\Delta t} = u^{**} + \hat{u} \tag{38}$$

$$v_{t+\Delta t} = v^{**} + \hat{v} \tag{39}$$

And the final positions of particles are calculated using a central difference scheme in time

$$x_t = x_{t-\Delta t} + \frac{\Delta t}{2}(u_t + u_{t-\Delta t}) \tag{40}$$

$$y_t = y_{t-\Delta t} + \frac{\Delta t}{2}(\mathrm{v}_t + \mathrm{v}_{t-\Delta t}) \tag{41}$$

## 6. Rigid Wall Boundary Conditions

Boundary conditions always receive special attention in SPH method; here we follow the treatment used by Koshizuka et al. [35] to model the wall boundaries by fixed wall particles, which are spaced according to the initial configuration. Here we employ a kernel with a compact support of $h = 1.5\,l_0$, where $l_0$ is the initial particle spacing. In addition, two lines of dummy particles with properties which are completely similar to boundary particles are also placed outside of solid walls at spacing $l_0$. The calculation procedure for both wall and dummy particles is completely similar to inner particles, except set the velocity to be zero to represent no-slip boundary conditions.

# 7. A Test Case

A test case that was performed by Hosseini and Amanifard(2007), is presented here to show the capability of the mentioned method, however it is not a test case of the porous media. Using the SPH for porous media is developing and current test case may provide a new idea for such a development.

### An Elastic Plate Subjected to Time-Dependent Water Pressure

In this experiment an elastic gate, is clamped at one end and free at the other one, interacts with a mass of water initially confined in a free-surface tank behind the gate.

A schematic view of this 2D problem is shown in figure 2. As it is shown, the left wall consists in an upper rigid part and in a lower deformable plate made of rubber. The rubber plate is free at its lower end, thus representing an elastic gate closing the tank. The geometric dimensions of the system and the physical characteristics of the elastic gate are reported in table 1.

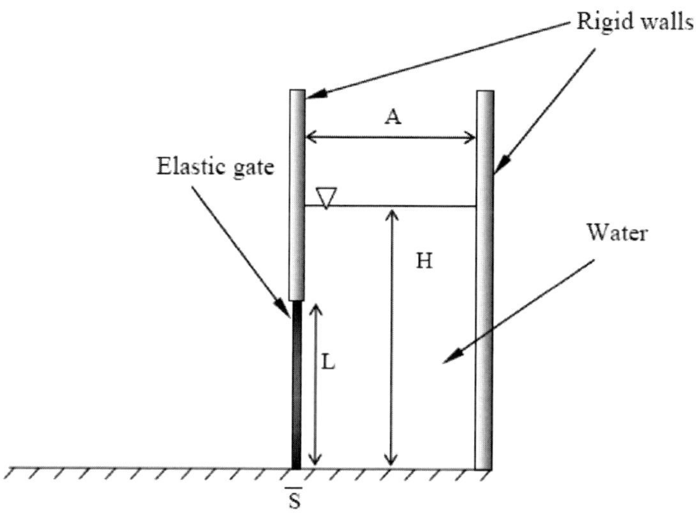


Figure 2. Schematic view of initial configuration.

**Table 1. Dimensions of the system and physical characteristics of the rubber plate**

| Dimensions | |
|---|---|
| A (m) | 0.1 |
| H (m) | 0.14 |
| L (m) | 0.079 |
| S (m) | 0.005 |
| Rubber | |
| $\rho$ (kg/m$^3$) | 1100 |
| G (N/m$^2$) | $4.27 \times 10^6$ |

In figure 3, the horizontal and vertical displacements computed for the plate (according to presented SPH algorithm and FSI model) are compared with those measured in the digitalized images acquired during the experiments and the results of old SPH model of FSI.

SPH particle positions at six different times are shown in figure 5. These Images are completely in good agreement with the frames from the experiment setup at corresponding times.

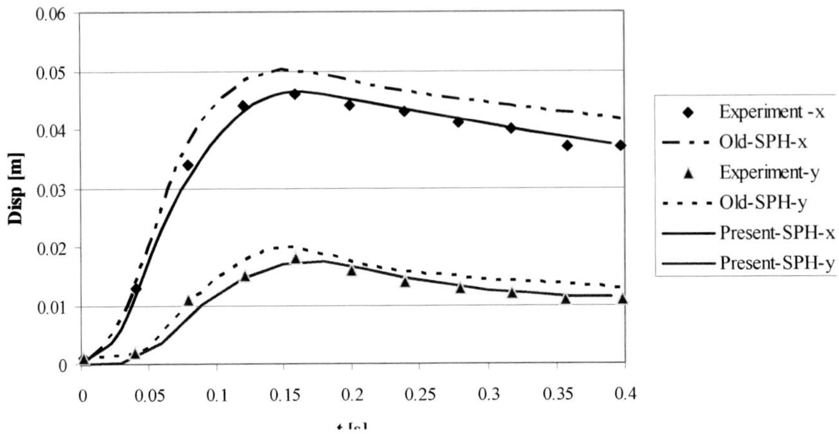


Figure 3. Horizontal and vertical displacements of the free end of the plate.

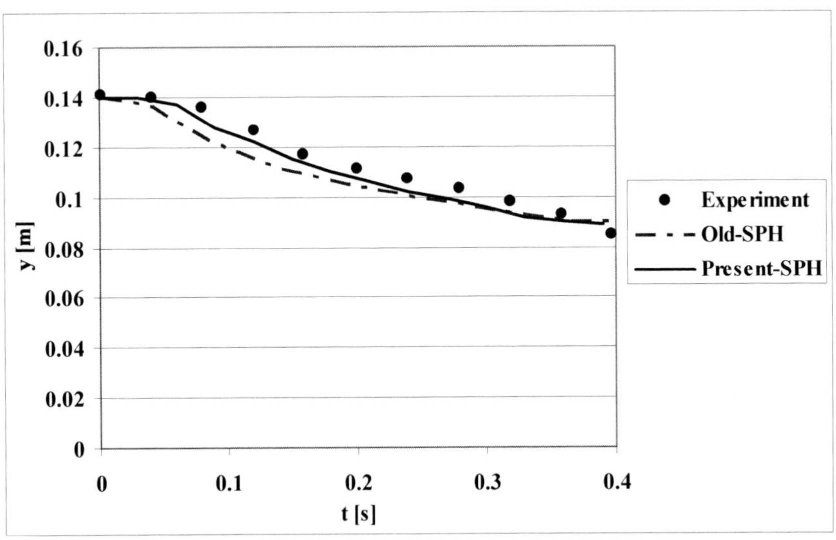


Figure 4. Water level (m) just behind the gate.

The evolution of the free-surface is also well reproduced by the simulation. Figure 5 shows the water level history immediately behind the gate.

## Conclusion

As it was observed, the accuracy of presented SPH simulation is completely greater than the old one. It is because of using a better algorithm for solid elastic simulation and no special treatment for interface particles (the interface modeled naturally).

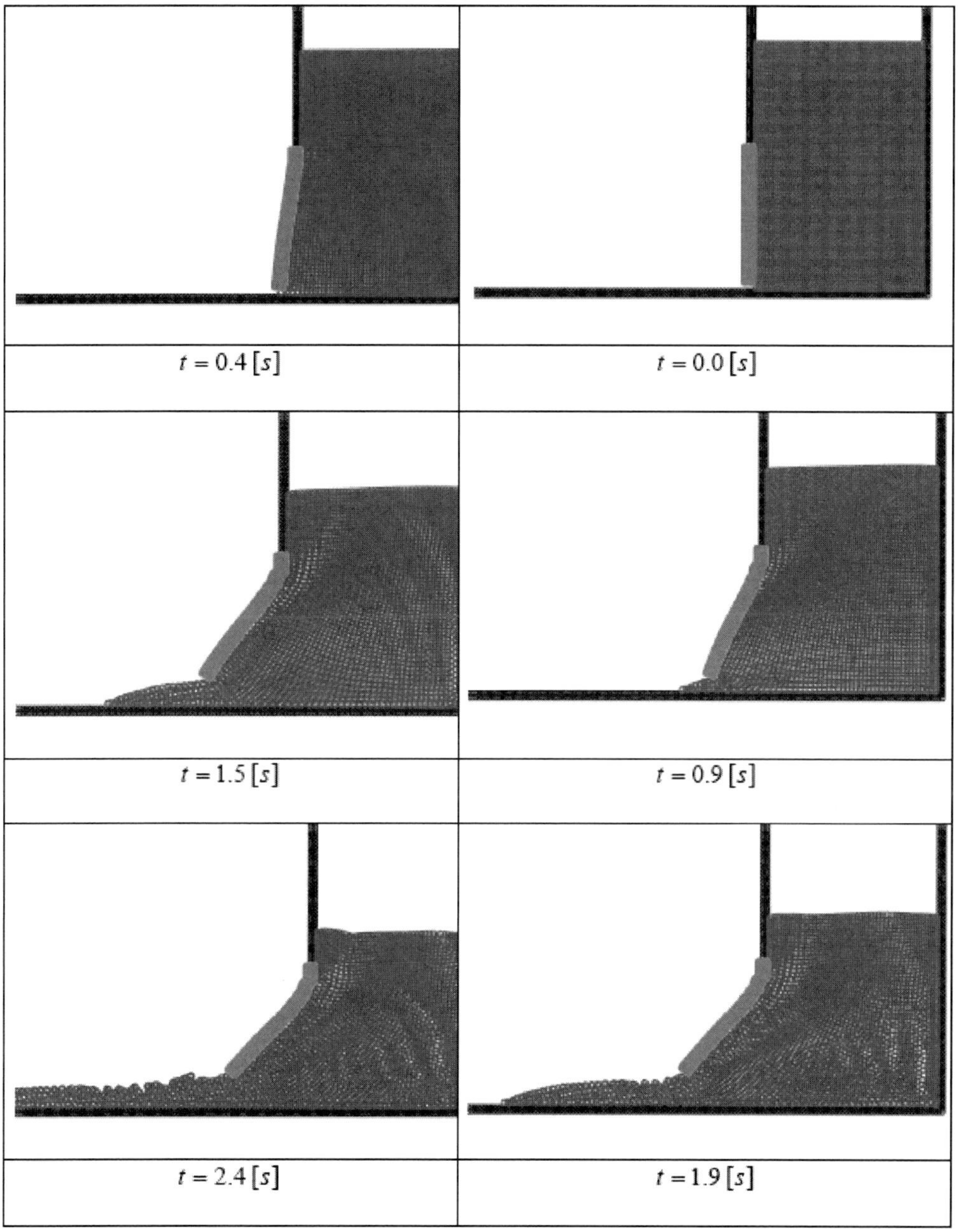


Figure 5. SPH particle positions at six different times.

## REFERENCES


[1] Schwartz LM, Martys N, Bentz DP, Garboczi EJ, Torquato S. (1993). Cross-property relations and permeability estimation in model porous media. *Phys. Rev. E.* 48:4584-91.

[2] Snyder LJ, Stewart WE. (1966). Velocity and pressure profiles for Newtonian creeping flow in regular packed beds of spheres. *A. I. Ch. E. Journal ;*12:167-73.

[3] Sorensen JP, Stewart WE. (1974). Computation of forced convection in slow flow through ducts and packed beds Đ II velocity profile in a simple cubic array of spheres. *Chem. Eng. Sci.;*29:8 19-25.

[4] Edwards DA, Shapiro M, Bar-Yoseph P, Shapira M. (1990). The influence of Reynolds-number upon the apparent permeability of spatially periodic arrays of cylinders. *Phys. Fluids*;2:45-53.

[5] Ghaddar CK. (1995). On the permeability of unidirectional fibrous media: a parallel computational approach. *Phys. Fluids* ;7:2563-86.

[6] Meegoda NJ, King IP, Arulanandan K. (1989). An expression for the permeability of anisotropic granular media. *Int. J. Numer Anal. Methods Geomech;*13:575-98.

[7] Larson RE, Higdon JJL. (1986). Microscopic flow near the surface of two-dimensional porous media. Part1. Axial flow. *J. Fluid Mech;*166:449-72.

[8] Larson RE, Higdon JJL. (1987). Microscopic flow near the surface of two-dimensional porous media. Part 2. Transverse flow'. *J. Fluid Mech*;178:119-36.

[9] Qian YH, d'Humires D, Lallemand P. (1992). Lattice BGK models for Navier-Stokes equation. *Europhys Lett;*17:479-84.

[10] Morris JP, Fox PJ, Zhu Y. (1997). Modeling low Reynolds number incompressible flows using SPH. *J. Comput. Phys*;136:214-26.

[11] Zhu Y, Fox PJ, Morris JP. (1997)`Smoothed particle hydrodynamics model for flow through porous media'. In: Proceedings of the 9th International Conference on Computer Methods and Advances in Geomechanics Vol. 2, China, Wuhan, , p. 1041-6.

[12] Zhu Y, Fox PJ, Morris JP. (1999)`A pore-scale numerical model for flow through porous media', *Int. J. Numer Anal Methods Geomech;*23:881-904.

[13] Morris J.P., Zhu Y., Fox P.J. (1999). Parallel simulations of pore-scale flow through porous media, *Comp. Geotech.* 25, 227-246.

[14] Lucy LB. (1977). A numerical approach to the testing of the fission hypothesis. *Astron. J;*83:1013-24.

[15] Gingold RA, Monaghan JJ. (1977). Smoothed particle hydrodynamics: theory and application to non- spherical stars. *Mon. Not. R Astron. Soc*;181:375-89.

[16] Owen JM, Villumsen JV, Shapiro PR, Martel H (1998) Adaptive smoothed particle hydrodynamics methodology ii. *Astrophys. J. Suppl. S.* 116:55–209

[17] Fulbright MS, Benz W (1995) A method of Smoothed Particle Hydrodynamics Using Spheroid Kernels. Astron. J. 440: 254–262

[18] Liu M. B. Liu G. R. (2005). Meshfree particle simulation of micro channel flows with surface tension. *Comput. Mech.*) 35: 332–341

[19] D. Vola and F. Babik and J.C. Latche, on a numerical strategy to compute gravity currents of non-Newtonian fluids, *J. Comp. Physics.* 201(2): 397-420, 2004.

[20] A.K. Chaniotis and D. Poulikakos and P. Koumoutasakos(2002), Remeshed smoothed particle hydrodynamics for the simulation of viscous and heat conducting flows, *J. Comp. Physics*. 182: 67-90.

[21] J. Michael Owen (2002). Remeshed smoothed particle hydrodynamics for the simulation of viscous and heat conducting flows. *J. Comp. Physics*. 182: 67-90.

[22] J.J. Monaghan (1994). Simulating free surface flows with SPH. *J. Comp. Physics*. 110: 399.

[23] H. Takeda and S.M Miyama and M. Sekiya (1994). Numerical Simulation of Viscous Flow by Smoothed Particle Hydrodynamics, *Progress of Theoretical Physics*. 92(5):939-960.

[24] J. P. Gray, J. J. Monaghan, R. P. Swift (2001). SPH elastic dynamics, *Comp. Methods Appl. Mech. Eng.*190:6641-6662.

[25] Carla Antoci , Mario Gallati, Stefano Sibilla (2007). Numerical simulation of fluid–structure interaction by SPH, Computers and Structures.

[26] J.J. Monaghan (1992), Smoothed particle hydrodynamics, *Annul. Rev. Astron. Astrophys.*, 30: 543-574.

[27] M. A. Hosseini and Nima Amanifard (2007). Presenting a New Modified SPH Algorithm for Numerical Fluid-Structure Interaction Studies. Accepted for publication in *Int. J. Eng.*

In: Monomers, Oligomers, Polymers, Composites… ISBN: 978-1-60456-877-6
Editors: R. A. Pethrick, G.E. Zaikov et al. 

*Chapter 10*

# ADVANCES IN HEAT AND FLUID FLOW COMPUTATIONAL TECHNIQUES WITH PARTICULAR REFERENCE TO MICROCHANNELS AS POROUS MEDIA

***N. Amanifard and A. K. Haghi****

University of Guilan, P. O. Box 3756, Rasht, Iran

## ABSTRACT

In this chapter advances in heat and fluid flow computational techniques with particular application in microchannels is introduced. Throughout the chapter, the flow and heat transfer development regions inside the channels are considered. The numerical results are then compared with the available experimental data. The effects of liquid velocity through channels and their effects on heat transfer and pressure drop along microchannels are investigated. Finally, the effects of aspect ratio on heat dissipation and pressure drop in microchannels are discussed and predicted.

## NOMECLATURE

$C_{pf}$ Specified heat capacity of the cooling liquid [j $kg^{-1}$ $k^{-1}$]
$D_h$ Hydraulic diameter [m]
e Elementary charge [C]
$E_x$ Streaming potential [v/m]
$\overline{E}_x$ Non-dimensional Streaming potential
$\overline{G}$ Non-dimensional parameter
$H_t$ Height of microchannel [m]
$H_c$ Depth of microchannel [m]

* Corresponding author e-mail: Haghi@Guila.ac.ir

$k$ Non-dimensional electrokinetic diameter
$k_b$ Boltzman constant [j mol$^{-1}$ k$^{-1}$]
$k_f$ Thermal conductivity of cooling liquid [w m$^{-1}$ k$^{-1}$]
$k$ Thermal conductivity of silicon[w m$^{-1}$ k$^{-1}$]
$k_f$ Thermal conductivity of fluid phase [w m$^{-1}$ k$^{-1}$]
$k_s$ Thermal conductivity of solid phase [w m$^{-1}$ k$^{-1}$]
$L$ Total length of microchannel [m]
$n_{io}$ Bulk concentration of type-I ion [m$^{-3}$]
$\overline{P}$ Non-dimensional pressure
$q$ Heat flux [w m$^{-2}$]
$R$ Thermal resistance [m$^2$ k w$^{-1}$]
Re Reynolds number
$T$ Absolute temperature [k]
$T_{out}$ Measured outlet temperature [k]
$T_{in}$ Measured inlet temperature [k]
$T_f$ Fluid temperature [k]
$\widetilde{T}_f$ Volume –averaged fluid temperature [k]
$\overline{T}_f$ Mean fluid temperature [k]
$T_s$ Solid phase temperature
$u$ Velocity in X-direction [ms$^{-1}$]
$\mathbf{u}$ Velocity vector [ms$^{-1}$]
$\widetilde{u}$ Volume-averaged velocity in x-direction[ms$^{-1}$]
$v$ Velocity in Y-direction [ms$^{-1}$]
$w$ Velocity in Z-direction [ms$^{-1}$]
$\overline{u}$ Non-dimensional Velocity in X-direction
$W_c$ Width of microchannel [m]
$W_s$ Wall thickness of microchannel [m]
$W_t$ Total width of microchannel [m]
$x_i$ General coordinate
$x$ x- coordinate
$y$ y- coordinate
$z$ z- coordinate
$Y$ Non-dimensional coordinate
$Z$ Non-dimensional coordinate
$Z_i$ Valence of type-i ion

**Greek Symbols**

$\alpha$ The aspect ratio of the channels
$\alpha_f$ Thermal diffusivity of the cooling liquid [wkm$^2$]

$\varepsilon$ Porosity

$\varepsilon_r$ Dielectric constant of the liquid

$\varepsilon_0$ Permittivity of vacuum $[CV^{-1}m^{-1}]$

$\theta$ Dimensionless temperature of cooling liquid [k]

$\aleph$ Debye-Huckel parameter $[m^{-1}]$

$\zeta$ Zeta potential [V]

$\bar{\zeta}$ Non-Dimensional zeta potential

$\rho_f$ Density of cooling liquid $[kg\ m^{-3}]$

$\rho_e$ Net volume charge density $[C\ m^{-3}]$

$\psi$ Electric potential [V]

$\Psi$ Non-Dimensional Electric potential

## 1. Introduction

With the advances in computing technology over the past few decades, electronics have become faster, smaller and more powerful. This results in an ever-increasing heat generation rate from electronic devices. In most cases, the chips are cooled using forced air flow. However, when dealing with a component that contains billions of transistors working at high frequency, the temperature can reach a critical level where standard cooling methods are not sufficient. In addition to high-performance electronic chips, high heat flux removal is also required in devices such as laser diode arrays and high-energy mirrors. In the last two decades, many cooling technologies have been pursued to meet the high heat dissipation rate requirements and maintain a low junction temperature. Among these efforts, the microchannel heat sink (MCHS) has received much attention because of its ability to produce high heat transfer coefficient, small size and volume per heat load, and small coolant requirements (Tsai and Chein (2007)). (See figure 1)

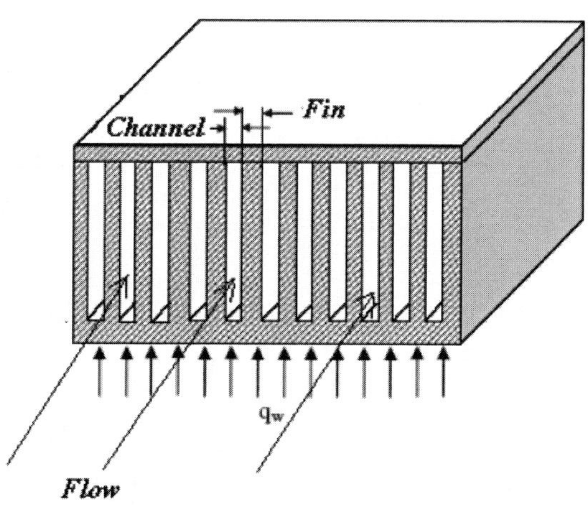


Figure 1. An schematic shape of a Microchannel.

The MCHS cooling concept was proposed by Tuckerman and Pease (1981). Since then, MCHS performances with different substrate materials and channel dimensions have been studied extensively in the past two decades. These studies can be categorized into theoretical (Knight et al., 1992; Ambatipudi and Rahman, 2000), numerical (Fedorov and Viskanta, 2000; Lee et al., 2005; Li et al., 2004; Li and Peterson, 2006; Amanifard and Haghi., 2006; Amanifard et al., 2007; Amanifard and Haghi, 2007), and experimental approaches (Qu and Mudawar, 2002; Tiselj et al., 2004).

In the theoretical approach, most studies employed the classical fin theory which models the solid walls separating microchannels as thin fins. The heat transfer process is simplified as one-dimensional, constant convection heat transfer coefficient and uniform fluid temperature, and these assumptions make validity of this approach in a limited range. The fin approach is effective for the analysis of micro-scale heat transfer in many practical applications, and has been used recently to investigate the efficiency of micro-cell honeycombs in compact heat exchangers (Lu, 1999), and the design of cellar metal system (Gu et al., 2001).However, the nature of the heat transfer process in MCHS is conjugated heat conduction in the solid wall and convection to the cooling fluid. The simplifications used in the theoretical approach usually under- or over predict MCHS performance. To overcome the shortcomings associated with MCHS thermal performance analysis using fin theory, several investigators proposed modeling the MCHS as a porous medium. Kohen and colony (1986) modeled the microchannels as a porous medium by using Darcy's law to describe the flow. Later, Tien and Kuo (1987) analyzed the convection heat transfer in microstructures by using the Brinkman-extended Darcy model for fluid flow to impose the boundary layer effects, and the Forchheimer-Brinkman-extended Darcy equation proposed by Vafai and Tien (1981) for porous mediums. More recently, Kim and Kim (1999) analyzed the laminar heat transfer in MCHS using a modified Darcy model for fluid flow and two-equation model for heat transfer. They found that their results agreed well with those predicted using fin theory models (Knight et al., 1992) and experimental measurements by Tuckerman and Pease (1981). Zhao and Lu (2002) further extended the model developed by Kim and Kim (1999) to study the channel geometries, effective thermal conductivities and porosities on MCHS thermal performance (figure 2 Shows the equivalent porous media).

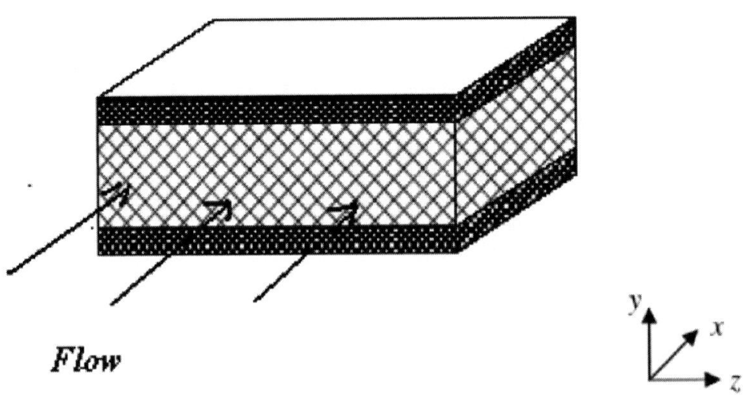


Figure 2. The equivalent Microchannel as a porous media.

Because conjugated heat transfer is involved in MCHS, it is believed that the porous medium model is better than the classical fin theory in describing MCHS thermal performance, and upon this basic idea, the microchannel studies become as a branch of porous media heat transfer problems.

# 2. The Continuity Approach of Microchannels

## 2.1. Flow without Electric Field

The fin physical approach relies on continuity concepts and the Navier-Stokes equations with the energy equation are solved with no-slip boundary condition and the viscous dissipation effects without any electric field on walls. Wu and Little (1983) measured the heat transfer characteristics for gas flows in miniature channels with inner diameter ranging from 134 to 164 μm. The tests involved both laminar and turbulent flow regimes. Their results showed that the turbulent convection occurs at Reynold's number of approximately 1000. They also found that the convective heat transfer characteristics depart from the predictions of the established empirical correlations for the macroscale tubes. They attributed these deviations to the large asymmetric relative roughness of the microchannel walls.

Harms et al. (1997) tested a 2.5 cm long, 2.5 cm wide silicon heat sink having 251μm wide and 1030 μm deep microchannels. A relatively low Reynolds number of 1500 marked transition from laminar to turbulent flow which was attributed to a sharp inlet, relatively long entrance region, and channel surface roughness. They concluded the classical relation for Nusselt number was fairly accurate for modeling microchannel flows.

Fedrov and Viskanta (2000) reported that the thermal resistance decreases when Reynolds number increases and approaches an asymptote at high Reynolds numbers.

Choi et al. (1991) measured the convective heat transfer coefficients for flow of nitrogen gas in microtubes for both laminar and turbulent regimes. They found that the measured Nusselt number in laminar flow exhibits a Reynolds number dependence in contrast with the conventional prediction for the fully developed laminar flow, in which Nusselt number is constant.

Adam et al. (1998)conducted single-phase flow studies in microchannel using water as the working fluid. Two diameters of the circular microchannels, namely 0.76 mm and 1.09 mm, were used in the investigation. It was found that the Nusselt numbers are larger than those encountered in macrochannels.

Peng and Peterson (1996) investigated water flows in rectangular microchannels with hydraulic diameters ranging from 0.133 to 0.336 mm. In laminar flows, it was found that the heat transfer depends on the aspect ratio and the ratio of the hydraulic diameter to the center-to-center distance of the microchannel.

***Computational Model***

The physical characteristics of a computational model for the fin approach are illustrated in figure 3.

The width and the height of the microchannels are $W_s$, and $H_c$ respectively. The thickness of the substructure is $H_t - H_c$; and the total length of the microchannels is $L$.

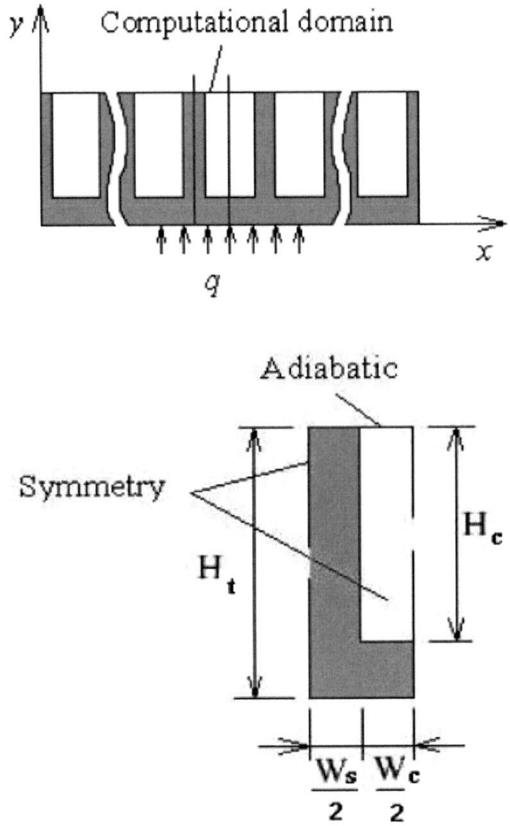


a

b

Figure 3. Dimensions of the computational model.

***Governing Equations***

Several simplifying assumptions are incorporated before establishing the governing equations for the fluid flow and heat transfer in the unit cell:

1. Steady fluid flow and heat transfer;
2. Incompressible fluid;
3. Laminar flow;
4. Negligible radiative heat transfer;
5. Constant solid and fluid properties except water viscosity.

However, the variation of water viscosity is significant. Therefore, a polynomial function in a limited temperature range is usually considered for the viscosity variations.

$$\mu(T) = a + bT + cT^2 + dT^3 + eT^4 + fT^5 \tag{1}$$

The coefficients of the above equation are given in table 1.

**Table 1. Coefficients of the polynomial function**

| a | b | c | d | e | f |
|---|---|---|---|---|---|
| 0.5779 | -0.0078 | 4.E-05 | -1E-07 | 2.E-10 | -9.E-14 |

Based on the above assumptions, the governing differential equations used to describe the fluid flow and heat transfer in the unit cell of a microchannel can be proposed on the continuity basis.

Continuity

$$\frac{\partial u_i}{\partial x_i} = 0 \tag{2}$$

Momentum

$$\rho\frac{\partial(u_i u_j)}{\partial x_j} = -\frac{\partial P}{\partial x_i} + \mu\frac{\partial}{\partial x_j}\left(\frac{\partial u_i}{\partial x_j} + \frac{\partial u_j}{\partial x_i}\right) - \frac{2}{3}\mu\frac{\partial}{\partial x_i}\left(\frac{\partial u_k}{\partial x_k}\right) \tag{3}$$

Energy

$$\rho C_p\frac{\partial u_i T}{\partial x_i} = k\frac{\partial^2 T}{\partial x_i^{\,2}} + \mu\Phi \tag{4}$$

where

$$\Phi = 2\left[\left(\frac{\partial u}{\partial x}\right)^2 + \left(\frac{\partial v}{\partial y}\right)^2 + \left(\frac{\partial w}{\partial z}\right)^2\right] + \left(\frac{\partial u}{\partial y} + \frac{\partial v}{\partial x}\right)^2 + \left(\frac{\partial u}{\partial z} + \frac{\partial w}{\partial x}\right)^2 + \left(\frac{\partial v}{\partial z} + \frac{\partial w}{\partial y}\right)^2 \tag{5}$$

***Flow Characteristics through a Specified Microchannel***

The general solution of the fin theory can not be achieved by analytical methods, and the only way is the numerical analysis. Amanifard and Haghi (2006) performed a numerical solution for a specified microchannel which is used for cooling of a supercomputer chip.

The heat supplies by a 1cm × 1cm heat source located at the entrance of the microchannels and is centered across the whole channel heat sink. A uniform heat flux of $q$ is provided to heat the microchannels. The heat is removed by flowing water through channels. The inlet temperature of the cooling water is $20°C$. The analysis is performed for four different cases. The dimensions related to each case are given in table 1.By these dimensions, there will be 150 microchannels for cases 0 and 1 and 200 microchannels for cases 2, 3, and 4.

**Table 2. Four different cases for the microchannels**

| | Case | | | | |
|---|---|---|---|---|---|
| | 0 | 1 | 2 | 3 | 4 |
| L(cm) | 2 | 2 | 1.4 | 1.4 | 1.4 |
| $W_s(\mu m)$ | 64 | 64 | 56 | 55 | 50 |
| $W_c(\mu m)$ | 36 | 36 | 44 | 45 | 50 |
| $H_c(\mu m)$ | 280 | 280 | 320 | 287 | 302 |
| $H_t(\mu m)$ | 489 | 489 | 533 | 430 | 458 |
| $\dot{Q}(cm^3/s)$ | 1.277 | 1.86 | 4.7 | 6.5 | 8.6 |
| $q(W/cm^2)$ | 34.6 | 34.6 | 181 | 277 | 790 |
| Number of channels | 150 | 150 | 200 | 200 | 200 |

***Boundary Conditions***

The whole substructure is made of silicon with thermal conductivity (*k*) of $148 W/m.K$. At the top of the channel $y = H_t$ is a pyrex plate to make an adiabatic condition (its thermal conductivity is two orders lower than silicon). There are two different boundary conditions at the bottom. For $z < L_h$ a uniform heat flux of $q$ is imposed over the heat sink and the rest is assumed to be adiabatic. Water flows through the channel from the entrance in *z* direction. The transverse velocities of the inlet are assumed to be zero. The axial velocity is considered to be evenly distributed through the whole microchannel. The velocities at the top and the bottom of channels are zero.

***Numerical Method***

The Finite Volume Method (FVM) is used to solve the continuity, momentum, and energy equations. A brief description of the method is explained here.

In this method the domain is divided into a number of control volumes such that there is one control volume surrounding each grid point. The grid point is located in the center of a control volume. The governing equations are integrated over the individual control volumes to construct algebraic equations for the discrete dependent variables such as velocities, pressure, and temperature. Then the discretized equations express the conservation principle for a finite control volume just as the partial differential equations express for an infinitesimal control volume. Amanifar and Haghi (2006) Performed the solution be converged when the mass imbalance in the continuity equation is less than $10^{-6}$.

***Temperature and Velocity Profiles***

Thermal specifications and the flow characteristics are the main parameters in microchannel design. The results of the mentioned problem are focused on these fields.

For solving the equations several grid structures were used. The grid density of 120×40×20 in *z*, *y*, and *x* directions is considered to be appropriate.

The thermal resistance is introduced as:

$$R(z) = \frac{T_{out} - T_{in}}{q} \tag{5}$$

Moreover, the Reynolds number is calculated as:

$$\mathrm{Re} \equiv \frac{\rho w_{ave} D_h}{\mu} \tag{6}$$

Where the hydraulic diameter is defined as:

$$D_h \equiv \frac{4A}{P} = \frac{2H_1W_1}{H_1 + W_1} \tag{7}$$

***Temperature Distribution and Thermal Resistance***

The temperature distributions at four *x-y* cross sections along the channel are shown in figure 4 for the Case 0. The four sections (z= 1, 3, 6, 9mm) are all in the heated area. The temperature contours are clustered at the entrance of the channels. This means that the heat removing (cooling) rate is higher at the entrance than the other parts of the channel.

The thermal resistance along the channel is shown in figure 5 for Cases 0 and 1. It can be seen that the thermal resistance has increased by increasing the *z* value and approaches to a maximum value at $z$ = 9mm. The adequate consistency with respect to Tuckerman's experimental results (1984) is achieved. The maximum thermal resistance occurred in $z$ = 9mm which is consistent with Tuckerman's results (1984). For the other Cases, The results are given in table 3. The computed resistances have acceptable values, except in Case 4, in which the error is noticeable. This may be due to the very large heat flux exposed to the heat sink, the prediction of large heat fluxes in micro-scales needs some modifications in fundamental approaches.

**Table 6. Thermal resistance comparison**

| Case | $q(\frac{W}{cm^2})$ | $R(cm^2K/W)$ | | Error (%) |
|---|---|---|---|---|
| | | Experimental | Numerical | |
| 0 | 34.6 | 0.277 | 0.253 | 8.5 |
| 1 | 34.6 | 0.28 | 0.246 | 12.1 |
| 2 | 181 | 0.110 | 0.116 | 5 |
| 3 | 277 | 0.113 | 0.101 | 8.1 |
| 4 | 790 | 0.090 | 0.086 | 3.94 |

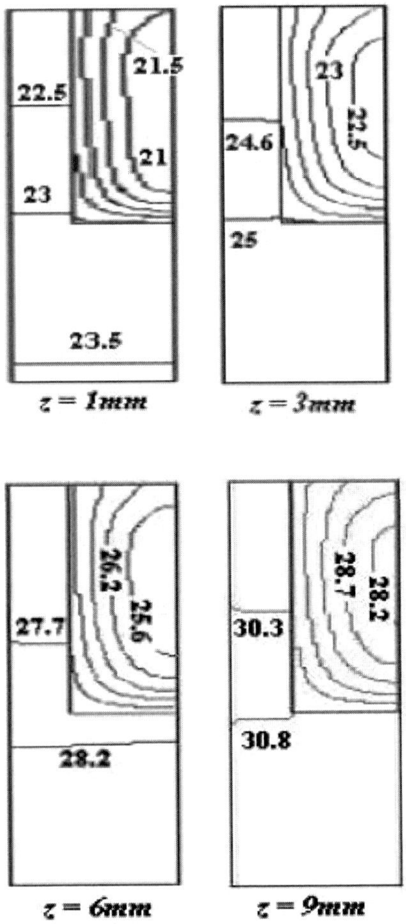


Figure 4. Temperature distribution along the channel for Case 0.

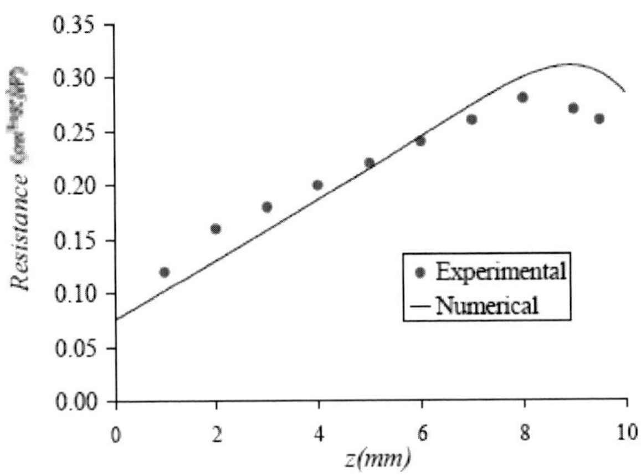


a

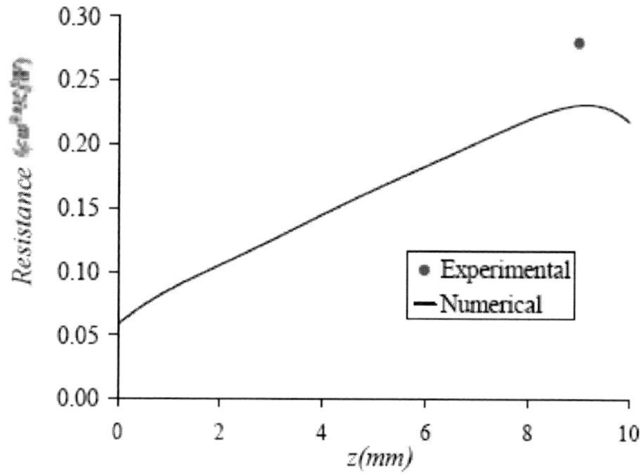


b

Figure 5. Numerical and experimental Thermal resistances for: (a) Case 0, and (b) Case 1.

***Velocity Distribution***

Figure 6 illustrates the velocity profiles in various *y-z* sections in Cases 0 and 1. It is obvious that, the velocity gradients near the channel walls are very large. This means that the wall shear stress is considerably large and consequently, the pressure drop along the channel becomes significant. Although the thermal boundary layer through the channel is fully developed, but the fully developed velocity is not valid and the velocity profile through whole channel does not reaches the developed form.

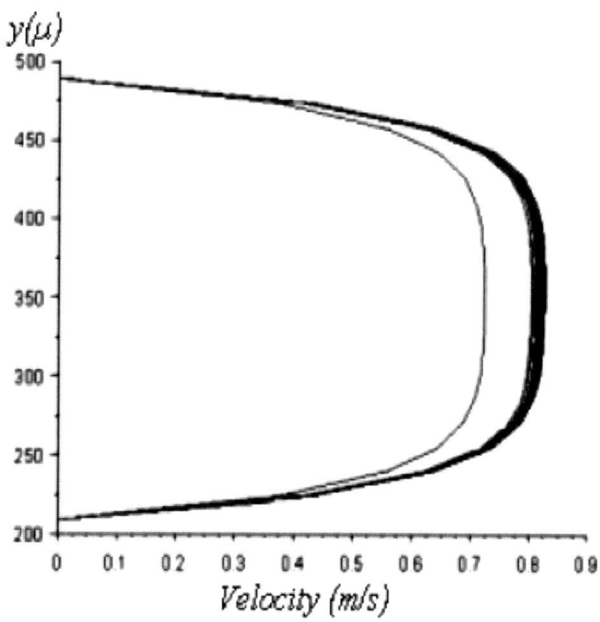


a

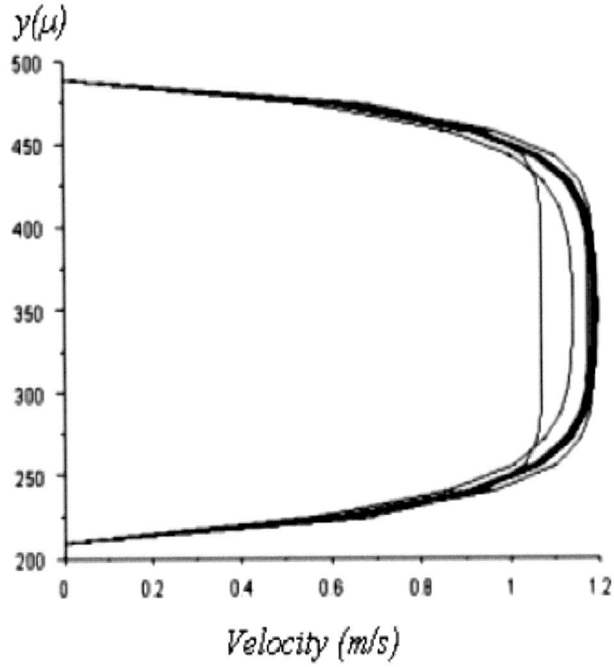


b

Figure 6. Velocity profiles in *y-z* sections, (a) Case 0, (b)Case 1.

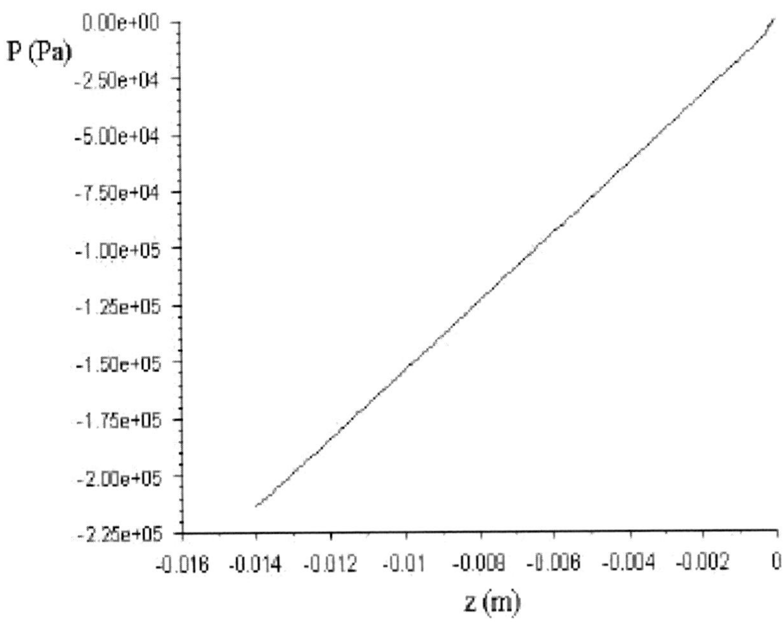


a

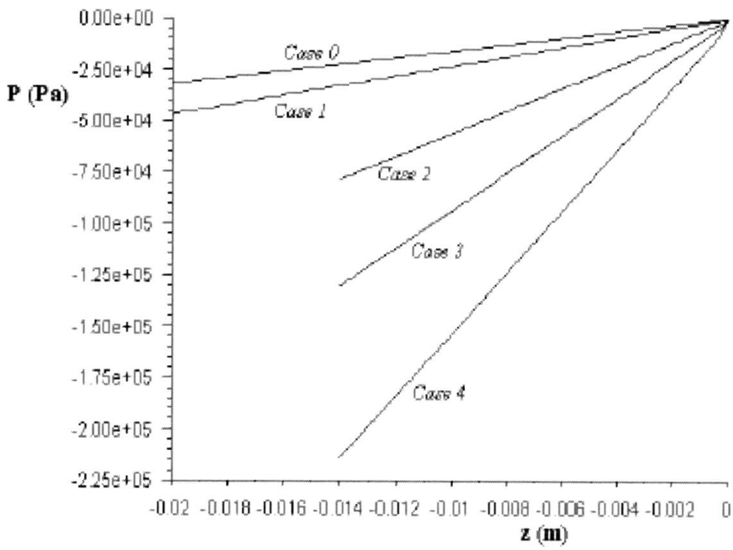


b

Figure 7. Pressure drop curves, (a) case 4, (b) all cases.

***Pressure Drop***

Figure 7(a) shows the pressure drop for the case 4 and figure 7(b) gives a comparison of the pressure drops for all study cases. As mentioned, the high level of the pressure drop relates to the high levels of the wall shear stresses.

These amounts are tabulated in table 4.

**Table 4. Pressure drop in 5 Cases**

| Case | Pressure Drop (bar) |
|---|---|
| 0 | 0.322 |
| 1 | 0.469 |
| 2 | 0.784 |
| 3 | 1.302 |
| 4 | 2.137 |

***The Effect of Velocity on Temperature Distribution and Pressure Drop***

The effects of velocity on the temperature rise and pressure drop in microchannels are illustrated in figures 8 and 9. The amount of heat dissipation increases by increasing the entrance velocity values. It also can be seen that the reduction in temperature values goes down drastically by augment in velocity levels.

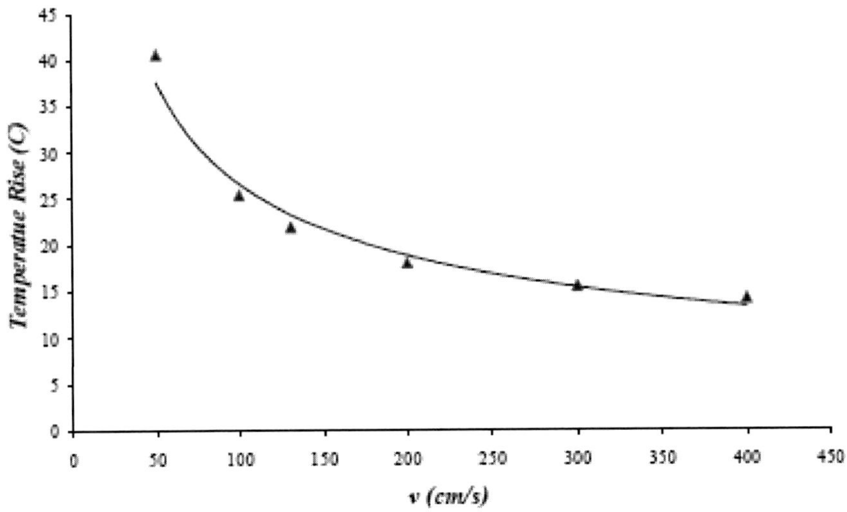


Figure 8. Temperature rise for different velocities.

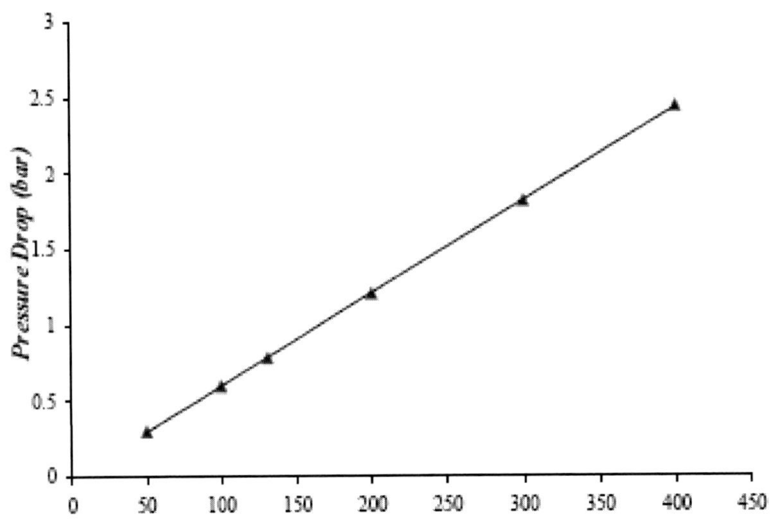


Figure 9. Pressure drop for different velocities.

A prediction empirical formula between the temperature and the velocity values was performed by Amanifard and Haghi (2006).

$$(T_{\max} - T_{in}) = 265.67u^{-0.4997} \tag{7}$$

The amounts of temperature rise and pressure drop are given in table 5.

**Table 5. Temperature rise and pressure drop**

| Velocity (cm/s) | Re | Temp. Rise (°C) | Pressure Drop (bar) |
|---|---|---|---|
| 50 | 47 | 40.6 | 0.298 |
| 100 | 95 | 25.4 | 0.598 |
| 131 | 124 | 21.9 | 0.784 |
| 200 | 190 | 18.0 | 1.206 |
| 300 | 285 | 15.5 | 1.817 |
| 400 | 380 | 14.2 | 2.439 |

***The Effect of Aspect Ratio on Temperature Distribution and Pressure Drop***

Aspect ratio is an important factor in microchannel design. In the current case the inlet heat flux of *181 W/cm*$^2$ imposed over the heat sinks and the hydraulic diameter changes between 85.8 and 104.2. Figure 10 shows the maximum temperature of each case. It may be seen that with an identical heat flux the heat dissipation of the largest aspect ratio is the lowest. But this case has the minimum pressure drop too (figure 11). When the aspect ratio grows the heat dissipation increases. The third geometry has the minimum temperature rise. The temperature rise of the second and the forth geometries are approximately similar. On the other hand, the pressure drop of these two cases has a noticeable difference. The pressure drop increases with growth of the aspect ratio. The temperature levels of all geometries are given in table 6.

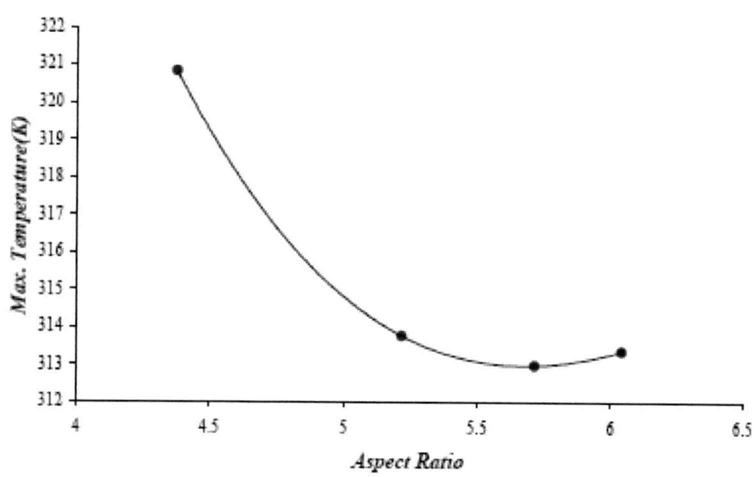


Figure 10. Max. temperature with respect to aspect ratio.

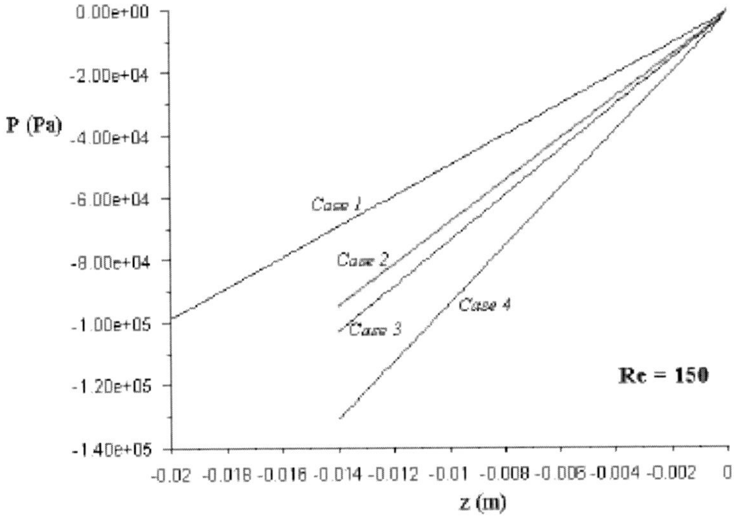


Figure 11. Pressure drop relating to each geometry in constant Reynolds of 150.

**Table 6. Temperature rise and pressure drop for section 3**

| Case | Aspect Ratio | Hydraulic Diameter (μm) | Velocity (m/s) | Temp. Rise (°C) | Pressure Drop (bar) |
|---|---|---|---|---|---|
| 1 | 4.375 | 104.19 | 1.447 | 27.84 | 0.98 |
| 2 | 5.714 | 95.32 | 1.58 | 19.97 | 0.95 |
| 3 | 5.218 | 92.31 | 1.633 | 20.76 | 1.03 |
| 4 | 6.040 | 85.80 | 1.76 | 20.35 | 1.31 |

The results gave the required assurance of using the full Navier-Stokes approach for the microchannels with aspect ratios about 0.1. However, for lower aspect ratios in the range of molecular mean free paths length scales, some modifications must be prepared for the boundary conditions and physical approaches (Mentioned in previous chapters on convection aspects).

## 2.2. The Continuity Approach with Electric Double Layer (EDL)

One possible explanation for these observed effects is the presence of the interfacial effects such as electric double layer, EDL. These effects are ignored in macro-scale fluid studies. Mala et al. (1996) considered the electrical body force resulting from the double layer field in the equations of motion. These equations were solved for the steady state flow. It was found that without the double layer a higher heat transfer rate is obtained. They proposed to consider the effects of the EDL on liquid flows and heat transfer in microchannels to prevent the overestimation of the heat transfer capacity of the system. Amanifard et al. (2007) completed their study (Amanifard and Haghi (2006)) by applying the EDL source terms in governing equations (2)-(5) and gave a complete illustration of EDL effects on microchannel cases.

### *EDL Equations*

Most of solid surfaces convey a static electric charge. In other words they will have superficial electric potential. If the liquid contains a very small number of ions (for instance, due to impurity), the electrostatic charges on the non-conducting solid surface will captivate the counter ions, and repel similar ions in cooling fluid. Consequently, as we can see in figure 12, the ions near the solid surface will found a new arrangement. The rearrangement of the charges on the solid surface and the balancing charges in the fluid is called the electric double layer. On the basis of electrostatic theorems,

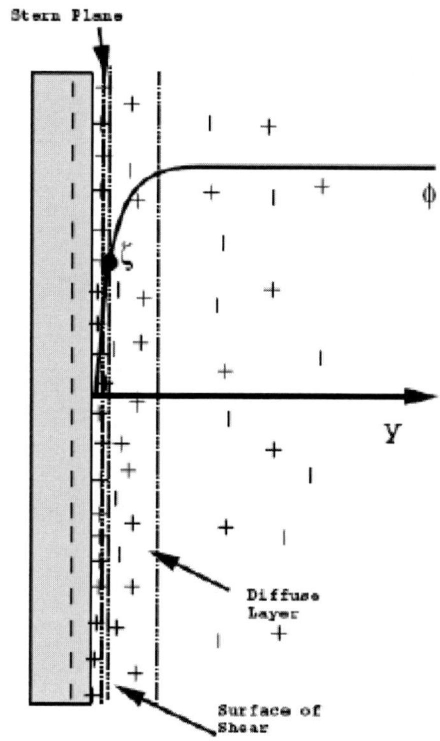


Figure 12. Electric Double layer.

The relation between electric potential Ψ, and net volume charge density $\rho_e$ in each point of fluid, is decrypted by Poisson equation as follows:

$$\frac{\partial^2 \psi}{\partial y^2} + \frac{\partial^2 \psi}{\partial z^2} = -\frac{\rho_e}{\varepsilon_r . \varepsilon_0} \tag{8}$$

$\varepsilon_r$, *and* $\varepsilon_0$ represent the dielectric constant of the solution and permittivity of vacuum respectively. Assuming the Boltzman distribution equation is applicable, the number concentration of the type-i ion in an electrolyte fluid can be considered as follows:

$$n_i = n_{i0}.\exp(\frac{-z_i e\psi}{k_b T}) \tag{9}$$

Where $n_{i0}$ and $z_i$ are the bulk concentration and the valence of type-i ion, respectively, e is the elementary charge, $k_b$ is the Boltzman constant, and T is the absolute temperature. This equation is established only when the system is in equilibrium state. The net volume charge density is commensurate to difference between concentration of cations and anions, via

$$\rho_e = z.e.(n_+ - n_-) = -2zen_0 \sinh(\frac{ze\psi}{k_b T}) \tag{10}$$

Substitution of equation (10) into equation (8) leads to the conspicuous Poisson-Boltzman equation:

$$\frac{\partial^2 \psi}{\partial y^2} + \frac{\partial^2 \psi}{\partial z^2} = \frac{2zen_0}{\varepsilon_r \varepsilon_0} \sinh(\frac{ze\psi}{k_b T}) \tag{11}$$

With considering the Debye-Huckel parameter and following dimensionless groups:

$$\aleph = (\frac{2z^2 e^2 n_0}{\varepsilon_r \varepsilon_0 k_b T})^{1/2} \tag{12}$$

$$D_h = \frac{2H_t W_c}{H_t + W_c} \tag{13}$$

$$Y = \frac{y}{D_h} \tag{14}$$

$$Z = \frac{z}{D_h} \tag{15}$$

$$K = \aleph.D_h \tag{16}$$

$$\Psi = ze\frac{\psi}{k_b T} \tag{17}$$

Where $D_h$ is hydraulic diameter of the rectangular channel, Y and Z are non-dimensional coordinates. By using above non-dimensionless groups equation (11) can be non-dimensionalized as:

$$\frac{\partial^2\psi}{\partial Y^2}+\frac{\partial^2\psi}{\partial Z^2}=K^2\sinh\psi \tag{18}$$

Here K, is the non-dimensional electro-kinetic diameter, depicted as the ratio of the hydraulic diameter to the electrical double layer thickness, and $\psi$ is the non-dimensional electrical potential standing for the ratio of the electrical energy $ze\psi$ to the thermal energy $k_b T$. Considering figure 3 we can issue following boundary conditions:

$$Y=0 \Rightarrow \psi=\bar{\zeta} \qquad Y=\frac{H}{D_h} \Rightarrow \psi=\bar{\zeta}$$
$$Z=0 \Rightarrow \frac{\partial\psi}{\partial Z}=0 \qquad Z=\frac{W}{D_h} \Rightarrow \psi=\bar{\zeta} \tag{19}$$

Where $\bar{\zeta}$ defined by $\bar{\zeta}=\frac{ze\zeta}{k_b T}$ is a non-dimensional zeta potential of the channel walls (here $\zeta$ is the potential of the channel wall). The zeta potential is an electric potential at the channel walls. After solving equation (18) and computing $\psi$, the net volume charge density can be obtained as follows:

$$\rho_e(Y,Z)=-2zen_0\sinh\psi(Y,Z) \tag{20}$$

This net volume charge density is needed for computing of body forces originating from EDL.

### *Modified Navier-Stokes Equations*

Assuming a laminar fully developed flow in rectangular channels in positive x-direction, the components of velocity satisfy $u=u(y,z)$ and $v=w=0$ in terms of Cartesian coordinate. The equation of motion is written as follows:

$$\frac{\partial^2 u}{\partial y^2}+\frac{\partial^2 u}{\partial z^2}=\frac{1}{\mu_f}\frac{dP}{dx}-\frac{1}{\mu_f}E_x\rho_e(y,z) \tag{21}$$

In this equation the final term at right hand, is the effects of body forces originating from EDL. Considering following dimensionless groups we can obtain the non-dimensional form of equation (21),

$$\mathrm{Re}_0=\frac{\rho_f D_h U}{\mu_f} \tag{22}$$

$$\bar{u} = \frac{u}{U} \tag{23}$$

$$\bar{P} = \frac{P - P_0}{\rho_f U^2} \tag{24}$$

$$\bar{X} = \frac{x}{D_h \operatorname{Re}_0} \tag{25}$$

$$\frac{d\bar{P}}{d\bar{X}} = \frac{D_h \operatorname{Re}_0}{\rho_f U^2} \frac{dP}{dx} \tag{26}$$

$$\bar{E}_x = \frac{E_x D_h \operatorname{Re}_0}{\zeta_0} \tag{27}$$

$$\bar{G}_1 = \frac{2zn_0\zeta_0}{\rho_f U^2} \tag{28}$$

Substitution of resent equations in equation (21), the non-dimensional for of this equation can be obtained:

$$\frac{\partial^2 \bar{u}}{\partial Y^2} + \frac{\partial^2 \bar{u}}{\partial Z^2} = \frac{d\bar{P}}{d\bar{X}} + \bar{G}_1 \bar{E}_x \sinh \Psi(Y, Z) \tag{29}$$

Related boundary conditions are as follows:

$$\begin{aligned} &Y = 0 \ \Rightarrow \ \bar{u} = 0 \qquad && Y = \frac{H_c}{D_h} \ \Rightarrow \ \bar{u} = 0 \\ &z = 0 \ \Rightarrow \ \frac{\partial \bar{u}}{\partial z} = 0 \qquad && Z = \frac{W_c}{D_h} \ \Rightarrow \ \bar{u} = 0 \end{aligned} \tag{30}$$

After numerical solution of equation (29) the velocity field will be obtained.

***Energy Equation***

As presented in figure 1, a silicon wafer plate with a large number of microchannels is connected to the chip. A liquid is forced to flow through these microchannels to remove the heat. All microchannels are assumed to have a uniform rectangular cross-section with geometric parameters as shown in table 2. For a steady-state, fully developed, laminar flow in a microchannel, the energy equation (with consideration of the axial thermal conduction in flow direction and the viscous dissipation) for the cooling liquid takes the specific form:

$$u\frac{\partial\theta}{\partial x}=\alpha_f\left(\frac{\partial^2\theta}{\partial x^2}+\frac{\partial^2\theta}{\partial y^2}+\frac{\partial^2\theta}{\partial z^2}\right)+\frac{\mu_f}{\rho_f C_{pf}}\left[\left(\frac{\partial u}{\partial y}\right)^2+\left(\frac{\partial u}{\partial z}\right)^2\right] \tag{31}$$

Where $\theta$ and $\alpha_f$ are the temperature and the thermal diffusivity of the cooling liquid, respectively, $C_{pf}$ is the specified heat capacity of the cooling liquid. Based on presented computational domain, the adiabatic condition can be used along the channel symmetric center line:

$$z=0 \quad \Rightarrow \quad \frac{\partial\theta}{\partial z}=0 \tag{32}$$

At the bottom of channels, a uniform heat flux of q" is imposed over the heat sink, and can be expressed as:

$$y=0 \quad \Rightarrow \quad q''=-k_f\left(\frac{\partial\theta}{\partial y}\right) \tag{33}$$

Hear $k_f$ is the thermal conductivity of the liquid coolant. Since the thermal conductivity of the glass is about two-order of magnitude lower than that the top boundary is insulted. This is a conservative assumption which will lead to slight underestimation of the overall heat transfer coefficient. This assumption yields:

$$y=H \quad \Rightarrow \quad \frac{\partial\theta}{\partial y}=0 \tag{34}$$

***Numerical Solution***

The power-low scheme is used to model the combined convection-diffusion effects in the transport equations with FVM approach. The SIMPLER scheme is employed to resolve the pressure-velocity coupling. The resulting algebraic equations are solved using a line-by-line Tri-Diagonal matrix Algorithm.

The study cases are the same given in table (2) and the computed results were compared in table (7) with the results of Toh et. al. (2002) (without EDL), and the experimental results of Tuckermann (1984). The thermal resistance was the basis of their comparisons. It is shown that sufficiently reasonable agreement exists in such comparison. The thermal resistance because of decreasing in volumetric flow rate and consequently increasing in $T_{\max}$ increases from case 1 to case 4.

**Table 7. Thermal resistance comparison**

| Case | Heat Flux ( $w/m^2$ ) ×10-4 | 104× ( $m^2K/W$ )Thermal Resistance | | |
|---|---|---|---|---|
| | | Tuckerman's Result | Toh's Result | Current numerical Result |
| 1 | 36.4 | 0.280 | 0.253 | 0.301 |
| 2 | 181 | 0.110 | 0.157 | 0.136 |
| 3 | 277 | 0.113 | 0.128 | 0.116 |
| 4 | 790 | 0.090 | 0.105 | 0.086 |

For considering of effects of EDL, the gradient of pressure, velocity profile, and temperature profile are inspected for both, existing of EDL with two amounts for zeta potential, 75 milivolt and 200 milivolt, and absence of EDL. The 75 milivolt zeta potential is considered for feeble EDL effects, and 200 milivolt zeta potential is selected for strong EDL effects. The effects of EDL on distribution of pressure along the microchannel are showed in figure 13 for case1.

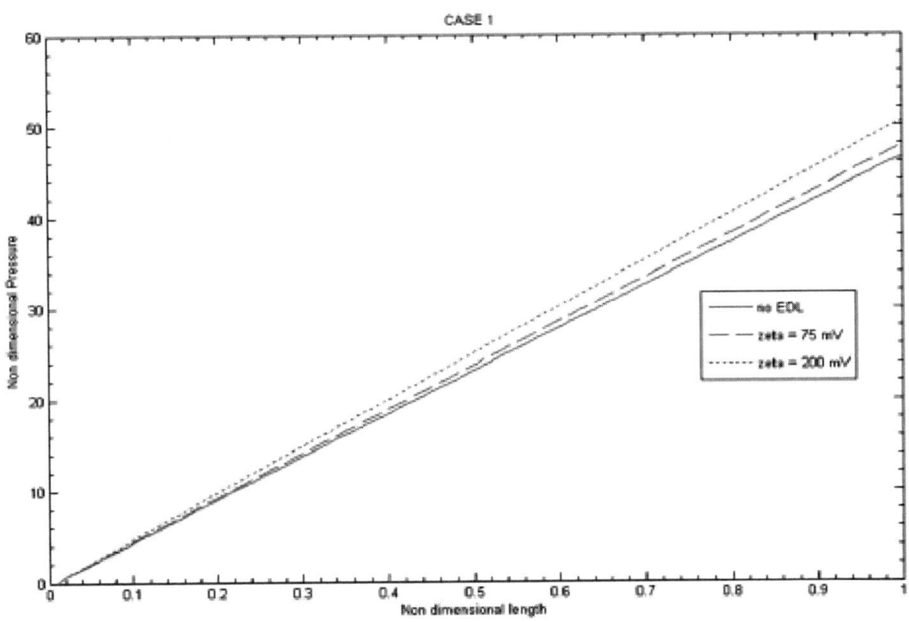


Figure 13. variation of dimensionless pressure drop with respect to dimensionless length of channel.

The percentage of effects of EDL on dimensionless distribution of pressure for case 1 to case 4 is presented in table (7).

**Table 7. The percentage of variation of dimensionless distribution of pressure**

| Case | The percentage of variation of dimensionless distribution affected by EDL | |
|---|---|---|
| | Zeta Potential (75mv) | Zeta Potential (200mv) |
| 0 | 9.4 | 24.7 |
| 1 | 2.6 | 8.5 |
| 2 | 4.1 | 10.2 |
| 3 | 4.4 | 10.9 |
| 4 | 6.2 | 13.7 |

The illustrated difference in alterations of pressure affected of EDL is because of coming into existence of apparent viscosity that is much more than viscosity in absence of EDL. As it can be seen in table (7), the percentage of alterations of pressure in 75 milivolt zeta potential for cases 1-4 is not so intense, but in 200 milivolt zeta potential for case 0 that has minimum channel width, exists a large mutation causes to considerable thickness of EDL with respect to channel width. Hence we can conclude that whit increasing of thickness of EDL in comparison with channel width the effects of EDL will be stronger. Pressure drop considerations will determine the required pumping power. The more pressure drop, the more required pumping power. Thus, with existing of a powerful EDL effect, we will require much pumping power, and this is not so suitable. The effects of EDL on velocity profile are depicted in figures 14 and 15 for case 0, and case 1 respectively, and percentage of alterations of velocity for cases 1-4 is shown in table (8). As presented in table (8), a large mutation in velocity alliteration exists for case 0 in 200 milivolt zeta potential as there is for pressure alliteration in this case. Adhesiveness of ions to channels walls and body forces originated from EDL causes to reducing of volumetric flow rate. Decreasing in characteristic length of microchannel, and increasing in body forces originating from EDL causes to increasing in velocity alliteration.

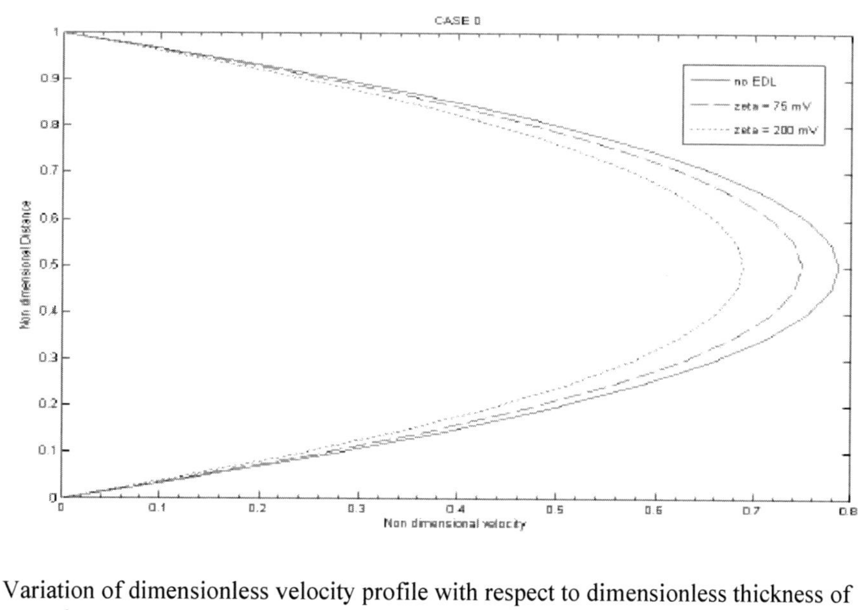


Figure 14. Variation of dimensionless velocity profile with respect to dimensionless thickness of channel for case 0.

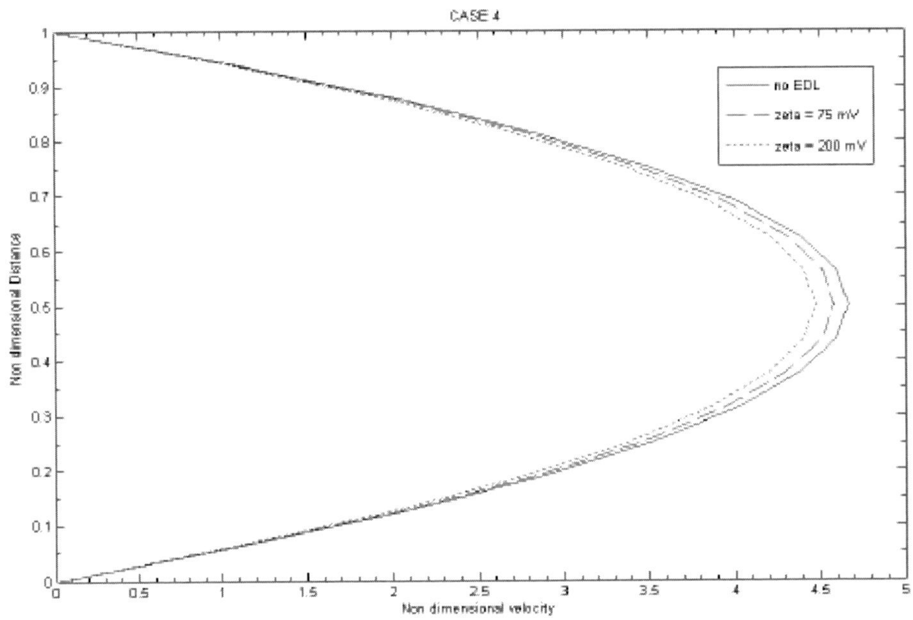


Figure 15. Variation of dimensionless velocity profile with respect to dimensionless thickness of channel for case 4.

**Table 8. The percentage of variation of dimensionless velocity**

| Case | The percentage of variation of dimensionless Velocity affected by EDL | |
|---|---|---|
| | Zeta Potential (75mv) | Zeta Potential (200mv) |
| 0 | 4.8 | 12.5 |
| 1 | 0.9 | 2.1 |
| 2 | 1.4 | 3.1 |
| 3 | 1.6 | 3.5 |
| 4 | 2.9 | 4.1 |

The EDL effects on dimensionless temperature profile with respect to dimensionless channel high are presented for cases 3, 4 in figures 16 and 17. Asshown, temperature of cooling fluid decreases from bottom wall that is exposed to uniform heat flux to top of the channel that is adiabatic.The gradient of temperature near the upper wall of the channel depicted that there is no heat flux near the wall. The effects of EDL on dimensionless temperature profile is shown for cases 0-4 in table (9) on percentage base. As it is obvious, the presence of EDL causes an augment of temperature, particularly in case 0, this growth is result of reduction in volumetric flow rate. The growth of temperature of causes the reduction in Nusselt number.

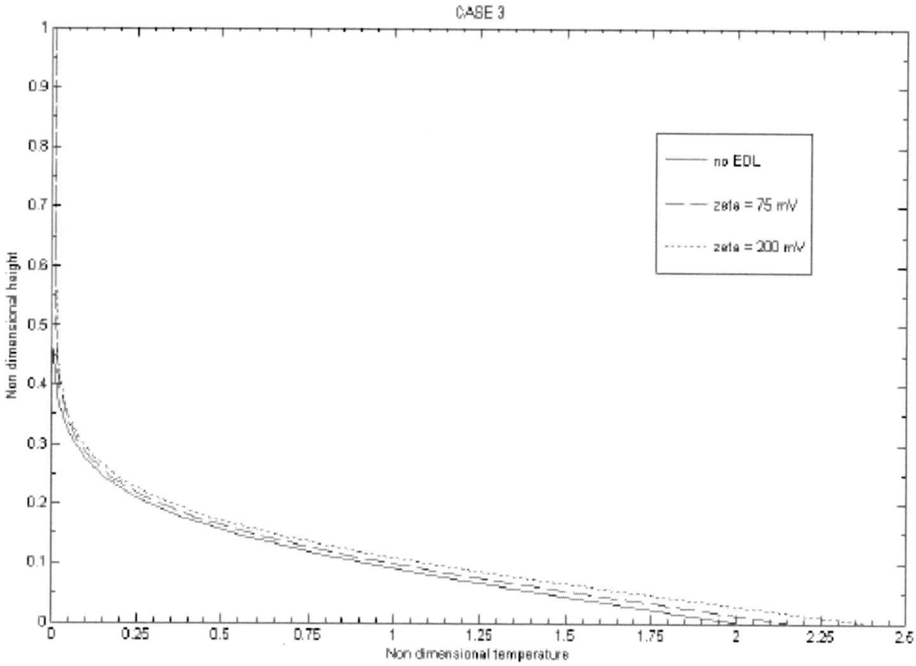


Figure 16. variation of dimensionless Temperature profile with respect to dimensionless height of channel for case 3.

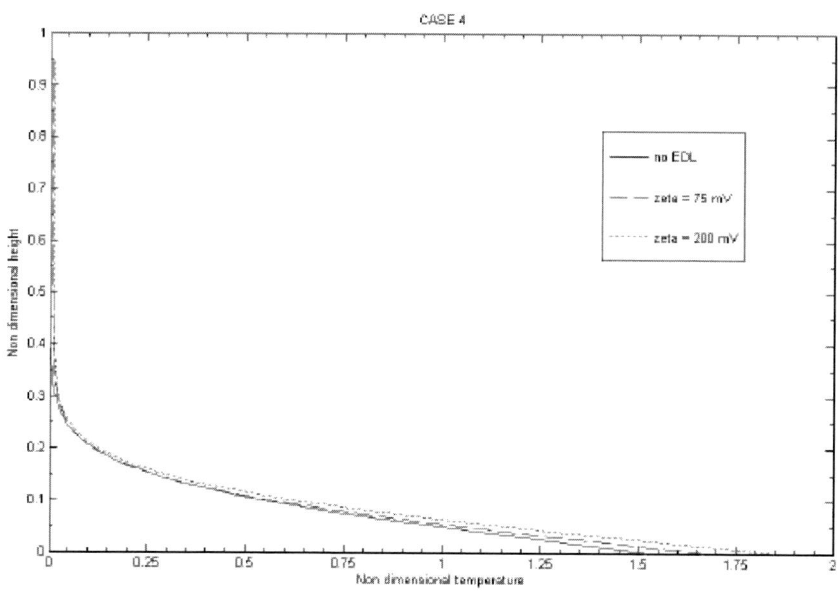


Figure 17. variation of dimensionless Temperature profile with respect to dimensionless height of channel for case 4.

**Table 9. The percentage of variation of dimensionless Temperature**

| Case | The percentage of variation of dimensionless Temperature affected by EDL | |
|---|---|---|
| | Zeta Potential (75mv) | Zeta Potential (200mv) |
| 0 | 8/8 | 26/6 |
| 1 | 3/9 | 9/8 |
| 2 | 4/7 | 12/6 |
| 3 | 5/1 | 13/1 |
| 4 | 6/4 | 14/8 |

We can say that; considering the effects of EDL is very necessary for exact solution of equations of motion in electrolyte fluid flow. Because of using of ionized liquid in practical manners, the effects of EDL are not negligible. For the cases that characteristic length of microchannel is comparable with the thickness of EDL or exist a high electric potential, the liquid flow and heat transfer characteristics are significantly affected by the presence of the EDL, and omission of the effects of EDL causes to a much deviates from the prediction of conventional theorems. Presence of EDL causes to an apparent viscosity that is much more than the viscosity of fluid, and increasing of zeta potential also causes to decreasing in volumetric flow rate. Similarly, presence of EDL leading to a large amount of pressure drop in microchannel heat sinks. As a result we can say that, existence of EDL causes to decreasing in efficiency of microchannel.

## 3. Porous Micro-Structure Approach

As proposed by Kim and Kim (1999), the MCHS can be modeled as a porous micro-structure. That is, the region between the cover plate and base plate(substructure) of the MCHS is modeled as a porous medium (figure 2). The governing equations for the fluid flow and heat transfer can be established by applying the volume-averaged technique. Since the one equation model is valid only when the fluid phase is in local equilibrium with the solid phase and is not suitable in the evaluation of MCHS thermal performance (Tsai and Chen, 2007), the two-equation model is employed. The governing equations for the fluid and energy transport under fully hydrodynamic and thermal development conditions can be written as

Fluid flow:

$$-\frac{dp_f}{dx}+\mu\frac{d^2\tilde{u}}{dy^2}-\frac{\mu}{K}\varepsilon\tilde{u}-\rho_f C\tilde{u}^2=0 \tag{35}$$

Energy in solid phase:

$$k_{se}\frac{\partial^2 T_s}{\partial y^2}=ha(T_s-\tilde{T}_f) \tag{36}$$

Energy in fluid phase:

$$\varepsilon(\rho c_p)_f u \frac{\partial \widetilde{T}_f}{\partial x} = ha(T_s - \widetilde{T}_f) + \frac{\partial}{\partial y}(k_{fe} \frac{\partial \widetilde{T}_f}{\partial y}) \tag{37}$$

In which the $k_{se}$ and $k_{fe}$ are defined as following

$$k_{se} = (1-\varepsilon)k_s \tag{39}$$

$$k_{fe} = \varepsilon k_f \tag{40}$$

In equations (35)–(37), $p$, $\widetilde{u}$, $\widetilde{T}_f$, $T_s$, $C$ and $h$ are the volume-averaged fluid pressure, fluid velocity, fluid temperature, solid temperature, inertia force and interfacial heat transfer coefficient between fluid and solid, respectively. The MCHS shown in figure 1, modeled as a porous medium, the porosity ε, permeability $K$ and wetted area per volume $a$ can be expressed as (Bejan, 1984).]

$$\varepsilon = \frac{w_c}{w_c + (H_t - H_c)}, \quad a = \frac{2}{w_c + (H_t - H_c)} \tag{41}$$

The boundary conditions for this problem are:

$$\widetilde{u} = 0 \text{ at } y = 0, H \tag{42-a}$$

$$\widetilde{T}_f = T_s = T_w \text{ at } y = 0 \tag{42-b}$$

$$\frac{\partial T_s}{\partial y} = \frac{\partial \widetilde{T}_f}{\partial y} = 0 \text{ at } y = H \tag{42-c}$$

To solve the governing equations (35)–(37), the permeability $K$ and the interstitial heat transfer coefficient $h$ should be determined in advance. For the present configuration, these parameters can be determined analytically through an approximation method proposed by Kim(2004). It is assumed that the characteristics of pressure drop across and heat transfer from the fins under consideration can be approximated as those found for the Poiseuille flow between two infinite parallel plates that are subject to a constant heat flux. Then the velocity and temperature distributions can be obtained easily as:

$$\widetilde{u} = \frac{w^2{}_c}{2\mu}(-\frac{dp}{dx})[(\frac{z}{w_c}) - (\frac{z}{w_c})^2] \tag{43}$$

$$\widetilde{T}_f = \frac{qw_c}{k_f}[-(\frac{z}{w_c})^4 + 2(\frac{z}{w_c})^3 - (\frac{z}{w_c})] + T_w \tag{44}$$

$$\widetilde{u} = \frac{1}{V_f}\int_{V_f} u dV = \frac{w_c^{\,2}}{12\mu}(-\frac{dp}{dx}) \tag{45}$$

$$\widetilde{T}_f = \frac{1}{V_f}\int_{V_f} T_f dV = T_w - \frac{w_c^{\,2}}{10\alpha}\widetilde{u}\frac{d\overline{T}_f}{dx} \tag{46}$$

In which the $\alpha$ is the aspect ratio of the channels.

## Conclusion

New developments on heat transfer and fluid flow computational techniques in different types of microchannel heat sinks was investigated. The numerical results of temperature distribution, thermal resistances and pressure drop were shown and the results had good consistency with experimental data.

## References

[1] Tsai, T., and Chein, R. (2007). Performance analysis of nanofluid-cooled microchannel heat sinks. *Int. J. Heat and Fluid Flow,* In press.

[2] Tuckerman, D.B. and Pease, R.F. (1981). High-performance heat sinking for VLSI. *IEEE Electronic Devices Letters EDL*, Vol. 2, pp. 126–129.

[3] Knight, R.W., Hall, D.J., Goodling, J.S., Jaeger, R.C. (1992). Heat sink optimization with application to microchannels. *IEEE Transactions on Components, Hybrids, and Manufacturing Technology,* Vol. 15, pp. 832–842.

[4] Ambatipudi, K.K., Rahman, M.M. (2000). Analysis of conjugate heat transfer in microchannel heat sinks. *Numerical Heat Transfer Part A, Vol.* 37, pp.711–731.

[5] Fedorov, A.G., Viskanta, R. (2000). Three-dimensional conjugate heat transfer in the microchannel heat sink for electronic packaging. *International Journal of Heat and Mass Transfer,* Vol. 43, pp.399–415.

[6] Lee, P., Garimella, S.V., Liu, D. (2005). Investigation of heat transfer in rectangular microchannels. *International Journal of Heat and Mass Transfer Vol.* 48, pp.1688–1704.

[7] Li, J., Peterson, G.P., Cheng, P. (2004). Three-dimensional analysis of heat transfer in a micro-heat sink with single phase flow. *Int. J. of Heat and Mass Transfer,* Vol. 47, pp.4215–4231.

[8] Li, J. and Peterson, G.P. (2006). Geometric optimization of a micro heat sink with liquid flow. *IEEE Transactions on Components and Packaging Technologies,* Vol. 29, pp.145–154.

[9] Amanifard, N. and Haghi, A. K. (2006). Numerical Investigation of fluid flow and heat transfer in microchannels. *Int. J. Heat and Technology,* Vol.1, pp.......

[10] Amanifard,N. Borji, M., Haghi, A. K. (2007). Heat Transfer in Porous Media. *Brazilian J. Chemical Engineering,* Vol.1, In press.

[11] Amanifard,N. and Haghi, A. K. (2007). Numerical Investigation of fluid flow and heat transfer in microchannels. *Int. J. Heat and Technology,* Vol.1, In press.

[12] Qu, W., Mudawar, I. (2002). Experimental and numerical study of pressure drop and heat transfer in a single-phase microchannel heat sink. *Int. J. of Heat and Mass Transfer,* Vol. 45, pp. 2549–2565.

[13] Tiselj, I., Hetsroni, G., Mavko, B., Mosyak, A., Pogrebnyak, E., Segal, Z. (2004). Effect of axial conduction on the heat transfer in microchannels. *Int. J. of Heat and Mass Transfer,* Vol.47, pp. 2551–2565.

[14] Lu, T. J. (1999). Heat transfer efficiency of metal honeycombs. *Int. J. Heat 7 Mass Transfer,* Vol.42, pp.2031-2040.

[15] Gu, T. J. and Lu, A. G., Evans. (2001). On the design of two dimensional cellar metals for combined heat dissipation and structural load capacity. *Int. J. Heat and Mass Transfer,* Vol. 44, pp. 2163-2175.

[16] Kohn, J.C.Y. and Colony, R. (1986). Heat transfer of microstructures for integrated circuits. *Int. J. Heat and Mass Transfer,* Vol. 13, pp. 89-98.

[17] Tien, C.L. and Kuo, S.M. (1987). Analysis of forced convection in microstructures for electronic system cooling. *Proceeding of Int. Symp. Cooling Technology for Electronic Equipment, Honolulu, HI,* pp. 217-226.

[18] Kim, S.J., Kim, D. (1999). Forced convection in microstructures for electronic equipment cooling. *ASME Journal of Heat Transfer.* 121, pp. 639–645.

[19] Knight, R.W., Hall, D.J., Goodling, J.S., Jaeger, R.C. (1992). Heat sink optimization with application to microchannels. *IEEE Transactions on Components, Hybrids, and Manufacturing Technology.* 15, pp. 832–842.

[20] Zhao, C.Y., Lu, T.J. (2002). Analysis of microchannel heat sinks for electronics cooling. *International Journal of Heat and Mass Transfer.* 45, pp. 4857–4869.

[21] Vafai, K. and Tien, C. L. (1981). Boundary and inertia effects on flow and heat transfer in porous media. *Int. J. Heat and Mass Transfer.* 43, pp. 195-203.

[22] Wu, P.Y. and Little, W.A. (1983). Measurement of friction factor for flow of gases in very fine channels used for microminiature, *Joule Thompson refrigerators,Cryogenics.* 24 (8), pp.273-277.

[23] Choi, S.B., Barren, R.R., Warrington, R.O. (1991). Fluid Flow and Heat Transfer in Microtubes. *ASME DSC.* 40 pp.89-93.

[24] Adams, T.M., Abdel-Khalik, S.I., Jeter, S.M., Qureshi, Z. H. (1998). AN Experimental investigation of single-Phase Forced Convection in Microchannels. *Int. J. Heat Mass Transfer.* 41 pp. 851-857.

[25] Peng, X.F., Peterson, G.P. (1996). Convective Heat Transfer and Flow Friction for Water Flow in Microchannel Structure. *Int. J. Heat Mass Transfer.* 36, pp. 2599-2608.

[26] Mala, G., Li, D., Dale, J.D. (1996). Heat Transfer and Fluid Flow in Microchannels. *J. Heat Transfer,* 40 pp.3079-3088.

[27] Tuckerman, D.B. (1984). Heat transfer microstructures for integrated circuits. Ph.D. thesis, Stanford University.

[28] Toh, K.C., Chen, X.Y., and Chai, J.C. (2002). Numerical Computation of Fluid Flow and Heat Transfer in Microchannels. *Int. J. Heat and Mass Transfer.* 45, pp. 5133–5141.

[29] Kim, S.J. (2004).Methods for thermal optimization of microchannel heat sinks. *Heat Transfer Eng.* 25, pp.37–49.

In: Monomers, Oligomers, Polymers, Composites...
Editors: R. A. Pethrick, G.E. Zaikov et al.

ISBN: 978-1-60456-877-6


*Chapter 11*

# IMAGE ANALYSIS OF PORE SIZE DISTRIBUTION IN ELECTROSPUN NANOFIBER WEBS: NEW TRENDS AND DEVELOPMENTS

***M. Ziabari, V. Mottaghitalab and A. K. Haghi****

The University of Guilan, P. O. Box 3756, Rasht, Iran

## ABSTRACT

Nanofibers produced by electrospinning method are widely used for drug delivery, as tissue scaffolding materials and filtration purposes where specific pore characteristics are required. For continued growth in these areas, it is critical that the nanofibers be properly designed for these applications to prevent failure. Most of the current methods only provide an indirect way of determining pore structure parameters and contain inherent disadvantages. In this study, we developed a novel image analysis method for measuring pore characteristics of electrospun nanofiber webs. 5 electrospun webs with different pore characteristics were analyzed by this method. The method is direct, so fast and presents valuable and comprehensive information regarding to pore structure parameters of the webs. Two sets of simulated images were generated to study the effects of web density, fiber diameter and its variations on pore characteristics. The results indicated that web density and fiber diameter significantly influence the pore characteristics whereas the effect of fiber diameter variations was insignificant.

## 1. INTRODUCTION

Fibers with a diameter of around 100 nm are generally classified as *nanofibers*. What makes nanofibers of great interest is their extremely small size. Nanofibers compared to conventional fibers, with higher surface area to volume ratios and smaller pore size, offer an opportunity for use in a wide variety of applications. To date, the most successful method of producing nanofibers is through the process of *electrospinning*. The electrospinning process

* Corresponding author E-Mail: Haghi@Guilan.ac.ir

uses high voltage to create an electric field between a droplet of polymer solution at the tip of a needle and a collector plate. When the electrostatic force overcomes the surface tension of the drop, a charged, continuous jet of polymer solution is ejected. As the solution moves away from the needle and toward the collector, the solvent evaporates and jet rapidly thins and dries. On the surface of the collector, a nonwoven web of randomly oriented solid nanofibers is deposited [1]-[5]. Figure 1 illustrates the electrospinning setup.

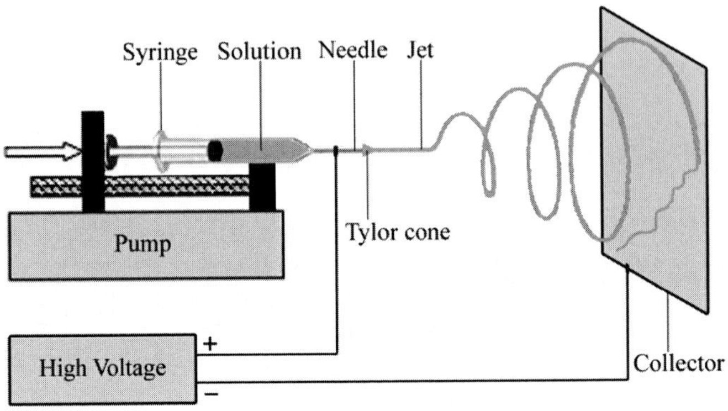


Figure 1. Electrospinning setup.

Material properties such as melting temperature and glass transition temperature as well as structural characteristics of nanofiber webs such as fiber diameter distribution, pore size distribution and fiber orientation distribution determine the physical and mechanical properties of the webs. The surface of electrospun fibers is important when considering end-use applications. For example, the ability to introduce porous surface features of a known size is required if nanoparticles need to be deposited on the surface of the fiber, if drug molecules are to be incorporated for controlled release, as tissue scaffolding materials and for acting as a cradle for enzymes [6]. Besides, filtration performance of nanofibers is strongly related to their pore structure parameters, i.e., percent open area (POA) and pore-opening size distribution (PSD). Hence, the control of the pore of electrospun webs is of prime importance for the nanofibers that are being produced for these purposes. There is no literature available about the pore size and its distribution of electrospun fibers and in this work, the pore size and its distribution was measured using an image analysis technique.

Current methods for determining PSD are mostly indirect and contain inherent disadvantages. Recent technological advancements in image analysis offer great potential for a more accurate and direct way of determining the PSD of electrospun webs. Overall, the image analysis method provides a unique and accurate method that can measure pore opening sizes in electrospun nanofiber webs.

## 2. METHODOLOGY

The porosity, $\varepsilon_V$, is defined as the percentage of the volume of the voids, $V_v$, to the total volume (voids plus constituent material), $V_t$, and is given by

$$\varepsilon_V = \frac{V_v}{V_t} \times 100$$

Similarly, the Percent Open Area (POA), $\varepsilon_A$, that is defined as the percentage of the open area, $A_o$, to the total area $A_t$, is given by

$$\varepsilon_A = \frac{A_o}{A_t} \times 100$$

Usually porosity is determined for materials with a three-dimensional structure, e.g. relatively thick nonwoven fabrics. Nevertheless, for two-dimensional textiles such as woven fabrics and relatively thin nonwovens it is often assumed that porosity and POA are equal [7].

The size of an individual opening can be defined as the surface area of the opening, although it is mostly indicated with a diameter called Equivalent Opening Size (EOS). EOS is not a single value, for each opening may differ. The common used term in this case is the diameter, $O_i$, corresponding with the equivalent circular area, $A_i$, of the opening.

$$O_i = (4A_i / \pi)^{1/2}$$

This diameter is greater than the side dimension of a square opening. A spherical particle with that diameter will never pass the opening (Figurea) and may therefore not be considered as an equivalent dimension or equivalent diameter. This will only be possible if the diameter corresponds with the side of the square area (Figureb). However, not all openings are squares, yet the equivalent square area of openings is used to determine their equivalent dimension because this simplified assumption results in one single opening size from the open area. It is the diameter of a spherical particle that can pass the equivalent square opening, hence the equivalent opening or pore size, $O_i$, results from

$$O_i = (A_i)^{1/2}$$

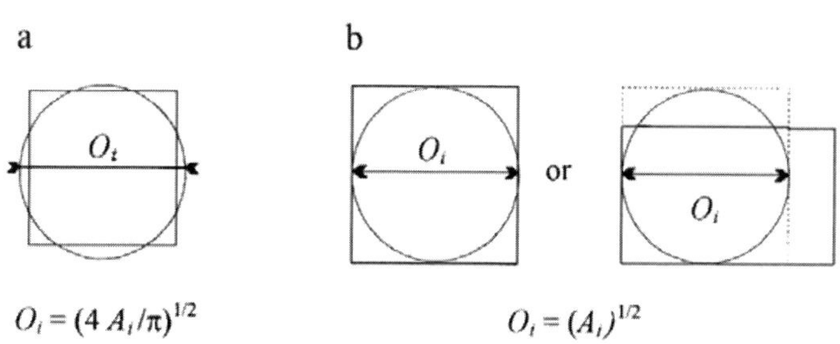


Figure 2. Equivalent opening size, $O_i$, based on (a) equivalent area, (b) equivalent size.

From the EOSs, Pore Size Distribution (PSD) and an equivalent diameter for which a certain percentage of the opening have a smaller diameter ($O_x$, pore opening size that $x$ percent of pores are smaller than that size) may be measured.

The PSD curves can be used to determine the uniformity coefficient, $C_u$, of the investigated materials. The uniformity coefficient is a measure for the uniformity of the openings and is given by

$$C_u = O_{60}/O_{10}$$

The ratio equals 1 for uniform openings and increases with decreasing uniformity of the openings [7].

Pore characteristic is one of the main tools for evaluating the performance of any nonwoven fabric and for electrospun webs as well. Understanding the link between processing parameters and pore structure parameters will allow for better control over the properties of electrospun fibers. Therefore there is a need for the design of nanofibers to meet specific application needs. Various techniques may be used to evaluate pore characteristics of porous materials including sieving techniques (dry, wet and hydrodynamic sieving), mercury porosimetry and flow porosimetry (bubble point method) [8], [9]. As one goes about selecting a suitable technique for characterization, the associated virtues and pitfalls of each technique should be examined. The most attractive option is a single technique which is non-destructive, yet capable of providing a comprehensive set of data [10].

### 2.1. Sieving Methods

In dry sieving, glass bead fractions (from finer to coarser) are sieved through the porous material. In theory, most of the glass beads from the first glass bead fraction should pass. As larger and larger glass bead fractions are sieved, more and more glass beads should become trapped within and on top of the material. The number of pores of a certain size should be reflected by the percentage of glass beads passing through the porous material during each glass bead fraction sieved; however, electrostatic effects between glass beads and between glass beads and the material can affect the results. Glass beads may stick to fibers making the pores effectively smaller and they may also agglomerate to form one large glass bead that is too large to pass through the any of the pores. Glass beads may also break from hitting each other and the sides of the container, resulting in smaller particles that can pass through smaller openings.

In hydrodynamic sieving, a glass bead mixture is sieved through a porous material under alternating water flow conditions. The use of glass bead mixtures leads to results that reflect the original glass bead mixture used. Therefore, this method is only useful for evaluating the large pore openings such as $O_{95}$. Another problem occurs when particles of many sizes interact, which likely results in particle blocking and bridge formation. This is especially a problem in hydrodynamic sieving because the larger glass bead particles will settle first when water is drained during the test. When this occurs, fine glass beads which are smaller than the pores are prevented from passing through by the coarser particles.

In wet sieving, a glass bead mixture is sieved through a porous material aided by a water spray. The same basic mechanisms that occur when using the hydrodynamic sieving method also take place when using the wet sieving method. Bridge formation is not as pronounced in the wet sieving method as in the hydrodynamic sieving method; however, particle blocking and glass bead agglomeration are more pronounced [8], [9].

The sieving tests are very time-consuming. Generally 2 hours are required to perform a test. The sieving tests are far from providing a complete PSD curve because the accuracy of the tests for pore sizes smaller than 90 μm is questionable [12].

## 2.2. Mercury Porosimetry

Mercury porosimetry is a well known method which is often used to study porous materials. This technique is based on the fact that mercury as a non-wetting liquid does not intrude into pore spaces except under applying sufficient pressure. Therefore, a relationship can be found between the size of pores and the pressure applied.

In this method, a porous material is completely surrounded by mercury and pressure is applied to force the mercury into pores. As mercury pressure increases the large pores are filled with mercury first. Pore sizes are calculated as the mercury pressure increases. At higher pressures, mercury intrudes into the fine pores and when the pressure reaches a maximum, total open pore volume and porosity are calculated.

The mercury porosimetry thus gives a PSD based on total pore volume and gives no information regarding the number of pores of a porous material. Pore sizes ranging from 0.0018 to 400 $\mu m$ can be studied using mercury porosimetry. Pore sizes smaller than 0.0018 $\mu m$ are not intruded with mercury and this is a source of error for porosity and PSD calculations. Furthermore, mercury porosimetry does not account for closed pores as mercury does not intrude into them. Due to applying high pressures, sample collapse and compression is possible, hence it is not suitable for fragile compressible materials such as nanofiber sheets. Other concerns would include the fact that it is assumed that the pores are cylindrical, which is not the case in reality. After the mercury intrusion test, sample decontamination at specialized facilities is required as the highly toxic mercury is trapped within the pores. Therefore this dangerous and destructive test can only be performed in well-equipped labs [6], [8], [9].

## 2.3. Flow Porosimetry (Bubble Point Method)

The flow porosimetry is based on the principle that a porous material will only allow a fluid to pass when the pressure applied exceeds the capillary attraction of the fluid in largest pore. In this test, the specimen is saturated with a liquid and continuous air flow is used to remove liquid from the pores. At a critical pressure, the first bubble will come through the largest pore in the wetted specimen. As the pressure increases, the pores are emptied of liquid in order from largest to smallest and the flow rate is measured. PSD, number of pores and porosity can be derived once the flow rate and the applied pressure are known. Flow porosimetry is capable of measuring pore sizes within the range of 0.013–500 $\mu m$.

As the air only passes through the through pores, characteristics of these pores are measured while those of closed and blind pores are omitted. Many times, 100% total flow is not reached. This is due to porewick evaporation from the pores when the flow rate is too high. Extreme care is required to ensure the air flow does not disrupt the pore structure of the specimen. The flow porosimetry method is also based on the assumption that the pores are cylindrical, which is not the case in reality. Finding a liquid with low surface tension which could cover all the pores, has no interaction with the material and does not cause swelling in material is not easy all the times and sometimes is impossible [6], [8], [9].

### 2.4. Image Analysis

Because of its convenience to detect individual pores in a nonwoven image, it seemed to be advantageous to use image analysis techniques for pore measurement. Image analysis was used to measure pore characteristics of woven [11] and nonwoven geotextiles [12]. In the former, successive *erosion* operations with increasing size of *structuring element* was used to count the pore openings larger than a given structuring element. The main purpose of the erosion was to simulate the conditions in the sieving methods. In this method, the voids connected to border of the image which are not complete pores are considered in measurement. Performing opening and then closing operations proceeding pore measurement cause the pore sizes and shapes deviate from the real ones. The method is suitable for measuring pore sizes of woven geotextiles with fairly uniform pore sizes and shapes and is not appropriate for electrospun nanofiber webs of different pore sizes.

In the later case, cross sectional image of nonwoven geotextile was used to calculate the pore structure parameters. A *slicing* algorithm based on a series of morphological operations for determining the mean fiber thickness and the optimal position of the uniform slicing grid was developed. After recognition of the fibers and pores in the slice, the pore opening size distribution of the cross sectional image may be determined. The method is useful for measuring pore characteristics of relatively thick nonwovens and cannot be applied to electrospun nanofiber webs due to extremely small size.

Therefore, there is a need for developing an algorithm suitable for measuring the pore structure parameters in electrospun webs. In response to this need, we have developed a new image analysis based method and presented in the following.

In this method, a binary image of the web is used as an input. First of all, voids connected to the image border are identified and cleared using *morphological reconstruction* [13] where mask image is the input image and marker image is zero everywhere except along the border. Total area which is the number of pixels in the image is measured. Then the pores are labeled and each considered as an object. Here the number of pores may be obtained. In the next step, the number of pixels of each object as the area of that object is measured. Having the area of pores, the porosity and EOS regarding to each pore may be calculated. The data in pixels may then be converted to *nm*. Finally PSD curve is plotted and $O_{50}$, $O_{95}$ and $C_u$ are determined.

#### *2.4.1. Real Webs*

In order to measure pore characteristics of electrospun nanofibers using image analysis, images of the webs are required. These images called micrographs usually are obtained by Scanning Electron Microscope (SEM), Transmission Electron Microscope (TEM) or Atomic Force Microscope (AFM). The images must be of high-quality and taken under appropriate magnifications.

The image analysis method for measuring pore characteristics requires the initial segmentation of the micrographs in order to produce binary images. This is a critical step because the segmentation affects the results dramatically. The typical way of producing a binary image from a grayscale image is by *global thresholding* [13]] where a single constant threshold is applied to segment the image. All pixels up to and equal to the threshold belong to object and the remaining belong to the background. One simple way to choose the threshold is picking different thresholds until one is found that produces a good result as judged by the observer. Global thresholding is very sensitive to any inhomogeneities in the gray-level distributions of object and background pixels. In order to eliminate the effect of inhomogeneities, *local thresholding* scheme [13]] could be used. In this approach, the image is divided into subimages where the inhomogeneities are negligible. Then optimal thresholds are found for each subimage. A common practice in this case, which is used in this study, is to preprocess the image to compensate for the illumination problems and then apply a global thresholding to the preprocessed image. It can be shown that this process is equivalent to segment the image with locally varying thresholds. In order to automatically select the appropriate thresholds, *Otsu's method* [14]] is employed. This method chooses the threshold to minimize interaclass variance of the black and white pixels. As it is shown in Figure 7, global thresholding resulted in some broken fiber segments. This problem was solved using local thresholding. Note that, since the process is extremely sensitive to noise contained in the image, before the segmentation, a procedure to clean the noise and enhance the contrast of the image is necessary.

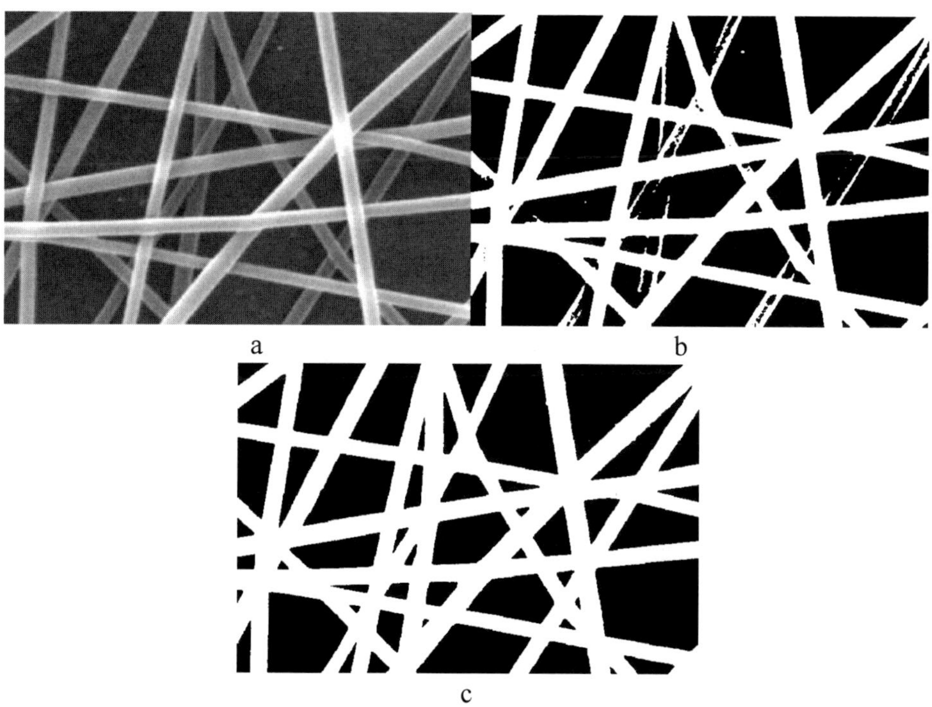


Figure 3. a) A real web, b) Global thresholding, c) Local thresholding.

### *2.4.2. Simulated Webs*

In is known that the pore characteristics of nonwoven webs are influenced by web properties and so are those of electrospun webs. There are no reliable models available for predicting these characteristics as a function of web properties [15]. In order to explore the effects of some parameters on pore characteristics of electrospun nanofibers, simulated webs are generated. These webs are images simulated by straight lines. There are three widely used methods for generating random network of lines. These are called S-randomness, μ-randomness (suitable for generating a web of continuous filaments) and I-randomness (suitable for generating a web of staple fibers). These methods have been described in details by Abdel-Ghani et al. [16] and Pourdeyhimi et al. [17]. In this study, we used *μ-randomness* procedure for generating simulated images. Under this scheme, a line with a specified thickness is defined by the perpendicular distance *d* from a fixed reference point *O* located in the center of the image and the angular position of the perpendicular *α*. Distance *d* is limited to the diagonal of the image. Figure 2 demonstrates this procedure.

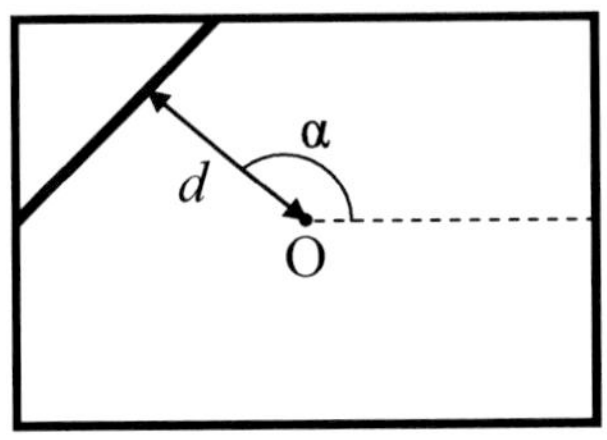


Figure 4. Procedure for μ-randomness.

One of the most important features of simulation is that it allows several structural characteristics to be taken into consideration with the simulation parameters. These parameters are: web density (controlled as line density), angular density (sampled from a normal or random distribution), distance from the reference point (sampled from a random distribution), line thickness (sampled from a normal distribution) and image size.

## 3. Experimental

Nanofiber webs were obtained from electrospinning of PVA with average molecular weight of 72000 g/mol (MERCK) at different processing parameters for attaining different pore characteristics. Table 1summarizes the electrospinning parameters used for preparing the webs. The micrographs of the webs were obtained using Philips (XL-30) environmental Scanning Electron Microscope (SEM) under magnification of 10000X after being gold coated. Figure 8 shows the micrographs of the electrospun webs.

**Table 1. Electrospinning parameters used for preparing nanofiber webs**

| No. | Concentration (%) | Spinning Distance (Cm) | Voltage (KV) | Flow Rate (ml/h) |
|---|---|---|---|---|
| 1 | 8 | 15 | 20 | 0.4 |
| 2 | 12 | 20 | 15 | 0.2 |
| 3 | 8 | 15 | 20 | 0.2 |
| 4 | 8 | 10 | 15 | 0.3 |
| 5 | 10 | 10 | 15 | 0.2 |

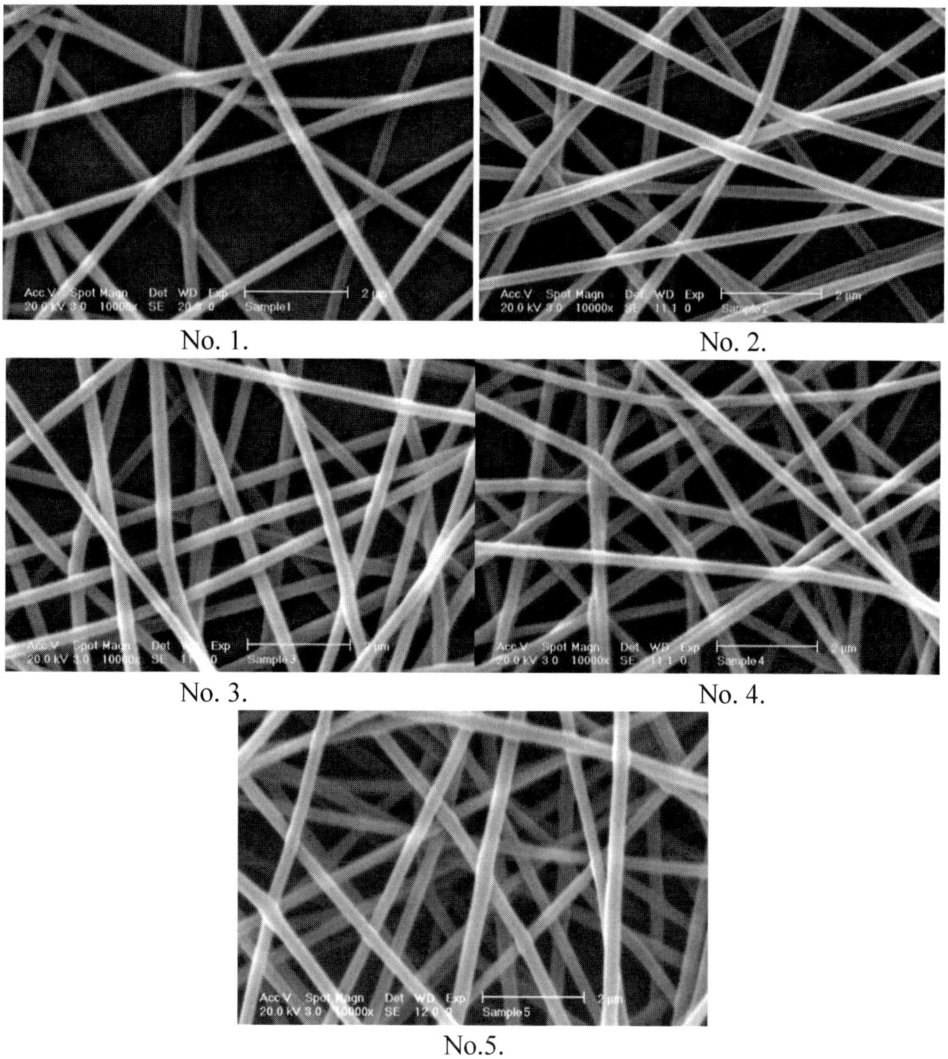


Figure 5. Micrographs of the electrospun webs.

## 4. Results and Discussion

Due to previously mentioned reasons, sieving methods and mercury porosimetry are not applicable for measuring pore structure parameters in nano-scale. The only method which seems to be practical is flow porosimetry. However, since in this study, the nanofibers were made of PVA, finding an appropriate liquid for the test to be performed is almost impossible because of solubility of PVA in both organic and inorganic liquids.

As an alternative, image analysis was employed to measure pore structure parameters in electrospun nanofiber webs. PSD curves of the webs, determined using the image analysis method, are shown in Figure 6. Pore characteristics of the webs ($O_{50}$, $O_{95}$, $C_u$, number of pores, porosity) measured by this method are presented in Table 2. It is seen that decreasing the porosity, $O_{50}$ and $O_{95}$ decrease. $C_u$ also decreases with respect to porosity, that's to say increasing the uniformity of the pores. Number of pores has an increasing trend with decreasing the porosity.

The image analysis method presents valuable and comprehensive information regarding to pore structure parameters in nanofiber webs. This information may be exploited in preparing the webs with needed pore characteristics to use in filtration, biomedical applications, nanoparticle deposition and other purposes. The advantages of the method are listed below:

1. The method is capable of measuring pore structure parameters in any nanofiber webs with any pore features and it is applicable even when other methods may not be employed.
2. It is so fast. It takes less than a second for an image to be analyzed (with a 3 GHz processor).
3. The method is direct and so simple. Pore characteristics are measured from the area of the pores which is defined as the number of pixels of the pores.
4. There is no systematic error in measurement (such as assuming pores to be cylindrical in mercury and flow porosimetry and the errors associated with the sieving methods which were mentioned). Once the segmentation is successful, the pore sizes will be measured accurately. The quality of images affects the segmentation procedure. High-quality images reduce the possibility of poor segmentation and enhance the accuracy of the results.
5. It gives a complete PSD curve.
6. There is no cost involved in the method and minimal technical equipments are needed (SEM for obtaining the micrographs of the samples and a computer for analysis).
7. It has the capability of being used as an on-line quality control technique for large scale production.
8. The results obtained by image analysis are reproducible.
9. It is not a destructive method. A very small amount of sample is required for measurement.

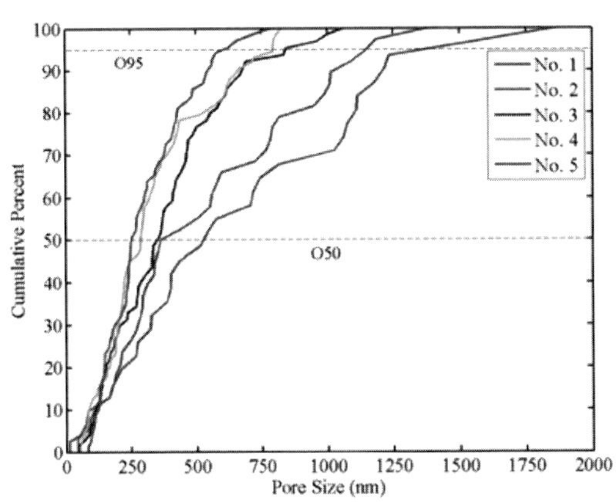


Figure 6. PSD curves of electrospun webs.

**Table 2. Pore characteristics of electrospun webs**

| No. | $O_{50}$ | | $O_{95}$ | | $C_u$ | Pore No. | Porosity |
|---|---|---|---|---|---|---|---|
| | pixel | nm | pixel | nm | | | |
| 1 | 39.28 | 513.9 | 94.56 | 1237.1 | 8.43 | 31 | 48.64 |
| 2 | 27.87 | 364.7 | 87.66 | 1146.8 | 5.92 | 38 | 34.57 |
| 3 | 26.94 | 352.5 | 64.01 | 837.4 | 3.73 | 64 | 26.71 |
| 4 | 22.09 | 289.0 | 60.75 | 794.8 | 3.68 | 73 | 24.45 |
| 5 | 19.26 | 252.0 | 44.03 | 576.1 | 2.73 | 69 | 15.74 |

In an attempt to establish the effects of some structural properties on pore characteristics of electrospun nanofibers, two sets of simulated images with varying properties were generated. The simulated images reveal the degree to which fiber diameter and density affect the pore structure parameters. The first set contained images with the same density varying in fiber diameter and images with the same fiber diameter varying in density. Each image had a constant diameter. The second set contained images with the same density and mean fiber diameter while the standard deviation of fiber diameter varied. The details are given in Table 3 and Table 4. Typical images are shown in figure 7 and figure 8..

**Table 3. Structural characteristics of first set images**

| No. | Angular Range | Line Density | Line Thickness |
|---|---|---|---|
| 1 | 0-360 | 20 | 5 |
| 2 | 0-360 | 30 | 5 |
| 3 | 0-360 | 40 | 5 |
| 4 | 0-360 | 20 | 10 |
| 5 | 0-360 | 30 | 10 |
| 6 | 0-360 | 40 | 10 |
| 7 | 0-360 | 20 | 20 |
| 8 | 0-360 | 30 | 20 |
| 9 | 0-360 | 40 | 20 |

**Table 4. Structural characteristics of second set images**

| No. | Angular Range | Line Density | Line Thickness | |
|---|---|---|---|---|
| | | | Mean | Std |
| 1 | 0-360 | 30 | 15 | 0 |
| 2 | 0-360 | 30 | 15 | 4 |
| 3 | 0-360 | 30 | 15 | 8 |
| 4 | 0-360 | 30 | 15 | 10 |

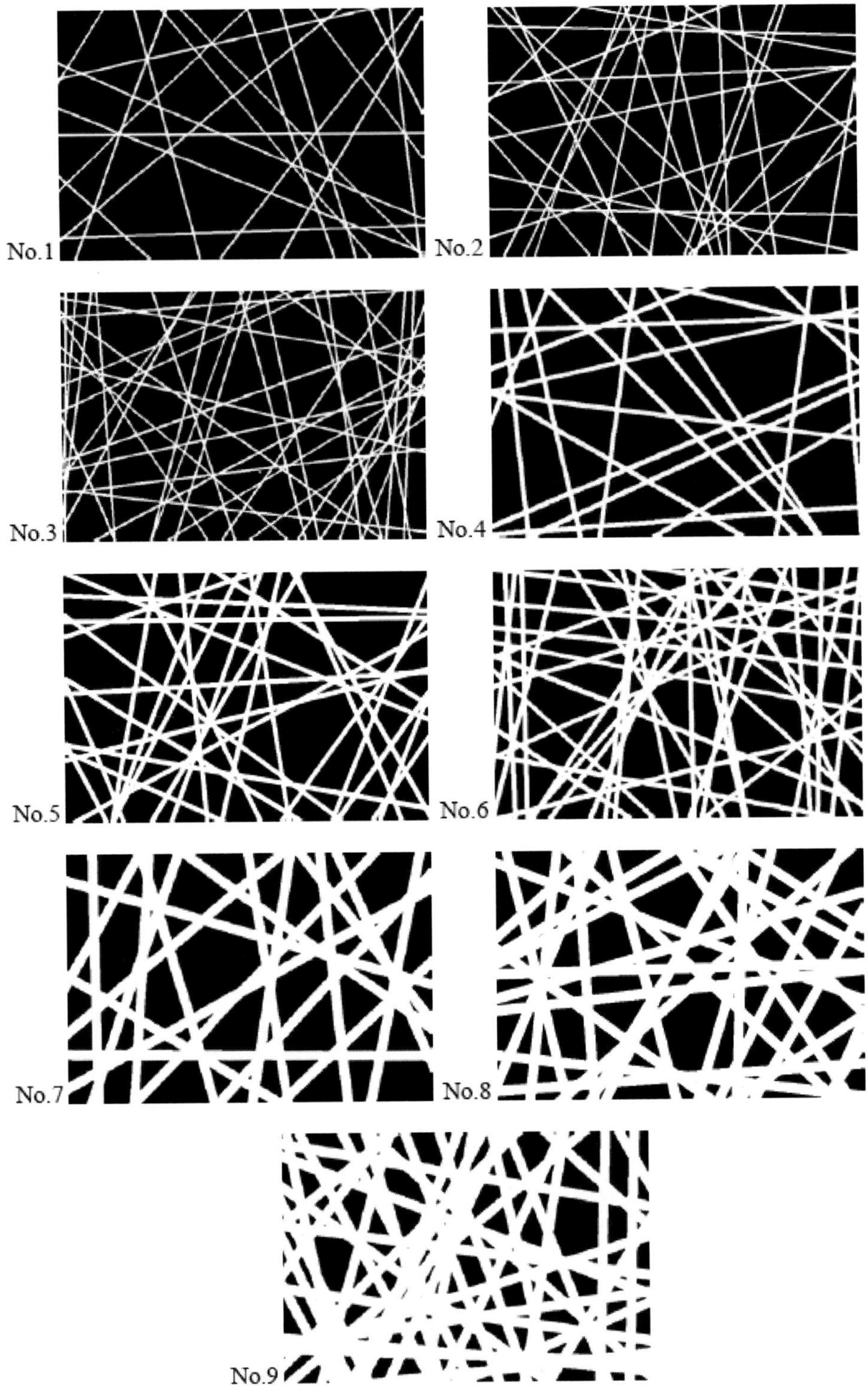


Figure 7. Simulated images of the first set.

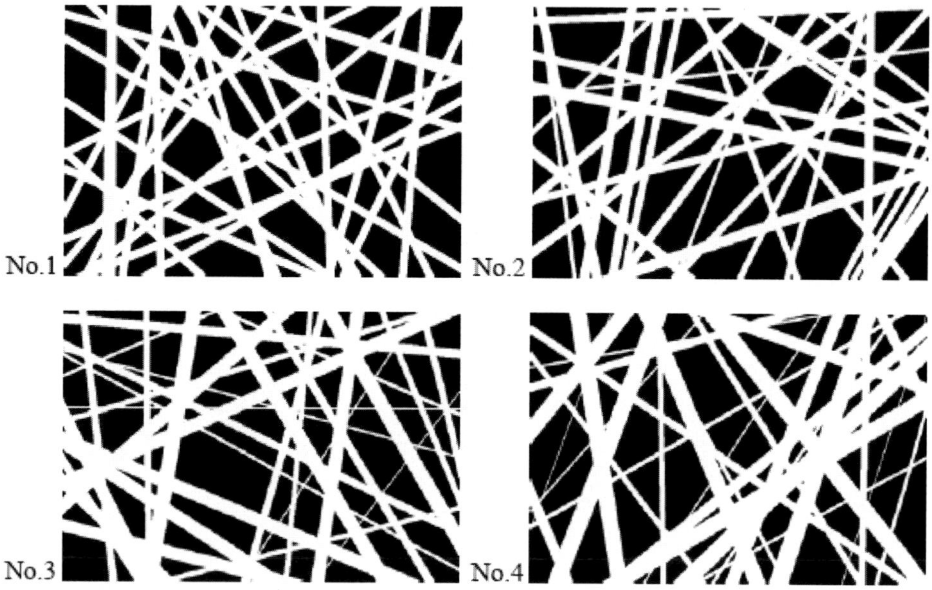


Figure 8. Simulated images of the second set.

Pore structure parameters of the simulated webs were measured using image analysis method. Table 5 summarizes the pore characteristics of the simulated images in the first set. For the webs with the same density, increasing fiber diameter resulted in a decrease in $O_{95}$, number of pores and porosity. No particular trends were observed for $O_{50}$ and $C_u$. Figure 9 and figure 10 show the PSD curves of the simulated images in the first set. As the web density increases, the effects of fiber diameter are less pronounced since the PSD curves of the webs become closer to each other. For the webs with the same fiber diameter, increasing the density resulted in a decrease in $O_{50}$, $O_{95}$, $C_u$ and porosity whereas number of pores increased with the density.

**Table 5. Pore characteristics of the first set of simulated images**

| No. | $O_{50}$ | $O_{95}$ | $C_u$ | Pore No. | Porosity |
|---|---|---|---|---|---|
| 1 | 27.18 | 100.13 | 38.38 | 84 | 79.91 |
| 2 | 15.52 | 67.31 | 22.20 | 182 | 71.78 |
| 3 | 13.78 | 52.32 | 18.71 | 308 | 69.89 |
| 4 | 36.65 | 94.31 | 43.71 | 67 | 66.10 |
| 5 | 17.89 | 61.64 | 22.67 | 144 | 53.67 |
| 6 | 12.41 | 51.60 | 16.70 | 245 | 47.87 |
| 7 | 24.49 | 86.90 | 33.11 | 58 | 41.05 |
| 8 | 16.31 | 56.07 | 21.66 | 108 | 32.53 |
| 9 | 13.11 | 45.38 | 17.75 | 126 | 22.01 |

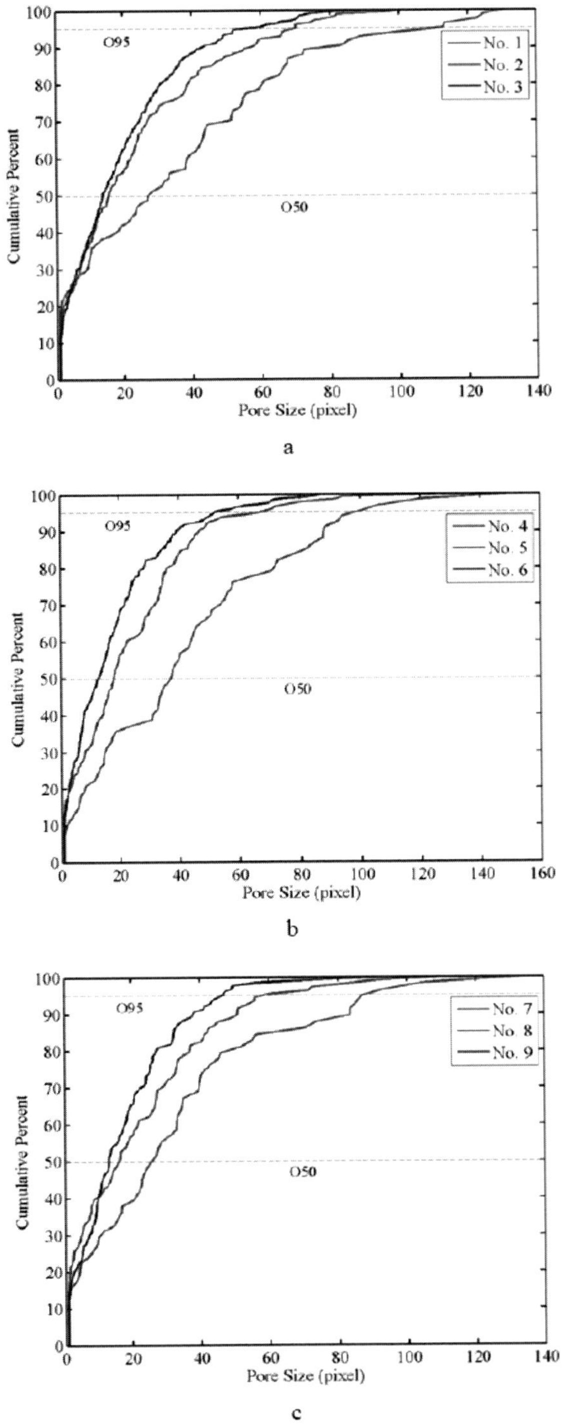


Figure 9. PSD curves of the first set of simulated images; effect of density, images with the diameter of a) 5, b) 10, c) 20 pixels.

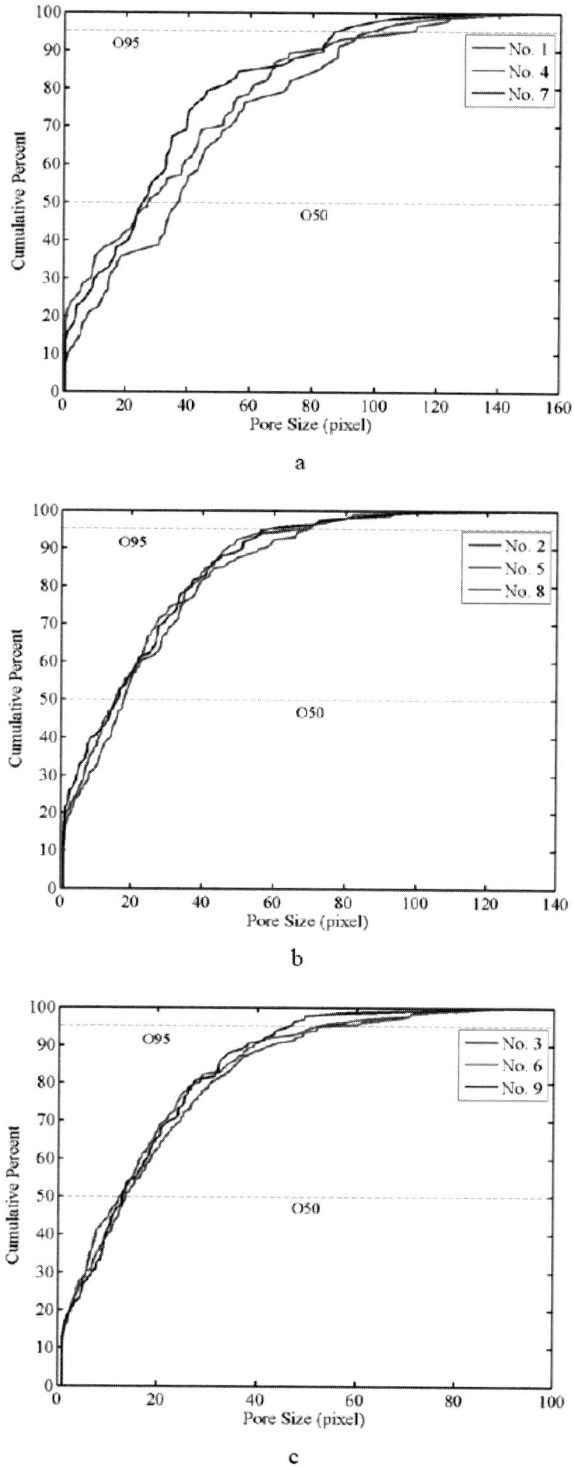


Figure10. PSD curves of the first set of simulated images; effect of fiber diameter, images with the density of a) 20, b) 30, c) 40 lines.

Table 6 summarizes the pore characteristics of the simulated images in the second set set. No significant effects for variation of fiber diameter on pore characteristics were observed. Suggesting that average fiber diameter is determining factor not variation of diameter. Figure 11 shows the PSD curves of the simulated images in the second set.

**Table 6. Pore characteristics of the second set of simulated images**

| No. | $O_{50}$ | $O_{95}$ | $C_u$ | Pore No. | Porosity |
|---|---|---|---|---|---|
| 1 | 14.18 | 53.56 | 18.79 | 133 | 35.73 |
| 2 | 13.38 | 61.66 | 20.15 | 136 | 41.89 |
| 3 | 18.14 | 59.35 | 22.07 | 121 | 41.03 |
| 4 | 15.59 | 62.71 | 20.20 | 112 | 37.77 |

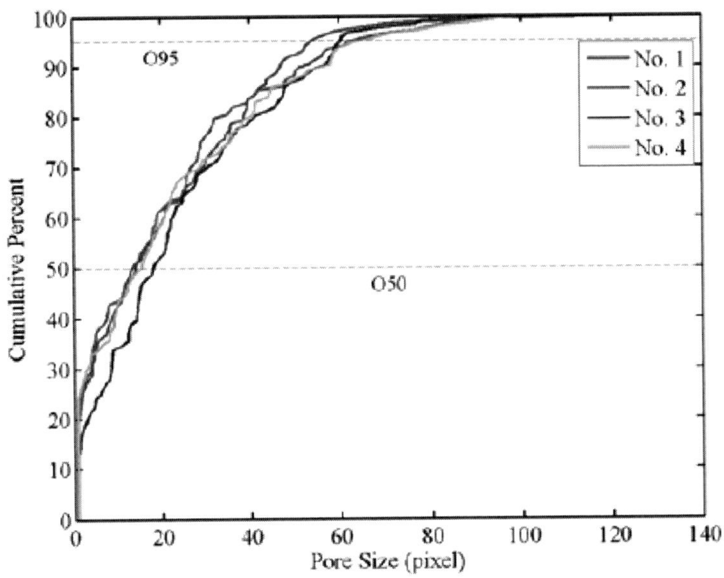


Figure 11. PSD curves of the second set of simulated images, the effect of fiber diameter variation.

## 5. Conclusion

The evaluation of electrospun nanofiber pore structure parameters is necessary as it facilitates the improvement of the design process and its eventual applications. Various techniques have been developed to assess pore characteristics in porous materials. However, most of these methods are indirect, have inherent problems and are not applicable for measuring pore structure parameters of electrospun webs. In this investigation, we have successfully developed an image analysis based method as a response to this need. The method is simple, comprehensive and so fast and directly measures the pore structure parameters.

The effects of web density, fiber diameter and its variation on pore characteristics of the webs were also explored using some simulated images. As fiber diameter increased, $O_{95}$, number of pores and porosity decreased. No particular trends were observed for $O_{50}$ and $C_u$. Increasing the density resulted in a decrease in $O_{50}$, $O_{95}$, $C_u$ and porosity whereas number of pores increased with the density. The effects of variation of fiber diameter on pore characteristics were insignificant.

## REFERENCES

[1] A. K. Haghi, M. Akbari, Trends in Electrospinning of Natural Nanofibers, *Physica Status Solidi (a),* 204, 1830-1834 (2007).

[2] D. H. Reneker, I. Chun, Nanometre Diameter Fibers of Polymer, Produced by Electrospinning, *Nonotechnology,* 7, 216-223 (1996).

[3] D. R. Salem, Structure Formation in polymeric Fibers, Hanser, Cincinnati , Chapter 6, H. Fong, , D. H. Reneker, *Electrospinning and the Formation of Nanofibers.* (2001).

[4] Th. Subbiah, G. S. Bhat, R. W. Tock, S. Parameswaran, S. S. Ramkumar, Electrospinning of Nanofibers, *Journal of Applied Polymer Science,* 96, 557-569 (2005).

[5] A. Frenot, I. S. Chronakis, Polymer Nanofibers Assembled by Electrsopinning, *Current Opinion in Colloid and Interface Science,* 8, 64-75 (2003).

[6] Ch. L. Casper, J. S. Stephens, N. G. Tassi, D. B. Chase, J. F. Rabolt, Controlling Surface Morphology of Electrospun Polystyrene Fibers: Effect of Humidity and Molecular Weight in the Electrospinning Process, *Macromolecules,* 37, 573-578 (2004).

[7] W. Dierickx, Opening Size Determination of Technical Textiles Used in Agricultural Applications, *Geotextiles and Geomembranes,* 17 (4), 231–245 (1999).

[8] S. K. Bhatia, J. L. Smith, Geotextile Characterization and Pore Size Distribution: Part II. *A Review of Test Methods and Results, Geosynthetics International,* 3 (2), 155–180 (1996).

[9] S. K. Bhatia, J. L. Smith, B. R. Christopher, Geotextile Characterization and Pore Size Distribution: Part III. Comparison of Methods and Application to Design, Geosynthetics International, 3 (3), 301–328 (1996).

[10] S. T. Ho, D. W. Hutmacher, A Comparison of Micro CT with Other Techniques Used in the Characterization of Scaffolds, *Biomaterials,* 27, 1362-1376 (2006).

[11] A. H. Aydilek, T. B. Edil, Evaluation of Woven Geotextile Pore Structure Parameters Using Image Analysis, *Geotechnical Testing Journal,* 27 (1), 1-12 (2004).

[12] A. H. Aydilek, S. H. Oguz, T. B. Edil, Digital Image Analysis to Determine Pore Opening Size Distribution of Nonwoven Geotextiles, *Journal of Computing in Civil engineering,* 280-290 (2002).

[13] R. C. Gonzalez, R. E. Woods, Digital Image Processing, Prentice Hall, New Jersey, Second Edition (2001).

[14] B. Jähne, Digital Image Processing, Springer, England, 5th Revised and Extended Edition (2002).

[15] H. S. Kim, B. Pourdeyhimi, A Note on the Effect of Fiber Diameter, Fiber Crimp and Fiber Orientation on Pore Size in Thin Webs, *International Nonwoven Journal,* 15-19 (Winter 2000).

[16] M. S. Abdel-Ghani, G. A. Davis, Simulation of Nonwoven Fiber Mats and the Application to Coalescers, *Chemical Engineering Science,* 117 (1985).

[17] B. Pourdeyhimi, R. Ramanathan, R. Dent, Measuring Fiber Orientation in Nonwovens, *Part I: Simulation, Textile Research Journal,* 66 (11), 713-722 (1996).

In: Monomers, Oligomers, Polymers, Composites... ISBN: 978-1-60456-877-6
Editors: R. A. Pethrick, G.E. Zaikov et al. 

*Chapter 12*

# INTERPOLYMERIC ASSOCIATIONS BETWEEN ALGINIC ACID AND POLY (N-ISOPROPYLACRYLAMIDE), POLY (ETHYLENE GLYCOL) AND POLYACRYLAMIDE

***Catalina Natalia Duncianu and Cornelia Vasile***
"Petru Poni" Institute of Macromolecular Chemistry,
41 A, Gr.Ghica Voda Alley, 700487, Iasi, Romania

## INTRODUCTION

It is well-known that hydrogen bonding and polyelectrolyte complexes are two categories of intermacromolecular (interpolymeric) associations depending on the type of the interacting forces between polymeric constituents. H-bonding complexes result by the interaction between a proton donating polymer (weak polyacid) and a proton accepting polymer (a weak Lewis polybase) via hydrogen bonds. They can be formed between the protons low density region and a high electron density charge region. The proton acceptor polymers can interact with proton donor polymers in aqueous solution or organic solvents [1, 2]. The interest for these interpolymeric complexes (IPC) can be explained because of their unique physical and chemical properties in comparison with pure components. The hydrogen-bonded IPCs have attracted a great attention of pharmaceutical scientists due to the wide possibilities of their use in the development of different drug formulations [3, 4, 5, 6, 7, 8]

It is well-known that IPC formation and stabilization are due to the cooperative hydrophilic and hydrophobic effects and depend on: the composition of the system, structure of the polymers, concentration of polymer solution, pH, temperature, ionic strength, etc. [1, 3, 9, 10].

Several authors reported IPC's formation between the proton donating polymer (e.g. poly(acrylic acid)- PAA, poly(methacrylic acid)-PMMA), and the proton accepting polymer (e.g. poly (ethylene glycol)- PEG, poly (acrylamide)- PAM, poly(N-isopropylacrylamide)-PNIPAM) in aqueous solutions. [11, 12, 13, 14] or polyacid copolymers with different polybases [15, 16]. Only few studies deal with the IPC formation using a polysaccharide as a weak polybase [17, 18, 19, 20]. They found that by using a polycarboxylic acid as a proton donor polymer, the complex formation will depend on pH of the solution and, of course, on the dissociation degree of the polyacid. The dissociation of the acid is crossed out in the presence of the Lewis base (e.g. PEO) and can be estimated via an apparent dissociation constant (Kd). The existence of an amount of undissociated carboxylic groups (COOH) is a condition for obtaining a stable complex through hydrogen bonds. Thus, this condition is accomplished at a certain pH when the complex can be irreversibly formed. During the last decade, the researchers interest have been caught by the intermacromolecular associations between non-complementary polymers containing variable number of functional groups on the mole unit like polysaccharides, random, alternant [21] or graft-copolymers as they can be better simulate the intermolecular interactions between natural complex macromolecules. In addition to H-bonding, hydrophobic interactions contribute to the formation and stabilization of the IPCs [22]

Using a natural polymer as weak polyacid e.g. alginic acid to obtain interpolymeric complexes mainly with chitosan represents an attractive target for many researchers especially due to its applications in the drug delivery systems [23, 24] and pharmaceutics [25]

In the present work, it has been studied the interpolymeric associations between alginic acid (AgA) with poly (N-isopropylacrylamide) (PNIPAM), poly (ethylene glycol) (PEG) or poly (acrylamide) (PAM) by using the following methods: viscosity, pH-metry and conductivity measurements.

## 2. Experimental

### 2.1. Materials

Alginic acid is a natural hydrophilic polysaccharide which is extracted from different brown seaweeds (*Macrocystis pyrifera, Ascophyllum nodosum*).

Alginic acid can be found in three different structures depending on the type of the seaweed extracted from. One of it's structure can have only blocks of L- Guluronic acid (G), a second one can contain only segments of D- Manuronic acid (M) and the last one can have an alternated structure between D- Manuronic and L- Guluronic units (scheme 1).

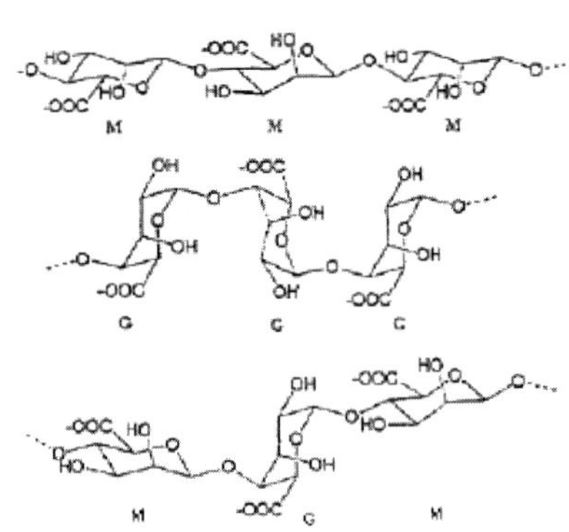

Scheme 1. Structural formulas of alginic acid (AgA) [26, 27].

The commercial alginic acid used in this study is a Fluka product with an average molecular weight of 48,000-186,000 [9005-32-7] (according to Fluka specifications), the reduced viscosity in water at 25 °C for an aqueous solution of c= 0.2 wt % is $\eta_{red}$ =2.41 ml*g$^{-1}$, a drying loss ≤ 10 wt % and ash ≤ 3%.

***Poly (N-Isopropylacrylamide), PNIPAM***

The polymers with a low critical solubility temperature (LCST) presents a great interest in green chemistry field, catalysts, drug delivery, sensors, permeability control in composites used as films. PNIPAM is the most representative of this class of polymers because it shows a phase transition at a low critical solubility temperature at about 31- 33 °C, close to physiological temperature [34]. Aqueous solutions based on PNIPAM, are the best examples of thermo-sensitive materials. It can be observed that it precipitates above LCST and it re-dissolves below it. It has been shown that, in the case of PNIPAM, the behavior is due to hydrophobic N-alkyl groups (see scheme 3) and solvent type.

LCST is the result of the variation of entropy at the dewatering of the amidic groups with the increase of the temperature and it can be influenced by the content of salts, alcohols, surfactants addition. In this way, pH of solution is varying also. LCST varies in a range close to human body temperature; so, the PNIPAM was proposed to be used in biological applications.

With the increase of the temperature, PNIPAM configuration suffers some changes regarding the shape of the chain leading to reversible swelling-deswelling properties. [35] These properties can be useful in the control process of the thickness of the gels layers and their composition. PNIPAM gels can show temperature sensitive phase changes and they can be used as sensors or actuators.

In this study PNIPAM was synthesized via free radical polymerization of N-isopropylacrylamide in water at 30 °C using the redox system of ammonium persulfate $(NH_4)_2$ $S_2O_8$, with potassium bisulfite, $K_2S_2O_5$. [36] The resulting product was purified by dialyse against water through a membrane (cut off ~12 000 Dalton, Sigma) and freeze drying. The average viscometric molecular weight of PNIPAM, measured at 20°C in 0.5 M $LiNO_3$ aqueous solution is 28 000 Dalton.

**Poly (ethylene glycol) (PEG)** is a linear, water-soluble polymer and it can be obtained by an addition reaction of ethylene oxide (or ethylene glycol). Its general formulae is:

$$H\text{-}(OCH_2CH_2)_n\text{-}OH$$

where “n" is the polymerization degree.

PEG is widely used in pharmaceutical industry and cosmetics; it is non-volatile and inert from physiologically point of view. It posses moderate swelling properties. According to its molecular weight it can show different viscosity values and different properties [28]. For example, PEG with a molecular weight of about 2000 can be used as thickening agent for shaving creams emulsions, teeth paste, different creams and hair conditioner production. The compound with a high molecular weight can be used as suspension and thickening agent for washing powders, soaps, cosmetics.

In pharmaceutics, PEG can be used for the different ointments, emulsions, pastes, lotions and suppositories production. [29]

In this study, a PEG with a molecular weight 35,000 and with a melting temperature of 60-65$^0$C was used.

***Poly (Acrylamide) (PAM)***

In last decades poly (acrylamide) (PAM) had been identified as an efficient polymer both from ecological point of view mainly for infiltrations and erosions preventing in agriculture. [30, 31] It has the following formulae (scheme2):

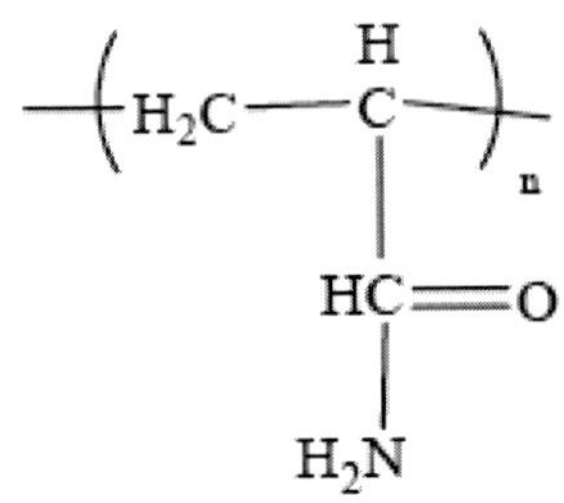

Scheme 2. Poly (acrylamide) structure.

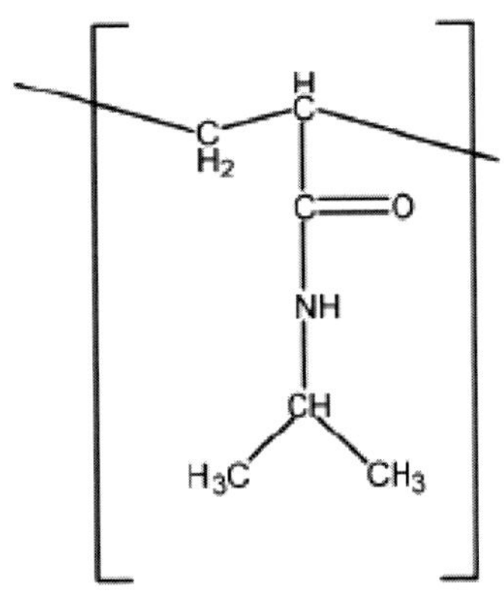

Scheme 3. Poly (N-isopropylacrylamide) (PNIPAM) structure.

PAM can be successfully used in efficient technologies for soil treatment and to improve the infrastructure (roads). [32]

PAM with high molecular weight is used as conditioning agent being added in the water used for irrigations.

PAM is used as flocculant agent in minerals processing and industrial wastes treatment. Copolymers based on acrylamide enhance the paper resistance.

The PAM used in our study was synthesized in a 5% water solution of acrylamide (Sigma) using hydrogen peroxide as initiator at 50 °C. [33] It was then precipitated in methanol, dissolved in water and freeze dried. Its molecular weight was viscometrically determined and it was found to be 1.5 * $10^5$ Dalton.

## 2.2. Preparation of the Aqueous Solution of PEG, PAM, PNIPAM, Aga in Twice Distilled Water

To minimize the polymer – polymer interactions effect, the diluted aqueous polymers solutions were prepared, all with the same concentration of 0.2 wt %. Twice distilled water was used for the preparation of the solutions. The pH of each solution was measured and they were of 3.21; 5.3; 5.4 and 5.5 for AgA, PNIPAM, PEG and PAM, respectively.

The pH of each polymer solution was adjusted at pH ≈ 4 by adding several drops of 2 M NaOH or HCl, corresponding with the minimum value for alginic acid to be dissolved.

Alginic acid solution adjusted at pH = 4 was filtered to remove undissolved particles. The real concentration of the polymer solution had been determined after filtration by evaporation of a certain volume of solution and weighing the dry mass. The alginic acid solutions was mixed with each polybase solution (PNIPAM, PEG or PAM to obtain binary mixtures in various ratios; so the entire range of composition of AgA/ polybase (PNIPAM, PEG or PAM) mixtures was covered from 5 wt % to 95 wt % AgA.

## 2.3. Investigation Methods

### *Viscometry*

Viscometric experiments have been performed by means of an Ubbelhode type viscometer with dilution and suspended level, at 25 °C ± 0.02 ° C and flow times were measured with an accuracy of ± 0.1s. The reduced viscosity ($\eta sp /c$) was determined for each partner solution and their binary blends in different weight ratios. The ideal value of the viscosity of the mixture solution can be calculated by using the equation (1) assuming the simple additivity of the viscosities of the solutions of components if there are no interactions between them. Equation (1) is applicable only for non-ionic (non polar) components.

$$\eta_{red,\ id} = \eta_1 * \omega_1 + \eta_2 * \omega_2, \quad (1)$$

$\eta_1, \eta_2$ are the reduced viscosities of the pure components from the system; $\omega_1, \omega_2$ are the weight fractions of the polyacid and polybases, respectively.

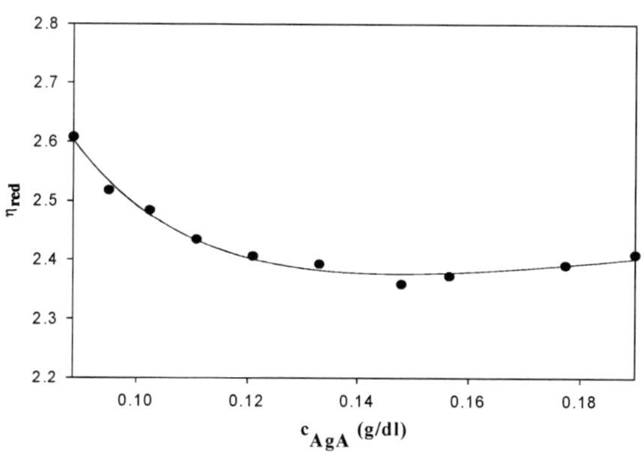


Figure 1. Dilution curve of alginic acid in twice distilled water.

In the study conditions, as it appears from dilution curve of AgA, in figure 1, the alginic acid behaves like a weak polyelectrolyte. Thus, the reduced viscosity records an increase with the decreasing of concentration of the polyacid.

In this case it is necessary to apply another equation for the evaluation of the ideal reduced viscosity of the system [11, 12, 37], namely:

$$(\eta_{sp} / c)_{calc} = w_1 * (\eta_{sp\,1} / c1) + w_2 * (\eta_{sp\,2} / c), \qquad (2)$$

$\eta_{sp\,1}/ c_1$ is the reduced viscosity of the polyacid at the concentration $c_1$; $\eta_{red1}$ as $(\eta_{sp1}/ c_1)$ can be determined from the dilution curve of AgA (figure 1) being dependent on the real concentration of AgA within the system ($c_1$).

$\eta_{sp2}/c$ is the reduced viscosity of the polybase at the total polymer concentration in system, c; $w_1$, $w_2$ are the weight ratios of the two polymers in the solution mixture.

To distinguish better the interactions in all three studied systems, it had been introduced the ratio between the experimental value and the calculated one, named viscosity ratio (equation 3) [38, 39]:

$$r_\eta = \eta_{exp} / \eta_{calc} \qquad (3)$$

$\eta_{exp}$ is experimental value of the reduced viscosity of the polymer mixture ; $\eta_{calc}$ is the calculated value of the reduced viscosity of the polymer mixture by using equation (2).

When $r_\eta$ takes values lower than unity it means that the formed interpolymeric complex via hydrogen bonds has a rather compact structure in comparison with that of the pure components. In this case the system shows a contraction of the polymers conformation due to the presence of the hydrogen bonds. [11, 12, 14, 18, 40]

A value higher than unity indicates an expansion of the complex conformation which leads to the appearance of a gel like structure. [12]

An ideal behavior is marked by a value equal with the unity. In this case the two polymers, practically, do not interact at all.

### *pH Measurements*

pH measurements were performed at 25 °C ± 0.02 ° C, in a thermostated bath, with a Consort C835 multimeter equipped with a separate pH glass electrode suitable for diluted solutions domain.

### *Conductometry*

The conductivity measurements were carried out by using a Consort C 835 multimeter equipped with a separate conductivity glass electrode specific to the measurements in 0.1μS-1000 mS range, characteristic for diluted polymer solutions.

To evidence better the formation of hydrogen bonds between the polymers in aqueous solutions, the "iso-pH" method was applied which consists in measurements of all characteristics starting with solutions of components having the same values of pH. [11, 12, 13, 14, 41, 42]. In such conditions, the binary solutions obtained by mixing of the weak polyacid with weak polybase solution will present or not deviations of experimental values in respect with additive ones only because of some intermacromolecular interactions between polymers.

# 3. RESULTS AND DISCUSSIONS

### 3.1. Viscosimetry

Studying the system AgA/ PNIPAM, it had been found the values of the viscosity ratio ($r_\eta$) lay below 1 (figure 1). This means that within this system some interpolymeric associations are formed, having a compact structure with a maximum of deviation at weight of 15- 70% AgA / 85- 30% PNIPAM.

For the systems AgA/ PEG and AgA/ PAM have been found different behaviors depending on the composition range. In the range 2- 38% AgA/ 98-62% PEG or 18-52 % AgA / 82-48% PAM there are positive values of the viscosity ratio. The maximum value of the ratio it was found at weight of 2-20 % AgA/ 98-80% PEG and 18-40 % AgA/ 82-60% PAM, respectively- figure 1. In the second region (62-100% AgA/ 38-2% PEG or PAM- figure 1) it can be noticed that the value of the viscosity ratio is tending to unity; it means that, in this composition range the polymers do not interacting each other anymore.

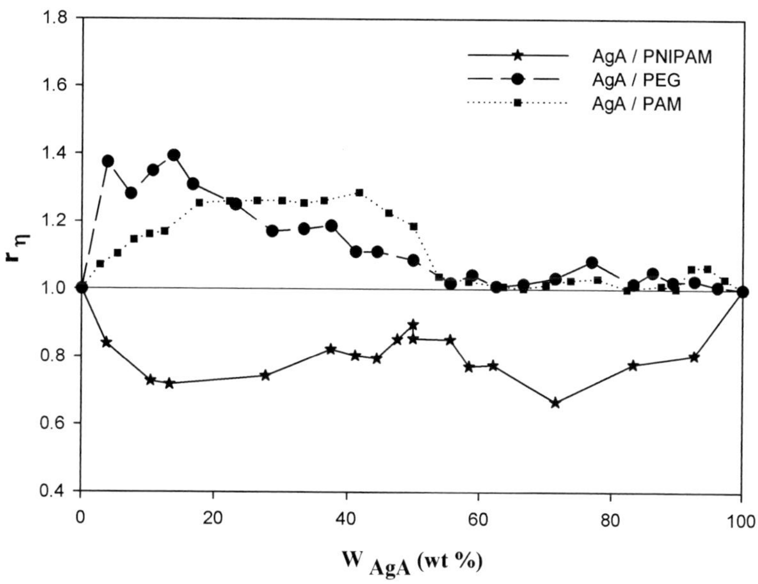


Figure 2. The dependence of the viscosity ratio on the composition of the systems: AgA/ PNIPAM (*), AgA/ PEG (•), AgA/ PAM( ) in twice-distilled water at 25 °C.

### 3.2. pH – Measurements

The pH-measurements represent a useful method for the identification of the presence of an interpolymeric complex in protic solvents. The complexation process due to the hydrogen bonds formed between COOH groups of the polyacid and basic groups (-OH, -O-, $-NH_2$) of the polybase can change the pH of the solution.

Addition of the polybase to polyacid is leading to the increase of the pH of the solution till a constant value when the complexation equilibrium is established. Therefore, the –COOH groups concentration will decrease and, according to the dissociation equilibrium of the polyacid, the concentration of the dissociated groups will decrease also at higher values of pH. [1, 14, 42]

Interpolymeric complexation between a polyacid (PA) like alginic acid (AgA) and a polybase (PB) like poly (N-isopropyl acrylamide) (PNIPAM) , poly(ethylene glycol) (PEG) or poly(acrylamide) (PAM) takes place according to a complexation equilibrium as following:

$$PA + PB \leftrightarrow C,$$

Equilibrium reaction is characterized by the apparent complexation constant:

$$Kc = [C] / [PA] [PB] \qquad (4)$$

where: [C] is complex concentration; [PA], [PB] are the uncomplexed polyacid and polybase concentrations.

The complexation process takes place via the successive hydrogen bonds between the reactive groups:

$$\text{-COOH} + B \leftrightarrow \text{-COOH...B}$$

The acidic groups are dissociated according to dissociation equilibrium

$\text{-COOH} \leftrightarrow COO^- + H^+$ characterized by dissociation constant Kd:

$$Kd = [COO^-]^* [H^+] / [COOH], \qquad (5)$$

The association degree can be determined according to the equation (6). [13]

$$\theta = 1 - ([H^+] / [H^+]_0)^2 \qquad (6)$$

where:

- $[H^+]$, $[H^+]_0$ are the hydrogen concentrations in the presence and absence of the acceptor polymer
- $\theta$ – association/dissociation degree of the –COOH groups of the polyacid.

$$pKd = a + b^* pH \qquad (7)$$

Dissociation constant, $K_d$ has been obtained by using Kern empiric equation (eq 7) [43] and some blank experiments which consist in measurements of the pH of the pure alginic acid at different concentrations and different temperatures in the absence of the polybase solutions and the linear dependences as in figure 3 were obtained for all temperatures tested.

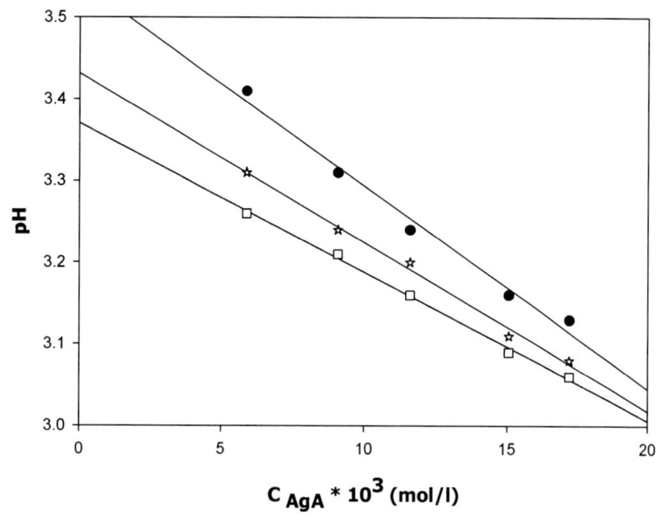


Figure 3. The dependence of the pH on the alginic acid concentration, $C_{AgA}$, at 25 °C (•), 30 °C (*) and 35 °C( ).

From the linear dependence of the calculated pKd according equation (7) pKd values of alginic acid at different pH values are shown in figure 4.

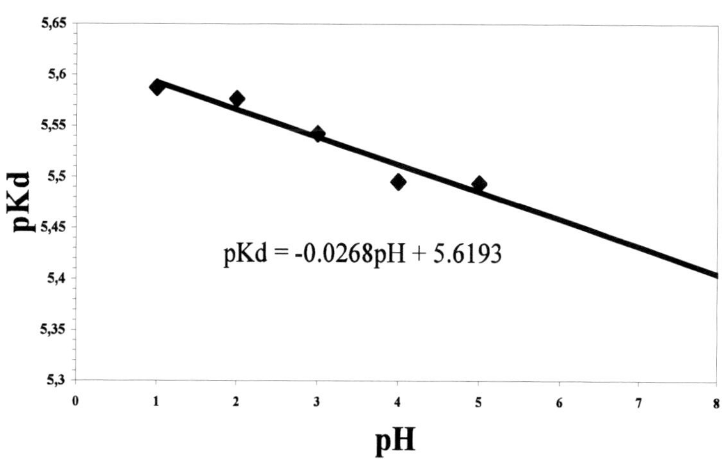


Figure 4. pKd - pH dependence for the alginic acid solution in twice- distilled water.

The behaviour of a polyelectrolyte can be decribed also by the Henderson-Hasselbalch equation according to the disociation equilibrium of the polyacid and eq. (5):

$$pKd = pH + \log_{10} [COO^-]/[COOH] \quad (8)$$

Kagawa and Tsumura [44- ref 3 therein], in their study on the carboxymethylcellulose dissociation found that Kern formula [43] has a good applicability in the individual experiments series under fixed conditions like dilution or neutralization studies but it lacks universal applicability. Kagawa transformed Kern's equation according to his opinion that the dissociability of polyelectrolytes should be dependent on the degree of dissociation [44- ref 6 therein].

Nagasawa and Rice [45] showed the equivalence between the two equations as being the equation of a titration curve obtained in the potentiometric titration of a polyacid with a polybase.

$$pKd = pH + n\log_{10}(1-\theta)/\theta \quad (9)$$

where n- empirical parameter,

θ dissociation degree, $\theta = [-COO^-] / [-COO^-] + [-COOH]$

Therefore in the present study we focused on the using of the Kern empirical equation (equation 7) to determine the dissociation constant of the alginic acid and thus the stability constants of the interpolymeric associations between alginic acid and PNIPAM, PEG and PAM. It was evaluated also the complexation degree (θ)-equation 6- and the thermodynamic characteristics-according to the equation 12.

Potentiometric curves from the figures 5 a, b, c show large maxima of pH which become even larger when temperature increases. These maxima are not very specific with type of system studied. They range between 16 – 70 wt% AgA for the system AgA/PNIPAM and 15 – 45 wt % in the case of AgA/PEG si AgA/PAM. This should mean that, if the intermacromolecular associations exist, they are not stoichiometrical. It is possible that temperature increase favors the solubility of the alginic acid the chain is extended increasing the number of intercontacts between components. This can be explained also by the fact that with higher temperature there are more accentuated interactions between components. [42] Another reason should be that in the working conditions we are at the limit between H-bonding or electrostatic interactions between components.

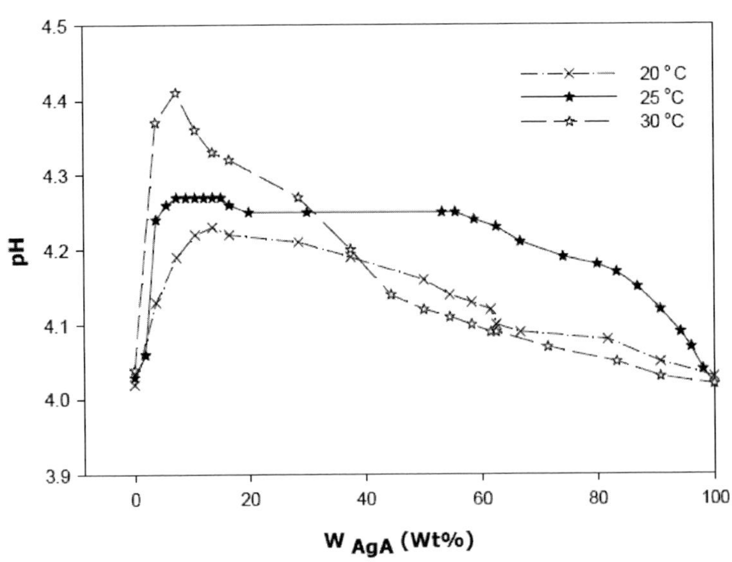


a

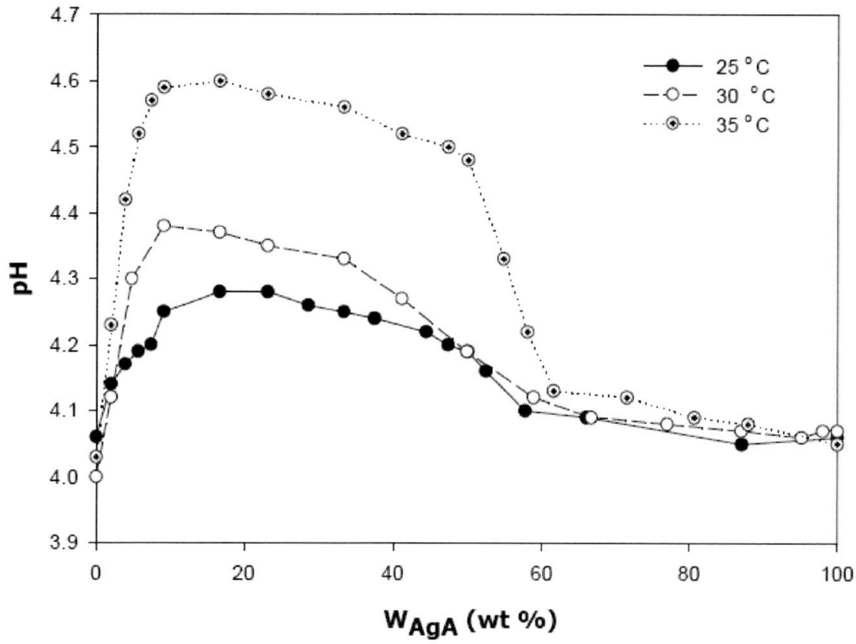


b

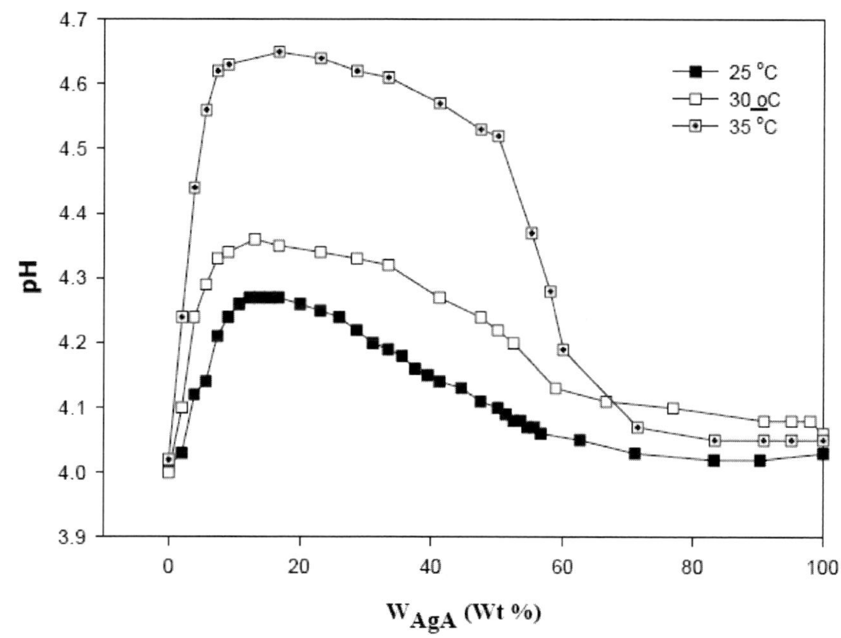


c

Figure 5. pH- values recorded at AgA titration with PNIPAM (a), with PEG (b) and PAM (c) at different temperatures.

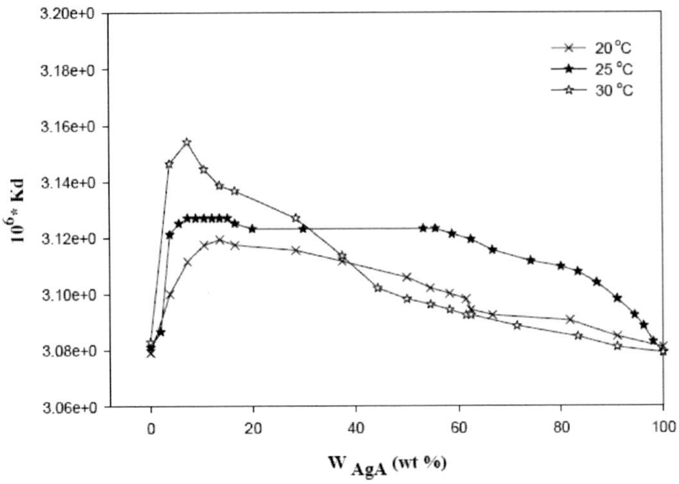


a

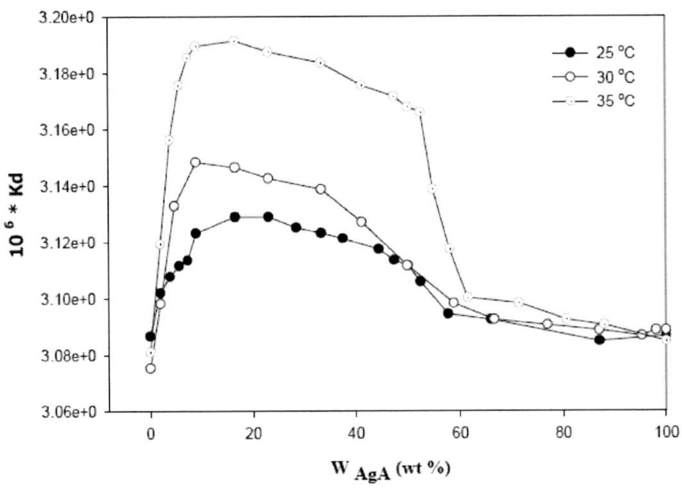


b

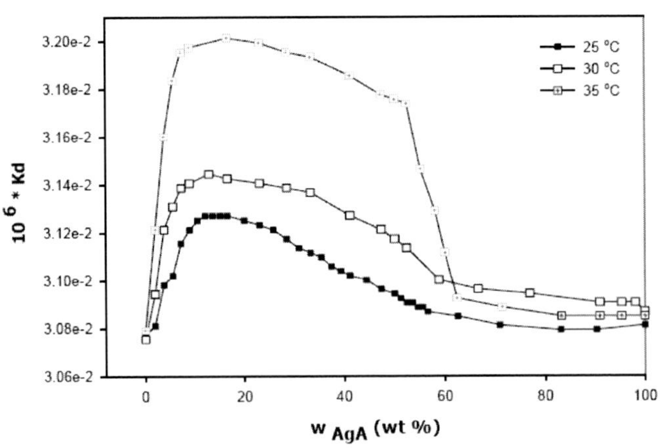


c

Figure 6. The dependence of the dissociation constant (Kd) on the system composition of AgA/ PNIPAM (a) AgA/ PEG (b ) and AgA/ PAM (c) at different temperatures.

A similar dependence was found for variation of the the dissociation constant (Kd) with composition of the system – figure 6 and also of the interpolymeric associations concentration – figure 7.

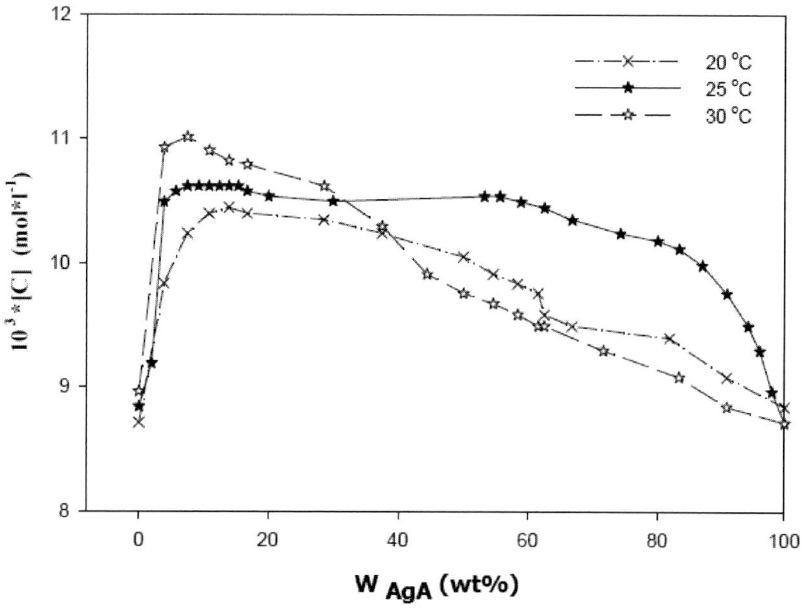


a

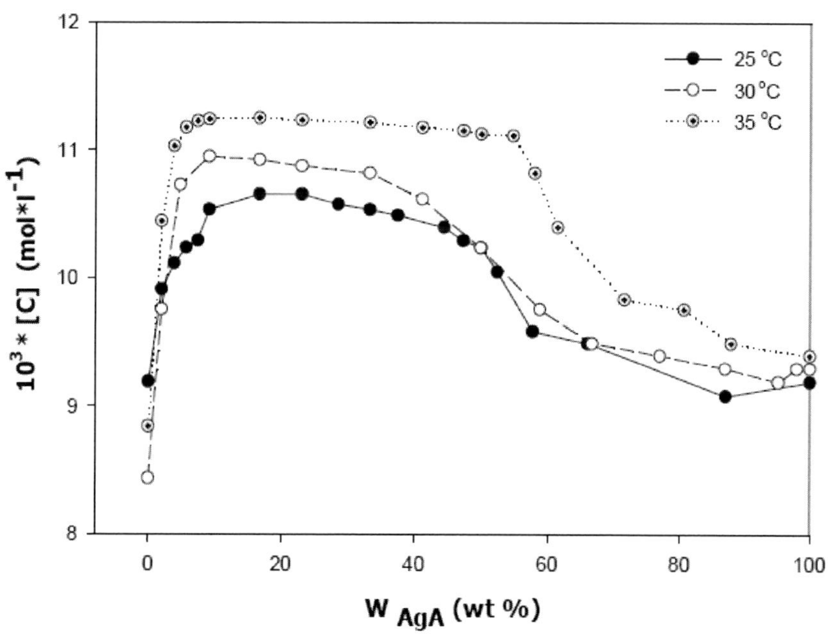


b

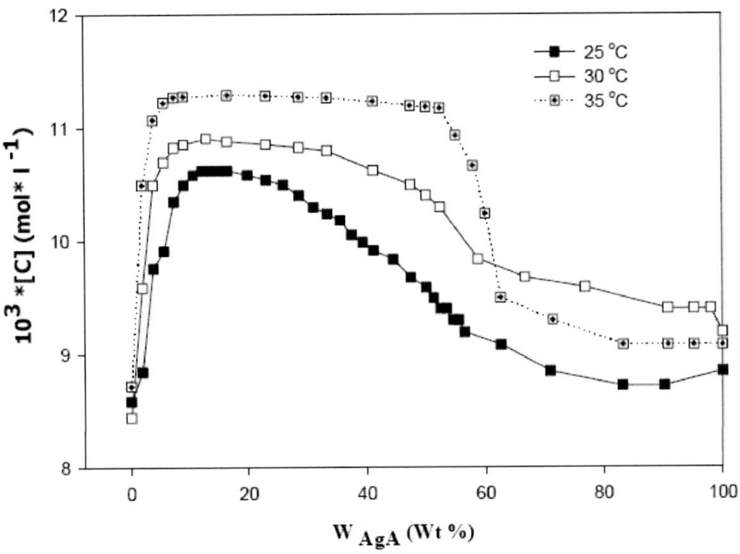


c

Figure 7. Variation of the complex concentration for the system AgA/ PNIPAM (a); AgA/ PEG (b) and AgA/ PAM (c) at different temperatures.

Interpolymeric complex concentration can be obtained by applying the mass conserving law for –COOH groups of the polyacid:

$$[C]= [AgA]_0- [COOH]- [COO^-], \quad (10)$$

Where:

[COOH] – molar concentration of –COOH uncomplexed groups;

$[COO^-]$- concentration of –$COO^-$ groups which are uncomplexable.

$[COO^-]$ was obtained from the equation which expresses solution the electroneutrality (equation 11), taking into account that the AgA solution pH had been adjusted to pH≅ 4 by using a solution of NaOH 0.2 M, and according isoionic dilution. [14, 42]

$$[COO^-]= [H^+] - [HO^-] \quad (11)$$

Similar procedures had been used for the determination of stability constants of the formed complexes and thermodynamic parameters of the complexation process of other systems. [9, 13, 46]

The concentrations of the interpolymeric associations are very low being of $1.062 * 10^{-2}$ (mol* $L^{-1}$) for AgA/ PNIPAM; $1.065* 10^{-2}$ (mol* $L^{-1}$) for AgA/PEG and $1.06 * 10^{-2}$ (mol* $L^{-1}$) for AgA/PAM system.

Association degree, θ, as a function of the system composition is plotted in figure 8. Its values are of 0.66, 0.63 and 0.69 for AgA/PNIPAM, AgA/PEG and AgA/PAM systems.

The dependence of the θ on the composition has the same shape as other characteristics of the intermacromolecular associations presented above as: pH, Kd, complex concentration.

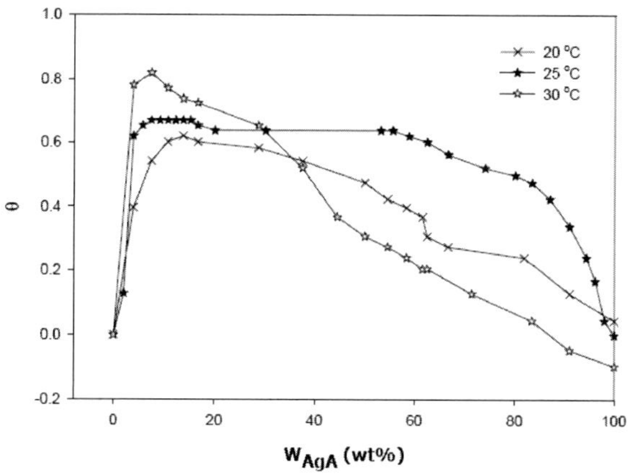


a

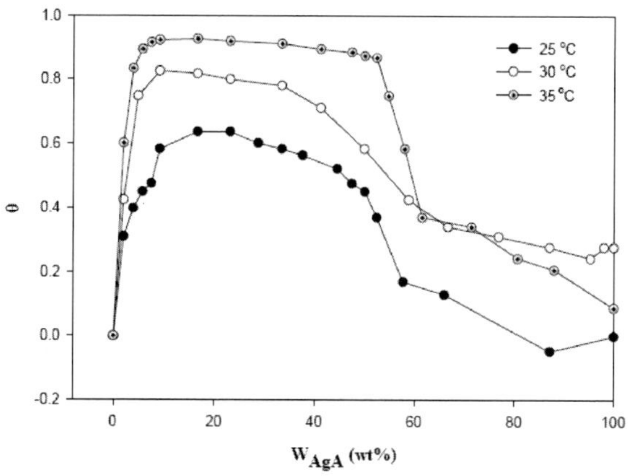


b

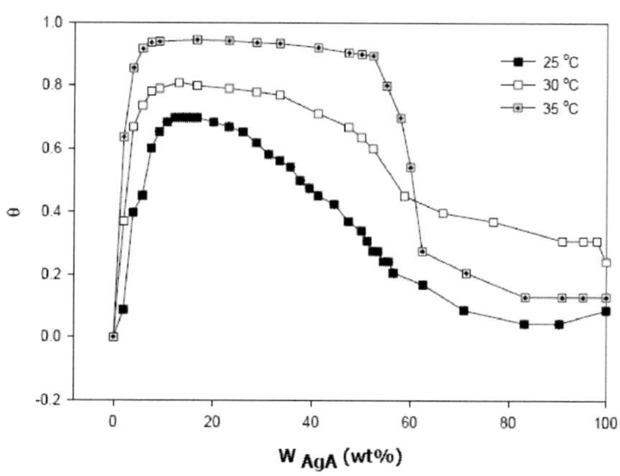


c

Figure 8. The dependence of the association degree on the system composition in the case of mixture AgA/PNIPAM (a); AgA/ PEG (b) AgA/ PAM (c) at different temperatures.

The average values of the equilibrium (stability) constants, Kc, calculated according equation 4, of the interpolymeric associations corresponding to the studied systems AgA/ PNIPAM, AgA/PEG, AgA/ PAM are given in the table 1 and figure 9.

Figure 9 displays the variation of the apparent complexation constant of AgA/ PNIPAM, AgA/ PEG, AgA/PAM with the weight fraction of alginic acid. It can be observed that the values of apparent complexation constant laid in close limits given above and are independent on polymer mixture composition.

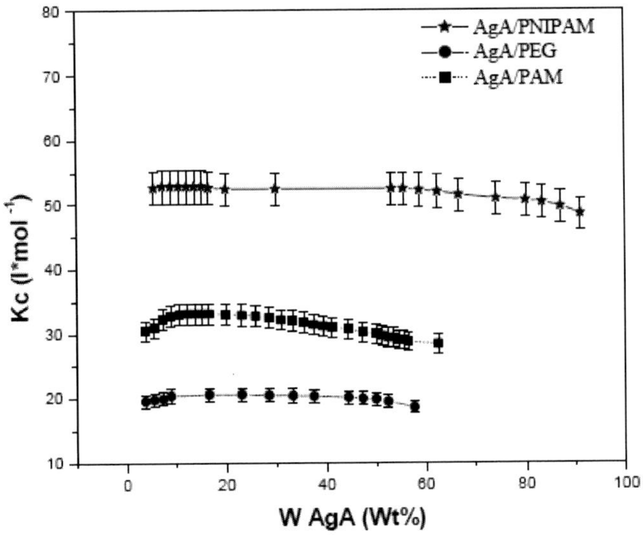


Figure 9. The dependence of the stability constant, Kc, of the system AgA/ PNIPAM, AgA/ PEG and AgA/ PAM on the composition (a).

**Table 1. Values of the stability constants, Kc, for the studied systems at different temperatures**

| T (°C) | $Kc_{AgA/PNIPAM}$ ($L^*mol^{-1}$) | $Kc_{AgA/PEG}$ ($L^*mol^{-1}$) | $Kc_{AgA/PAM}$ ($L^*mol^{-1}$) |
|---|---|---|---|
| 20 | 51.94 | - | - |
| 25 | 52.80 | 20.64 | 33.14 |
| 30 | 54.76 | 21.20 | 34.02 |
| 35 | - | 21.78 | 35.22 |

The average values have been evaluated for the composition range where maxima in pH values and other characteristics were found for each system. At different temperatures there is a slight increase of Kc from 51.9 till 54.76 for AgA/PNIPAM system, from 20.64 till 21.78 for AgA/PEG system and from 33.1 till 35.2 for AgA/PAM system, this being an indication for the presence of the hydrophobic forces within the systems in the case of AgA/PNIPAM for which the Kc takes the highest values.

The van't Hoff dependence of ln Kc versus 1/T estimated according to eq. 12 and plotted in figure 10.

$$-\ln Kc = \Delta H - T\,\Delta S \qquad (12)$$

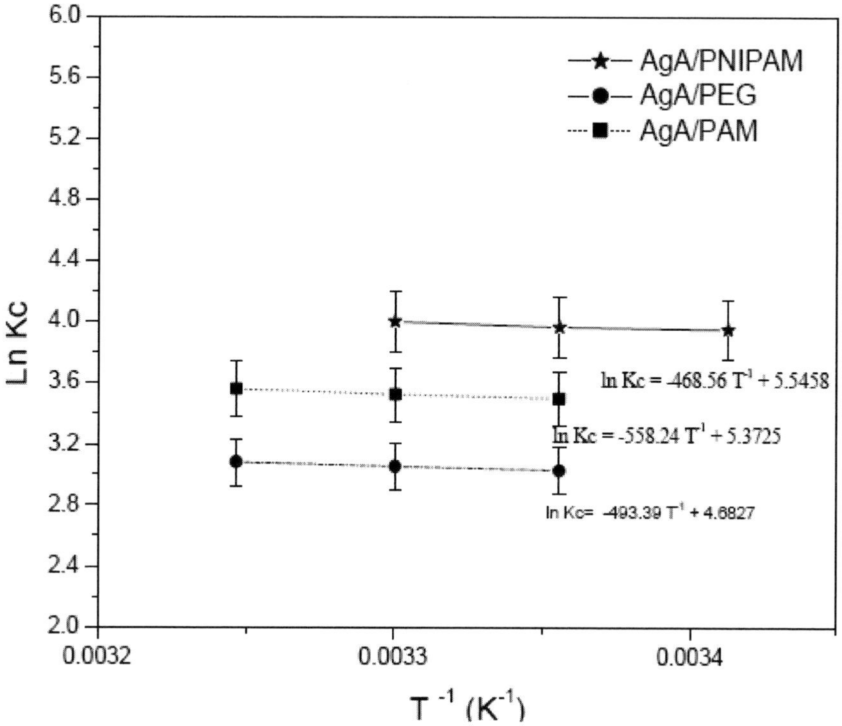


Figure 10. The dependence ln Kc= f (1/T) for the mixtures AgA/PNIPAM, AgA/PEG, AgA/PAM.

The values of the stability constants determined above and its variation with temperature show that the stability of the associations formed in the case of AgA/ PNIPAM, AgA/PEG and AgA/ PAM are not significantly influenced by the temperature variation.

It allows also the evaluation of the enthalpy and entropy of hydrogen-bonded associations from the slope and the intercept of the van't Hoff plot yielding for ΔH the values of - 3.89 KJ * $mol^{-1}$ for AgA/ PNIPAM, -4.1 KJ*$mol^{-1}$ for AgA/ PEG and - 4.6 KJ*$mol^{-1}$ for AgA/PAM systems. (table 2).

**Table 2. Estimated values for the enthalpy and entropy for the studied systems**

| System | ΔH (KJ*$mol^{-1}$) | ΔS (J*$mol^{-1}$*$K^{-1}$) |
|---|---|---|
| AgA/ PNIPAM | - 3.89 | 46 |
| AgA/ PEG | - 4.1 | 38.9 |
| AgA/ PAM | - 4.6 | 44.6 |

The obtained values of enthalpy are comparable with those evaluated by Tsushida [47] within the study of interpolymeric complexes based on poly (methacrylic acid), (PMMA) and poly (N-vinyl-2-pyrolidone), (PVPo) and poly (ethylene oxide). In comparison with the general hydrogen bond enthalpy (which is about 5 kcal/mol), the obtained experimental values are low and it can be attributed to the fact that only some of the active sites are involved in the formation of the complexes and the employed conditions are at the limit of H-bond formation.

## 3.3. CONDUCTOMETRY

In the last years, conductometry had become an important analysis instrument of the counterions distribution in aqueous diluted polyelectrolytes solutions. The counterions influence the polyelectrolyte solution properties by their size, valence and polarizability. Conductometry measures the transport of all charged species within studied system. [48] Most of conductometric studies showed a slight increase of the conductivity with the decrease of concentration of polyion followed by an important increase of conductivity at high dilution. [49, 50, 51, 52, 53] Another studies reported that the conductivity can be independent on concentration [54] or it can show a minimum value especially in the case of cationic polyelectrolytes with quaternary ammonium salts groups. [55, 56] Studying the conductometric behavior of some cationic polysaccharides, Ghimici [57] had observed an almost linear increase of conductivity with the dilution over a wide concentration range. Wandrey had taken into account the influence of macromolecular parameters and the chemical structure on the polyion- counterion interactions of flexible polyelectrolytes in diluted solutions. [58] Conductometric measurements of Arabic gum [59] showed dependence with a minimum value of conductivity. Nelson explained this behavior on the base of the increase of the ions mobility due to the intermacromolecular interactions and the coiled backbones at high concentrations leading to the decrease of the number of the condensed conterions.

Electrolytic conductivity behavior of the systems of AgA/PNIPAM, AgA/PEG, AgA/ PAM in twice-distilled water is presented in figure 11. A curve with a minimum value of conductivity at a mixture composition of 36 % AgA / 64 %PNIPAM, 18 %AgA/ 82 % PEG and 18 %AgA / 82 %PAM was found that means at these compositions the H-bonding associations are formed. Conductometric results are in accordance with the viscometric and potentiometric results and there are comparable with results obtained by Betktutov [9] at titration of poly (acrylic acid) (PAA) with poly (vinyl pyrolidone) (PVP) in DMSO or DMF or by Ghimici [57] in the study of the conductivity dependence on polyelectrolyte concentration, charge density, substituent at the ionic group and solvent polarity.

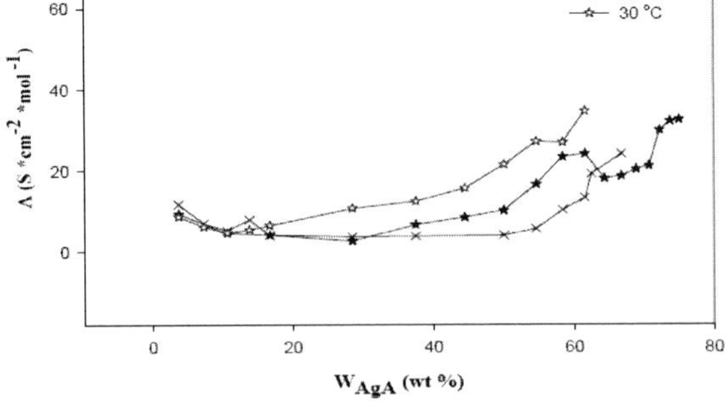


a

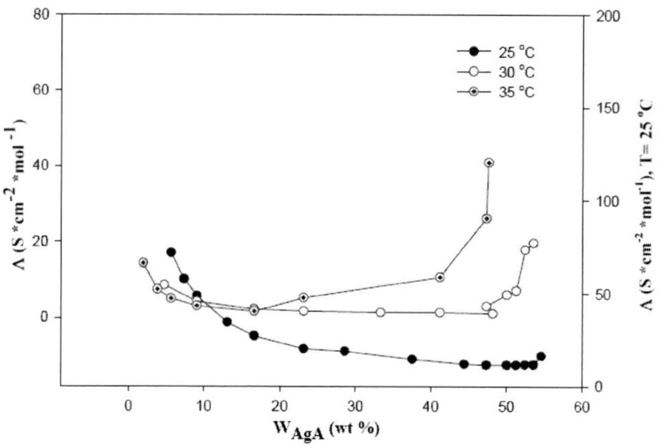


b

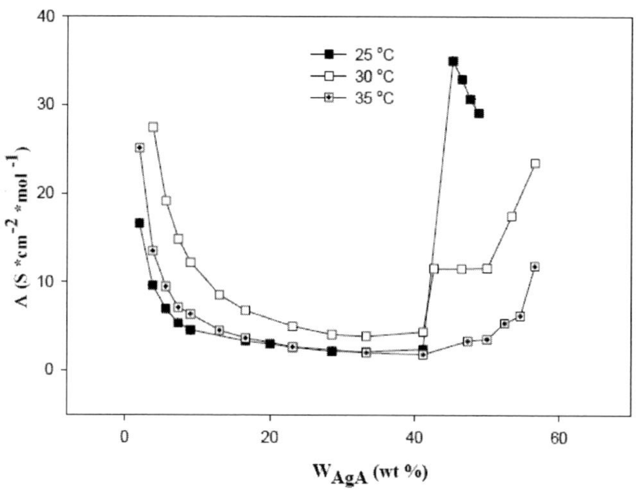


c

Figure 11. Molar conductivities of the systems AgA/PNIPAM (a) AgA/ PEG (b) in twice - distilled water, at different temperatures in function of composition.

## Conclusions

It was established via viscometric measurements, pH determinations, and conductometry experiments that between alginic acid (AgA) and PNIPAM, PEG and PAM, interpolymeric associations with different conformations and different stabilities were formed.

It was studied the influence of temperature on the formation and stability of these associations and their thermodynamic characteristics like association degree, association (stability) constant, thermodynamic parameters have been evaluated.

pH studies had been performed at pH=4 which means the solubility limit of alginic acid (AgA). Therefore it can be appreciated that this limit is also available for the formation of the interpolymeric associations via hydrogen bonds however there are not excluded the presence of some ionic bonds. The most probable composition ranges found for the existence of these weak intermacromolecular associations is as follows:

| System | Viscometry | Potentiometry | Conductometry |
|---|---|---|---|
| AgA/ PNIPAM | 15- 70 % AgA/ 85-30% PNIPAM | 16-70% AgA/ 84-30% PNIPAM | 36 % AgA/ 64 %PNIPAM |
| AgA/ PEG | 2-38 % AgA/ 98-62% PEG | 15-45% AgA/ 90-55% PEG | 18% AgA/ 82% PEG |
| AgA/ PAM | 18-52 % AgA/ 82-48% PAM | 15-45% AgA/ 90-55% PAM | 18 %AgA/ 82% PAM |

A good accordance between the results of the three used methods can be remarked.

## References

[1] G. Staikos, G.Bokias, G. G. Bumbu, Water soluble polymer systems- phase behaviour and complex formation, chapter 5 in *Handbook of Polymer Blends and Composites*, Vol 3A, C. Vasile, A. K. Kulshreshtha (Eds) , Rapra Technology Limited, Shawbury, 2003, 135-178.

[2] M. Jiang, M. Li, M. Xiang, H. Zhou, Interpolymer Complexation and Miscibility Enhancement by Hydrogen Bonding, *Adv. in Polym. Sci.*, 1999, 146: 121-196

[3] V. V. Khutoryanskiy, G. A Mun, Z. S Nurkeeva, A. V Dubolazov, pH and salt effects on interpolymer complexation via hydrogen bonding in aqueous solutions, *Polym. Int.*, 2004, 53:1382–1387.

[4] Z. S. Nurkeeva, G. A. Mun, V. V. Khutoryanskiy, Interpolymer Complexes of Water-Soluble Nonionic Polysaccharides with Polycarboxylic Acids and Their Applications, *Macromol. Biosci.* 2003, 3: 283-295

[5] T. Ozeki, H. Yuasa, Y. Kanaya, Controlled release from solid dispersion composed of poly(ethylene oxide)–Carbopol interpolymer complex with various cross-linking degrees of Carbopol, *J. Controlled Release.* 2000, 63: 287-295.

[6] M. K. Chun, C. S. Cho , H. K. Choi, Mucoadhesive drug carrier based on interpolymer complex of poly(vinyl pyrrolidone) and poly(acrylic acid) prepared by template polymerization. *J. Controlled Release*, 2002, 81: 327-334.

[7] B. S. Lele, A. S. Hoffman, Mucoadhesive drug carriers based on complexes of poly (acrylic acid) and PEG-ylated drugs having hydrolysable PEG–anhydride–drug linkages. *J. Controlled Release*, 2000, 69:237-248.

[8] G. Colo, S. Falchi, Y. Zambito, In vitro evaluation of a system for pH-controlled peroral delivery of metformin, *J. Controlled Release*, 2002, 80:119-128.

[9] E. A. Betktutov, L.A. Bimmendina, Interpolymer Complexes, *Adv. Polym Sci.,* 1981, 41:99-145.

[10] V. A. Prevysh, B. C. Wang, R. J. Spontak, Effect of added salt on the stability of hydrogen- bonded interpolymer complexes, *Colloid Polym. Sci.*, 1996, 274:532-538.

[11] G. Staikos, K. Tsitsilianis, Viscometric investigation of the poly (acrylic acid)-polyacrylamide interpolymer association, *J. Appl. Polym. Sci.*, 1991, 42:867- 872.

[12] G. Staikos, G. Bokias , C. Tsitsilianis, The viscometric methods in the investigation of the polyacid-polybase interpolymer complexes, *J. Appl. Polym. Sci.*, 1993, 48:215 – 217.

[13] E. Tsuchida, Y. Osada, H. Ohno, Formation of Interpolymer Complexes, *J. Macromol. Sci.- Phys., B.* 1980, 17:683-714.

[14] C. Vasile, G. G. Bumbu, Y. Mylonas, I. Cojocaru, G. Staikos, Hydrogen - Bonding interaction of an alternating maleic acid- vinyl acetate copolymer with poly (ethylene glycol), polyacrylamide and poly (N-isopropylacrylamide): a comparative study, *Polym. Int*, 2003, 52:1887-1891.

[15] G. G. Bumbu, J. Eckelt, C. Vasile, Interpolymeric Complexes Containing Copolymers in Hydrogen Bonded Interpolymer Complexes: Formation, Structure and Applications.V. V. Khutoryanskiy and G. Staikos Eds. *Elsevier* in press

[16] G. G. Bumbu, C. Vasile, Stimuli responsive copolymers of poly (maleic acid-alt-vinyl acetate) grafted with poly(N-isopropyl acrylamide). Influence of copolymer composition and of pH, *Memoriile Stiintifice ale Academiei Romane*:In press.

[17] O. Nikolaeva, T. Budtova, V. Alexeev, S. Frenkel, Interpolymer association between polyacrylic acid and cellulose ethers: formation and properties, *J. Polym. Sci: Part B: Polym. Phys*., 2000, 38:1323-1330.

[18] C. Vasile, G. G. Bumbu, R. Dumitriu, G. Staikos, Comparative study of the behavior of carboxymethyl cellulose-g- poly (N-isopropylacrylamide) copolymers and their equivalent physical blends, *Eur. Polym. J*., 2004, 40:1209- 1215.

[19] C. Vasile, G. G. Bumbu, G. Staikos Carboxymethyl cellulose grafted poly (N-isopropylacrylamide) II) Influence of temperature and pH on the solution behaviour. *Cell. Chem. Technol*. In press.

[20] G. G. Bumbu J. Eckelt , C. Vasile, Interpolymer Complexes containing maleic copolymers. rheological behaviour, *European Polymer Congress*, *Portoroz, Slovenia, July* 2-6, 2007, P7.4.28.

[21] G. G. Bumbu, C. Vasile, G.C. Chitanu, G. Staikos, Interpolymer complexes between hydroxypropylcellulose and copolymers of maleic acid: A comparative study, *Macromol. Chem. Phys*., 2004, 206:540-546.

[22] G. Staikos, K. Karayanni, Y. Mylonas, Complexation of polyacrylamide and poly(N-isopropylacrylamide) with poly(acrylic acid). The temperature effect, *Macromol. Chem. Phys*. 1997, 198:2905

[23] C. Tapia, E. Costa, M. Moris, J. Sapag-Hagar, F. Valenzuela, C. Basualto , Study of the Influence of the pH Media Dissolution, Degree of Polymerization, and Degree of Swelling of the Polymers on the Mechanism of Release of Diltiazem from Matrices Based on Mixtures of Chitosan/Alginate, *Drug Development and Industrial Pharmacy*, 2002, 28:217-224.

[24] K. Y. Lee, W. H. Park, W. S. Ha, Polyelectrolyte complexes of sodium alginate with chitosan or its derivatives for microcapsules, *J. of Appl. Polym. Sci*, 1998, 63:425 – 432.

[25] A. Iruín, M. Fernández-Arévalo, J. Álvarez-Fuéntes, A. Fini, M. A. Holgado, Elaboration and "In Vitro" Characterization of 5-ASA Beads, *J. Drug Development and Industrial Pharmacy*, 2005, 31:231-239.

[26] J. I. Kadokawaa, S. Saitoub, S. I. Shodab, *Carbohydrate Polymers*, 2005, 60:253–258.

[27] Kenneth Clare, Algin, chapter 6 in *Industrial Gums*, Third edition,. R. L. Whistler, J.N. BeMiller (Ed), Academic Press, San Diego, 1993, 105

[28] http://www.arpc-ir.net/ PDF/ catalogue/ ChemicalSpec/Ethoxylates/PEG-Chemical %20Grade.pdf

[29] http://www.arpc-ir.net/PDF/catalogue/ChemicalSpec/Ethoxylates/PEG-Chemical%20 Grade.pdf

[30] R. D. Lentz, I. Shainberg, R. E. Sojka , D. L. Carter, Preventing irrigation furrow erosion with small applications of polymers, *Soil Sci. Soc. Am. J.*, 1992, 56:1926-1932.

[31] C. C. Shock, B. M. Shock, Comparative effectiveness of polyacrylamide and straw mulch to control erosion and enhance water infiltration in A. Wallace (Ed), *Handbook of Soil Conditioners*, M. Dekker Inc. New York, 1997, 429 – 444.

[32] C.C. Shock, E. B. G. Feibert, L. D. Saunders, S. Wimpy, H. Cox Treatment of soil with Bright Sun Soil Booster and polyacrylamide as soil conditioners for improved seedling emergence, *Malheur Experiment Station Special Report*, 1993, 924:199-200.

[33] W. M. Kulicke , J. Klein, *Angew Makromol. Chem.* 1987, 69: 169.

[34] H. Feil, Y. H. Bae, J. Feijen, S. W. Kim Effect of co-monomer hydrophilicity and ionization on the lower critical solution temperature of N-isopropylacrylamide copolymers, *Macromolecules,* 1993, 26: 496-250.

[35] X. Ma, J. Xi, X. Zhao, X. Tang, Deswelling comparison of temperature-sensitive poly (N-isopropylacrylamide) microgels containing functional OH groups with different hydrophilic long side chains, *J. Polym. Sci. Part B*: *Polym. Phys*, 2005, 43: 3575 – 3583.

[36] M. Heskins, J. E. Guillet, Solution properties of poly (N-isopropylacrylamide), *J. Macromol Sci A*, 1968, 2: 1441-1445.

[37] H. Ohno, K. Abe, E. Tsuchida, Solvent effect on the formation of poly(methacrylic acid)-poly(N-vinyl-2-pyrrolidone) complex through hydrogen bonding, *Makromol. Chem.*, 1978, 179: 755–763.

[38] G. R. Williams, B. Wright, Interactions in binary polymer systems, *J. Polym. Sci., Part A*, 1965, 3: 3885-3891.

[39] D. Staszewska, M. Bohdanecky, A viscometric study of dilute aqueous solutions of poly(vinylalcohol)-polyacrylamide mixtures, *Eur. Polym. J.*, 1981, 17: 245-248.

[40] N. G. Belnikevich, T. V. Budtova, N. S. Nesterova, Y. N. Panov, S. Y. Frenkel, Applying the Viscosimetric Method of Determination of Intermolecular Interaction Constant for Probing Complex Formation, *Polym. Sci.*, 1992, 34: 3.

[41] G. Bokias, G. Staikos, I. Iliopoulos, R. Audebert, Interpolymer association between acrylic acid copolymers and poly(ethylene glycol) : effects of the copolymer nature, *Macromolecules*, 1994, 27: 427-431.

[42] G. G. Bumbu, C. Vasile, J. Eckelt, B. Wolf, Investigation of the Interpolymer Complex between Hydroxypropyl Cellulose and Maleic Acid-Styrene Copolymer, Dilute solutions studies, *Macromol, Chem. and Phys*, 2004, 205: 1869-1876.

[43] W. Kern, *Z. Phys. Chem.*, A 1938, 181: 268.

[44] H. C. Trivedi, C. K. Patel, R. D. Patel, Studies on carboxymethylated cellulose: Potentiometric Titrations, *Makromol. Chem.*, 1980, 182: 3561 – 3567.

[45] M. Nagasawa, S.A. Rice, A Chain Model for Polyelectrolytes. V. A Study of the effects of local charge density, *J. Amer.Chem. Soc.* 1960, 82: 5070.

[46] M. Koussathana, P. Lianos, G. Staikos, Investigation of Hydrophobic Interactions of Hydrogen- Bonding Interpolymer Complexes, *Macromol.*, 1997, 30: 7798-7802.

[47] E. Tsuchida, K. Abe Interactions between Macromolecules in Solution and Intermacromolecular Complexes, *Adv. Polym. Sci.*, 1982, 45: 1–119.

[48] A. Marc, G. T. Van den Hoop, J. C. Benegas, Improvement in Conductometric Analysis of Metal/ Polyelectrolyte Systems, *Macromol.*, 1997, 30: 3930-3932.
[49] H. P. V. Leeuwen, R. F. M. J. Cleven, P. Valenta, Conductometric analysis of polyelectrolytes in solution, *Pure Appl. Chem.*, 1991, 63: 1251-1268.
[50] H. E. Ricos, R.G. Barraza, I. C. Gamboa, Polyelectrolyte solutions. Electrical conductivity and counterion condensation, *Polym. Int.*, 1993, 31: 213-216.
[51] H. L. Cheng, K.F. Lin, Molecular Weight Dependence of Chain Conformation and Counterion Dissociation of Poly(xylylene tetrahydrothiophenium chloride) in the Salt-Free Semidilute Aqueous Solutions, *Laugm.*, 2002, 18: 7287- 7290.
[52] H. Vink, Conductivity of polyelectrolytes in very dilute solutions, *J. Chem. Soc. Trans.*, 1981, 77:2439.
[53] Z. Hu, S. Zhang, J. Yang, Y.Chen, Some properties of aqueous-solutions of poly(vinylamine chloride), J. *Appl. Polym. Sci.*, 2003, 89: 3889.
[54] H. Vink, Electrolytic conductivity of polyelectrolyte solutions, *Macromol. Chem.*, 1982, 183:2272-2283.
[55] L. Ghimici, S. Dragan, F. Popescu, Interaction of the low-molecular weight salts with cationic polyelectrolytes, *J. Polym. Sci. B*, 1997, 35: 2571- 2581.
[56] L. Ghimici, S. Dragan, Behaviour of cationic polyelectrolytes upon binding of electrolytes: effects of polycation structure, counterions and nature of the solvent, *Colloid Polym. Sci.*, 2002, 280: 130-134.
[57] L. Ghimici, M. Nichifor, Electrical conductivity of some cationic polysaccharides. Effects of polyelectrolyte concentration, charge density, substituent at the ionic group and solvent polarity, *J.Polym. Sci., B*, 2005, 43: 3584-3590.
[58] C. Wandrey, D. Hunkeler, Study of Polyion Counterion Interaction. by Electrochemical Methods, chapter 5 in S. K. Tripathy, J. Kumar, H. S. Nalwa (Eds), *Handbook of Polyelectrolyte and Their Applications*, ACS Publisher, USA, 2002, 2: 147-172.
[59] R. Nelson, P. Ander, Electrical Conductivities of Salts of Gum Arabic and Carrageenan in Aqueous Solutions, *J. Phys. Chem.*, 1971, 75: 1691-1697.

In: Monomers, Oligomers, Polymers, Composites... ISBN: 978-1-60456-877-6
Editors: R. A. Pethrick, G.E. Zaikov et al. 

*Chapter 13*

# A THEORETICAL APPROACH FOR PREDICTION OF YARN STRENGTH IN TEXTILE INDUSTRY

***A.Shams-Nateri and A.K.Haghi****
University of Guilan, P. O. Box 3756, Rasht, Iran

## ABSTRACT

This paper describes advance techniques that can be used to predict ring spun yarn strength from HVI/FMT measured properties of cotton fibers such as span length, bundle strength, fineness, uniformity ratio and maturity of cotton fibers. Neural networks, neuro-fuzzy, multiple-linear regression techniques and Sitra's expressions were used to predict yarn strength from cotton fibers properties. The results of intelligence system were better than multiple-linear regression techniques and Sitra's expressions. The best results were obtained by neuro-fuzzy method.

**Keywords:** Cotton, Fiber, Yarn, Relationship, Regression, Neural Network, Neuro-fuzzy

## INTRODUCTION

One of the most important production processes in the textile industry is the spinning process, which started with cotton fiber. Yarn has been produced in rotor or ring spinning machine. The quality of the resulting yarns is very important in determining their application possibilities. The two most important characteristics of yarn are its tenacity and elongation. Prediction formulae to related fiber properties to yarn quality have been an interesting field of work for several research workers for more than three decades [1, 2].

Sitra was developed expressions for prediction yarn strength from fiber properties in the year 1989[3]. In the prediction expressions fiber properties measured with conventional instrument. In next step, they establish prediction expressions for strength of spun yarns from

* Corresponding author: Haghi@Guilan.ac.ir

cotton fiber properties using HVI test system. The following expression was derived for yarn Lea CSP under good working condition for carded count:

$$Lea \cdot CSP = 280 \times \sqrt{FQI} + 700 - 13C \tag{1}$$

In addition, And for combed count:

$$Lea \cdot CSP = (280 \times \sqrt{FQI} + 700 - 13C) \times (1 + \frac{W}{100}) \tag{2}$$

Where FQI is fiber quality index and arrived from the following formula:

$$FQI = \frac{L \times S \times m}{F} \tag{3}$$

Where

L: 50% span length (mm)
S: Bundle strength ( g/tex)
F: Fineness (micronnaire value)
m: Maturity
C: Yarn count (Ne)
W: Percentage waste extracted during combing

If the maturity coefficient values are not readily available, as a rough approximation the yarn CSP values may be arrived from the following formula:

$$Lea \cdot CSP = 250 \times \sqrt{\frac{LS}{F}} + 590 - 13C \tag{4}$$

For yarn strength, RKm value defined as:

$$RKm = \frac{Single\ Yarn\ strength}{Tex\ Count\ of\ Yarm} \tag{5}$$

RKm value can be predicted from CSP value by following formula:

$$RKm = \frac{Lea.CSP}{150} \tag{6}$$

Chellamani et al. reported a similar relationship between fiber properties and yarn strength [4]. The relationship between the yarn strength and fiber properties was studied and a

Fiber Quality Index was derived to overcome inconsistencies within the basic fiber properties. Then, it was used to predict the strength of the yarn and significant correlations were obtained.

This work reports a new method for prediction ring spun yarn strength from cotton fiber properties by intelligence system such as neural network and neuro-fuzzy methods.

## NEURAL NETWORK

Recently, many researches have utilized a parallel processing structure that has a large number of simple processing structures that has a large number of simple processing with many interconnections between them [5-8]. The use of these processors is much simpler and faster than one central processing unit (CPU). Because of recent advantages in VLSI technology, the neural network has emerged as a new technology and has found wide application in many areas (as well).

In this work, the multi-layer perceptron was used to process data by using the modified back-propagation algorithm. This algorithm attempt to minimize an error function $\Phi$ by modification of network connection weights and bias. The parameters of $\Phi$ are the weights of network and its value is error measure.

In each iteration an input vector is presented to the network and propagated forward to determine the output signal. The output vector is then compared with the target vector resulting an error signal, which is backed propagated through the network in order to adjust the weights and bias. This learning process is repeated until the network respond for each input vector with an output vector that is sufficiently close to the desired one. The general formula for the output of each unit in the network (except for the input units) is given by:

$$y_{i,l} = \varphi\left(\sum_{j=1} \omega_{ij,l} \times y_{j,l-1} + b_{i,l}\right) \tag{7}$$

Where $j$ runs over all nodes of $(l\text{-}1)^{th}$ layer and $\omega_{ij,l}$ is the strength of the coupling between unit $I$ in $l^{th}$ layer and unit j in the previous layer, $y_{j,l-1}$ is the activation of $j^{th}$ unit in $(l\text{-}1)^{th}$ layer, and $b_{i,l}$ is the bias for unit $I$ in $l^{th}$ layer. $\varphi(.)$ Is the nonlinear activation function which can be log-sigmoid (logistic sigmoid), hard limiting, etc., but usually the log-sigmoid function is used, $\varphi(s)=1/91+e^{-s})$.

At each iteration, the values of the weights are modified in the direction in which the error function should decrease most rapidly. The direction and magnitude of the modification is given by the gradient of the error function with proportionality commonly referred to as the learning rate or step size. The formula is, [8]:

$$\omega_{ij}^{n=1} = \omega_{ij}^{n} + \Delta\omega_{ij}^{n} \tag{8}$$

$$\omega_{ij}^{n=1} = \omega_{ij}^{n} - \eta \frac{\partial \phi^{n}}{\partial \omega_{ij}^{n}} \tag{9}$$

Where $\omega_{ij}^{n}$ is the interconnection weight between the $i^{th}$ unit in any layer and $j^{th}$ unit in the previous layer in the $n^{th}$ iteration. $\phi^{n}$ is the error function of network at the $n^{th}$ iteration, and $\eta$ is the learning rate.

The computation of partial derivatives is described in what follows in some detail. In the general back-propagation method, $\Phi$ is defined as the sum of the squared error for all output nodes:

$$\phi = E_{\Sigma} = \|E\|^{2} = \|O - T\|^{2} = \sum_{i=1}^{N0} (O_i - t_i)^2 \tag{10}$$

where $N_0$ is the number of the output nodes. Therefore, the partial derivative of the sum of the squared error, $E_{\Sigma}$, with respect to w is given by:

$$\frac{\partial E_{\Sigma}}{\partial W} = \frac{\partial E_{\Sigma}}{\partial O} \times \frac{\partial O}{\partial X} \times \frac{\partial X}{\partial W} = 2 \times E \times O \times (1 - O) \times Y' \tag{11}$$

Where $w$ is the weights between the last hidden layer and the output layer.

Since, the original back-propagation method converges slowly, the new method used to speed convergence. In this method, the new function $\Phi$ is designed that is given by [9,10]:

$$\phi_{new} = (1-T)^{n} \times [\log_e(s) - \log_e(1-s)] + \sum_{j=1}^{p} \left\{ \begin{array}{c} (\prod_{i=0}^{j-1}(m-i)) \frac{(1-T)^{m-j}}{j!} .(-1)^{j} \\ \left[ \log_e(s) - \binom{j}{1} S + \binom{j}{2} \frac{s^2}{2} + \ldots + \right. \\ \left. (-1)^{r} . \binom{j}{r} \frac{s^{r}}{r} + \cdots + (-1)^{j} . \frac{s^{j}}{j} \right] \end{array} \right\} \tag{12}$$

$$\forall \begin{cases} s = O \quad \text{if} \quad E \rangle 0 \\ s = 2T - O \quad \text{if} \quad E < 0 \end{cases}$$

To minimize where followed to:

$$\frac{\partial \phi}{\partial W} = \frac{\partial \phi}{\partial O} \times \frac{\partial O}{\partial X} \times \frac{\partial X}{\partial W} = sign(E).\left|E\right|^{m} \times Y' \tag{13}$$

The approach of this algorithm is much faster than the other ones designed and provided the better performance

## ANFIS

The fuzzy inference system (FIS) is a popular computing framework based on the concepts of fuzzy set theory, fuzzy IF/THEN rule and fuzzy reasoning to transform an input space into an output space. The basic structure of fuzzy inference system consists of three conceptual components:

1. A rule base
2. Database or dictionary which defines the membership functions used in the fuzzy rules
3. Reasoning mechanism, which performs the inference procedure upon the rule and a given condition to derive a reasonable output or conclusion.

A fuzzy system can be created to match any set input/output data. This can be done with an adaptive neuro-fuzzy inference system (ANFIS). ANFIS is about taking a fuzzy inference system and training it with a backpropagation algorithm, well known in the artificial neural network (ANN) theory, based on some collection of input/output data[11,12,13,14].

ANFIS consist of a Takagi Sugeno FIS and has a five layered as shown in figure 1. The first hidden layer is for fuzzification of the input and T-norm operators are positioned in the second hidden layer to compute the rule antecedent part. The third hidden layer normalizes the rule strengths followed by the fourth hidden layer where the resultant parameters of the rule are determined. Output layer computes the overall input as the summation of all incoming signals. ANFIS uses backpropagation learning algorithm to determine premise parameters (to learn the parameters related to membership functions) and least mean square estimation to determine the consequent parameters. A step in the learning procedure has two parts: In the first part, the input data are propagated, and the best consequent parameters are estimated by an iterative least mean square method, while the premise parameters are assumed to be fixed for the current cycle during the training set. In the second part, the patterns are propagated again, and in this epoch, backpropagation is used to modify the argument parameters, while the resulting parameters remain fixed. This method is then repeated. The fuzzy inference system is known by numerous other names, such as fuzzy-rule-based system, fuzzy expert system, fuzzy model, fuzzy associative memory, fuzzy logic controller and simply fuzzy system [15, 16, and 17].

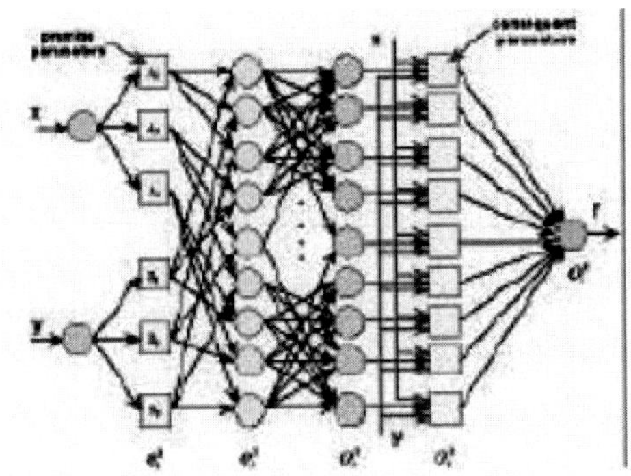

Figure 1. Structure of ANFIS.

## EXPERIMENTAL

In this work, 22 variety quality cotton fibers samples were selected for making relation between cotton fibers properties and yarn quality. The cotton fiber properties such as spun length, fiber bundle strength, maturity; micronnaire was measured by HVI/FMT system. Ring spun yarn was spun in Rieter spinning system and their quality was measured. The properties of cotton fiber and quality of spun yarn are shown in tables 1.

**Table 1. Cotton fibers Properties and yarns quality**

| Sample No. | Yarns quality | | Cotton fibers properties | | | |
|---|---|---|---|---|---|---|
| | Count (Ne) | RKm | 50% span length (mm) | Micronnair | Maturity ratio | Fiber bundle Strength ( g/tex) |
| 1 | 41 | 14 | 14 | 2.69 | 0.46 | 20.21 |
| 2 | 39 | 16.95 | 13.54 | 4.6 | 0.86 | 23.38 |
| 3 | 40.2 | 16.56 | 13.83 | 4.04 | 0.82 | 18.47 |
| 4 | 40.2 | 17.6 | 13.76 | 3.79 | 0.87 | 17.99 |
| 5 | 39.5 | 14.57 | 11.57 | 3.79 | 0.88 | 20.88 |
| 6 | 39.8 | 15.37 | 13.03 | 3.47 | 0.75 | 21.11 |
| 7 | 40.6 | 16.24 | 13.48 | 4.53 | 0.86 | 22.86 |
| 8 | 40 | 14.47 | 13.38 | 3.39 | 0.66 | 16.65 |
| 9 | 40.7 | 17 | 13.17 | 3.34 | 0.66 | 16.28 |
| 10 | 40 | 16.3 | 14.77 | 3.69 | 0.81 | 19.77 |
| 11 | 39.9 | 15.4 | 13.78 | 3.55 | 0.82 | 18.86 |
| 12 | 39 | 15.8 | 12.6 | 3.21 | 0.72 | 19.58 |
| 13 | 39.2 | 15.7 | 14.37 | 5.19 | 0.9 | 18.54 |
| 14 | 40 | 15.07 | 13.38 | 3.61 | 0.72 | 18.58 |
| 15 | 29.8 | 16.04 | 13.3 | 3.04 | 0.54 | 19.99 |
| 16 | 24 | 13.9 | 13.97 | 3.58 | 0.66 | 19.26 |
| 17 | 29.5 | 14.78 | 13.41 | 3.06 | 0.6 | 19.42 |
| 18 | 29.3 | 15.96 | 13.26 | 3.33 | 0.63 | 19.36 |
| 19 | 38.8 | 15.47 | 13.72 | 3.48 | 0.69 | 19.64 |
| 20 | 40.4 | 14.36 | 14.76 | 3.56 | 0.76 | 20.2 |
| 21 | 40.6 | 14.43 | 13.82 | 3.84 | 0.64 | 19.69 |
| 22 | 28.9 | 16.7 | 14.72 | 4.31 | 0.86 | 20.4 |

## RESULTS AND DISCUSSIONS

At first, the relationship between fiber properties and strength of spun yarn was study by using Sitra's expression (Equation 1). The predicted value of yarn RKm for various samples is given in table 2. in second method, two multiple linear regressions was used to correlate yarn strength to cotton fiber properties. In first method of regression, Lea CSP of yarn was calculated from 50% span length (L), Bundle strength(S), Fineness ( F), Maturity(M) of cotton fibers and yarn count (C) by Equation 14:

$$RKm = 9.73 + 0.229 \times L - 0.08 \times F + 3.7 \times M - 0.027 \times S + 0.0249 \times C \quad (14)$$

In second method, RKm value of yarn was calculated from fiber quality index (FQI) and yarn count (C) by equation 15:

$$RKm = 12 + 0.0527 \times C + 0.0308 \times FQI \quad (15)$$

In neural network technique, network has two input nods, one output nod and two hidden layers respectively with 4 and 2 nods. Two input nods referred to fiber quality index (FQI) and yarn count and one output nod referred to yarns strength. Neural network training was continued over 1000 epochs by back propagation algorithm. After training, the neural network was tested with all sample. The results of yarns strength prediction are given in tables 2 and 3.

In neuro-fuzzy method, ANFIS system applied to prediction yarns strength from cotton fibers properties. ANFIS system was designed with two input, 81 IF-THEN rule and one output (figure 2 ). Two input nods referred to fiber quality index (FQI) and yarn counts. One output nod referred to spun yarn strength. First input had three Gaussian shape membership function and second input had four Gaussian shape membership function. Network was trained by back-propagation algorithm. After training, the ANFIS was tested with all samples. The predicted results are shown in tables 2 and 3.

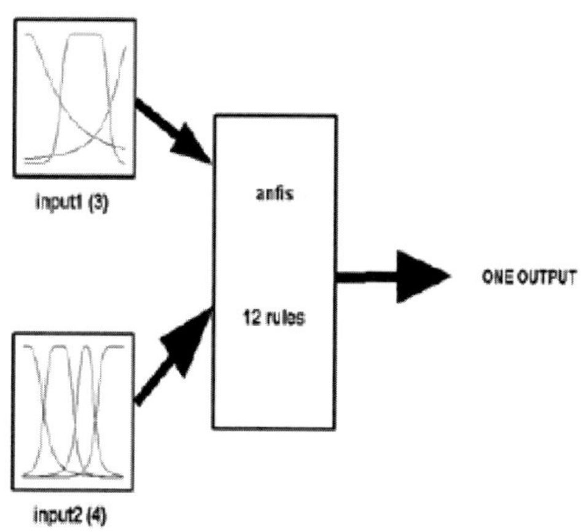


Figure 2. The ANFIS system structure.

**Table 2. Actual and predicted RKm value of ring spun yarns**

| Sample No. | Actual Value | Regression Method I | Regression Method II | Sitra Method | Neural network Method | Neuro-fuzzy Method |
|---|---|---|---|---|---|---|
| 1 | 14 | 14.90 | 15.65 | 14.76 | 13.88 | 14.00 |
| 2 | 16.95 | 15.98 | 15.88 | 16.31 | 16.83 | 16.95 |
| 3 | 16.56 | 16.11 | 15.72 | 15.29 | 16.66 | 16.56 |
| 4 | 17.6 | 16.31 | 15.87 | 15.92 | 16.96 | 17.60 |
| 5 | 14.57 | 15.75 | 15.81 | 15.89 | 15.87 | 14.57 |
| 6 | 15.37 | 15.63 | 15.93 | 16.28 | 15.38 | 15.41 |
| 7 | 16.24 | 16.03 | 15.94 | 16.09 | 16.25 | 16.24 |
| 8 | 14.47 | 15.51 | 15.44 | 14.16 | 14.46 | 14.47 |
| 9 | 17 | 15.49 | 15.45 | 13.96 | 17.02 | 17.00 |
| 10 | 16.3 | 16.28 | 16.08 | 16.81 | 16.31 | 16.30 |
| 11 | 15.4 | 16.12 | 15.95 | 16.34 | 15.3 | 15.36 |
| 12 | 15.8 | 15.47 | 15.76 | 15.84 | 15.21 | 15.78 |
| 13 | 15.7 | 16.41 | 15.49 | 14.62 | 15.75 | 15.70 |
| 14 | 15.07 | 15.66 | 15.64 | 15.01 | 14.99 | 15.07 |
| 15 | 16.04 | 14.73 | 15.03 | 15.58 | 16.12 | 16.05 |
| 16 | 13.9 | 15.16 | 14.79 | 16.40 | 13.92 | 13.90 |
| 17 | 14.78 | 14.99 | 15.13 | 16.12 | 14.84 | 14.78 |
| 18 | 15.96 | 15.04 | 15.04 | 15.80 | 15.96 | 15.95 |
| 19 | 15.47 | 15.58 | 15.69 | 15.61 | 15.2 | 15.48 |
| 20 | 14.36 | 16.10 | 16.09 | 16.72 | 14.37 | 14.00 |
| 21 | 14.43 | 15.43 | 15.54 | 14.39 | 14.41 | 16.95 |
| 22 | 16.7 | 16.11 | 15.37 | 17.28 | 16.73 | 16.56 |

**Table 3. The absolute error of yarns strength prediction**

| No. | Methods | Mean | SD | Max | Min |
|---|---|---|---|---|---|
| 1 | Sitra | 5.960 | 5.551 | 17.986 | 0.253 |
| 2 | Regression | 5.594 | 3.449 | 12.047 | 0.253 |
| 3 | Regression | 5.108 | 3.205 | 12.117 | 0.123 |
| 4 | Neural network | 1.069 | 2.043 | 8.922 | 0 |
| 5 | Neuro-fuzzy | 0.984 | 3.721 | 17.464 | 0 |

## CONCLUSIONS

In this chapter, Intelligence System was used to prediction ring spun yarns strength from cotton fiber properties and compared with conventional method. The ring spun strength was predicted from HVI/FMT measured properties of cotton fibers such as span length, bundle strength, fineness, uniformity ratio and maturity. The fiber quality index (FQI) was defined from 50% span length, Bundle strength , Fineness (micronnaire value), Maturity of cotton fibers. The FQI, as a quality parameter of cotton fiber, was used in making relationship between yarns strength and cotton fibers properties. Several methods such as neural networks, neuro-fuzzy and multiple-linear regression techniques and Sitra's expressions were used to predict yarns strength. The results conclusively prove the superiority of intelligence system over multiple-linear regression techniques and Sitra's expressions. The best results was achieved by neuro-fuzzy method.

## REFERENCES


[1] Sette, S.; L. Boullart ; L. Van Langenhove and P. Kiekens 1997. " Optimizing the Fiber-to-Yarn Production Process With a Combined Neural Network/Generic Algorithm Approach". *Textile Research Journal,* 67(2), 84-92.

[2] Ramey H.H. ; R. Lawson R. and S.Worley 1977. " Relationship of Cotton Fiber Properties to Yarn Tenacity". *Textile Research Journal,* 47(10), 685-691.

[3] Chellmani K.P. ; K. Gnanasekar ; M.S. Ravindran and T.V. Ratnam 1995. " Fiber Yarn Relationships Using HVI/FMT Measured Fiber Properties". The South India Textile Research Association (SITRA), Vol.41, No. 1.

[4] Chellamani P. ; Indra Doraiswamy; T.V. Ratnam 1990. " Fiber Quality and Yarn Strength Relationships". *Indian Journal of Fiber & Textile Research,* Vol. 15, 1-5.

[5] Bishop J.M. , Bushnel M.J. and Westland S., " Application of Neural Networks to computer recipe Prediction", *Color Res. Appl. J.* 16,(1991)

[6] Westland S.,"Advances in Artificial Intelligent for the Color Industry", *J. Soc. Dyers Color,* 110, (1994), 370-375.

[7] Amirshahi S.H., Roushan_Zamir J.M. and Torkamani-Azar F.,"An Attempt to Application of Neural Networks in Recipe Prediction", *Int. J. Eng. Science,* 11, (200), 51-59.

[8] Rumelhart G. E., Hilton G.E. and Williams R.G.," Learning Internal Representation by Error Propagation, in: Parallel Distributed Processing", Vol.1, Chp. 8, Cambridge, MA, MIT. Press, (1986).

[9] Torkamani-Azar F. "Comparative studies of Diffusion Equation Image Recovery Methods with an Improved Neural Network Embedded Technique" PhD Thesis, The University of New South Wales, (Jan. 1995).

[10] Torkamani-Azar F.,"A Modified Back Propagation Algorithm", Second Annual Computer Society of Iran Conference, Tehran,(Dec. 1996), 217-226.

[11] Marjoniemi M. and E. Mantysalo 1997. " Neuro-Fuzzy Modeling of Spectroscopic Data. Part A: Modeling of Dye Solutions". *J.S.D.C.,* Vol.113, 13-17.

[12] Marjoniemi M. and E. Mantysalo 1997. " Neuro-Fuzzy Modeling of Spectroscopic Data. Part B: Dye Concentration Prediction", *J.S.D.C.,* Vol.113, 64-67.

[13] Jang J.S.R. 1993. " ANFIS: Adaptive-network-based fuzzy Inference System". *IEEE Trans. On Sys., Man.* Cyb., 23.

[14] Nariman-Zadeh N. and a. Darvizeh 2001. " Design of Fuzzy System for the Modeling of Explosive Cutting Process of Plates Using Singular Value Decomposition". WSES 2001 Conf. On fuzzy sets and fuzzy systems (FSFS, 01) , spain( Feb).

[15] Jang R, *Neuro-Fuzzy Modeling: Architectures, Analyses and Applications*, PhD Thesis, University of California, Berkeley, July 1992.

[16] A. Abraham, "Neuro Fuzzy Systems: State-of-the-art Modeling Techniques", http://ajith.softcomputing.net,2007.

[17] J.S.R. Jang, C.T. Sun, E. Mizutani, "Neuro fuzzy and Soft Computing" , The Prentice-Hall, Inc. USA, 1997.

In: Monomers, Oligomers, Polymers, Composites... ISBN: 978-1-60456-877-6
Editors: R. A. Pethrick, G.E. Zaikov et al. 

*Chapter 14*

# TECHNOLOGICAL ADVANCES IN GEOTEXTILES

***A.H. Tehrani and A. K. Haghi****
University of Guilan, P. O. Box 3756, Rasht, Iran

## ABSTRACT

According to the need and ascendant approach of technical textiles and the absence of its technical application as respect to the construction industry, this chapter is intended to provide a comprehensive and critical review on geotextiles. This should be of value to those interested in civil, chemical, textile and polymer engineering.

## 1. ABRASION RESISTANCE OF THERMALLY BONDED 3D NONWOVEN FABRICS

Employing the established standard Martindale abrasion method, the abrasion resistance of the three-dimensional nonwoven filter sample produced using the recently developed air laid web formation process and through-air thermal bonding process and the commercially available polypropylene (PP)/polyester (PET) (sheath/core) bi-component staple fiber as raw material, has been evaluated. Results obtained indicated that there are two failure forms for the thermally bonded nonwoven samples during the abrasion testing, the peeling of the fiber PP sheath and the pilling forming and breaking off. The process parameters, including the bonding temperature, dwell time and hot air velocity and fabric weight clearly affect the abrasion resistance of the thermally bonded nonwoven filter samples. These effects could be correlated with the thermal oxidative degradation of the fiber PP sheath during the thermal bonding process and the compactness of the resulting samples.

---

* Corresponding author e-mail: Haghi@Guilan.ac.ir

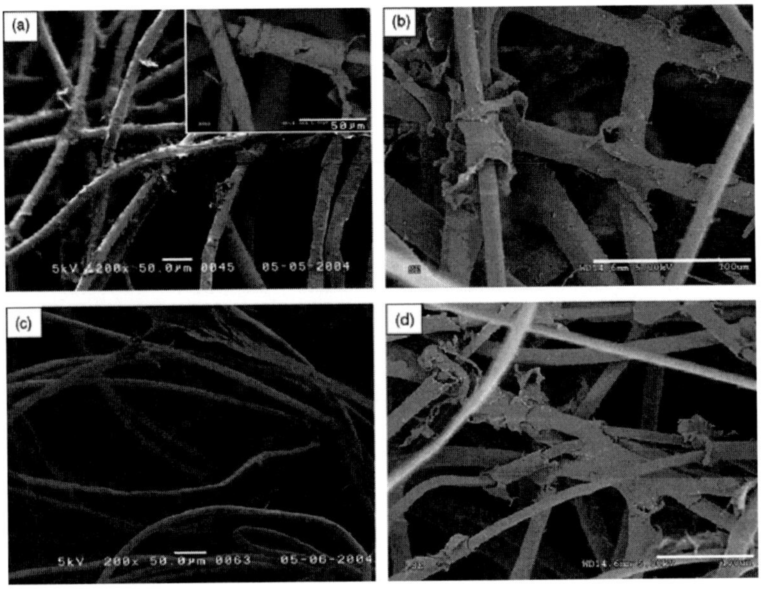


Figure 1. SEM images showing the development of the peeling and wearing away of the PP sheath of the bi-component fiber during the abrasion testing for the thermally bonded nonwoven samples: (a) BT4, surface wear and lifting, the inset shows the details of fiber peeling; (b) AD1, broken fragments of fibers, surface peeling and breakdown of fibers; (c) IV1, fiber failure by multiple splitting or breakage; (d) DT3, peeling away of large pieces.

Technological advances (figures 1-4) in the textile industry have resulted in an ongoing demand for cost-effective processing techniques and textiles. In the past 40 years, nonwoven techniques and products have been developing and growing at a phenomenal rate, corresponding to such a demand [1-6]. The most significant feature of nonwoven fabrics, and also the one that contributes most to their economical appeal, is that the fabrics are usually made directly from raw materials in a continuous production line, thus partially or completely eliminating conventional textile operations, such as carding, roving, and spinning, weaving or knitting. The simplicity of fabric formation, coupled with high productivity, allows nonwovens to compete favorably with woven's and knits on a performance per cost basis in many industrial applications, from simple low cost replacements for more expensive textiles to high-quality textiles and many functions that could never be filled by regular textiles.

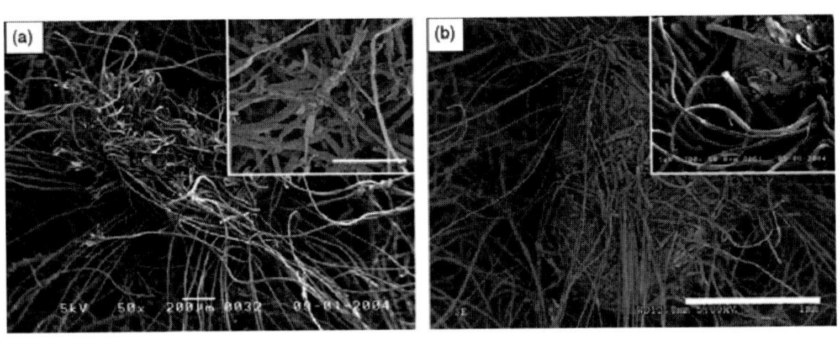


Figure 2. SEM images show (a) the early stage of pill development; (b) the fiber entanglement, splitting and well-developed pill during the abrasion testing for the thermally bonded nonwoven sample DT3. The insets in both images present the details under higher magnification.

Over the past few decades, the overwhelming majority of the research is related to the manufacture and use of nonwovens as essentially two-dimensional (2D) sheet structures. In many applications where three-dimensional (3D) fibrous web structures are needed, they have to be constructed from the flat sheet fabrics. If 3D nonwoven shell structures are produced in a single process, directly from fibers, the packaging, freight and labor costs and the cost of wastage inevitably generated during panel cutting can be saved possibly up to 70% of the production cost.

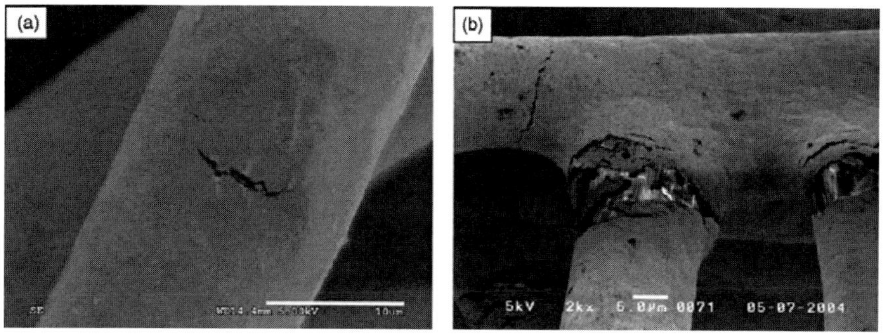


Figure 3. SEM images showing the cracks on the fiber surface of the thermally bonded nonwoven sample: (a) BT1 and (b) BT4.

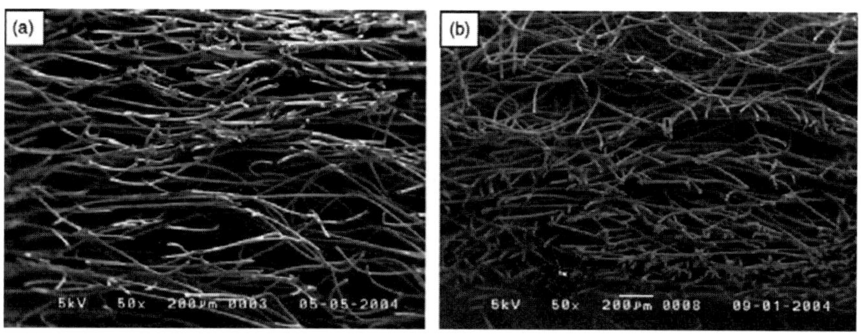


Figure 4. SEM images taken from the cross sections of the thermally bonded nonwoven samples: (a) IV1 produced using the hot air velocity 1.5 m/s and (b) IV3produced using the hot air velocity 4 m/s.

It can also shorten the process and save equipment investment, space and energy. However, 3D shell structure nonwovens are still at the very beginning stage of development. Major difficulties for manufacturing of 3D shell structure nonwovens (figure 5) arise in web formation and web solidification. This is because it is much more difficult to manipulate and control the fiber distribution and bonding of a 3D shell structure web than do those of a 2D web, aiming to achieve nearly uniform and orientation random fiber distribution, and, hence, uniformity and isotropic properties of the final product[7-11]. Up to date, there have been only a few reports on the production of 3D nonwoven structures. Recently, a pilot process to air-form 3D nonwoven fabrics, using the air laid web formation process and through-air bonding technique, has been developed. Samples of the hats prepared using this process have been proved to be remarkably even and strong, and are about to be commercialized by the only chef's hat producer in the UK who previously sponsored the research work. The principles of this process were promising and further studied in the present work for a variety of other products, especially filters, finger-shaped towel and garment interlinings, e.g., bra cups.

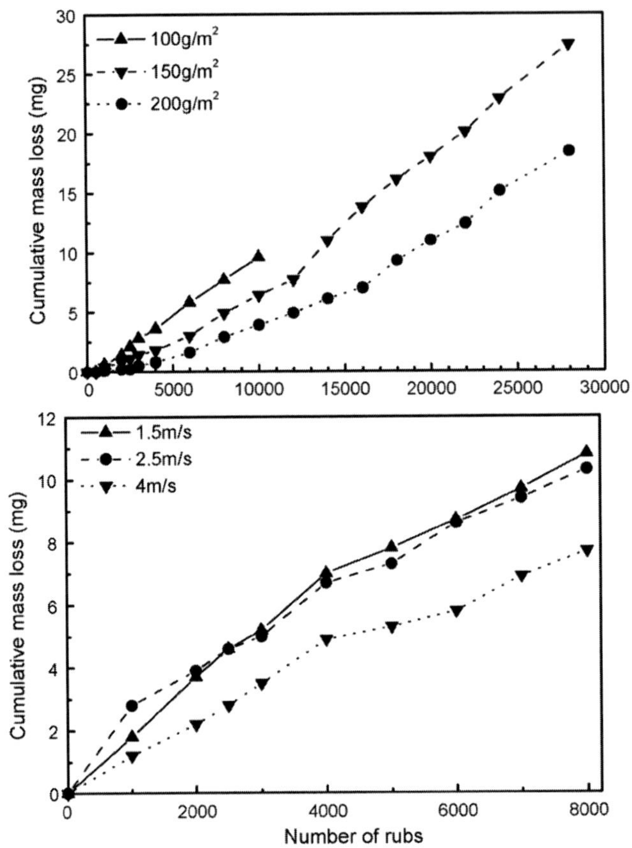


Figure 5. Cumulative mass losses of nonwoven fabrics produced using different hot air velocities and Cumulative mass losses of nonwoven fabrics with and different fabric area densities.

As the conclusion the abrasion resistance of the 3D nonwoven filter samples, produced using the air laid web formation process and through air thermal bonding process, has been evaluated using the established standard Martindale abrasion method. The effects of the bonding temperature, dwell time, air velocity and fabric weight on the abrasion resistance were considered and the following conclusions can be drawn:

1. SEM examination reveals that there are two failure forms for the thermally bonded nonwoven samples during the abrasion tests, the peeling of the fiber PP sheath, and the pilling forming and breaking off.
2. The abrasion resistance of the thermally bonded nonwoven fabrics decreases with increasing the thermal bonding temperature. This can be attributed to the fact that more mechanical defects, such as cracks, may be formed resulting from a higher lever of the thermal oxidative degradation and a higher amount shrinkage of the PP sheath of the fiber at a higher bonding temperature.

3. The abrasion resistance of the nonwoven fabrics increases up to a maximum, and, then, decreases with further increasing the dwell time. The poor abrasion resistance of a nonwoven fabric produced using a shorter dwell time can be ascribed to the under-bonding between the fibers. The poor abrasion resistance of a nonwoven fabric produced using a longer dwell time may be due to a higher lever of the thermal oxidative degradation of the PP sheath, resulting in the peeling and multiple splitting of the PP sheath more serious.
4. The abrasion resistance of the thermally bonded nonwoven samples is also significantly affected by the hot air velocity and the fabric weight. It increases with the increasing both the hot air velocity and the nonwoven fabric weight. This can be due to the fact that the thermally bonded nonwoven samples become more compact with the increasing both the hot air velocity and the fabric weight, hence leading to the higher abrasion resistance.
5. The effects of the bonding temperature, dwell time, air velocity and fabric weight on the abrasion resistance indicate that the various process parameters are all critical for achieving the most abrasion-resistant fabric.

## 2. Internal Deformation Behavior of Geosynthetic-Reinforced Soil Walls

Local deformation of geosynthetics, such as Geogrids, and nonwoven and woven geotextiles, was measured to analyze the stability of geosynthetic-reinforced soil (GRS) structures. To analyze the deformation behavior of geosynthetics applied to a reinforced soil structure, the tensile load–elongation properties of the geosynthetic and local deformation measurement data are required. However, local deformation of nonwoven geotextile (NWGT), which is permeable, is difficult to measure with strain gauges. This study proposes a new, more convenient, method to measure the deformation behavior of NWGTs using a strain gauge and examines its suitability via laboratory tests and field trials on two GRS walls. A wide-width tensile test, conducted under a confining pressure of 70 kPa, showed that local deformation of NWGT, measured with strain gauges of type AE-11-S80N-120-EL, was similar to total deformation measured with linear variable deformation transformer (LVDT). In field trials, NWGT showed a larger deformation range than woven geotextile or Geogrids. However, the deformation patterns of the three materials were similar. The strain gauges attached to NWGT in the walls worked normally for 16 months. Therefore, the method proposed in this study for measuring NWGT deformation using a strain gauge was effective and valuable. Pore water pressure in the GRS wall can be ignored since the backfill remains unsaturated regardless of rainfall. However, it should be noted for design purposes that horizontal earth pressures at the wall face are greater at the bottom and top of the wall than at rest.

Reinforced walls have been the subject of considerable research, and a number of recent papers have examined different aspects of their design and behavior Owing to the increasing need for clayey soil (CL) as backfill in reinforced soil walls, nonwoven geotextiles (NWGTs) with drainage capability have received attention. NWGTs have the merit of high drainage capability and low cost, but also have a drawback of low tensile stiffness and higher deformability than Geogrids or woven geotextiles. To analyze the deformation behaviors of reinforcements, load–elongation properties and local deformation measurement data are needed; however, measuring local deformations of GRS walls in the field is problematic. It is suggested using silicon to attach strain gauges to the woven geotextile.

Strain gauges can be attached directly to woven geotextiles and Geogrids (however, it is not easy to measure deformation of NWGTs by direct attachment of strain gauges because the gauges separate from the surface of NWGTs to which they are attached as NWGTs are elongated by a tensile force. A method of attaching strain gauges to the surface of NWGT composed of a core layer of knitted textile and needle-punched double layer of NWGT, by using gauge cement (Kyowa, EC-30). This study examines an easy method, using an adhesive, to attach strain gauges to NWGTs, and the applicability of the technique via laboratory and field tests. To analyze deformation behavior of reinforcements within GRS walls, two 5-m walls were constructed on a weak, shallow-layered foundation and fitted with a compound arrangement of NWGTs/woven geotextiles and nonwoven/Geogrids. Deformation behavior inside the GRS walls was analyzed using data collected from four earth-pressure gauges, four pore-water pressure gauges, and 124 strain gauges attached to NWGTs and woven geotextiles and Geogrids reinforcements over a period of about 1.5 years.

As a conclusion a laboratory wide-width tensile test conducted under a confining pressure of 70 kPa showed that the pattern of local deformation on NWGT measured with strain gauges resembled that of the total deformation measured with LVDT. In GRS walls, NWGT showed a larger deformation range than the woven geotextile or Geogrids. However, deformation patterns of these three reinforcement materials were similar and the strain gauges attached to the geosynthetics functioned normally for 16 months. Therefore, the method of measuring a NWGT deformation by using a strain gauge, as suggested by this study, was effective. The backfill material probably remained unsaturated regardless of rainfall because there were no signs of drainage through NWGT from the backfill, and the pore water pressures throughout the measurement period showed negative values. Therefore, pore water pressures in the wall can be ignored. However, horizontal earth pressures at the wall face were larger at the bottom and top of the wall than earth pressures at rest. Therefore, when a GRS wall with a flexible wall face is constructed on a shallow, weak foundation, as in this study, precautions must be taken during the design and construction of the wall, since the horizontal earth pressure can be larger than earth pressure at rest at the bottom of the wall.

## 3. Comparative Study between Needle Punch Nonwoven Geotextile Structures Made from Flax and Polyester Fibers

Geotextiles are widely used in civil and geotechnical engineering applications. In this study, a comparison is made between the properties of the needle punched nonwoven geotextiles produced from polyester and flax fibers. The properties of geotextiles including density, pore size and permeability have been investigated. It has been found that large inherent variation in flax fiber length and fineness can result in loss of tensile strength and cause large variation in smallest detected pore diameter. Nevertheless, flax fiber-based geotextiles have a great potential in various civil engineering applications as they are found to be less anisotropic, more compact and have produced an "open structure". Furthermore, the influence of process parameters, namely feed rate, stroke frequency and depth of needle penetration, on the properties of the geotextiles has been observed. Geotextiles are woven, knitted or nonwoven structures, widely used in civil engineering applications and are a core member of geosynthetic family. Recently, there has been an ever-increasing interest in nonwoven geotextiles as these structures are simple, flexible and can be produced at a lower cost. It is essential that the geotextile structures should be able to fulfill more than one function, i.e. separation, drainage and filtration. The design of a geotextile accounts for soil–geotextile interaction, internal stability of the soil, isotropic characteristics of the geotextile structure and flow conditions, i.e. seepage velocities and hydraulic gradients.

Comparison of tensile properties of geotextiles produced from flax and polyester fibres

| Sample ID | Tensile strength in machine direction (TMD) (N/50 mm) | | Tensile strength in cross-machine direction (TXMD) (N/50 mm) | | TXMD/TMD | |
|---|---|---|---|---|---|---|
| | Polyester | Flax | Polyester | Flax | Polyester | Flax |
| G1 | 107.05 | 58.20 | 164.88 | 83.0 | 1.54 | 1.43 |
| G2 | 168.08 | 75.20 | 287.04 | 118.2 | 1.71 | 1.57 |
| G3 | 156.44 | 43.80 | 220.56 | 52.2 | 1.41 | 1.19 |
| G4 | 241.90 | 99.00 | 386.78 | 91.0 | 1.60 | 0.92 |
| G5 | 139.32 | 70.68 | 225.48 | 68.56 | 1.62 | 0.97 |
| G6 | 197.48 | 99.84 | 348.32 | 124.8 | 1.76 | 1.25 |
| G7 | 167.82 | 57.40 | 248.54 | 68.8 | 1.48 | 1.20 |
| G8 | 284.48 | 104.16 | 455.92 | 135.4 | 1.60 | 1.30 |
| G9 | 289.10 | 113.04 | 467.8 | 148.1 | 1.62 | 1.31 |
| G10 | 147.04 | 55.80 | 227.04 | 65.2 | 1.54 | 1.17 |
| G11 | 234.24 | 76.80 | 344.22 | 114.4 | 1.47 | 1.49 |
| G12 | 116.40 | 85.40 | 208.94 | 134.6 | 1.80 | 1.58 |
| G13 | 253.34 | 84.60 | 388.76 | 93.91 | 1.53 | 1.11 |
| G14 | 128.08 | 91.92 | 198.72 | 113.06 | 1.55 | 1.23 |
| G15 | 224.62 | 83.04 | 367.02 | 119.57 | 1.63 | 1.44 |

However, the selection of fiber and fabric types is of paramount importance for any geotextile-based application. The advantages of natural fibers are robust, superior strength/durability properties, good drape ability and biodegradability/ environment friendliness but these fibers are biocompatible. Nevertheless, several researchers have reported the use of natural fibers including jute, coir, wood and bamboo in various applications such as soil erosion control, vertical drains, road bases, bank protection and slope stabilization.

Therefore, the overall objective of the present work is to compare and analyze the properties of needle punched nonwoven geotextiles produced from flax and polyester fibers under the similar process conditions. Fifteen samples of needle punched nonwoven geotextile structures were produced from each polyester and flax fiber by employing a central composite experimental design as shown in table 1. The fineness and length of fiber samples were normalized to 6 dtex and 60 mm, respectively. However, the coefficients of variation in fineness and length of the flax fiber were found to be 25.35% and 26.85%, respectively. The needle punched nonwoven geotextile structures were produced by initially opening the staple fiber bale by carding and subsequently orientated to cross-machine direction using a cross-lapper to form a web of required area density. Following the production of the needle punched nonwoven fabrics, standard tests were performed on the fabrics to determine their area density (ASTM D3776, 1996), thickness (ASTMD5729, 1995), the pore size (ASTM E1294, 1999), and water permeability (EN ISO 11058, 1999).

Comparison of dimensional properties of geotextiles produced from flax and polyester fibres

| Sample ID | Fabric weight (g/m²) | | | Fabric thickness (mm) | | | Fabric density (g/cm³) | | |
|---|---|---|---|---|---|---|---|---|---|
| | Polyester | Flax | Polyester/flax | Polyester | Flax | Polyester/flax | Polyester | Flax | Polyester/flax |
| G1 | 284.2 | 128.3 | 2.22 | 3.32 | 1.66 | 2.00 | 0.0856 | 0.0773 | 1.11 |
| G2 | 407.6 | 170.4 | 2.39 | 4.46 | 2.08 | 2.14 | 0.0914 | 0.0819 | 1.12 |
| G3 | 261.6 | 93.8 | 2.79 | 2.94 | 1.17 | 2.51 | 0.0890 | 0.0804 | 1.11 |
| G4 | 372.3 | 165.0 | 2.26 | 3.77 | 2.59 | 1.46 | 0.0987 | 0.0890 | 1.11 |
| G5 | 259.6 | 117.8 | 2.2 | 2.7 | 1.32 | 2.05 | 0.0962 | 0.0637 | 1.51 |
| G6 | 366.3 | 166.4 | 2.2 | 3.55 | 1.91 | 1.86 | 0.1032 | 0.0872 | 1.18 |
| G7 | 236.1 | 112.8 | 2.09 | 2.37 | 1.25 | 1.9 | 0.0996 | 0.0869 | 1.15 |
| G8 | 339.1 | 173.6 | 1.95 | 3.00 | 1.99 | 1.51 | 0.1130 | 0.0903 | 1.25 |
| G9 | 363.9 | 188.4 | 1.93 | 3.85 | 2.15 | 1.79 | 0.0945 | 0.0874 | 1.08 |
| G10 | 228.7 | 93.0 | 2.46 | 2.48 | 0.99 | 2.51 | 0.0922 | 0.0942 | 0.98 |
| G11 | 263.7 | 128.0 | 2.06 | 2.85 | 1.48 | 1.93 | 0.0925 | 0.0864 | 1.07 |
| G12 | 296.3 | 145.0 | 2.04 | 3.47 | 1.74 | 1.99 | 0.0854 | 0.0835 | 1.02 |
| G13 | 303.6 | 141.0 | 2.15 | 2.77 | 1.86 | 1.49 | 0.1096 | 0.0757 | 1.45 |
| G14 | 312.4 | 153.2 | 2.04 | 3.61 | 1.99 | 1.81 | 0.0865 | 0.0767 | 1.13 |
| G15 | 307.6 | 138.4 | 2.22 | 3.26 | 1.83 | 1.78 | 0.0944 | 0.0755 | 1.25 |

As a conclusion a comparison is made between the properties of the needle punched nonwoven geotextiles produced from flax and polyester fibers using the same experimental matrix. Polyester fiber-based geotextiles are found to be dense corresponding to the flax fiber-based geotextiles due to high elastic recovery of polyester fiber. Although the fibers in the nonwoven structure were preferentially orientated in the cross-machine direction, the flax fiber-based geotextiles resulted in loss of tensile strength in the cross-machine direction (sample IDs G4 and G5) due to large inherent variation in length and fineness of the flax fiber. This has also led to the large variation in the smallest detected pore diameter in flax fiber geotextiles in comparison to the polyester fiber-based geotextiles. Nevertheless, flax-based geotextiles are found to be more compact and less anisotropic in nature. In addition, flax fiber geotextiles have produced an "open structure" at a lower FR, moderate SF and depth of NP (sample ID G10).

## 4. Dynamic Studies of Polypropylene Nonwovens in Environmental Scanning Electron Microscope

Environmental scanning electron microscopy (ESEM) provides new tools to examine the dynamic behavior of various materials under different conditions. The dynamic experiments of water wetting, oil sorption and loading deformation of polypropylene (PP) nonwovens in the ESEM were studied in this paper. Water wetting tests were performed by controlling the temperature of the specimens and chamber pressure in favor of water condensation at 100% relative humidity. The wetting by oil was made using a micro-injector to add oil droplets onto specimens being observed. The ESEM observations revealed the contrast in the wetting behavior of the PP nonwovens towards water and oil. Tensile testing experiments were performed in the ESEM using a tensile stage. The dynamic studies gave new insight into microscopic behavior of PP nonwovens.

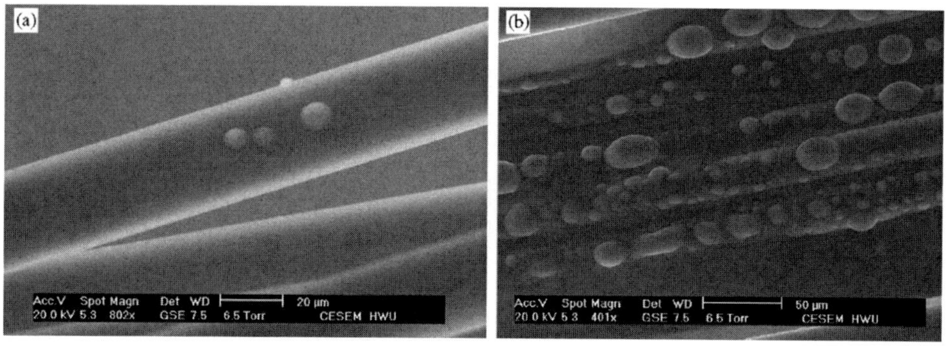


Figure 6. Wetting of needle punched polypropylene nonwoven by water.

The nonwoven industry is one of the fastest growing industries in the world [1]. Technological innovation and commercial development have been driving the industry into a sophisticated and diverse market with versatile products for a wide spectrum of applications in many industries, such as agricultural, automotive, building and construction, medical and hygiene, packaging, protective clothing, sportswear, transport, defense, leisure and safety.

One of the most important factors fuelling the growth of nonwovens is the development and application of man-made fibers, dominated by polyolefins [2].

For these increasing applications, nonwoven products are specially engineered to create the structure that gives the product its characteristic properties. The characterization of a nonwoven material under varying or dynamic conditions is of importance in understanding how the particular structure of the material is engineered, and, therefore, how it relates to the properties of the product.

Microscopy technology has provided the tools for the observation, analysis, and explanation of phenomena occurring at a micrometer scale of textile materials [3]. Scanning electron microscopes (SEMs) have long been important instruments in these studies. However, the high vacuum and the imaging process in SEMs impose special requirements for specimen preparation. Specimens that are not naturally conductive must be coated with a thin layer of a conductive material to bleed off any charge on the specimens imposed by the incident electron beam. One major disadvantage of the SEM is that it is normally not possible to examine wet specimens or dynamic processes of specimens in a wet state. Environmental scanning electron microscopy (ESEM) is a newer development in microscope technology, which is specifically suited to dynamic experimentation on the micron scale [4].

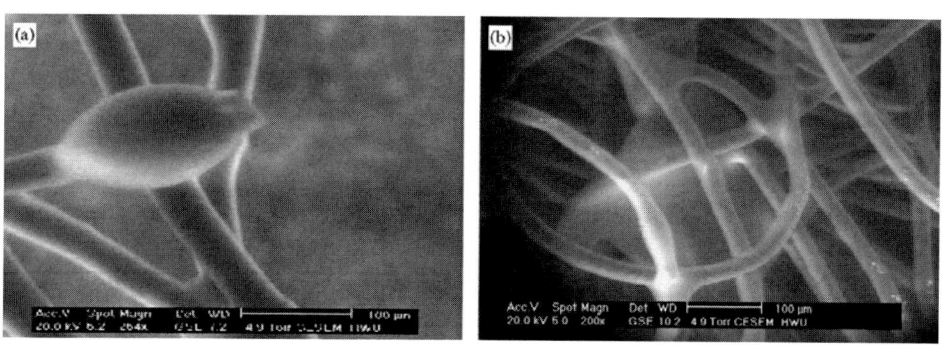


Figure 7. Oil adsorption of needle punched nonwoven.

The Philips XL30 ESEM was used for the dynamic experiments of water wetting, oil sorption and tensile deformation of polypropylene (PP) nonwovens in this study.

Figure 8. Loading deformation of needle punched nonwoven: (a) before tensile deformation; (b) deformation at 50% strain; (c) deformation at 100% strain and (d) deformation at 150% strain.

## Conclusion

As a conclusion this study has explored the use of the ESEM for the examination and observation of PP nonwovens under varying conditions. The ability of the ESEM to follow dynamic events under a variety of conditions gives new insight into the dynamic wetting of water on fiber surface, the oil sorption in nonwovens and the tensile behavior of nonwoven materials. Direct observations of water droplets on the PP nonwovens reveal the high contact angles, indicating the hydrophobic properties of the fibers. The oil contact angles are lower than 201, indicating the affinity of PP fibers for oil. The ESEM studies also show the oil sorption on individual fibers, fiber intersections and fibrous pores in the PP nonwovens. The tensile testing in the ESEM gives some further evidence on the dynamic process of fiber orientation, rearrangement and breakdown in the PP nonwovens. ESEM provides a new and powerful approach to imaging of textile materials in connection with such applications as filtration, medical and hygiene, composites and biomedical products. The potential of use of ESEM in textile research and development is promising and significant.

## References

[1] X.Y. Wang＊, R.H. Gong, Z. Dong, I. Porte Textiles and Paper, School of Materials, the University of Manchester, P.O. Box 88, Sackville Street, Manchester, M60 1QD, UK Received 8 September 2005; received in revised form 23 May 2006; accepted 14 June 2006

[2] Myoung-Soo Won_, You-Seong KimDepartment of Civil Engineering, Chonbuk National University, 664-14 Deogjin-dong 1Ga, Deogjin-gu, Jeonju, Jeollabuk-do 561-756, South Korea Received 16 April 2006; received in revised form 29 September 2006; accepted 16 October 2006

[3] Amit Rawala,_, Rajesh Anandjiwalaa,b aCSIR, National Fiber, Textile and Clothing Centre, Port Elizabeth 6000, South Africa Department of Textile Science, Faculty of Science, Nelson Mandela Metropolitan University, Port Elizabeth, South Africa Received 10 February 2006; received in revised form 21 August 2006; accepted 27 August 2006

[4] Qufu Weia,b,_, Ya Liua, Xueqian Wangb, Fenglin Huanga aKey Laboratory of Science and Technology of Eco-textile Ministry of Education, Southern Yangtze University, Wuxi 214122, PR China bAnhui University of Technology and Science, Wuhu 241000, PR China Received 11 June 2006;

[5] Abdelmalek Bouazza_, Michelle Freund1, Hani Nahlawi Department of Civil Engineering, Building 60, Monash University, Melbourne, Vic. 3800, Australia Received 5 May 2006;

[6] Cho-Sen Wua,_, Yung-Shan Honga, Yun-Wei Yanb, Bow-Shung Changb aDepartment of Civil Engineering, Tamkang University, Tamsui, Taipei 25137, Taiwan bDepartment of Civil Engineering, Tamkang University, Taipei 106, Taiwan Received 18 September 2004; received in revised form 6 September 2005; accepted 7 September 2005

[7] Muhammet V. Akpinara,_, Craig H. Bensonb,1 aDepartment of Civil Engineering, Mustafa Kemal Universities, Tayfur Sokmen Campus, Hatay, Turkey bDepartment of Civil and Environmental Engineering, University of Wisconsin-Madison, Madison, WI 53706, USA

[8] G.L. Hebelera, J.D. Frosta,_, A.T. Myersb aSchool of Civil and Environmental Engineering, Georgia Institute of Technology, 790 Atlantic Drive, Atlanta, GA, USA bGeorgia Institute of Technology, Atlanta, GA, USA

[9] T. Iryoa, R. Kerry Roweb,*Department of Civil and Environmental Engineering, University of Western Ontario, London, Ont., Canada N6A 5B9 bGeoEngineering Centre of Queen's-RMC, Department of Civil Engineering, Queen's University, Ellis Hall, Kingston, Ont., Canada K7L 3N6 Received 10 March 2003; received in revised form 30 May 2003;

[10] D.T. Bergadoa,*, S. Youwaib, C.N. Haic, P. Voottipruexd a Geotechnical Engineering Program, School of Civil Engineering, Asian Institute of Technology, P.O. Box 4, Klong Luang, Pathumthani 12120, Thailand b School of Civil Engineering, Asian Institute of Technology, P.O. Box 4, Klong Luang, Pathumthani 12120, Thailand cThe Polytechnic University of HoChiMinh City, HoChiMinh City, Viet Nam d King Mongkut's Institute of Technology, North Bangkok, Piboonsongkram Rd., Bangsue District, Bangkok, Thailand Received 28 May 2000;

[11] J. Prabakara,*, R.S. Sridharb a*Regional Research laboratory (CSIR), Bhopal, India* b*Coimbatore Institute of Technology, Coimbatore, India* Received 21 July 2000; received in revised form 23 July 2001; accepted 9 January 2002

In: Monomers, Oligomers, Polymers, Composites… ISBN: 978-1-60456-877-6
Editors: R. A. Pethrick, G.E. Zaikov et al. 

*Chapter 15*

# SOME ASPECTS OF HEAT FLOW DURING DRYING OF POROUS STRUCTURES

*A. K. Haghi*[*]
University of Guilan, P. O.Box 3756, Rasht, Iran

## ABSTRACT

In The first part of this chapter a detailed study on different aspects of heat flow in porous structures is presented. In the second part, a mathematical model was developed for optimization of heat and mass transfer in capillary porous media during drying process to predict the drying constants.

## 1.1. INTRODUCTION

For Heat flow analysis of wet porous materials, the liquid is water and the gas is air. Evaporation or condensation occurs at the interface between the water and air so that the air is mixed with water vapor. A flow of the mixture of air and vapor may be caused by external forces, for instance, by an imposed pressure difference. The vapor will also move relative to the gas by diffusion from regions where the partial pressure of the vapor is higher to those where it is lower.

Heat flow in porous media is the study of energy movement in the form of heat which occurs in many types of processes. The transfer of heat in porous media occurs from the high to the low temperature regions. Therefore a temperature gradient has to exist between the two regions for heat transfer to happen. It can be done by conduction (within one porous solid or between two porous solids in contact), by convection (between two fluids or a fluid and a porous solid in direct contact with the fluid), by radiation (transmission by electromagnetic waves through space) or by combination of the above three methods.

The general equation for heat transfer in porous media is:

---

[*] Haghi@Guilan.ac.ir

$$\begin{pmatrix}\text{rate of}\\ \text{heat in}\end{pmatrix} + \begin{pmatrix}\text{rate of generation}\\ \text{of heat}\end{pmatrix} = \begin{pmatrix}\text{rate of}\\ \text{heat out}\end{pmatrix} + \begin{pmatrix}\text{rate of accumulation}\\ \text{of heat}\end{pmatrix}$$

When a wet porous material is subjected to thermal drying two processes occur simultaneously, namely:

a. Transfer of heat to raise the wet porous media temperature and to evaporate the moisture content.
b. Transfer of mass in the form of internal moisture to the surface of the porous material and its subsequent evaporation.

The rate at which drying is accomplished is governed by the rate at which these two processes proceed. Heat is a form of energy that can across the boundary of a system. Heat can, therefore, be defined as "the form of energy that is transferred between a system and its surroundings as a result of a temperature difference". There can only be a transfer of energy across the boundary in the form of heat if there is a temperature difference between the system and its surroundings. Conversely, if the system and surroundings are at the same temperature there is no heat transfer across the boundary.

Strictly speaking, the term *"heat"* is a name given to the particular form of energy crossing the boundary. However, heat is more usually referred to in thermodynamics through the term "heat transfer", which is consistent with the ability of heat to raise or lower the energy within a system.

There are three modes of heat flow in porous media:

- convection
- conduction
- radiation

All three are different. Convection relies on movement of a fluid in porous material. Conduction relies on transfer of energy between molecules within a porous solid or fluid. Radiation is a form of electromagnetic energy transmission and is independent of any substance between the emitter and receiver of such energy. However, all three modes of heat flow rely on a temperature difference for the transfer of energy to take place.

The greater the temperature difference the more rapidly will the heat be transferred. Conversely, the lower the temperature difference, the slower will be the rate at which heat is transferred. When discussing the modes of heat transfer it is the rate of heat transfer Q that defines the characteristics rather than the quantity of heat.

As it was mentioned earlier, there are three modes of heat flow in porous structures, convection, conduction and radiation. Although two, or even all three, modes of heat flow may be combined in any particular thermodynamic situation, the three are quite different and will be introduced separately.

The coupled heat and liquid moisture transport of porous material has wide industrial applications. Heat transfer mechanisms in porous textiles include conduction by the solid material of fibers, conduction by intervening air, radiation, and convection. Meanwhile, liquid and moisture transfer mechanisms include vapor diffusion in the void space and moisture sorption by the fiber, evaporation, and capillary effects. Water vapor moves through porous textiles as a result of water vapor concentration differences. Fibers absorb water vapor due to their internal chemical compositions and structures. The flow of liquid moisture through the textiles is caused by fiber-liquid molecular attraction at the surface of fiber materials, which is determined mainly by surface tension and effective capillary pore distribution and pathways. Evaporation and/or condensation take place, depending on the temperature and moisture distributions. The heat transfer process is coupled with the moisture transfer processes with phase changes such as moisture sorption/desorption and evaporation/condensation.

## 1.2. Heat Flow and Drying of Porous Structures

All three of the mechanisms by which heat is transferred- conduction, radiation and convection, may enter into drying. The relative importance of the mechanisms varies from one drying process to another and very often one mode of heat transfer predominates to such extent that it governs the overall process.

As an example, in air drying the rate of heat transfer is given by:

$$q = h_s A(T_a - T_s) \tag{1.1}$$

Where $q$ is the heat transfer rate in $Js^{-1}$, $h_s$ is the surface heat-transfer coefficient in $Jm^{-2}\, s^{-1}$ $°C^{-1}$, A is the area through which heat flow is taking place, $m^{-2}$, $T_a$ is the air temperature and $T_s$ is the temperature of the surface which is drying, °C.

To take another example, in a cylindrical dryer where moist material is spread over the surface of a heated cylinder, heat transfer occurs by conduction from the cylinder to the porous media, so that the equation is

$$q = UA(T_i - T_s) \tag{1.2}$$

Where $U$ is the overall heat-transfer coefficient, $T_i$ is the cylinder temperature (usually very close to that of the steam), $Ts$ is the surface temperature of textile and A is the area of the drying surface on the cylinder. The value of $U$ can be estimated from the conductivity of the cylinder material and of the layer of porous solid.

Mass transfer in the drying of a wet porous material will depend on two mechanisms: movement of moisture within the porous material which will be a function of the internal physical nature of the solid and its moisture content; and the movement of water vapour from the material surface as a result of water vapour from the material surface as a result of external conditions of temperature, air humidity and flow, area of exposed surface and supernatant pressure.

Some porous materials such as textiles exposed to a hot air stream may be cooled evaporatively by bleeding water through its surface. Water vapour may condense out of damp air onto cool surfaces. Heat will flow through an air-water mixture in these situations, but water vapour will diffuse or convect through air as well. This sort of transport of one substance relative to another called mass transfer. The moisture content, $X$, is described as the ratio of the amount of water in the materials, $m_{H2O}$ to the dry weight of material, $m_{material}$:

$$X = \frac{m_{H2O}}{m_{material}} \tag{1.3}$$

There are large differences in quality between different porous materials depending on structure and type of material. A porous material such as textiles can be hydrophilic or hydrophobic. The hydrophilic fibres can absorb water, while hydrophobic fibers do not. A textile that transports water through its porous structures without absorbing moisture is preferable to use as a first layer. Mass transfer during drying depends on the transport within the fiber and from the textile surface, as well as on how the textile absorbs water, all of which will affect the drying process .

As the critical moisture content or the falling drying rate period is reached, the drying rate is less affected by external factors such as air velocity. Instead, the internal factors due to moisture transport in the material will have a larger impact. Moisture is transported in textile during drying through

- capillary flow of unbound water
- movement of bound water and
- vapour transfer

Unbound water in a porous media such as textile will be transported primarily by capillary flow .

As water is transported out of the porous material, air will be replacing the water in the pores. This will leave isolated areas of moisture where the capillary flow continues .

Moisture in a porous structure can be transferred in liquid and gaseous phases. Several modes of moisture transport can be distinguished:

- transport by liquid diffusion
- transport by vapour diffusion
- transport by effusion (Knudsen-type diffusion)
- transport by thermodiffusion
- transport by capillary forces
- transport by osmotic pressure and
- transport due to pressure gradient.

## 1.3. Convection Heat Flow in Porous Media

A very common method of removing water from porous structures is convective drying. Concevtion is a mode of heat transfer that takes place as a result of motion within a fluid. If the fluid, starts at a constant temperature and the surface is suddenly increased in temperature to above that of the fluid, there will be convective heat transfer from the surface to the fluid as a result of the temperature difference. Under these conditions the temperature difference causing the heat transfer can be defined as:

$\Delta T$ = surface temperature-mean fluid temperature

Using this definition of the temperature difference, the rate of heat transfer due to convection can be evaluated using Newton's law of cooling:

$$Q = h_c A \Delta T \tag{1.4}$$

where A is the heat transfer surface area and $h_c$ is the coefficient of heat transfer from the surface to the fluid, referred to as the "convective heat transfer coefficient".

The units of the convective heat transfer coefficient can be determined from the units of other variables:

$$\begin{aligned} Q &= h_c A \Delta T \\ W &= (h_c) m^2 K \end{aligned} \tag{1.5}$$

so the units of $h_c$ are $W / m^2 K$.

The relationships given in equations (1.4 and 1.5) are also true for the situation where a surface is being heated due to the fluid having higher temperature than the surface. However, in this case the direction of heat transfer is from the fluid to the surface and the temperature difference will now be

$\Delta T$ = mean fluid temperature-surface temperature

The relative temperatures of the surface and fluid determine the direction of heat transfer and the rate at which heat transfer take place.

As given in previous equations, the rate of heat transfer is not only determined by the temperature difference but also by the convective heat transfer coefficient $h_c$. This is not a constant but varies quite widely depending on the properties of the fluid and the behaviour of the flow. The value of $h_c$ must depend on the thermal capacity of the fluid particle considered, i.e. $mC_p$ for the particle. So the higher the density and $C_p$ of the fluid the better the convective heat transfer.

Two common heat transfer fluids are air and water, due to their widespread availability. Water is approximately 800 times more dense than air and also has a higher value of $C_p$. If the argument given above is valid then water has a higher thermal capacity than air and should have a better convective heat transfer performance. This is borne out in practice because typical values of convective heat transfer coefficients are as follows:

| Fluid | $h_c(W/m^2K)$ |
|---|---|
| water | 500-10000 |
| air | 5-100 |

The variation in the values reflects the variation in the behaviour of the flow, particularly the flow velocity, with the higher values of $h_c$ resulting from higher flow velocities over the surface.

When a fluid is in forced or natural convective motion along a surface, the rate of heat transfer between the solid and the fluid is expressed by the following equation:

$$q = h.A(T_W - T_f) \tag{1.6}$$

The coefficient $h$ is dependent on the system geometry, the fluid properties and velocity and the temperature gradient. Most of the resistance to heat transfer happens in the stationary layer of fluid present at the surface of the solid, therefore the coefficient $h$ is often called film coefficient.

Correlations for predicting film coefficient $h$ are semi empirical and use dimensionless numbers which describe the physical properties of the fluid, the type of flow, the temperature difference and the geometry of the system.

The Reynolds Number characterizes the flow properties (laminar or turbulent). $L$ is the characteristic length: length for a plate, diameter for cylinder or sphere.

$$N_{\mathrm{Re}} = \frac{\rho L v}{\mu} \tag{1.7}$$

The Prandtl Number characterizes the physical properties of the fluid for the viscous layer near the wall.

$$N_{\mathrm{Pr}} = \frac{\mu c_p}{k} \tag{1.8}$$

The Nusselt Number relates the heat transfer coefficient $h$ to the thermal conductivity $k$ of the fluid.

$$N_{Nu} = \frac{hL}{k} \tag{1.9}$$

The Grashof Number characterizes the physical properties of the fluid for natural convection.

$$N_{Gr} = \frac{L^3 \Delta \rho g}{\rho \gamma^2} = \frac{L^3 \rho^2 g \beta \Delta T}{\mu^2} \tag{1.10}$$

## 1.4. Conduction Heat Flow in Porous Materials

If a fluid could be kept stationary there would be no convection taking place. However, it would still be possible to transfer heat by means of conduction. Conduction depends on the transfer of energy from one molecule to another within the heat transfer medium and, in this sense, thermal conduction is analogous to electrical conduction.

Conduction can occur within both porous solids and fluids. The rate of heat transfer depends on a physical property of the particular porous solid of fluid, termed its thermal conductivity k, and the temperature gradient across the porous medium. The thermal conductivity is defined as the measure of the rate of heat transfer across a unit width of porous material, for a unit cross-sectional area and for a unit difference in temperature.

From the definition of thermal conductivity k it can be shown that the rate of heat transfer is given by the relationship:

$$Q = \frac{kA\Delta T}{x} \tag{1.12}$$

where $\Delta T$ is the temperature difference $T_1 - T_2$, defined by the temperature on the either side of the porous solid. The units of thermal conductivity can be determined from the units of the other variables:

$$\begin{aligned} Q &= kA\Delta T / x \\ W &= (k) m^2 K / m \end{aligned} \tag{1.13}$$

so the unit of k are $W / m^2 K / m$, expressed as W/mK.

Fourier's Law can be integrated through a flat wall of constant cross section $A$ for the case of steady-state heat transfer when the thermal conductivity of the wall $k$ is constant.

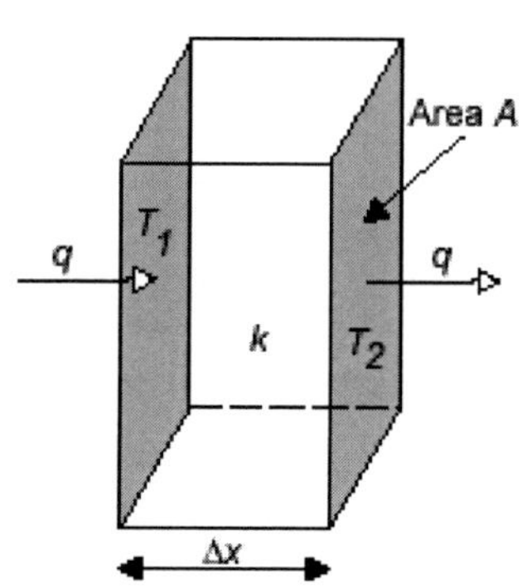

$$\frac{q}{A}\int_{x_1}^{x_2} dx = -k\int_{T_1}^{T_2} dT \rightarrow \frac{q}{A} = \frac{k}{\Delta x}(T_1 - T_2) \quad (1.14)$$

At any position $x$ between $x_1$ and $x_2$, the temperature $T$ varies linearly with the distance:

$$\frac{q}{A} = \frac{k}{x - x_1}(T_1 - T) \quad (1.15)$$

## 1.5. Radiation Heat Flow In Porous Solids

The third mode of heat flow, radiation, does not depend on any medium for its transmission. In fact, it takes place most freely when there is a perfect vacuum between the emitter and the receiver of such energy. This is proved daily by the transfer of energy from the sun to the earth across the intervening space.

Radiation is a form of electromagnetic energy transmission and takes place between all matters providing that it is at a temperature above absolute zero. Infra-red radiation form just part of the overall electromagnetic spectrum. Radiation is energy emitted by the electrons vibrating in the molecules at the surface of a porous body. The amount of energy that can be transferred depends on the absolute temperature of the porous body and the radiant properties of the surface.

A porous body that has a surface that will absorb all the radiant energy it receives is an ideal radiator, termed a "black body". Such a porous body will not only absorb radiation at a maximum level but will also emit radiation at a maximum level. However, in practice, porous bodies do not have the surface characteristics of a black body and will always absorb, or emit, radiant energy at a lower level than a black body.

It is possible to define how much of the radiant energy will be absorbed, or emitted, by a particular surface by the use of a correction factor, known as the "emissivity" and given the symbol $\varepsilon$. The emmisivity of a surface is the measure of the actual amount of radiant energy that can be absorbed, compared to a black body. Similarly, the emissivity defines the radiant energy emitted from a surface compared to a black body. A black body would, therefore, by definition, have an emissivity $\varepsilon$ of 1.

Since World War II, there have been major developments in the use of microwaves for heating applications. After this time it was realized that microwaves had the potential to provide rapid, energy-efficient heating of materials. These main applications of microwave heating today include food processing, wood drying, plastic and rubber treating as well as curing and preheating of ceramics. Broadly speaking, microwave radiation is the term associated with any electromagnetic radiation in the microwave frequency range of 300 MHz-300 Ghz. Domestic and industrial microwave ovens generally operate at a frequency of 2.45 Ghz corresponding to a wavelength of 12.2 cm. However, not all materials can be heated rapidly by microwaves. Porous materials may be classified into three groups, *i.e.* conductors insulators and absorbers. Porous materials that absorb microwave radiation are called dielectrics, thus, microwave heating is also referred to as dielectric heating. Dielectrics have two important properties:

- They have very few charge carriers. When an external electric field is applied there is very little change carried through the material matrix.
- The molecules or atoms comprising the dielectric exhibit a dipole movement distance.

An example of this is the stereochemistry of covalent bonds in a water molecule, giving the water molecule a dipole movement. Water is the typical case of non-symmetric molecule. Dipoles may be a natural feature of the dielectric or they may be induced. Distortion of the electron cloud around non-polar molecules or atoms through the presence of an external electric field can induce a temporary dipole movement. This movement generates friction inside the dielectric and the energy is dissipated subsequently as heat.

The interaction of dielectric materials with electromagnetic radiation in the microwave range results in energy absorbance. The ability of a material to absorb energy while in a microwave cavity is related to the loss tangent of the material.

This depends on the relaxation times of the molecules in the material, which, in turn, depends on the nature of the functional groups and the volume of the molecule. Generally, the dielectric properties of a material are related to temperature, moisture content, density and material geometry.

An important characteristic of microwave heating is the phenomenon of "hot spot" formation, whereby regions of very high temperature form due to non-uniform heating. This thermal instability arises because of the non-linear dependence of the electromagnetic and thermal properties of material on temperature. The formation of standing waves within the microwave cavity results in some regions being exposed to higher energy than others. This result in an increased rate of heating in these higher energy areas due to the non-linear dependence. Cavity design is an important factor in the control, or the utilization of this "hot spots" phenomenon.

Microwave energy is extremely efficient in the selective heating of materials as no energy is wasted in "bulk heating" the sample. This is a clear advantage that microwave heating has over conventional methods. Microwave heating processes are currently undergoing investigation for application in a number of fields where the advantages of microwave energy may lead to significant savings in energy consumption, process time and environmental remediation.

Compared with conventional heating techniques, microwave heating has the following additional advantages:

- higher heating rates;
- no direct contact between the heating source and the heated material;
- selective heating may be achieved;
- greater control of the heating or drying process;
- reduced equipment size and waste.

As mentioned earlier, radiation is a term applied to many processes which involve energy transfer by electromagnetic wave (x rays, light, gamma rays ...). It obeys the same laws as light, travels in straight lines and can be transmitted through space and vacuum. It is an important mode of heat transfer encountered where large temperature difference occurs between two surfaces such as in furnaces, radiant driers and baking ovens.

The thermal energy of the hot source is converted into the energy of electromagnetic waves. These waves travel through space into straight lines and strike a cold surface. The waves that strike the cold body are absorbed by that body and converted back to thermal energy or heat. When thermal radiations falls upon a body, part is absorbed by the body in the form of heat, part is reflected back into space and in some case part can be transmitted through the body.

The basic equation for heat transfer by radiation from a body at temperature $T$ is:

$$q = A\varepsilon\sigma T^4 \tag{1.16}$$

where $\epsilon$ is the emissivity of the body. $\epsilon = 1$ for a perfect black body while real bodies which are gray bodies have an $\epsilon < 1$

## 1.6. Porosity and Pore Size Distribution in a Body

Porosity refers to volume fraction of void spaces. This void space can be actual space filled with air or space filled with both water and air. Many different definitions of porosity are possible. For non-hygroscopic materials, porosity does not change with change in moisture content. For hygroscopic materials, porosity changes with moisture content. However, such changes during processing are complex due to consideration of bound water and are typically not included in computations.

The distinction between porous and capillary-porous is based on the presence and size of the pores. Porous materials are sometimes defined as those having pore diameter greater than or equal to $10^{-7}$ m and capillary-porous as one having diameter less than $10^{-7}$ m. Porous and capillary porous materials were defined as those having a clearly recognizable pore space.

In non-hygroscopic materials, the pore space is filled with liquid if the material is completely saturated and with air if it is completely dry. The amount of physically bound water is negligible. Such a material does not shrink during heating. In non-hygroscopic materials, vapour pressure is a function of temperature only. Examples of non-hygroscopic capillary-porous materials are sand, polymer particles and some ceramics. Transport materials in non-hygroscopic materials do not cause any additional complications as in hygroscopic materials.

In hygroscopic materials, there is large amount of physically bound water and the material often shrinks during heating. In hygroscopic materials there is a level of moisture saturation below which the internal vapour pressure is a function of saturation and temperature. These relationships are called equilibrium moisture isotherms. Above this moisture saturation, the vapour pressure is a function of temperature only and independent of the moisture level. Thus, above certain moisture level, all materials behave non-hygroscopic.

Transport of water in hygroscopic materials can be complex. The unbound water can be in funicular and pendular states. This bound water is removed by progressive vaporization below the surface of the solid, which is accompanied by diffusion of water vapour through the solid.

Examples of porous materials are to be found in everyday life. Soil, porous or fissured rocks, ceramics, fibrous aggregates, sand filters, snow layers and a piece of sugar or bread are but just a few. All of these materials have properties in common that intuitively lead us to classify them into a single denomination: *porous media.*

Indeed, one recognizes a common feature to all these examples. All are described as "solids" with "holes", i.e. presenting *connected void spaces*, distributed - randomly or quite homogeneously - within a *solid matrix.* Fluid flows can occur within the porous medium, so that we add one essential feature: this *void space* consists of a complex tridimensional network of *interconnected* small empty volumes called *"pores"*, with several continuous paths linking up the porous matrix spatial extension, to enable flow across the sample.

If we consider a porous medium that is not consolidated, it is possible to derive the *particle-size distribution* of the constitutive solid grains. The problem is obvious when dealing with spherical shaped particles, but raises the question of what is meant by particle size in the case of an irregular shaped particle. In both cases, a first intuitive approach is to start with a *sieve analysis.* It consists to sort the constitutive solid particles among various sieves, each one having a calibrated mesh size. The most common type of sieve is a woven cloth of stainless steel or other metal, with wire diameter and tightness of weave controlled to produced roughly rectangular openings of known, uniform size. By shaking adequately the raw granular material, the solid grains are progressively falling through the stacked sieves of decreasing mesh sizes, i.e. a *sieve column.* We finally get separation of the grains as function of their *particle- size* distribution that is also denoted by the porous medium *granulometry.* This method can be implemented for dry granular samples. The *sieve analysis* is a very simple and inexpensive separation method, but the reported granulometry depends very much on the shape of the particles and the duration of the laboratory test, since the sieve will let in theory pass any particle with a smallest cross-section smaller than the nominal mesh opening. For example, one gets very different figure while comparing long thin particles to spherical particles of the same weight.

The definition of a porous medium can be based on the objective of describing flow in porous media. A porous medium is a heterogeneous system consisting of a rigid and stationary solid matrix and fluid filled voids. The solid matrix or phase is always continuous and fully connected. A phase is considered a homogeneous portion of a system, which is separated from other such portions by a definitive boundary, called an interface. The size of the voids or pores is large enough such that the contained fluids can be treated as a continuum. On the other hand, they are small enough that the interface between different fluids is not significantly affected by gravity.

The topology of the solid phase determines if the porous medium is permeable, i.e. if fluid can flow through it, and the geometry determines the resistance to flown and therefore the permeability. The most important influence of the geometry on the permeability is through the interfacial or surface area between the solid phase and the fluid phase. The topology and geometry also determine if a porous medium is isotropic, i.e. all parameters are independent of orientation or anisotropic if the parameters depend on orientation. In multi-phase flow the geometry and surface characteristics of the solid phase determine the fluid distribution in the pores, as does the interaction between the fluids. A porous medium is homogeneous if its average properties are independent of location, and heterogeneous if they depend on location. An example of a porous medium is sand. Sand is an unconsolidated porous medium, and the grains have predominantly point contact. Because of the irregular and angular nature of sand grains, many wedge-like crevices are present. An important quantitative aspect is the surface area of the sand grains exposed to the fluid. It determines the amount of water which can be held by capillary forces against the action of gravity and influences the degree of permeability.

The fluid phase occupying the voids can be heterogeneous in itself, consisting of any number of miscible or immiscible fluids. If a specific fluid phase is connected, continuous flow is possible. If the specific fluid phase is not connected, it can still have bulk movement in ganglia or drops. For single-phase flow the movement of a Newtonian fluid is described. For two-phase immiscible flow, a viscous Newtonian wetting liquid together with a non-viscous gas are described. In practice these would be water and air.

## 1.7. Pore-Size Distribution in Porous Structure

A detailed description of the complex tri-dimensional network of pores is obviously impossible to derive. For consolidated porous media, the determination of a *pore-size distribution* is nevertheless useful. For those particular media, it is indeed impossible to handle any particle-size distribution analysis.

One approach to define a *pore size* is in the following way: the *pore diameter* $\delta$ at a given point within the pore space is the diameter of the largest sphere that contains this point, while still remaining entirely within the pore space. To each point of the *pore space* such a "diameter" can be attached rigorously, and the *pore-size distribution* can be derived by introducing the *pore-size density function* $\theta(\delta)$ defined as the fraction of the total void space that has a pore diameter comprised between $\delta$ and $\delta + d\delta$. This distribution is normalized by the relation:

$$\int_0^\infty \theta(\delta)d\delta = 1 \tag{1.17}$$

A *porous structure* should be:

- A material medium made of heterogeneous or multiphase matter. At least one of the considered phases is not solid. The solid phase is usually called the *solid matrix*. The space within the porous medium domain that is not part of the *solid matrix* is named *void space* or *pore space*. It is filled by gaseous and/or liquid phases.
- The solid phase should be distributed throughout the porous medium to draw a network of pores, whose characteristic size can vary greatly. Some of the pores comprising the void space must enable the flow across the solid matrix, so that they should then be interconnected.
- The interconnected pore space is often denoted as the *effective pore space*, while unconnected pores may be considered from the hydrodynamic point of view as part of the solid matrix, since those pores are ineffective as far as flow through the porous medium is concerned. They are *dead-end pores* or *blind pores*, that contain stagnant fluid and no flow occurs through them.

A porous material is a set of pores embedded in a matrix of mostly solid material. The pores are the voids in the material itself. Pores can be isolated or interconnected. Furthermore, a pore can contain a fluid or a vapor, but it can also be empty. If the pore is completely filled with the fluid, it will be called saturated and if it is partially filled, it will be called non-saturated. So the porous material is primarily characterized by the content of its voids and not by the properties of the material itself. Figure 1 gives a sketch of a porous material.

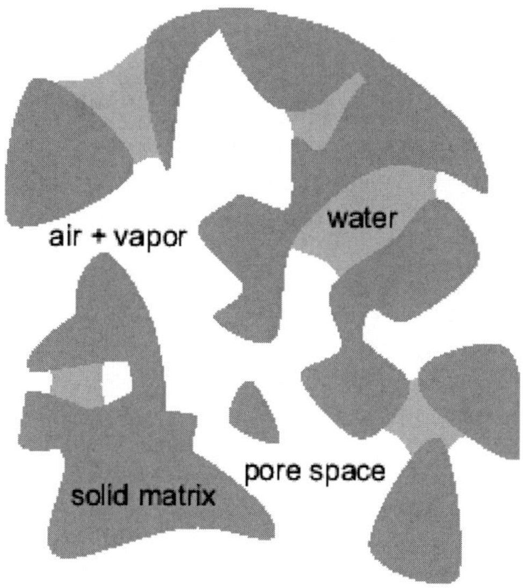


Figure 1.1. A 2D sketch of a non-saturated porous material.

If the pores are not interconnected very well, the relaxation-time distribution of an NMR(Nuclear Magnetic Resonance) spin-echo measurement can be interpreted in terms of a pore-size distribution (PSD). For magnetically doped materials like clay and .red-clay this so-called relaxometry technique gives a pore-size distribution between 100 nm and 100 $\mu$m, which is also the range of the majority of the pores in these materials. NMR (Nuclear Magnetic Resonance) can be used for spectroscopy, because different nuclei resonate at different frequencies and can therefore be distinguished from each other. Not only nuclei, but also different isotopes can be distinguished. Since also the surrounding of the nucleus has an effect on the exact resonance frequency, NMR spectroscopy is also used to distinguish specific molecules. By manipulating the spatial dependence of the magnetic field strength and the frequency of the RF excitation, the NMR sensitive region can be varied. This enables a noninvasive measurement of the spatial distribution of a certain nucleus and is called NMR Imaging (MRI).

In many NMR experiments it was noticed that liquids confined in porous materials exhibit properties that are very different from those of the bulk fluid. The so-called longitudinal ($T1$) and transverse ($T2$) relaxation time of bulk water, *e.g.*, are on the order of seconds, whereas for water in a porous material these times can be on the order of milliseconds. The measurement of $T1$ and $T2$ in an NMR experiment is often called NMR relaxometry. The transverse relaxation time is more sensitive to local magnetic field gradients inside the porous material than the longitudinal relaxation time. This sensitivity can be used to measure the self-di.usion coe.cient of the liquid. The interpretation of the measured self-diffusion coefficient of a confined liquid is often called NMR diffusometry.

Nuclear Magnetic Resonance is based on the following principle. When a nucleus is placed in a static magnetic field, the nuclear spin _$I$ will start to presses around this field, since the magnetic moment $\overline{\mu}$ of the nucleus is related to the nuclear spin $\overline{I}$ (figure 2).

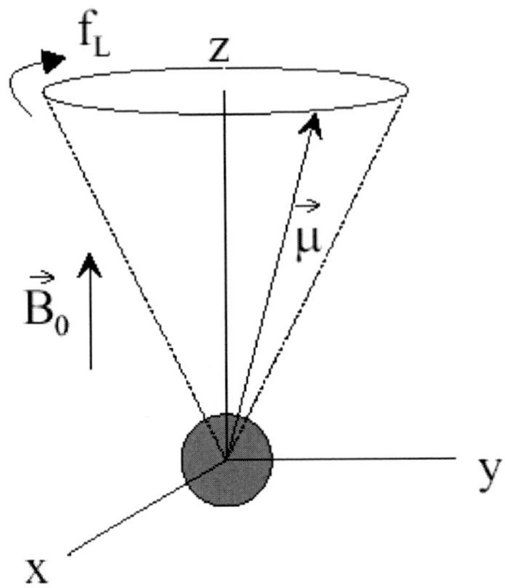


Figure 1.2. Larmor precession of a nuclear magnetic moment in a magnetic field.

The frequency of this precession motion is called the Larmor frequency:

$$f_L = \frac{\gamma}{2\pi} B_0 \tag{1.18}$$

where $B_0$ is the magnitude of the static magnetic field, which is usually taken aligned with the *z*-axis, $f_L$ is the Larmor frequency and $\gamma$ is the gyromagnetic ratio of the nucleus.

The NMR resonance condition (Eq. 1.18) states that the Larmor frequency depends linearly on the magnetic field. Normally one starts to assume that the magnetic field in the porous material is equal to the magnetic field generated by the experimental setup. This can be either the magnetic field emerging from a permanent magnet, an electromagnet, or a superconducting magnet. Frequently, an extra magnetic field gradient is added to the main magnetic field. This magnetic field gradient is used to discriminate spins at a certain position from spins at other positions. It is the basic principle of NMR Imaging (MRI). However, the magnetic field inside the porous sample can deviate largely from the magnetic field applied externally.

Because the magnetic susceptibility of the porous material differs from that of the surrounding air, the magnetic field inside the porous sample will deviate from the magnetic field that is present in the sample chamber or insert. Apart from this, the magnetic field in the pores of the material may differ from that in the bulk matrix. Consider two media with a different susceptibility. If the magnetic susceptibility of the sphere is larger (figure 3 on the left) than that of the environment, the magnetic field inside this sphere is larger than the external magnetic field and the sphere is called paramagnetic. If, on the other hand, the susceptibility of the sphere is smaller (figure 3 on the right) than that of the environment, the magnetic field inside the sphere is smaller than the external magnetic field and the sphere is called diamagnetic.

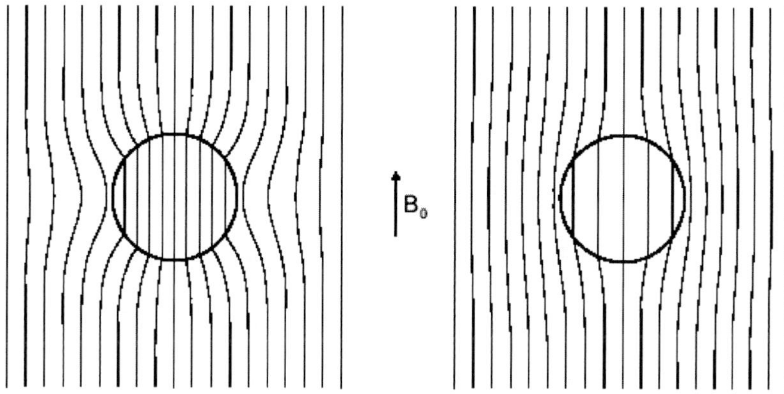


Figure 1.3. Disturbance of homogeneous magnetic field $B_0$ by an object with different susceptibility. Plotted are the magnetic field lines. On the left: a paramagnetic sphere; on the right: a diamagnetic sphere.

The amount of water in a porous body such as the textiles at the EMC is defined as bound water and it is absorbed by the textile fibers. When the textile is unable to absorb more water, all excess water is defined as unbound moisture. The unbound moisture is often found as a continuous liquid within the porous material.

Drying of porous media is accomplished by vaporizing the water and to do this the latent heat of vaporization must be supplied. There are, thus, two important process-controlling factors that enter into the process of drying:

(a) transfer of heat to provide the necessary latent heat of vaporization,
(b) movement of water or water vapour through textiles and then away from it to effect separation of water.

## 1.8. Basic Flow Relations in Porous Body

The motion of a fluid is described by the basic hydrodynamic equations, the continuity equation

$$\partial_t \rho + \nabla.(\rho u) = 0 \tag{1.19}$$

which expresses the conservation of mass, and the momentum equation

$$\partial_t(\rho u) + \nabla.(\rho u) = -\nabla p + \nabla.\tau + \rho g \tag{1.20}$$

which expresses the conservation of momentum. Here $\rho$ is the fluid density, $u$ the fluid velocity, $p$ the hydrostatic pressure, $\tau$ the fluid stress tensor, and $g$ the acceleration due to external forces including e.g. the effect of gravity on the fluid.

The equation for energy conservation can be written as

$$\rho\frac{d\hat{u}}{dt} + p(\nabla.u) = \nabla.(k\nabla T) + \Phi \tag{1.21}$$

where $T$ is temperature, $k$ the coefficient of thermal conductivity of the fluid, $\Phi$ the viscous dissipation function, and the density of thermal energy $\hat{u} = \hat{u}(p,T)$ is often approximated such that $d\hat{u} \approx c_v dT$ , where $c_v$ is the specific heat.

At low Reynolds numbers, the most important relation describing fluid transport through porous media is Darcy's law

$$q = -\frac{k}{\mu}\nabla p \tag{1.22}$$

where $q$ is the volumetric fluid flow through the (homogeneous) medium and $k$ is the permeability coefficient that measures the conductivity to fluid flow of the porous material.

## 1.9. Transport Mechanisms in Porous Media

The study of flow systems which compose of a porous medium and a homogenous fluid has attracted much attention since they occur in a wide range of the industrial and environmental applications. Examples of practical applications are: flow past porous scaffolds in bioreactors, drying process, electronic cooling, ceramic processing, and overland flow during rainfall, and ground-water pollution.

In the single-domain approach, the composite region is considered as a continuum and one set of general governing equations is applied for the whole domain. The explicit formulation of boundary conditions is avoided at the interface and the transitions of the properties between the fluid and porous medium are achieved by certain artifacts. Although this method is relatively easier to implement, the flow behavior at the interface may not be simulated properly, depending on how the code is structured.

In the two-domain approach, two sets of governing equations are applied to describe the flow in the two regions and additional boundary conditions are applied at the interface to close the two set of equations. This method is more reliable since it tries to simulate the flow behavior at the interface. Hence, in the present study, the two-domain approach, and the implementation of the interface boundary conditions, will be considered.

Fluid flow in a porous medium is a common phenomenon in nature, and in many fields of science and engineering. Important everyday flow phenomena include transport of water in living plants and trees, and fertilizers or wastes in soil. Moreover, there is a wide variety of technical processes that involve fluid dynamics in various branches of process industry. The importance of improving our understanding of such processes arises from the high amount of energy consumed by them. In oil recovery, for example, a typical problem is the amount of unrecovered oil left in oil reservoirs by traditional recovery techniques. In many cases the porous structure of the medium and the related fluid flow are very complex, and detailed studies of these flows pose demanding tasks even in the case of stationary single-fluid flow. In experimental and theoretical work on fluid flow in porous materials it is typically relevant to find correlations between material characteristics, such as porosity and specific surface area, and flow properties. The most important phenomenological law governing the flow properties, first discovered by Darcy, defines the permeability as conductivity to fluid flow of the porous material. Permeability is given by the coefficient of linear response of the fluid to a non-zero pressure gradient in terms of the flux induced.

Some of the material properties that affect the permeability, e.g. tortuosity, are difficult to determine accurately with experimental techniques, which have been, for a long time, the only practical way to study many fluid-dynamical problems. Improvement of computers and the subsequent development of methods of computational fluid dynamics (CFD) have gradually made it possible to directly solve many complex fluid-dynamical problems. Flow is determined by its velocity and pressure fields, and the CFD methods typically solve these in a discrete computational grid generated in the fluid phases of the system. Traditionally CFD has concentrated on finding solution to differential continuum equations that govern the fluid flow. The results of many conventional methods are sensitive to grid generation which most often can be the main effort in the application. A successfully generated grid is typically an irregular mesh including knotty details that follow the expected streamlines.

Transport in a porous media can be due to several different mechanisms. Three of these mechanisms are often considered most dominant: molecular diffusion, capillary diffusion, and convection (*Darcy flow*).

The *Darcy law* has been derived as follows: we consider a macroscopic porous medium which has a cross section $A$ and overall length $L$, and we impose an oriented fluid flow rate $\vec{Q}$ , to flow through it .When a steady state is reached, the induced hydrostatic pressure gradient $\vec{\nabla} p$ is related to $\vec{Q}$ by the vectorial formula :

Fluid dynamics (also called fluid mechanics) is the study of moving (deformable) matter, and includes liquids and gases, plasmas and, to some extent, plastic solids. From a 'fluid-mechanical' point of view, matter can, in a broad sense, be considered to consist of fluid and solid, in a one-fluid system the difference between these two states being that a solid can resist shear stress by a static deformation, but a fluid can not . Notice also that thermodynamically a distinction between the gas and liquid states of matter cannot be made if temperature is above that of the so-called critical point, and below that temperature the only essential differences between these two phases are their differing equilibrium densities and compressibility.

$$\frac{\vec{Q}}{A} = \frac{\overline{K}}{\mu_f} . \left( \vec{\nabla} p - \rho_f . \vec{g} \right) \Leftrightarrow \vec{v}_m = \frac{\overline{K}}{\mu_f} . \left( \frac{\vec{\Delta} p}{L} - \rho_f . \vec{g} \right) \tag{1.23}$$

where $\vec{g}$ is the acceleration of the gravity field, $\rho_f$ and $\mu_f$ are respectively the *specific mass* and the *dynamic viscosity* of fluid, $\vec{v}_m$ the *filtration velocity* over the cross section *A*. Formula (1.23) defines a second order symmetrical tensor $\overline{K}$, the *permeability*. It takes into account the macroscopic influence of the porous structure from the "resistance to the flow" point of view. The more permeable a porous medium is, the less it will resist to an imposed flow. The *permeability* is an intrinsic property of the porous matrix, based only on geometrical considerations, and is expressed in [$m^2$]. The tensorial character of $\overline{K}$ reflects the porous matrix *anisotropy*.

At the surface of the textile, two processes occur simultaneously in drying: heat transfer from the air to the drying surface and mass transfer from the drying surface to the surrounding air. The energy transfer between a surface and a fluid moving over the surface is traditionally described by convection. The unbound moisture on the surface of the material is first vaporised during the constant drying rate period.

Heat transfer by convection is described as

$$\frac{dQ}{dt} = \bar{h}A(T_A - T_S) \tag{1.24}$$

where *dQ/dt* is the rate of heat transfer, $h[W/m^2K]$ is the average heat transfer coefficient for the entire surface, *A*. $T_S$ is the temperature of the material surface and $T_A$ is the air temperature. The temperature on the surface is close to the wet bulb temperature of the air when unbound water is evaporated .

A similar equation describes the convective mass transfer. The total molar transfer rate of water vapour from a surface, $dN_v/dt$ [kmol/s], is determined by

$$\frac{dN_v}{dt} = \bar{h}_m A(C_{v,A} - C_{v,S}) \tag{1.25}$$

where $\bar{h}_m$ [m/s] is the average convection mass transfer coefficient for the entire surface, $C_{v,A}$ is the molar concentration of water vapour in the surrounding air and $C_{v,S}$ is the molar concentration on the surface of the solid with the units of [kmol/m3]. During the constant drying rate period the drying rate is controlled by the heat and/or mass transfer coefficients, the area exposed to the drying medium, and the difference in temperature and relative humidity between the drying air and the wet surface of the material (Bejan *et al.* 2004).

The average convection coefficients depend on the surface geometry of the material and the flow conditions. The heat transfer coefficient, $h$, can be determined by the average Nusselt number, $\overline{Nu}$:

$$\overline{Nu} = \frac{\bar{h}L}{k_A} = f(\mathrm{Re}, \mathrm{Pr}) \tag{1.26}$$

where $k_A$ is the heat conductivity for the air and $L$ is the characteristic length ofthe surface of interest. $\overline{Nu}$ shows the ratio of the heat transfer that depends on convection to the heat transfer that depends on conduction in the boundary layer. The Nusselt number is a function of the Reynold number,*Re*, and the Prantdl- number,*Pr*. *Pr* is the relation between the thickness of the thermal and the velocity boundary layers. If *Pr*=1, the thickness of the thermal and velocity boundary layers are equal. For air *Pr*=0.7. To determine the mass transfer coefficient, $\overline{h}_m$ , the average Sherwood-number, $\overline{Sh}$ is used:

$$\overline{Sh} = \frac{\overline{h}_m L}{D_{AS}} = f(\mathrm{Re}, Sc) \tag{1.27}$$

where $D_{AS}$ is the diffusion coefficient. $\overline{Sh}$ is a function of the Reynold number, *Re*, and the Schmidt number, *Sc*, which is the relation between the thickness of the concentration and the velocity boundary layers.

## 1.10. Molecular Diffusion in Porous Structures

Water vapour in the porous media can move by molecular or Fickian diffusion if the pores are large enough. Molecular diffusion is described by Fick's law

$$J = -D\frac{\partial c}{\partial x} \tag{1.28}$$

Where D is the molecular diffusivity.

Flow in porous media plays an important role in many areas of science and engineering. Examples of the application of porous media flow phenomena are as diverse as flow in human lungs or flow due to solidification in the mushy zone of liquid metals.

Flow in porous media is difficult to be accurately modeled quantitatively. Richards equation can give good results, but needs constitutive relations. These are usually empirically based and require extensive calibration. The parameters needed in the calibration are amongst others: capillary pressure and pressure gradient, volumetric flow, liquid content, irreducible liquid content, and temperature. In practice it is usually too demanding to measure all these parameters.

The description of the behavior of fluids in porous media is based on knowledge gained in studying these fluids in pure form. Flow and transport phenomena are described analogous to the movement of pure fluids without the presence of a porous medium. The presence of a permeable solid influences these phenomena significantly. The individual description of the movement of the fluid phases and their interaction with the solid phase is modeled by an up-scaled porous media flow equation. The concept of up-scaling from small to large scales is widely used in physics. Statistical physics translates the description of individual molecules into a continuum description of different phases, which in turn is translated by volume averaging into a continuum porous medium description.

## 2. Case Study

A mathematical model was developed for optimization of heat and mass transfer in capillary porous media during drying process to predict the drying constants. The modeling equations verified the experimental results and proved to be an important tool in predicting the drying rate under different drying conditions. The importance of heat and mass transfer in capillary porous materials like wood has increased in the last few decades due to its wide industrial as well as research applications. In order to reduce moisture content in woods to a level low enough, to prevent undesirable biochemical reactions and microbiological growth, prolonged drying time and high temperature must often be used. In practice, several different techniques are used; natural drying, vacuum drying, convectional convective drying, high temperature convective drying, and more recently microwave drying [1].

Several physical mechanisms contribute to moisture migration during the process. For a porous solid matrix, with free water, bound water, vapor, and air, moisture transport through the matrix can be in the form of either diffusion or capillary flow driven by individual or combined effects of moisture, temperature and pressure gradients. The predominant mechanisms that control moisture transfer depend on the hygroscopic nature and properties of the materials, as well as the heating conditions and the way heat is supplied. In this regard, there is a need to assess the effects of the heat and mass transfer within the wood on the transfer in the fluid adjacent to it.

There are three stages of drying: In the first stage when both surface and core MC are greater than the FSP. Moisture movement is by capillary flow. Drying rate is evaporation controlled. In the second stage when surface MC is less than the FSP and core MC is greater than the FSP. Drying is by capillary flow in the core and by bound water diffusion near the surface as fiber saturation line recedes into wood, resistance to drying increases. Drying rate is controlled by bound water diffusion and finally in the third stage when both surface and core MC is less than the FSP. Drying is entirely by diffusion. As the MC gradient between surface and core becomes less, resistance to drying increases and drying rate decreases.

For wood, model developments have been based on either a mechanistic approach with the transfer phenomena derived from Fick's and Fourier's laws, or on the principles of thermodynamics and entropy production. These models may be divided into three categories: (a) diffusion models [2], (b) models based on transport properties [3,4] and (c) models based on both the transport properties and the physiological properties of wood related to drying [5,6].

Drying adds value to timber but also costs money. Working out the complete cost of drying is a complex process. Timber drying is a critical and costly part of timber processing. Comparing the cost and effectiveness of drying systems and technology is an important exercise, before drying systems are commissioned or are upgraded. Reduction in drying time and energy consumption offers the wood industries a great potential for economic benefit. But the reduction in drying time often results in an increase in drying related defects such as checks, splits and warp.

In previous work drying curves were fitted to four drying models and the goodness of fit of each model (Correlation Coefficient and Standard Error) was evaluated [7]. The main aim of this work is to find out a model for drying time and to predict the required time for drying samples to desired moisture content. In the second part the forecast time is compared with the theoretical approach. The predicted values by the theoretical model are compared with experimental data taken under actual drying conditions to demonstrate the efficiency of the predictive model.

A software tool “Trend Analysis” for analysis the time series was applied. Trend analysis fits a general trend model to time series data and provides forecasts. S-curve is best fitted to our drying case. The S-curve model fits the Pearl-Reed logistic trend model. This accounts for the case where the series follows an S-shaped curve. The model is:

$$MC = \frac{10^a}{b_0 + b_1 b_2^t} \tag{2.1}$$

This tool is useful when we have dried the wood to moisture content not near to 30% and then predict the time needed to dry it completely.

Minitab computes three measures of accuracy of the fitted model: MAPE, MAD, and MSD for each of the simple forecasting and smoothing methods. For all three measures, the smaller the value, the better the fit of the model. These statistics are used to compare the fits of the different methods.

Mean Absolute Deviation (MAD) measures the accuracy of fitted time series values. It expresses accuracy in the same units as the data, which helps conceptualize the amount of error:

$$MAD = \frac{\sum_{t=1}^{n} \left| y_t - \hat{y}_t \right|}{n} \tag{2.2}$$

Where $y_t$ equals the actual value at time $\hat{y}_t$ equals the fitted value, and n equals the number of observations.

Mean Absolute Percentage Error (MAPE) measures the accuracy of fitted time series values. It expresses accuracy as a percentage.

$$MAPE = \frac{\sum \left| \frac{(y_t - \hat{y}_t)}{y_t} \right|}{n} \times 100 \ (y_t \neq 0) \tag{2.3}$$

Where $y_t$ equals the actual value at time $\hat{y}_t$ equals the fitted value, and n equals the number of observations.

MSD stands for Mean Squared Deviation. MSD is always computed using the same denominator, n, regardless of the model, so we can compare MSD values across models. MSD is a more sensitive measure of an unusually large forecast error than MAD.

$$MSD = \frac{\sum_{t=1}^{n} \left| y_t - \hat{y}_t \right|^2}{n} \tag{2.4}$$

Where $y_t$ equals the actual value, t equals the forecast value, and n equals the number of forecasts.

## 2.1. Governing Equations

Heat and mass transfer in a body take place simultaneously during the drying process. The time required to go from an initial moisture content, $U_0$, to a certain value $\overline{U}$ is given in[8]:

$$t = \frac{1.6 \times 10^{-4} S_x^2 S_y^2}{\left(\mu_{x1}^2 D_x S_y^2 + \mu_{y1}^2 D_y S_x^2\right)} Log\left(\Gamma_{x1}\Gamma_{y1}\left(\frac{U_0 - U_{eq}}{\overline{U} - U_{eq}}\right)\right) \tag{2.5}$$

$\mu_{l1}^2$ can be defined as:

$$\mu_{l1}^2 = \frac{1}{\frac{4}{\pi^2} + \frac{1}{B_l}} \tag{2.6}$$

Where $B_l$ is the dimensionless constant called the "bio-criterion "of the sample:

$$B_l = \frac{\alpha_l R_l}{D_l} \tag{2.7}$$

Where $R_l$ is half of the length of the rod, $l$ is any of the two coordinates x,y, $S_x \times S_y$ is the width and thickness of sample, $\alpha_l$ is the coefficient of moisture exchange(m/s), $D_l$ is the moisture diffusion coefficient( $m^2$ /s) which can vary in each of the different directions for the wood sample.

The value $\Gamma_{l1}$ is determined as:

$$\Gamma_{l1} = \frac{2B_l^2}{\mu_{l1}^2\left(B_l^2 + B_l + \mu_{l1}^2\right)} \tag{2.8}$$

and an average dimensionless moisture content $\overline{E}_\Sigma$ is:

$$\overline{E}_\Sigma = \frac{\overline{U} - U_{eq}}{U_0 - U_{eq}} \tag{2.9}$$

$U_{eq}$ is the equilibrium moisture content of the wood.

Another theoretical approach is presented by [9]:

$$t = \frac{65S^2}{\overline{D}10^6}\left(1 + \frac{\pi^2 D}{2\alpha s}\right)\log\frac{U_0 - U_{eq}}{\overline{U} - U_{eq}} \tag{2.10}$$

Where $\overline{D}$ is the average diffusion coefficient and $\overline{S}$ is the average length of the dimensions of specimens.

## 2.2. Experimental

Experimental material was obtained from two types of wood species, Guilan spruce and pine. The wood specimens were selected from Guilan region which is located in the north of Iran. The experiments were performed in a programmable domestic microwave drying system (Deawoo, KOC-1B4k) with a maximum power output of 1000 W at 2450MHz. Samples were dried in four methods: convection drying (150°C), microwave drying (270 W), infrared drying (100% power) and combination of microwave and convection drying. The dryer was run without the sample placed in, for about 30 min to set the desired drying conditions before each drying experiment. Throughout the experimental run the sample weights were continuously recorded at predetermined time intervals until wood reached to 30% of its moisture content.

## 2.3. Results and Discussion

Figure 1-8 show the graphs moisture content variation against drying time, the model and the forecasted time for the four methods of drying on pine and Guilan spruce. Drying time is estimated to a moisture content of 14%. Results are relatively in a good agreement with drying curves. Just in some cases in heating up period this model didn't fit the experimental data closely. Heat is transferred by convection from heated air to the product to raise the temperatures of both the solid and moisture that is present. Moisture transfer occurs as the moisture travels to the evaporative surface of the product and then into the circulating air as water vapor. The heat and moisture transfer rates are therefore related to the velocity and temperature of the circulating drying air. Moreover, the momentum transfer may take place simultaneously coupled with heat and moisture transfer. Convective drying at intermediate temperatures has proved to be very effective from the economical point of view, thanks to the short drying time, the reduced sizes of the kilns, and the better control of the energy consumption and the possibility of a good integration in the production line.

Infrared energy is transferred from the heating element to the product surface without heating the surrounding air. When infrared radiation is used to heat or dry moist materials, the radiation impinges the exposed material, penetrates it and the energy of radiation converts into heat. Since the material is heated intensely, the temperature gradient in the material reduces within a short period the depth of penetration of radiation depends upon the property of the material and wavelength of radiation. Further by application of intermittent radiation, wherein the period of heating the material is followed by cooling, intense displacement of moisture from core towards surface can be achieved.

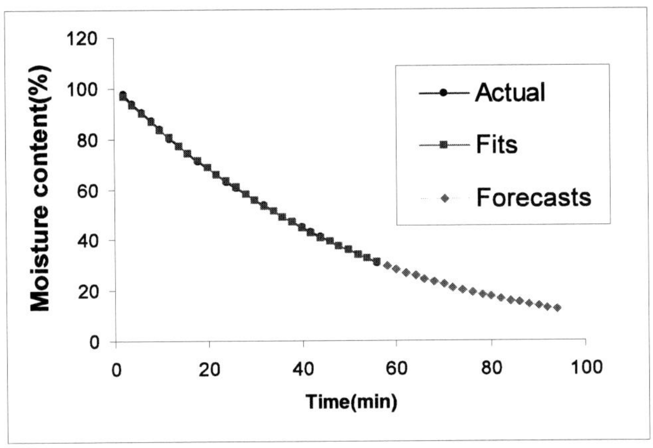


Figure 2.1. Moisture content vs. time for pine, (Convection drying).

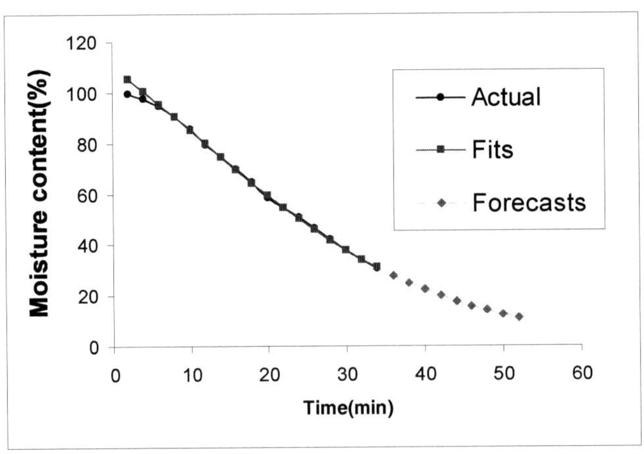


Figure 2.2. Moisture content vs. time for pine, (Infrared drying).

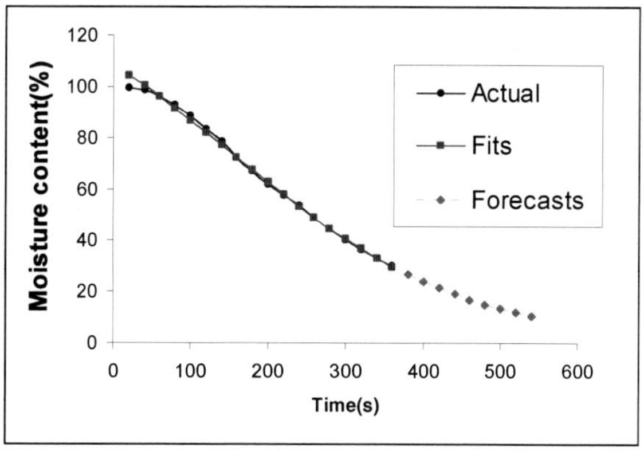


Figure 2.3. Moisture content vs. time for pine, (Microwave drying).

Microwave drying generate heat from within the grains by rapid movement of polar molecules causing molecular friction and help in faster and more uniform heating than does conventional heating. It should be pointed out that by variation of drying conditions (i.e. air temperature, humidity and air velocity) within a lumber stack, it is expected that the drying rate and the moisture content distribution varies as well [10].

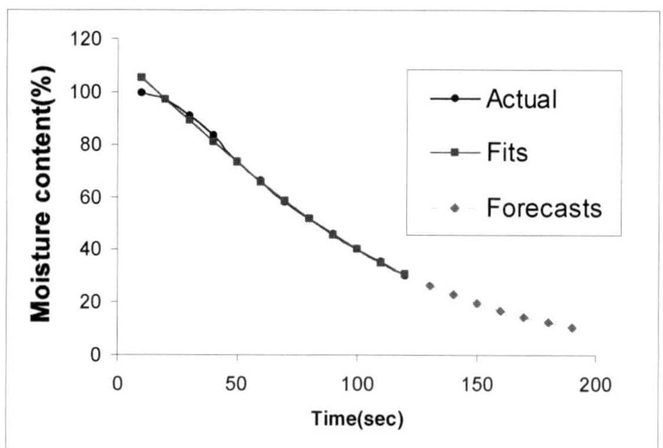


Figure 2.4. Moisture content vs. time for pine, (Combined dryer).

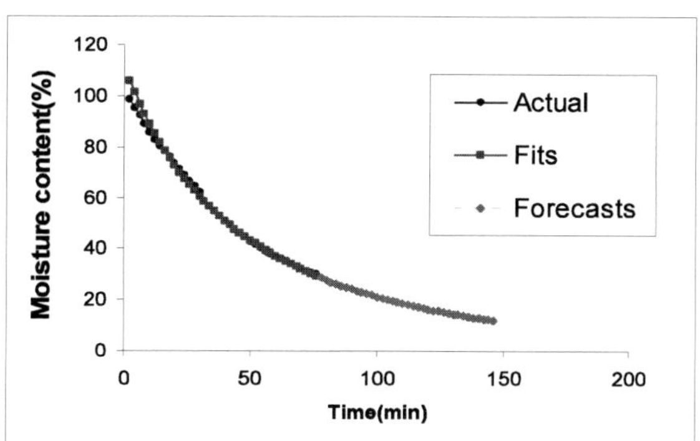


Figure 2.5. Moisture content vs. time for spruce, (Convection drying).

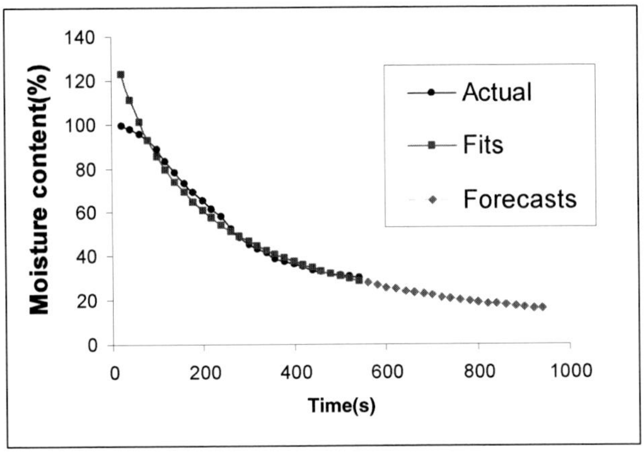


Figure 2.6. Moisture content vs. time for spruce, (Microwave drying).

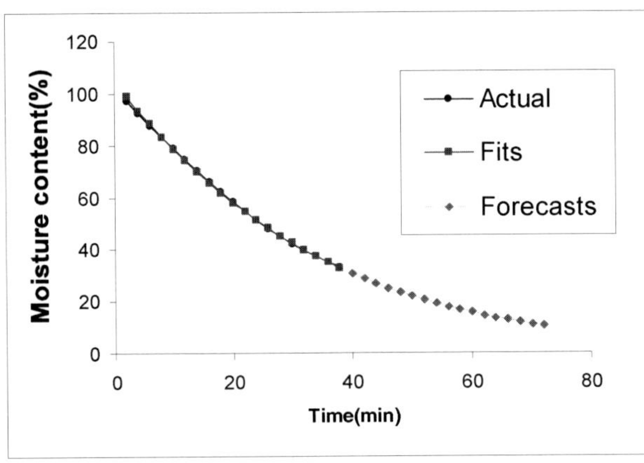


Figure 2.7. Moisture content vs. time for Spruce, (Infrared drying).

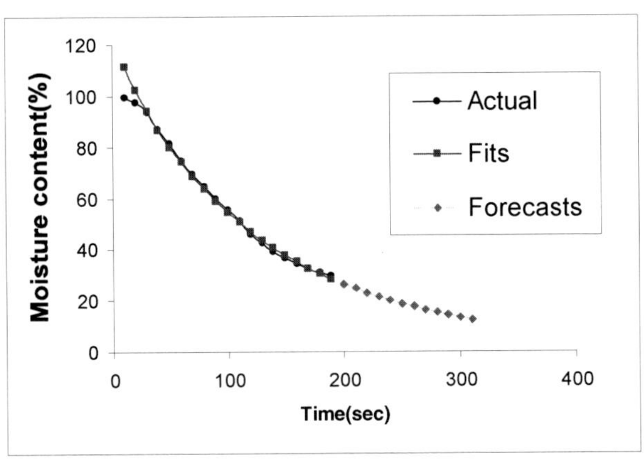


Figure 2.8. Moisture content vs. time for spruce, (Combined dryer).

The method of drying, type of samples, Mean Absolute Deviation, Mean Absolute Percentage Error, Mean Squared Deviation of these models used for moisture content change with time are presented in table 2.1.

**Table 2.1. Results of fitness**

| Type of Samples | Drying methods | MAPE | MAD | MSD |
|---|---|---|---|---|
| pine | Convection<br>Microwave<br>Infrared<br>Combined | 0.341876<br>1.08315<br>1.07610<br>1.26813 | 0.221418<br>0.86600<br>0.83372<br>1.00335 | 0.080966<br>2.08191<br>2.51506<br>3.72067 |
| spruce | Convection<br>Microwave<br>Infrared<br>Combined | 1.61692<br>4.8156<br>0.638023<br>2.46335 | 1.16996<br>3.3411<br>0.420579<br>1.63377 | 4.21973<br>33.2286<br>0.342695<br>9.40387 |

It is clear that the MAPE, MAD, MSD values of this model were changed between 0.34-4.8, 0.22-1.63 and 0.08-33.22 respectively. As it can be seen for pine samples the convection method has a better fitness to the model and for spruce infrared drying model fitted the experimental data properly.

The estimated values are based on data from [11] and can be conveniently used for theoretical approach are shown in table 2.2.

It was assumed that the diffusion coefficient bellow FSP can be represented by [11]:

$$D = A.e^{\frac{-5280}{T}}.e^{\frac{Bu}{100}} \tag{2.11}$$

Where $T$ is the temperature in Kelvin, $u$ is percent moisture content, $A$ and $B$ are experimentally determined.

Drying time is calculated from theoretical approach and evaluated model. Results show that real time had best agreement with which was obtained from equation (10) while there was a significant difference between real time and the one obtained from equation (5). Some authors have assumed that the diffusion coefficient depends strongly on moisture content [12-14] while others have taken the diffusion coefficient as constant [15-18]. Also, different boundary conditions have been assumed by different authors [19-22]. But Liu. et al concluded that the diffusion coefficient is a function of time, position, moisture content, and moisture gradient, which is at variance with assumptions in the literature that the diffusion coefficient is either a constant or a function of moisture content only [23].The difference in drying time may be due to the fact that diffusion coefficient was assumed to be the same in tangential and radial direction. So this assumption can't be used for equation (5). The same calculation can be done for other drying methods to predict the drying time.

**Table 2.2. Set of data selected for this study**

| Specifications | value | Reference |
|---|---|---|
| $S_x$ | 2.9cm | [11] |
| $S_y$ | 10.2cm | [11] |
| $u_0$ | 82.5% | [11] |
| $u_{eq}$ | 16.2% | [11] |
| $\bar{u}$ | 19% | [11] |
| T | 316.15K | [11] |
| $\alpha$ | $0.787\times10^{-5}\,cm/s$ | [11] |
| D | $8.711\times10^{-6}\,cm/s$ | Equation(11) |
| $\beta_x$ | 1.3099 | Equation(7) |
| $\beta_y$ | 4.6072 | Equation(7) |
| $\mu_x$ | 0.925 | Equation(6) |
| $\mu_y$ | 1.2676 | Equation(6) |
| $\Gamma_x$ | 0.99 | Equation(8) |
| $\Gamma_y$ | 0.985 | Equation(8) |
| A | $11.7cm^2/s$ | [11] |
| B | $3.14cm^2/s$ | [11] |
| t | 213hr | Equation(5) |
| t | 557.32hr | Equation(10) |
| t | 420hr | Trend analysis |
| t (real time) | 550hr | [11] |

## 2.4. CONCLUSION

Selection of the optimum operating conditions to obtain good quality dried products requires knowledge of the effect of the process parameters on the rate of internal-external mass transfer. High temperature heat treatment of wood is a complex process involving simultaneous heat, mass and momentum transfer phenomena and the effective models are necessary for process design, optimization, energy integration, and control.

Infrared heating offers many advantages over conventional drying under similar drying conditions. These results in high rate of heat transfer compared to conventional drying and the product is more uniformly heated rendering better quality characteristics. Microwave drying offers a number of advantages such as rapid heating, selective heating and self-limiting reactions which in turn can lead to improved quality and product properties, reduced processing time, and energy consumption and labor savings.

For pine samples the convection method has accurate result to the model and for spruce infrared drying model fitted the experimental data properly, thus their model was found to be adequate in predicting drying time of wood samples under different drying methods. The principle reason for drying wood at higher temperatures is because the rate of diffusion increases with the temperature. Water molecules generally diffuse from a region of high moisture content to a region of low moisture content, which reduces the moisture gradient and equalizes the moisture content. Diffusion plays an important role in the drying of lumber, at all moisture content with impermeable timbers and in permeable timber wherever the moisture content is too low for hydrodynamic flow of water through the lumens. Diffusion coefficient is influenced by the drying temperature, density and moisture content of timber. Other factors affecting the diffusion coefficient that are yet to be quantified are the species (specific gravity) and the growth ring orientation.

## REFERENCES


[1] Perre. P., Turner. I.W., The use of numerical simulation as a cognitive tool for studying the microwave drying of softwood in an over-sized waveguide, *Wood Science and Technology,* 33, 1999, 445–446.

[2] Rosen H.N., Drying of wood and wood products. In: Mujumdaar A.S. (ed.): Handbook of Industrial Drying. Marcel Dekker Inc., New York: 1987, 683-709.

[3] Plumb O.A., Spolek G.A., Olmstead B.A., Heat and mass transfer in wood during drying. *Intern. J. Heat Mass Transfer,* 28(9), 1985: 1669-1678.

[4] Stanish M.A., Schajer G.S., Kayihan F. A mathematical model of drying for porous hygroscopic media. *AIChE J.* 32(8): 1986, 1301-1311.

[5] Pang S. Moisture content gradient in softwood board during drying: simulation from a 2-D model and measurement. *Wood Science and Technology.* 30, 1996, 165-178.

[6] Pang S., Relationship between a diffusion model and a transport model for softwood drying. *Wood and Fiber Science.* 29(1), 1997, 58-67.

[7] Naghashzadegan. M., Haghi. A.K., Amanifard. N., Rahrovan. Sh., Microwave drying of wood: Prductivity improvement, *Wseas Trans. on Heat and Mass Transfer,* Issue 4, Vol.1, 2006, pp. 391-397.

[8] Pavlo Bekhta, Igor Ozarkiv , Saman Alavi, Salim Hiziroglu, A theoretical expression for drying time of thin lumber, *Bioresource Technology.* 97, 2006, 1572–1577.

[9] Sergovskii, P.S., Heat Treatment and Preservation of Timber, unpublished report, Moscow, Russia, 1975, p 400.

[10] Pang, S. "Airflow reversals for kiln drying of softwood lumber: Application of a kiln-wide drying model and a stress model", Proceedings of the 14th International Drying Symposium, vol. B, 2004, pp. 1369-1376.

[11] Baronas,F. Ivanauskas,M. Sapagovas, R., Modelling of wood drying and an influence of lumber geometry on drying dynamics, *Nonlinear Analysis: Modelling and Control, Vilnius,* IMI, No 4, 1999, pp.11-22.

[12] Meroney, R.N., The State of Moisture Transport Rate Calculations in Wood Drying, *Wood Fiber,* 1(1), 1969, pp. 64–74.

[13] Simpson, W.T., Determination and Use of Moisture Diffusion Coefficient to Characterize Drying of Northern Red Oak, Wood Science and Technology, 27(6), 1993, pp. 409–420.

[14] Skaar, C., Analysis of Methods for Determining the Coefficient of Moisture Diffusion in Wood, *Journal of Forest Products Research Society,* 4(6), 1954, pp. 403–410.

[15] Avramidis, S. and Siau, J.F., An Investigation of the External and Internal Resistance to Moisture Diffusion in Wood, *Wood Science and Technology,* 21(3), 1987, pp. 249–256.

[16] Droin, A., Taverdet, J.L. and Vergnaud, J.M., Modeling the Kinetics of Moisture Adsorption by Wood, *Wood Science and Technology,* 22(1), 1988, pp. 11–20.

[17] Mounji, H., Bouzon, J. and Vergnaud, J.M., Modeling the Process of Absorption and Desorption of Water in Two Dimension (Transverse) ina Square Wood Beam, *Wood Science and Technology,* 26(1), 1991, pp. 23–37.

[18] Soderstro¨ m, O. and Salin, J.G., On Determination of Surface Emission Factors in *Wood Drying, Holzforschung,* 47(5), 1993, pp. 391–397.

[19] Crank, J., The Mathematics of Diffusion, Chap. 9, 2nd ed., ClarendonPress, Oxford. 1975.

[20] Plumb, O.A., Spolek, G.A. and Olmstead, B.A., Heat and Mass Transfer in Wood during Drying, *International Journal of Heat and Mass Transfer,* 28(9), 1985, pp. 1669–1678.

[21] Salin, J.-G., Mass Transfer from Wooden Surface and Internal Moisture Non-equilibrium, *Drying Technology,* 14(10), 1996, pp. 2213–2224.

[22] Hukka, A., The Effective Diffusion Coefficient and Mass Transfer Coefficient of Nordic Softwoods as Calculated from Direct Drying Experiments, Holzforschung, 53(5), 1999, pp. 534–540.

[23] Jen Y. Liu, William T. Simpson, and Steve P. Verrill, An inverse moisture diffusion algorithm for the determination of diffusion coefficient, *Drying Technology,* 19(8), 2001, 1555–1568.

In: Monomers, Oligomers, Polymers, Composites... ISBN: 978-1-60456-877-6
Editors: R. A. Pethrick, G.E. Zaikov et al. 

*Chapter 16*

# "GLASSCRETE" CONTAINING POLYMER AGGREGATE AND POLYAMIDE FIBERS

*A. Sadrmomtazi and A. K. Haghi**
University of Guilan; P.O. Box: 3756; Rasht, Iran

## ABSTRACT

In this chapter the behavior of lightweight EPS glasscrete containing polymeric fibers are studied. The results are compared together and with the results observed by other authors. The comparison revealed the superiority of this new composite in improving the compressive and tensile strength. The results are indicative of strong potentials for application of waste polymeric fibers for the reinforcement of lightweight EPS glasscrete products.

## INTRODUCTION

Concrete is the most widely used construction material in the world. Its low cost, ease of application and compressive strength are the principle reasons for its universal acceptance. However it has few shortcomings most of which are attributable to the Portland cement binder. Shortcomings include poor tensile strength, high porosity freeze thaw deterioration, and destruction by corrosive chemicals, etc. [1]

The development of new composite materials possessing increased strength and durability when compared with conventional types is a major requirement of applications in repairs and in the improvement of infrastructure materials used in the civil construction industry. Polymer concrete (PC) is an example of a relatively new material with such high performance [2].

The demand for lightweight concrete in many applications of modern construction is increasing, owing to the advantage that lower density results in a significant benefit in terms

* Corresponding author e-mail: Haghi@Guilan.ac.ir

of load-bearing elements of smaller cross sections and a corresponding reduction in the size of the foundation [3].

Lightweight aggregate concrete, popular through the ages, was reported to have a comparable or some times better durability even in severe exposure conditions.

Lightweight aggregates are broadly classified in to two types: natural (pumice, diatomite, volcanic cinders, etc.) and artificial (perlite, expanded shale, clay, slate, sintered PFA, etc.). Expanded polystyrene (EPS) beads are a type of artificial ultra-lightweight, non-absorbent aggregate [4, 5]. It can be used to produce low-density concretes required for building applications like cladding panels, curtain walls, composite flooring systems, and load-bearing concrete blocks [6, 7].

Expanded polystyrene (EPS) concrete is a lightweight concrete with good energy-absorbing characteristics, consisting a discrete air voids in a polymer matrix. However, polystyrene beads are extremely light, with a density of only 12 to 20, which can easily cause segregation in mixing. Hence some chemical treatment of surface on this hydrophobic material is needed. Other investigators also reported that EPS tends to float and can result in a poor mix distribution and segregation, necessitating the use of admixtures [8, 9].

Our main aim is to contribute to the understanding of the behaviour of EPS Glasscrete reinforced with polyamide fibers. The quantity of fibers used is small, typically 1 to 5 per cent by volume.

## BACKGROUND

In general, waste glass produced can be sorted out as:

- building/automobile windows and doors;
- glassware and bottles;
- television tubes and light bulbs;
- others such as mirror and clock covers.

Among them, the first two are the major sources. Several domestic glass manufacturers have attempted to recycle the glass waste.

Waste recycling has become increasingly important for western society over the past few years. The European desire to increase recycling of glass, particularly that of glass bottles, has led to a large collect system in containers and a target of 75% of the collected glass (2.2 million tons) recycled by the year 2005 in France . Nevertheless, there will still be 5% of collected glass(0.1 to 0.15 million tons) that is not recycled and must be disposed of, as glass can only be melted down after the removal of non-ferrous metals and other contaminants, which are mixed with the smallest glass fragments . For the long period of disposal, it is predicted that the waste glass will corrode very slowly by contact with groundwater, and certain quantities of radionuclides will be released from the glass. Therefore, the waste glass corrosion and associated radionuclide release for the long-term are one of the most important phenomena to be evaluated for safety assessment of the disposal system .

Glass is a thermodynamically metastable, highly viscous liquid phase. When glass is placed in aqueous environments, various reactions, such as ion exchange, hydrolysis and transformation to more stable phases, occur simultaneously, and the effects of these reactions are generally referred to as "corrosion". However, the glass corrosion allows the release of radionuclides from glass, and the glass corrosion and associated radionuclide release for the long- term must be evaluated sufficiently .

Concrete, the composite material consisting of aggregate held together by a hydraulic cementing agent, has been known to ancient civilizations. In spite of its worldwide popularity, the proliferation of concrete has been a mixed blessing. If mixed or placed improperly or maintained inadequately, concrete structures can deteriorate prematurely and thereby contribute to the problems referred to generally as our "crumbling infrastructure". Also the indiscriminate use of concrete without concern for esthetic appearance has led to the partially deserved reputation of concrete as being ugly. More significantly, the increased worldwide concern about environmental issues and the need to change our way of life for the sake of sustainable development has led to the identification of the concrete industry as a major user and abuser of natural resources and energy and as an important contributor to the release of greenhouse gases. These issues pose formidable challenges for the concrete industry for years to come. The construction community as well as the public at large will demand increased emphasis on environmentally friendly high-performance building materials at affordable cost. This implies not only excellent mechanical properties but durability as well. Fortunately, concrete materials science has emerged as a tool well suited to face these issues .

The introduction and development of advanced composite material opened the door to new and innovative application in civil and structural engineering. Key points of this investigation are:

1. To convert glass and carpet waste into useful product.
2. To consume glass and carpet wastes which would otherwise gone to landfill.
3. Protection of Environment from being heavily contaminated

In developing concrete products with crushed waste glass aggregate, the economics is controlled by the price the product can fetch on the open market. Commodity products, by definition, are characterized by low values, which exert strong pressures on the production and manufacturing technology. The value added by the glass is marginal to nonexistent in those cases. But by utilizing the special properties of glass, chemical, physical, or esthetic, novel products can be developed, for which the prices fetched in the open market are much less exposed to competitive pressures.

What makes glass such a special ingredient for concrete becomes apparent by summarizing its special properties:

- Because it has basically zero water absorption, it is one of the most durable materials known to man. With the current emphasis on durability of high-performance concrete, it is only natural to rely on extremely durable ingredients.
- The excellent hardness of glass gives the concrete an abrasion resistance that can be reached only with few natural stone aggregates.
- For a number of reasons, glass aggregate improves the flow properties of fresh concrete so that very high strengths can be obtained even without the use of superplasticizers .

- The esthetic potential of color-sorted post-consumer glass, not to mention specialty glass, has barely been explored at all and offers numerous novel applications for design professionals.
- Very finely ground glass has pozzolanic properties and therefore can serve both as partial cement replacement and filler.

Waste glass can be used as partial replacement of coarse aggregate, fine aggregate, cementitious materials or ultra fine filler in concrete, depending on its chemical composition and particle size. Previous efforts have been made shown that replacement of glass as a part of coarse aggregate was not satisfactory because of chemical reaction between the alkali in the cement and the silica in the glass. This strongly expansive alkali-silica reaction (ASR) creates a gel, which swells in the presence of moisture, causes cracking and unacceptable damage of the concrete. Recent studies have shown that if the waste glass finely ground, could be used in mortars and concrete as a very fine addition without introducing problems concerning ASR. In fact, a general conclusion of literatures shows that if the waste glass is finely ground under 75μm, ASR does not occur and mortar durability is guaranteed because of its pozzolanic properties. Also on a market price basis, it would be much more profitable to use the glass in powder form as cement replacement to make a value added composite cement.

## EXPERIMENTAL PROCEDURE

### Materials

Ordinary Portland cement (OPC) was used. The main chemical composition of (OPC) used in this study are listed in table 1.

Table 7. Chemical and physical compositions of ordinary Portland cement

| | |
|---|---|
| $Sio_2$ | 20.40% |
| $Al_2o_3$ | 6.12% |
| $Fe_2O_3$ | 3.051% |
| CaO | 63.16% |
| MgO | 2.32% |
| $SO_3$ | 2.40% |
| Specific gravity | 3.13 |

Two types of commercially available spherical EPS beads were used (table 2). As indicated in table 2, the grading of EPS shows that type A has mostly 6.0-mm-size beads and type B has mostly 3.0-mm-size beads.

**Table 8. EPS beads characteristics**

| Type | Size (mm) | Bulk density ( $kg/m^3$ ) | Specific gravity |
|---|---|---|---|
| A | 6 | 16 | 0.014 |
| B | 3 | 20 | 0.029 |

The physical properties of PA 66 used in this study are summarized in table 3.

**Table 9. Physical properties of Polyamide fibers used**

| | |
|---|---|
| Specific gravity | 1.16 |
| Tensile strength | 965 MPa |
| Elastic modulus | 5.17 GPa |
| Ultimate elongation | 20% |

## Glasscrete Preparation

The cement replacement by the ground waste glass was 10%, 20%, 30% and 40% by total weight.

The composite cement paste containing ground waste glass were compared to the cement having the same percent replacement by silica fume and rice husk ash as well as to the control specimen without any mineral additives. The five batches were defined as follows:

- Ordinary Portland cement paste: no mineral additives
- Waste glass type I: 10%, 20%, 30% and 40% by weight of the Portland cement replaced by waste glass type I
- Waste glass type II: 10%, 20%, 30% and 40% by weight of the Portland cement replaced by waste glass type II
- Silica fume: 10%, 20%, 30% and 40% by weight of the Portland cement replaced by waste silica fume
- Rice husk ash: 10%, 20%, 30% and 40% by weight of the Portland cement replaced by husk rice ash.

## Specimen Preparation

A number of standard test specimens of different sizes were selected for investigating the various parameters. For studying the compressive strengths at 3, 7, and 28 days and also for investigating the absorption tests cubes of 100 mm size were used. Meanwhile, the split tensile strength test was conducted on 100 mm diameter × 200 mm cylinders at 28 days.

Lightweight EPS glasscrete were made with PA and without PA reinforcement. Initially, the EPS beads were wetted with 25% of the mixing water and superplasticizer before adding the remaining materials. It is expected that by addition of EPS in place of normal aggregate the weight of glasscrete can be reduced significantly and the interaction of Polyamide 66 fibers with EPS concrete can improve the strength. In this study, the estimated component additions were measured by volume in order to simplify mixing process. The non-absorbent, hydrophobic and closed cellular aggregates (expanded polystyrene beads) were mixed in a planetary mixer in compliance with the recommendation of ASTM C 305. To obtain a very uniform and flowing mixture, the mixing operation was continued on a regular bias. The molds are then filled with fresh EPS concrete and then with PA reinforced EPS concrete. Afterward, the mixture is firmly compacted by hand.

Based on ASTM C 143-98 we measured the slump values of the fresh concrete. Afterward, the compressive strength test was carried out in a testing machine with capacity of 2000 KN. The loading rate used during this test was 2.5 KN/s. According to ASTM C 496-89 we conducted the split tensile strength test on cylinders at 28 days. The absorption test was then carried out as per ASTM C 642-82.

## Results and Discussion

It appeared that the workability of the glasscrete in terms of the slump measurements were about 45–80 mm in non-reinforced cases and around 40-75 for polyamide EPS concretes. In addition, it should be noted that the mixes having the higher percentage silica fume can present higher slump values. During the experimental study, all the concretes specimens shown a interesting flexibility and provided a very easy condition to work with. Meanwhile, the specimens could be compacted using just hand compaction and they could be easily finished. In order to be able to solve the problems associated with the hydrophobic nature of the EPS beads, the silica fume and the superplasticizer were added. This will help to improve the cohesiveness of the mix significantly.

Figure 1 displays the development of compressive strength with the age for reinforced EPS concrete. This illustration clearly states that the compressive strength of reinforced EPS concrete in almost all mixes shows a continuous increase with age. It is seen that the rate of strength development was greater initially and decreased as the age increased. Comparison of strengths at 7 days revealed that non-reinforced concretes developed almost 40–50% of its 28-day strength. While it is seen that reinforced concretes developed almost 20–40% of the corresponding 28-day strength.

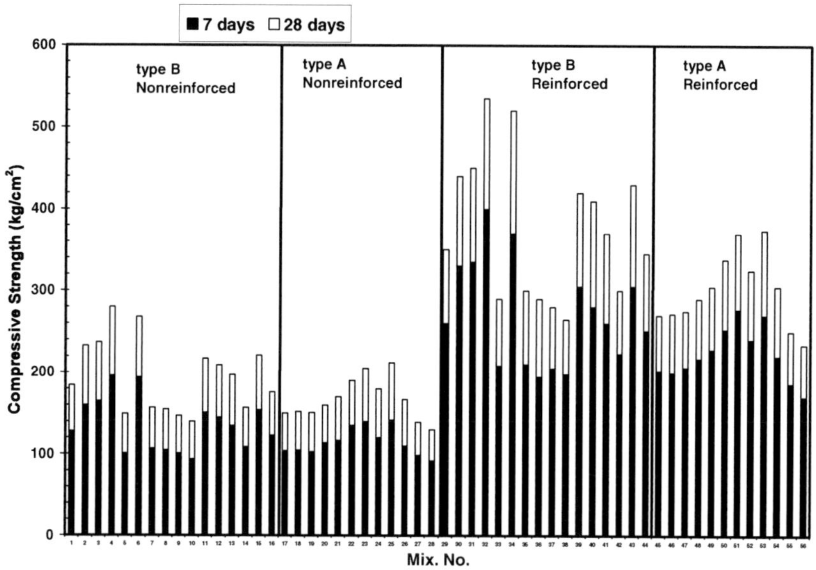


Figure 12. Effects of age on compressive strength.

Figures 2 and 3 illustrate the compressive strength of EPS concretes with different plastic densities for reinforced and non-reinforced concrete and the volume content of EPS.

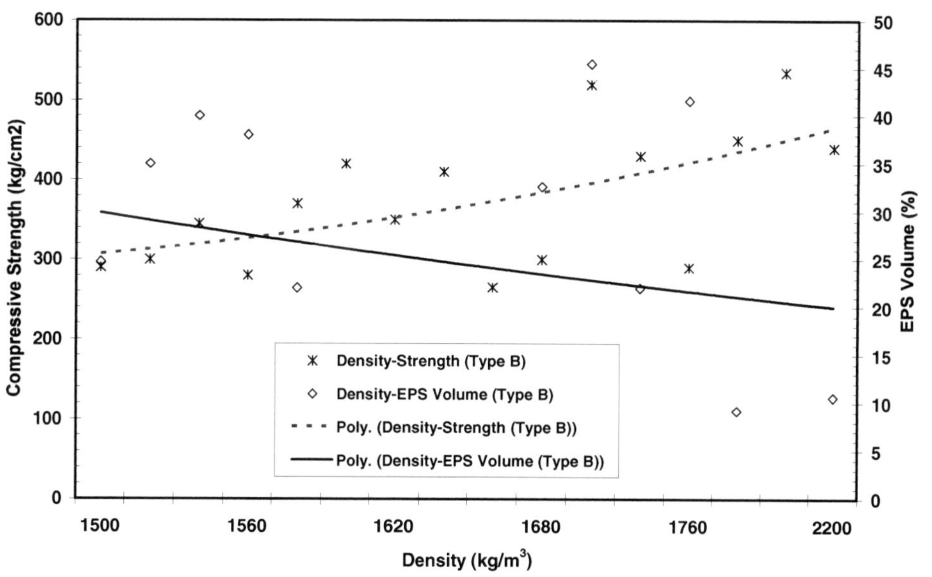


Figure 13. Variation of composite density with EPS volume and Compressive strength (type B).

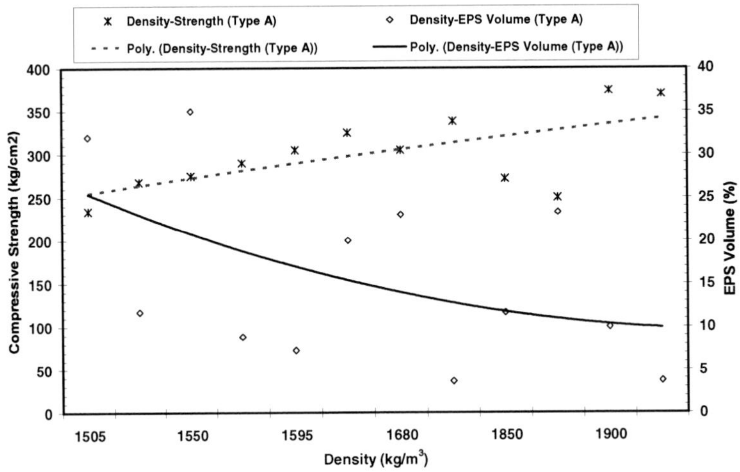


Figure 14. Variation of composite density with EPS volume and Compressive strength (type A).

It is appeared that the strength of non-reinforced and reinforced EPS concrete to increase linearly with a decrease in the EPS volume. Furthermore, it is noted that the density of EPS concretes decreased significantly with an increase in the EPS volume. It is clearly observed that the polyamide EPS concrete have high strength in compression, comparing to that non-reinforced concretes.

The plot represented in figure 4 show an inverse relationship between the concrete strength and the bead size. That means the strength in EPS concrete increased with a decrease in the EPS bead size for the same mix proportions.

From figure 5 it can be clearly appreciated that in the mixes content fine silica fume the compressive strength can increase significantly (i.e. mixes no. 4, 6,...). This represents the effects of fine silica fume and polyamide fiber on the compressive strength of EPS concrete. This is clearly visible that fine silica fume can improve the dispersion of EPS in the cement paste and interfacial bonding between EPS and cement paste. Moreover, the compressive strength of polyamide EPS concrete is almost 1.5 to 2 times non-reinforced EPS concretes.

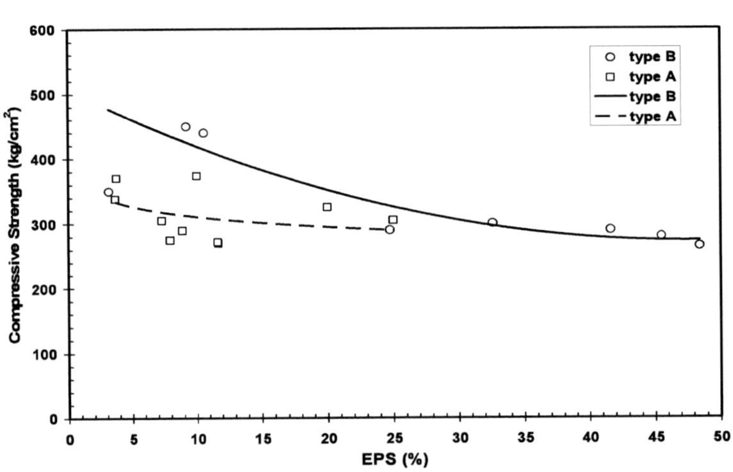


Figure 15. Effects of EPS bead size on compressive strength.

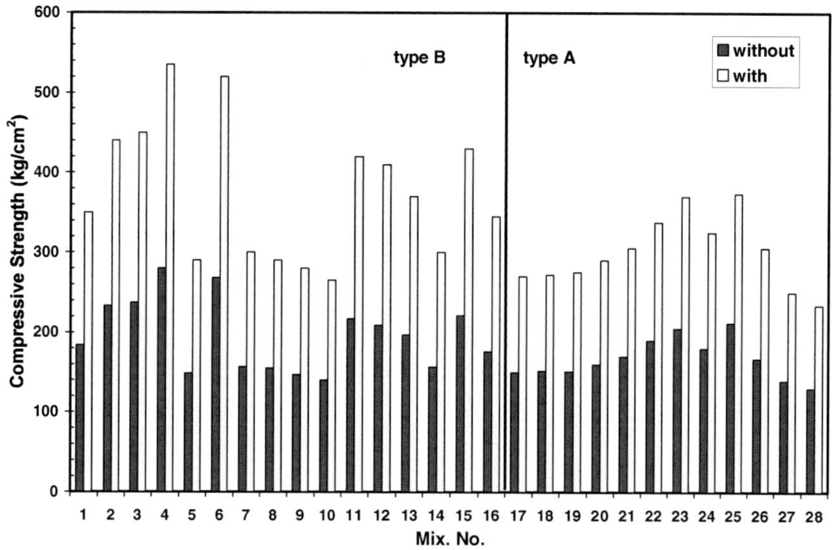


Figure 16. Effects of silica fume and polyamide on compressive strength.

The failure modes of different specimens are presented in figure 6. Figure (6a) shows the EPS concrete specimen before loading. The failure mode observed in figure (6b) is rather gradual (more compressible), and the specimen is capable of retaining the load after failure, without full disintegration. Figure (6c) illustrates failure mode in polymer concretes containing EPS with polyamide yarns. This is attributed to the fact that the failure mode of this specimen is near to a typical brittle failure.

(a) (b) (c)

Figure 17. The EPS concrete samples with and without polyamide fibers; (a) Compressive samples before failure, (b) Observed failure mode in polymer concretes containing EPS without polyamide fibers and c) Observed failure mode in polymer concretes containing EPS with polyamide fibers.

Figure 7 illustrates the effects of volume content of EPS on the tensile strength of polyamide EPS concretes. This attribute shows the fact that there is an inverse relationship between the concrete tensile strength and the EPS volume. This means that the tensile strength of EPS concrete appeared to increase with a decrease in the EPS volume.

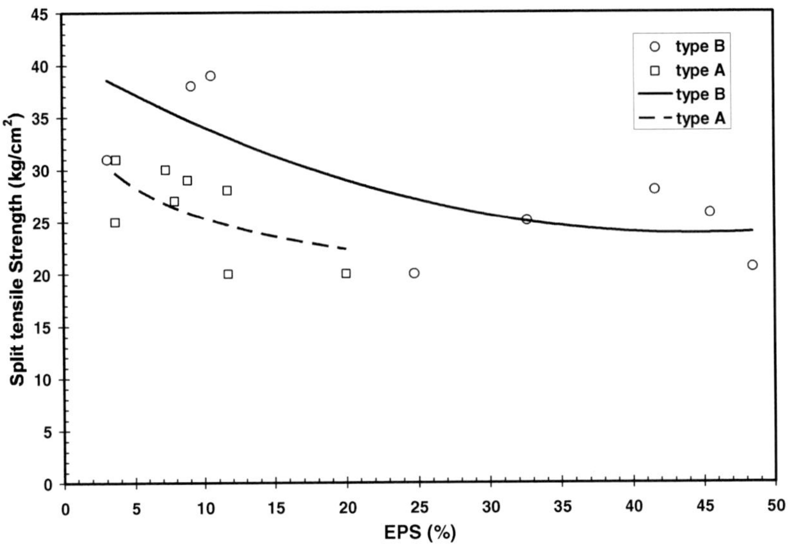


Figure 18. Variation of tensile strength with EPS volume.

From figure 8, the tensile strength of polyamide EPS concrete (type B) is 40 kg/cm2 whereas, the maximum tensile strength for polyamide EPS concrete (type A) is only 31 kg/cm2. But in general, we can expect an increase of 70% in the tensile strength for the lower bead sizes.

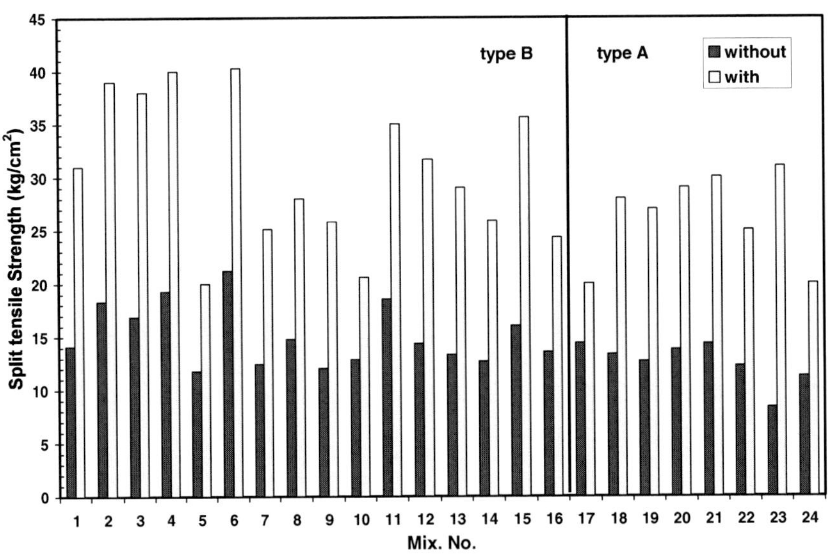


Figure 19. Effects of EPS bead size on tensile strength.

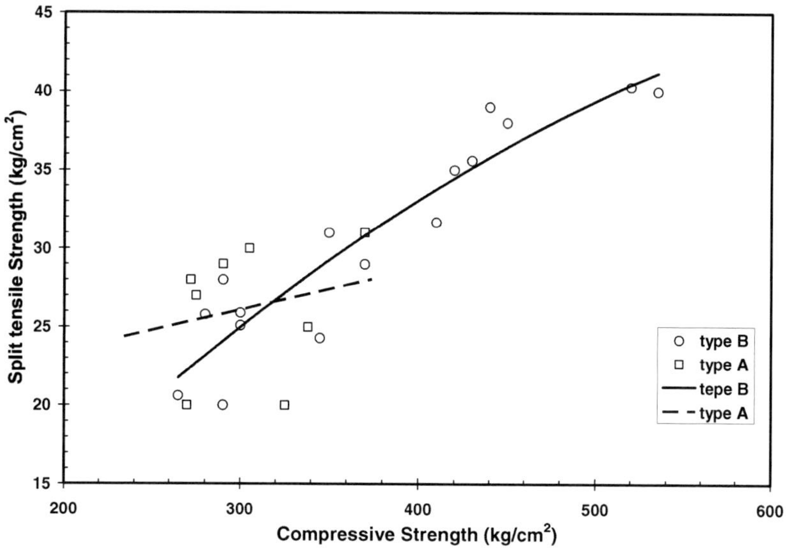


Figure 20. Variation of tensile strength with compressive strength.

Efforts were made to study the variation of tensile strength with the compressive strength (figure 9). It appears that the tensile strength increased with an increase in compressive strength. The rate of strength increasing in polyamide-EPS glasscrete containing small beads (type B) is higher than other.

The splitting failure mode of the concrete specimens containing non-reinforced EPS aggregates also did not exhibit the typical brittle failure principally expected in conventional concrete as in compressive strength. Consequently, the tensile failure mode observed in the polymer glasscrete containing polyamide fibers was rather a typical brittle failure which is normally resembles to conventional concrete (figure 10).

Figure 21. The failure mode in EPS glasscrete.

**Shrinkage**

The most significant factor affecting the shrinkage of concrete is the degree of restraint by the aggregate (i.e. its elastic properties and the volumetric proportion of the paste in the mix). It should be noted that the EPS beads principally offer little hindrance to the shrinkage of the paste. Therefore it is expected that as the volumetric proportion of the EPS is increased, the shrinkage would increase as well. This is clearly represented in figure 11. In this study, the drying shrinkage of normal concrete at 90 days was 610 microstrain. Inversely, when the volume content of EPS in the concrete specimen was 20%, the drying shrinkage at 90 days raised to 1000 microstrain, which represents a disadvantage for the application of EPS.

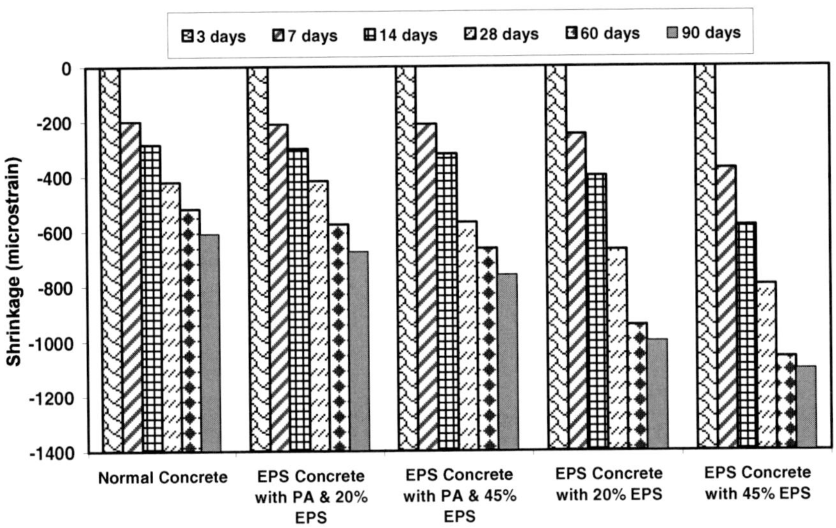


Figure 22. Relationship between drying shrinkage strain of EPS concrete (with polyamide and without polyamide) and reference concrete with age.

In a view of the above, it can be seen that polyamide yarns improved the drying shrinkage of EPS concrete significantly. Even for the case of EPS glasscrete which contained 20% of its volume by EPS, the drying shrinkage at 90 days was 670 microstrain, which seems to be rather close to normal concrete range.

# CONCLUSION

The data presented in this paper show that there is a great potential for the utilization of waste glass and waste fibers in EPS concrete. It is considered this utilization would provide much greater opportunities for value adding and cost recovery as it could be used as a replacement for expensive materials.

The use of recycled materials in glasscrete has been on the increase throughout the world due to conserving resources. The results of this study have shown that use of this locally available low-cost PA waste may provide an economical and effective alternative to concrete aggregates.

## REFERENCES

[1] M. Muthukumar and D. Mohsen, "Studies on polymer concrete based on optimized aggregate mix proportion", *European Polymer Journal,* Volume 40, 2004, pp. 2167-2177.

[2] J. P. Gorninski, D. C. Dal Mobin and C. S. Kazmierczak, Study of the modulus of elasticity of polymer concrete compounds and comparative assessment of polymer concrete and Portland cement concrete, *Cement and Concrete Research,* Volume 34, 2004, pp. 2091-2095.

[3] ACI Committee 213 R-87. Guide for Structural Lightweight Aggregate Concrete, ACI Manual of Concrete Practice, Part 1, American Concrete Institute, Farmington Hills (1987).

[4] A. Short and W. Kinniburgh, Lightweight Concrete. (3rd ed.),, *Applied Science Publishers*, London (1978).

[5] V. Sussman, Lightweight plastic aggregate concrete. *ACI J.* (1975 (July)), pp. 321–323.

[6] C. Bagon and S. Frondistou-Yannas, Marine floating concrete made with polystyrene expanded beads. *Mag. Concr. Res.* 28 (1976), pp. 225–229

[7] K. G. Babu and D. S. Babu, Behaviour of lightweight expanded polystyrene concrete containing silica fume, *Cement and Concrete Research,* Volume 33, Issue 5 , May 2003, pp. 755-762.

[8] A.A. Al-Manaseer and T.R. Dalal, Concrete containing plastic aggregates. *Concr. Int.* (1997 (August)), pp. 47–52.

[9] CEB-FIP. Diagnosis and assessment of concrete structures—State of the art report, CEB Bulletin 1989.

In: Monomers, Oligomers, Polymers, Composites... ISBN: 978-1-60456-877-6
Editors: R. A. Pethrick, G.E. Zaikov et al. 

*Chapter 17*

# Electrospun Nanofibers and Image Analysis

***M. Ziabari, V. Mottaghitalab and A. K. Haghi****
University of Guilan, P. O. Box 3756, Rasht, Iran

## Abstract

In the first part of this chapter electrospinning process of nanofiber is introduced. In the second part, a new image analysis technique for measuring the diameters of electrospun nanofibers is developed.

## 1.1. Introduction

When the diameters of polymer fiber materials are shrunk from micrometers (for example 10-100 $\mu m$) to sub-microns or nano meters (for example $10 \times 10^{-3}$ - $100 \times 10^{-3}$ $\mu m$), there appear several amazing characteristics such as very large surface area to volume ratio, flexibility in surface functionalities, and superior mechanical performance compared with any other known form of material. These outstanding properties make the polymer nanofibers to be optimal candidates for many important applications. The electrospinning process is not a new technology for polymer fiber production seems to be the only method which can be further developed for mass production of one-by-one continuous nanofibers from various polymers. It has been known since the 1930's; however, it did not gain significant industrial importance due to the low output of the process, inconsistent and low molecular orientation, poor mechanical properties and high diameter distribution of the electrospun fibers. Although special needs of military, medical and filtration applications have stimulated recent studies and renewed interest in the process, quantitative technical and scientific information regarding process and product characterization are extremely limited. Electrospinning has gained renewed interest as a method to produce ultra-thin fibers with diameters in the

* Corresponding author E-Mail: Haghi@Guilan.ac.ir

nanometer to micrometer scale range. Nanofibers exhibit special properties mainly due to extremely high surface to weifht ratio compared to conventional nonwovens [1-4]. Low density, large area to mass, high pore volume, and tight pore size make the nanofiber nonwoven appropriate for a wide range of the filtration applications. Figure 1 shows how much smaller nanofibers are compared to a human hair, which is 50-150 $\mu m$.

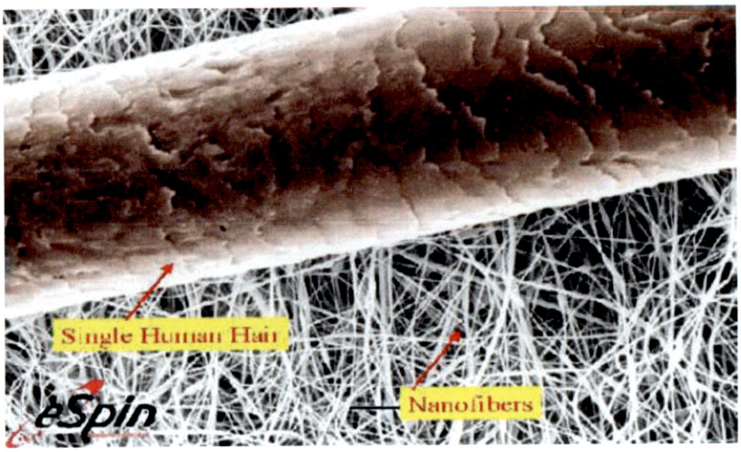


Figure 1. Comparison between human hair and nanofiber web.

Nanofibers have significant applications in the area of filtration since their surface area is substantially greater and have smaller micropores than melt blown (MB) webs. High porous structure with high surface area makes them ideally suited for many filtration applications. Nanofibers are ideally suited for filtering submicron particles from air or water. Electrospun fibers have diameters three or more times smaller than that of MB fibers. This leads to a corresponding increase in surface area and decrease in basis weight. Table 1 shows the fiber surface area per mass of nanofiber material compared to MB and SB fibers.

**Table 1. Fiber surface area per mass of fiber material for different fiber size**

| Fiber Type | Fiber size, in Micrometer | Fiber surface area per mass of fiber material m2/g |
|---|---|---|
| Nanofibers | 0.05 | 80 |
| Spunbond fiber | 20 | 0.2 |
| Melt blown fiber | 2.0 | 2 |

Nanofibers are widely used in medical applications, which include, drug and gene delivery, artificial blood vessels, artificial organs, and medical facemasks. For example, carbon fiber hollow nano tubes, smaller than blood cells, have potential to carry drugs in to blood cells (for more information refer to: www.donaldson.com , www.ecmjournal.org and www.zapmeta.com). A Comparison of red blood cell with nanofibers web is shown in figure 2. With electrospaun fibers we can even make bandages that are absorbed by the body (figure 3).

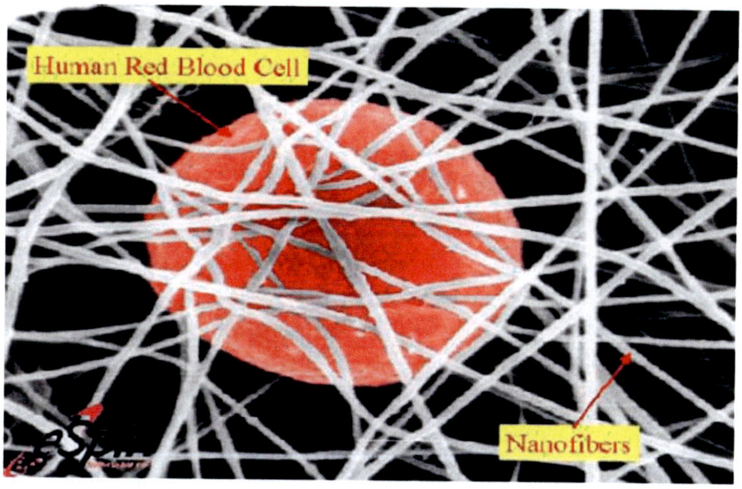


Figure 2. Comparison of red blood cell with nanofibers web.

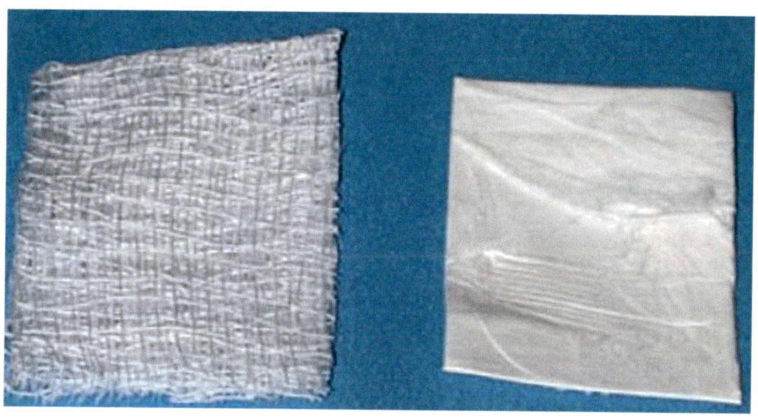

Figure 3. Photograph of a cotton gauze (left) and an electrospun bovine fibrinogen mat (right) produced from 0.167 g/ml fibrinogen in HFP/MEM that is approximately 6 x 6 cm with a thickness of 0.7 mm (dry mass is approximately 0.08 grams) for potential use as a tissue engineering scaffold or wound dressing.

Electrospinning is a fiber spinning technique driven by a high-voltage electrostatic field using polymeric solution or liquid that produces polymer nanofibers. The variables controlling the behavior of the electrified fluid jet during electrospinning can be divided into fluid properties and operating parameters. The relevant fluid properties are viscosity, conductivity, dielectric constant, boiling point, and surface tension. The operating parameters are flow rate, applied electric potential, and the distance between the tip and the collector called air gap.

Electrospinning is a method of producing nanofibers by accelerating a jet of charged polymer solution in an electric field (figures 4 and 5). A high-voltage generator generates an electric field between syringe with a capillary tip and grounded collector. A polymer solution is charged by the high-voltage generator and is ejected from the capillary tip. The grounded collector can be a screen or rotating drum and is placed at a fixed distance from the capillary tip. The ejected polymer solution forms a so-called "Taylor cone" at the end of the tip and is drawn toward the grounded collector. During the motion of the jet in traveling to the collector, the solvent evaporates and deposits a non-woven fiber mat on the collector. Homogeneously dissolved polymer solution can be taken in the glass syringe and mounted on the syringe pump. Initially, the polymer solution is held by its surface tension in the form of a droplet at the end of the capillary tip. The positive electrode from the high voltage source should be connected to the syringe needle tip by means of an alligator clip. The negative terminal of the power source and the collector screen should be grounded. As the voltage is increased, the droplet becomes distorted by the induced electrical charge on the liquid surface, and a stable jet of polymer solution is then ejected from the cone. Above a given critical voltage, the jet breaks up into droplets as a result of surface tension. The final products of fibers and beads are determined by the electrospinning parameters. The break-up of the jet depends on the magnitude of the applied electric current. The more electric current the jet carries, the less likely it is to form droplets. A higher net charge density of the polymer solution could, therefore, yield thinner fibers with no beads (figures 6-8). Nanofibres can be collected on a smooth aluminum foil.

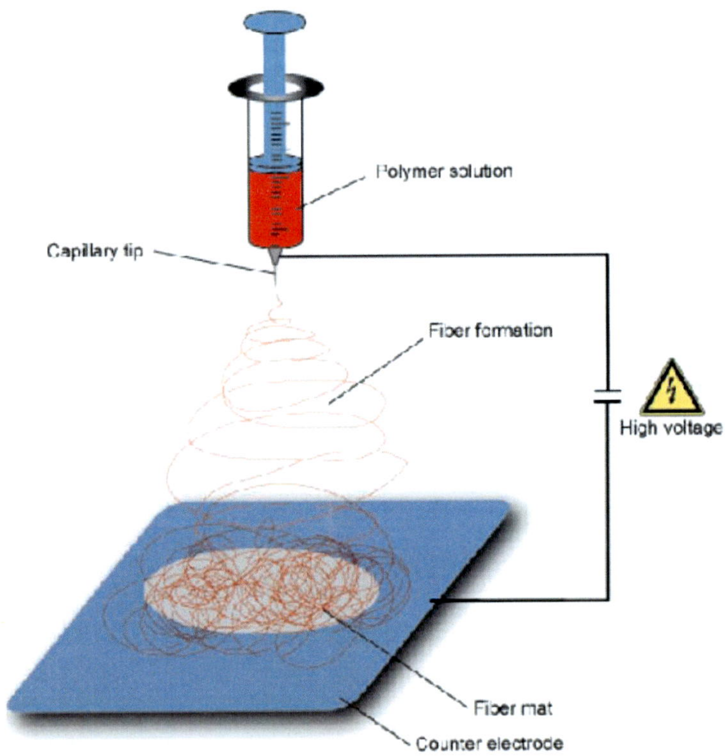


Figure 4. Electrospinning process.

The uniformity and non-uniformity in structure of nanofibers spun can be characterized using Scanning Electron Microscopy (SEM) as it is shown in figure 9. The morphology of nanofibres spun can be characterized using Atomic Force Microscopy (AFM). Sample preparation for the SEM characterization requires the usage of sputter coater. The samples to be analyzed should be placed over a carbon tape on an aluminum pedestal and sputter coated with gold-platinum alloy to a thickness of 200Ao-300Ao.Surface morphology of the nanofibre samples can be characterized using the AFM images. Structure and morphology of nanofibres produced by the electrospinning process depends on the process parameters like the applied voltage, needle tip-collector distance, solution concentration and conductivity and solvent volatility. Other parameters like needle tip or capillary diameter, surrounding gas stream, conductivity of the collector screen, etc. could also influence the fiber morphology and orientation.

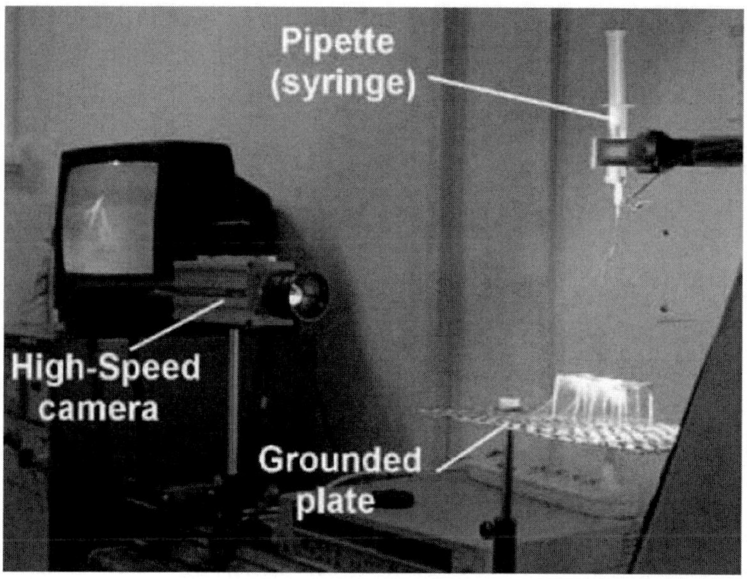


Figure 5. Photo of the experimental setup of the electrospinning process.

Figure 6. SEM Photograph of Electrospun Polyacrylonitrile Fiber.

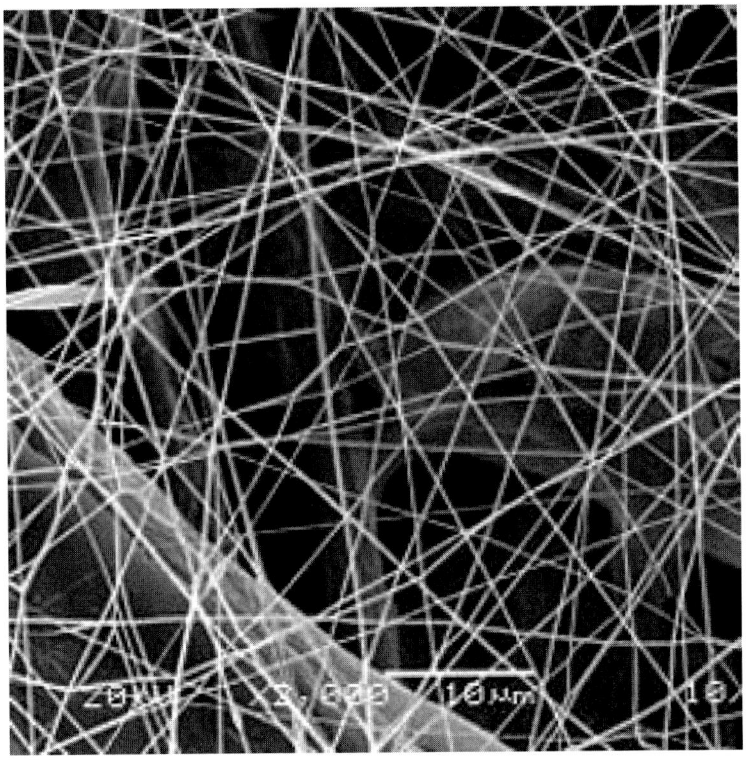

Figure 7. A typical SEM images of electrospun nanofibers.

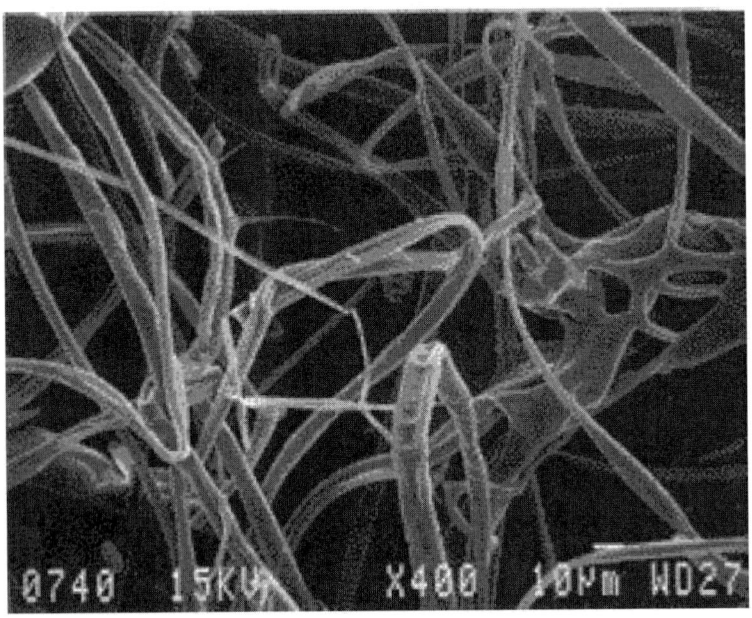


Figure 8. A typical SEM images of electrospun nanofibers.

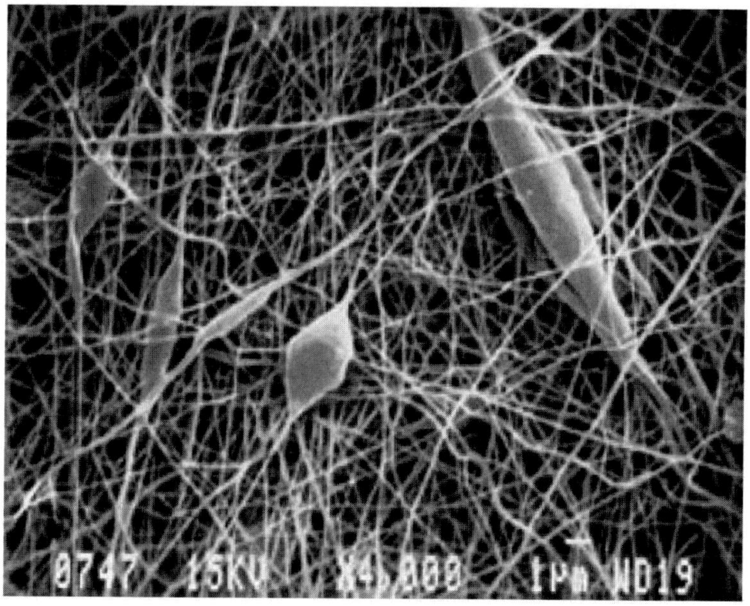


Figure 9. SEM images of nonuniform electrospun nanofibers.

In figure 8 it can be seen that how for certain experiments that we are collecting electrospun fibers onto an insulating film we can measure the charge to mass ratio by NanoCoulomb Meter.

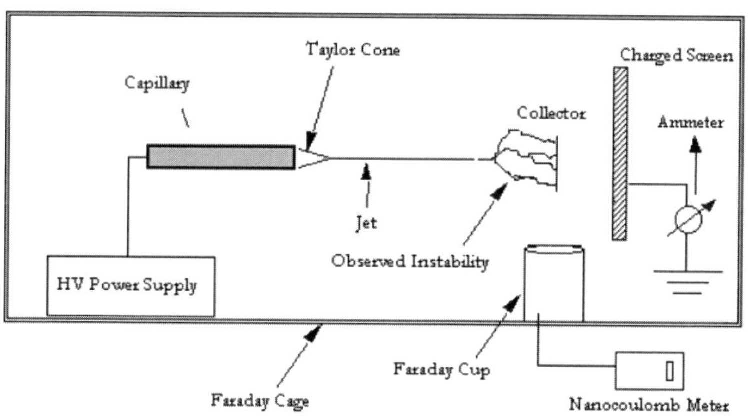


Figure 10. Charge Measurement Setup.

## 1.2. Polymer-Solvents Used in Electrospinning

The polymer is usually dissolved in suitable solvent and spun from solution. Nanofibers in the range of 10-to 2000 nm diameter can be achieved by choosing the appropriate polymer solvent system. Table 2 gives list of some of polymer solvent systems used in electrospinning.

**Table 2. Polymer solvent systems for electrospinning**

| Polymer | Solvents |
|---|---|
| Nylon 6 and nylon 66 | Formic Acid |
| Polyacrylonitrile | Dimethyl formaldehyde |
| PET | Trifluoroacetic acid/Dimethyl chloride |
| PVA | Water |
| Polystyrene | DMF/Toluene |
| Nylon-6-co-polyamide | Formic acid |
| Polybenzimidazole | Dimethyl acetamide |
| Polyramide | Sulfuric acid |
| Polyimides | Phenol |

## 1.3. Quantitive Analysis of Electrospinning Process

Quantitive analysis of the electrospinning process consists of the following:

- *Solution properties.* The relevant fluid properties are density, viscosity, surface tension, permittivity, conductivity and viscoelasticity. Of these, viscosity and conductivity appear to play the greatest role in the electrospinning of dilute solutions.
- *Operating parameters.* The relevant operating parameters are flow rate, electric field strength, and electric current flow between the spinnerette and collector. The volumetric flow rate is closely controlled through the use of the syringe pump. Field strength may be varied by changing either the applied voltage or the distance over which the voltage drop to ground occurs.
- *Online process monitoring.* Macrophotography is used to image the thinning of the fluid jet as it leaves the spinnerette.
- *Fiber characterization.* Fiber diameter distributions are measured using scanning electron microscopy.

## 1.4. Polymeric Nanocomposites

Polymeric nanocomposites made by incorporating small amount of nanoscale inclusions into polymer matrices exhibit dramatic changes in thermomechanical properties over the pure polymers. Because the properties of the nanoscale fillers can be extraordinary, even small volume fractions can result in significant changes. Enhancing the effect is the extremely significant role that the interphase plays in these systems. Given the enormous surface to volume ratio for nanoparticles, the interphase volume fraction can dwarf that of the inclusions themselves. Nanocomposites are a new class of composites, that are particle-filled polymers for which at least one dimension of the dispersed particles is in the nanometer range. Over the last decade, the utility of layered silicate nanoparticles as additives to enhance polymer performance has been established. Nanoscale fillers result in physical behaviour that is dramatically different from that observed for conventional microscale counterparts. For instance, increased moduli, gas barrier, increased strength and reduced thermal expansion coefficients are observed with only a few percent additions of nanofiller; thus maintaining polymeric processability, cost and clarity.

The reinforcement of polymers using fillers, whether inorganic or organic, is common in the production of modern plastics. Polymeric nanocomposites (PNCs) (or polymer nanostructured materials) represent a radical alternative to conventional-filled polymers or polymer blends. In contrast to the conventional systems where the reinforcement is on the order of microns, discrete constituents on the order of a few nanometers (~10,000 times finer than a human hair) exemplify PNCs.

Uniform dispersion of these nanoscopically sized filler particles (or nanoelements) produces ultra-large interfacial area per volume between the nanoelement and host polymer. This immense internal interfacial area and the nanoscopic dimensions between nanoelements fundamentally differentiate PNCs from traditional composites and filled plastics. Thus, new combinations of properties derived from the nanoscale structure of PNCs provide opportunities to circumvent traditional performance trade-offs associated with conventional reinforced plastics, epitomizing the promise of nano-engineered materials.

Main feature of polymeric nanocomposite, in contrast to conventional composites, is the reinforcement is on the order of nanometer deeply affected final macroscopic properties.

Several types of polymeric nanocomposites can be obtained with different particle nanosize, nature and shape, such as clay/polymer nanocomposites, and metal/polymer nanocomposites.

Nanocomposites have been formed with a wide variety of polymers including: epoxy, polyurethane, polyetherimide, poybenzoxazine, polypropylene, polystyrene, polymethyl methacrylate, polycaprolactone, polyacrylontrile, polyvinyl pyrrolidone, polyethylene glycol, polyvinylidene fluoride, polybutadiene, copolymers and liquid crystalline polymers.

### 1.4.1. Clay/Polymer Nanocomposites

Clay is a natural material that has been used by man from ancient times in pottery and building materials. Common clays are naturally occurring minerals and are thus subject to natural variability in their constitution. The purity of the clay can affect final nanocomposite properties. Many clays are aluminosilicates, which have a sheet-like (layered) structure, and consist of silica $SiO_4$ tetrahedra bonded to alumina $AlO_6$ octahedra in a variety of ways. A 2:1 ratio of the tetrahedra to the octahedra results in smectite clays, the most common of which is montmorillonite. Other metals such as magnesium may replace the aluminium in the crystal structure. Depending on the precise chemical composition of the clay, the sheets bear a charge on the surface and edges, this charge being balanced by counter-ions, which reside in part in the inter-layer spacing of the clay. The thickness of the layers (platelets) is of the order of 1 nm and aspect ratios are high, typically 100-1500. The clay platelets are truly nanoparticulate. In the context of nanocomposites, it is important to note that the molecular weight of the platelets (*ca.* $1.3 \times 10^8$) is considerably greater than that of typical commercial polymers, a feature which is often misrepresented in schematic diagrams of clay-based nanocomposites. In addition, platelets are not totally rigid, but have a degree of flexibility. The clays often have very high surface areas, up to hundreds of $m^2$ per gram. The clays are also characterised by their ion (e.g. cation) exchange capacities, which can vary widely. One important consequence of the charged nature of the clays is that they are generally highly hydrophilic species and therefore naturally incompatible with a wide range of polymer types. A necessary prerequisite for successful formation of polymer-clay nanocomposites is therefore alteration of the clay polarity to make the clay 'organophilic'. An organophilic clay can be produced from a normally hydrophilic clay by ion exchange with an organic cation such as an alkylammonium ion. For example, in montmorillonite, the sodium ions in the clay can be exchanged for an amino acid such as 12-aminododecanoic acid (ADA):

$$Na^{+}\text{-CLAY} + HO_2C\text{-R-}NH_3^{+}Cl^{-} \rightarrow .HO_2C\text{-R-}NH_3^{+}\text{-CLAY} + NaCl$$

Clay can be described as being a layered material. In layered materials the bonds between the atoms in the layer are very strong, but the bonds between the layers are much weaker. This enables the layers to be easily separated. Figure 11 shows the structure of a typical clay layer.

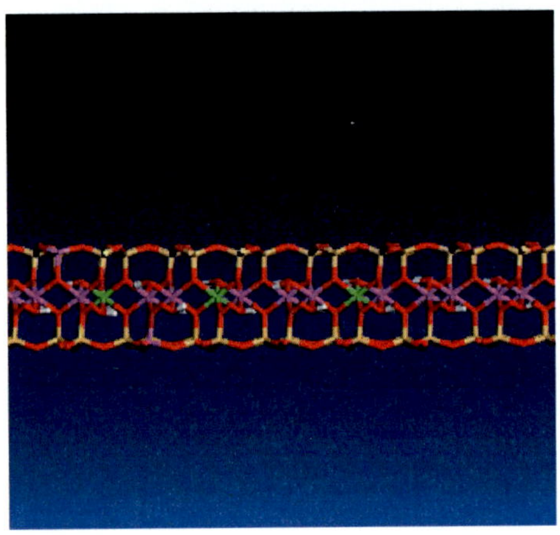

Figure 11. Structure of a typical clay layer.

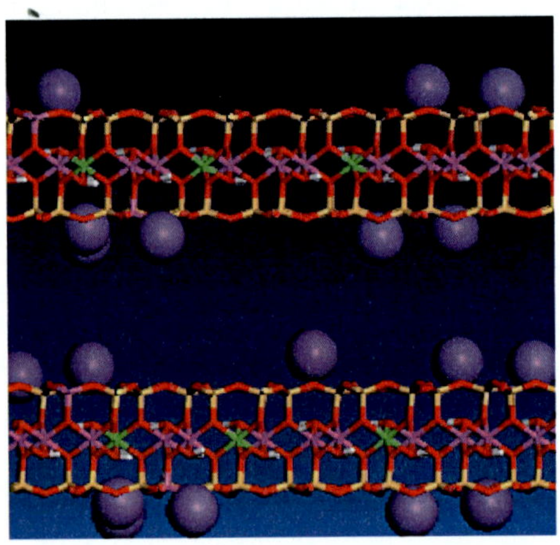

Figure 12. Sheets stack on top of one another.

The clay sheet shown consists of three sub-units: a octahedral center layer consisting of mainly aluminium (Al) cations; two tetrahedral layers consisting mainly of silica (Si) and oxygen (O) atoms. In the octahedral layer some of the $Al^{3+}$ atoms are subsituted by $magnesium^{2+}$ atoms which gives rise to a net negative charge on the layer. Similarly, some of the $Si^{4+}$ may be substituted by $Al^{3+}$ resulting in negative charge in the tetrahedral layers.

Figure 12 shows how the clay sheets stack on top of one another. The resulting negative charge has to be counter-balanced by the introduction of positive ions (cations) in the region between the clay sheets (interlayer or gallery) and at the clay surface. The amount of charge on the clay layer is known as the Cation Exchange Capacity of the clay. In natural clays the charge balancing cations are mainly $Na^+$, $K^+$ and $Ca^{2+}$. These cations tend to have water molecules associated with them in the interlayer region, a fact which makes clay soils swell in wet weather and shrink or crack in dry weather.

It is also possible to incorporate other molecules into the interlayer regions of the clay system. These molecules may be charged or neutral. Charged molecules, such as organo-ammonium species are incorporated by exchanging them for the small inorganic cations - a process known as cation exchange. Neutral molecules may also be intercalated into the clay through favourable entropic and enthalpic interactions. This enables the uptake of molecules such as poly(ethylene glygol) and related species, phenols, etc. Often water is displaced from the clay interlayer during this process. An example of a clay-polymer system is shown in figure 13.

When a clay-polymer nanocomposite is formed there are three ways in which the clay can be incorporated in the polymer. Figure 14 is a simple schematic showing the 3 forms.

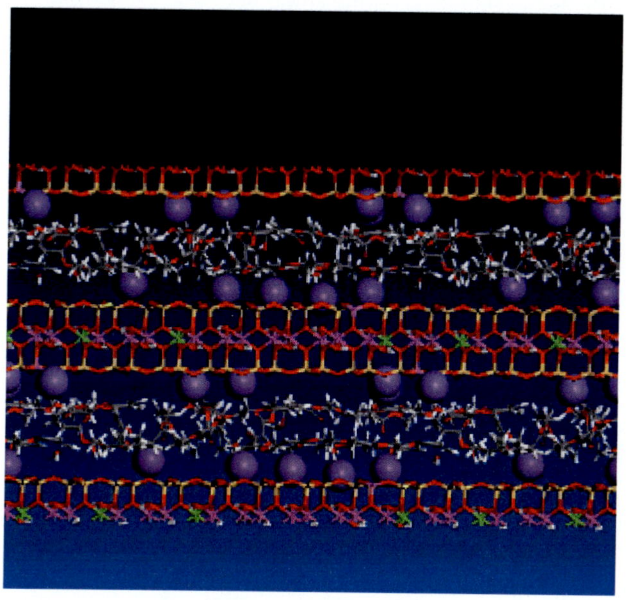

Figure 13. Clay-polymer system.

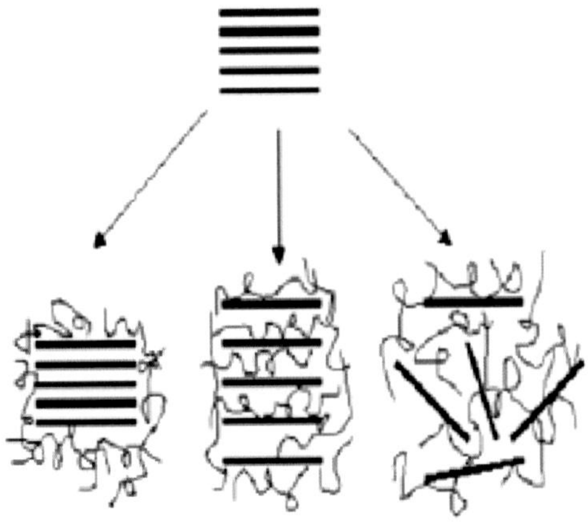

Figure 14. Clay-polymer nanocomposite.

In figure 14a the clay layers do not separate and the clay tactoid is incorporated wholly within the polymer. At the other extreme, figure 14c shows a fully exfoliated clay where all the layers are separated and incorporated within the polymer as individual sheets. Figure 14b shows an intercalated clay, where the clay sheets have separated enough to let some polymer between the sheets, but still retain the layered structure.

When low levels of clay are added to a polymer it is desirable to fully exfoliate the clay so that an even (homogeneous) distribution of clay sheets through the polymer is achieved.

Metal-polymer nanocomposites are obtained using nano-particles (nanostructured metal)as additives in polymer matrix. Nano-sized metals has different properties from bulk metals originating from nanocrystals size. Nanocrystals measure a few nanometers containing few hundred atoms. In this way, nanomaterials shows unique properties (electronic, magnetic, structural) depending on nano-size structure.

Physical and chemical properties of metal nanoparticles are as follow:

The specific-size dependence of these properties becomes evident when they:

- no longer follow classical physical laws but rather are described by quantum mechanical ones;
- are dominated by particular interface effects;
- exhibit properties due to a limited number of constituents, since the usual term "material";

## 2.1. Image Analysis

An image analysis based method was proposed by Pourdeyhimi et al. [5] for measuring fiber diameter in nonwoven textiles. In this method, a binary image of the textile is used to create a distance map and skeleton. The fiber diameter may be determined from the values of the distance map at any pixel location on the skeleton. However, the occurrence of a broken skeleton at intersection points is a main challenging area within the use of this method. Since two or more fibers cross each other at these intersections, the value of the center of the object in the distance map doesn't coincide with the fiber diameter at these points because it isn't associated with a single fiber. The problem becomes more serious as fibers get thicker and for points where more fibers cross each other. Hence, the method fails in measuring fiber diameter at intersections.

### 2.2.1. New Distance Transform Algorithm

We established a new method based on image analysis in which the problem associated with the intersections was solved. The method uses a binary image as an input. Then, the distance transformed image and the skeleton are created. It can be noted that the skeleton which is obtained by the process of *skeletonization* or *thinning* often contains short spurs which may be cleaned up through the use of a *pruning* procedure [2]. In order to solve the problems associated with measuring fiber diameter at the intersections, we first use a *sliding neighborhood* operation [6] to identify the location of these points. Then the thickness of each intersection is recorded from the distance transformed image. Finally the intersections are deleted from the skeleton image based on their measured thickness.

Figure 15 shows a simple simulated web and the resulting skeleton superimposed on the distance transformed image. The obtained skeleton is used as a guide for tracking the distance transformed image and the diameters are computed from the intensities of this image at all points along the skeleton. The data in pixels may then be converted to *nm* and the histogram of fiber diameter distribution is plotted.

### 2.2.2. Web Simulation

In order to validate the method, test samples with known characteristics were required. Algorithms for simulation of nonwoven mats have been proposed by Abdel-Ghani et al. [3] and Pourdeyhimi et al. [4]. Lately it has been discovered that the best way to simulate nonwoven mats of continuous fibers is through *μ-randomness* procedure [8] which was used in this study for generating a simulated image with known characteristics.

### 2.2.3. Real Web Treatment

Fiber diameter determination by the use of image analysis requires the initial segmentation of the micrographs in order to produce binary images. The typical way of producing a binary image from a grayscale image is by *global thresholding* [5, 6] where a single constant threshold, usually selected by trial and error, is applied to segment the image. Global thresholding is very sensitive to any inhomogeneities in the gray-level distributions of the object and background pixels. This effect can be eliminated through the use of a *local thresholding* [7, 8] scheme. Automatic selection of the appropriate thresholds can be carried out based on, *Otsu's method* [9]. Note that, since the process is extremely sensitive to noise, before the segmentation, a procedure to clean the noise and enhance the contrast of the image is necessary.

## 3. Results and Discussion

A simulated image with the diameter sampled from a normal distribution with the mean (M) of 15 and standard deviation (STD) of 4 pixels was used to test the validity of the method. It is noteworthy that the true M and STD of the simulated image (15.35 and 4.47) varies slightly with those used as simulation parameters.

Figure 16 shows the simulated image and its diameter distribution obtained from the new distance transform method. The M and STD of fiber diameter obtained by this method were 15.02 and 4.80 respectively, showing a good correlation between the calculated and true M and STD of the simulated image.

Using this method, there can be up to half a pixel error in either direction when measuring the fiber diameter, resulting in a total measurement error of up to 1-pixel. The slight differences observed between the calculated and true values could be attributed to this 1-pixel measurement error, some parts of branches remaining after pruning and other slight variations in the skeleton adjacent to the deleted intersections. Furthermore, the fiber diameters at the deleted intersections were not counted within the measurement and may contribute slightly to the variation observed.

To prove that this process is suitable for determination of fiber diameter on real samples, a real nanofiber web was obtained from electrospinning of polyvinyl alcohol (PVA) with average molecular weight of 72000 g/mol (MERCK). The micrograph of the electrospun web (figure 17a), was taken using a Philips (XL-30) Environmental Scanning Electron Microscope after gold sputter coating.

Figure 17b shows the diameter distribution for the real web. The respective M and STD of the fiber diameter obtained by this new method were 24.74 and 3.85 in terms of pixel and 323.7 and 50.4 in term of *nm* which are in good agreement with the values 24.36 and 3.19 pixels and 318.7 and 41.8 *nm* obtained from manual methods. The differences here can also be attributed to the different number of measurements taken between the methods used (over 2000 for our method versus 100 for the manual method). Nevertheless, in each case presented, the difference observed was within 1-pixel measurement error suggesting the main limitation with the process is with the resolution of the taken image.

## 4. Conclusion

In this study, a new image analysis based method for assessing nanofibers diameters was successfully developed. The validity of the method was tested using a simulated image as well as an image of a real electrospun nanofiber web. In the case of the real web, local thresholding was applied on the micrograph of the web taken from SEM to attain the necessary binary image. The M and STD of fiber diameter which were obtained using this new method were extremely close to true values on the simulated image. For the real web, M and STD of fiber diameter measured by the method were also in good agreement with those obtained from the manual method. The results show the effectiveness of the method for diameter measurement. The method is automated, accurate, and much faster than manual method and has the capability of being used as an on-line technique for quality control.

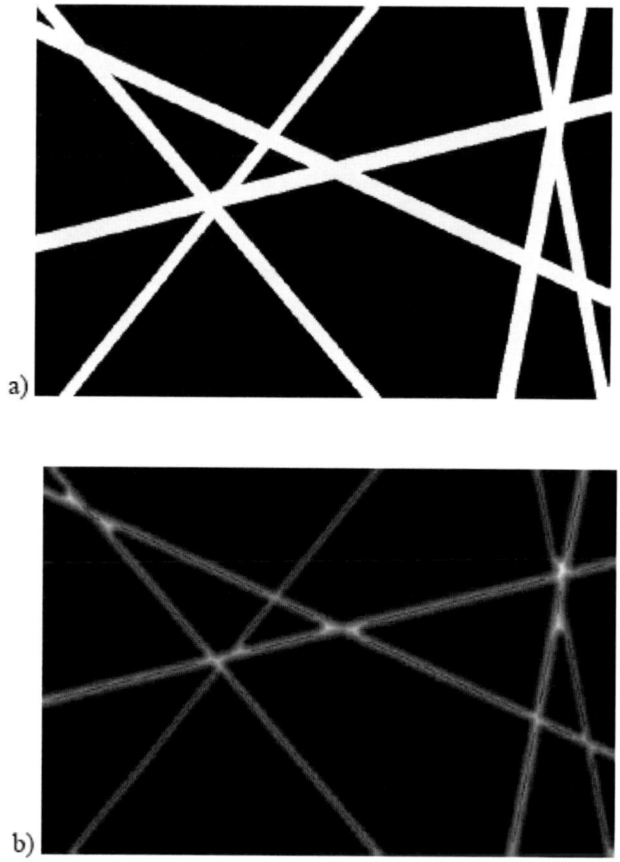

Figure 15. a) A simple simulated image, b) Resulting skeleton overlaid on distance transformed image.

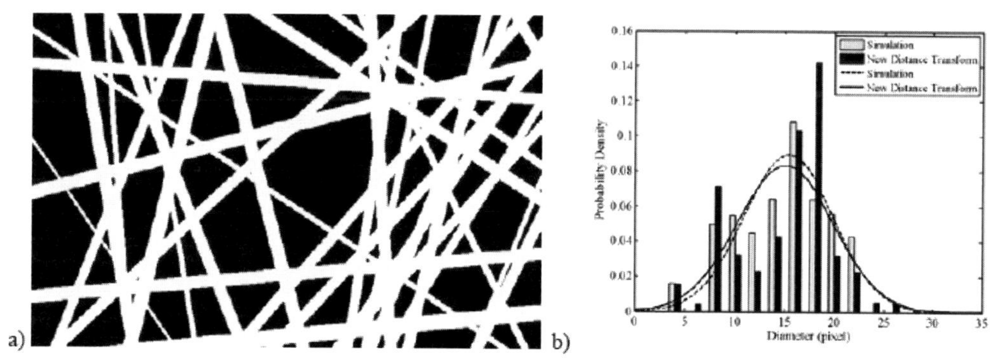


Figure 16. a) A simulated image, b) Its diameter distribution.

a) 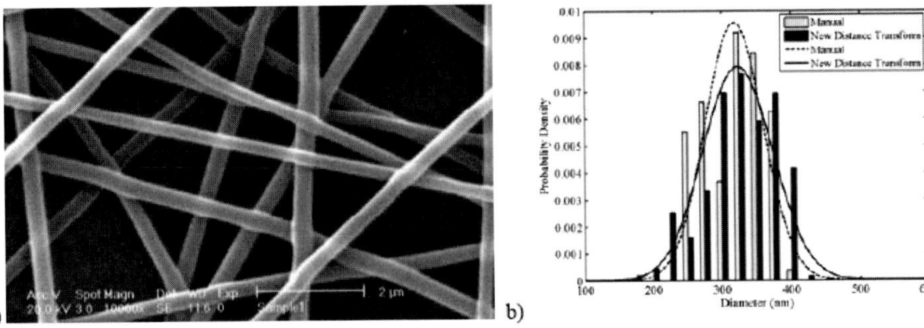 b)


Figure 17. a) Micrographs of an electrospun web, b) Its diameter distribution.

## REFERENCES


[1] A. K. Haghi, M. Akbari, *Phys. Stat. Sol.* (a) 204, 1830 (2007).

[2] D. H. Reneker, I. Chun, *Nanotechnology.* 7, 216 (1996).

[3] H. Fong, D. H. Reneker, Electrospinning and the Formation of Nanofibers, in: Structure Formation in polymeric Fibers, ed. by D. R. Salem (Hanser, Cincinnati, 2001), chap. 6, pp. 225-246.

[4] Th. Subbiah, G. S. Bhat, R. W. Tock, S. Parameswaran, S. S. Ramkumar, *J. Appl. Polym. Sci.* 96, 557 (2005).

[5] B. Pourdeyhimi, R. Dent, *Text. Res. J.* 69, 233 (1999).

[6] R. C. Gonzalez, R. E. Woods , Digital Image Processing (Prentice Hall, Second Edition, New Jersey, 2001).

[7] M. S. Abdel-Ghani, G. A. Davis, *Chem. Eng. Sci.* 40, 117 (1985).

[8] B. Pourdeyhimi, R. Ramanathan, R. Dent, *Text. Res. J.* 66, 713 (1996).

[9] M. Petrou, P. Bosdogianni, Image Processing the Fundamentals (John Wiley and Sons, England, 1999).

In: Monomers, Oligomers, Polymers, Composites…
Editors: R. A. Pethrick, G.E. Zaikov et al.
ISBN: 978-1-60456-877-6


*Chapter 18*

# INDUSTRIAL DRYING OF WOOD: TECHNOLOGY LIMITATION AND FUTURE TRENDS

***A.K. Haghi[*] and R. K. Haghi***
University of Guilan, P. O. Box 3756, Rasht, Iran

## ABSTRACT

In this paper, a comprehensive review is presented on the researches and developments related to drying processes of wood. Other issues regarding the technology limitation, research challenges, and future trends are also discussed.

## LIST OF SYMBOLS

| | |
|---|---|
| A, B, | constants |
| Bi | Biot number for moisture loss |
| $d_h$ | hydraulic diameter |
| D | moisture diffusivity |
| $f_D$ | friction factor from Moody chart |
| $F_O$ | Fourier number for moisture loss |
| k | moisture transfer coefficient, |
| L | half thickness, m |
| Re | Reynolds number |
| t | time |
| $u_m$ | average speed |
| W | moisture content |

[*] Corresponding author e-mail: Haghi@Guilan.ac.ir

| | |
|---|---|
| $\phi$ | moisture content difference |
| $\Phi$ | dimensionless moisture content |
| $\mu$ | root of the transcendental characteristic equation |
| $\rho$ | mass density of fluid |
| $\nu$ | Newtonian fluid viscosity |
| $\sigma$ | surface roughness |
| $\Gamma$ | dimensionless distance |

## 1. INTRODUCTION

Porous materials such as wood have microscopic capillaries and pores which cause a mixture of transfer mechanisms to occur simultaneously when subjected to heating. Transfer of vapor and liquids occurs in porous bodies in the form of diffusion. In essence, transfer of liquids can occur by means of diffusion arising from hydrostatic pressure gradient. Heat and mass transfer in porous media is a complicated phenomenon and a typical case is the drying of moist porous materials. Scheidegger [1] claimed 47 years ago that the structure of porous media is too complex to be described precisely either in macro-scale or micro-scale, not to mention the combination of water with matrix. To date, there is no credible work proving that Scheidegger was wrong.

Convective drying is usually encountered in wood industry. The study of this type of drying has attracted the attention of several authors. Among the works relating to this question we cite the works of Plumb et al [2] and Basilico and Martin [3]. Convective drying of timber is one of the oldest and time-consuming methods to prepare the wood for painting and chemical treatments. The drying method can obviously have significant effect on the mechanical properties of wood. Major disadvantages of hot air drying are low energy efficiency and lengthy drying time during the falling rate period. The desired to achieve fast thermal processing has resulted in the increasing use of radiation heating. In this case, not only the removal of moisture is accelerated but also a smaller floor space is required, as compared to conventional heating and drying equipment.

It has also been recognized that dielectric heating could perform a useful function in drying of porous materials in the leveling out moisture profiles across wet sample. This is not surprising because water is more reactive than any other material to dielectric heating so that water removal is accelerated. This leads to giving a temperature gradient inside the wood sample with opposite directions to that in conventional drying processes. The Fickian diffusion theory for wood results in a mass transfer coefficient (surface emission factor) which is not in accordance with classical heat and mass transfer boundary layer theory. The mass transfer coefficient is one order of a magnitude lower than the ones obtained from classical theory. Checking susceptibility is closely related to stress development across the grain during drying. The tension stress at the beginning of the drying period develops rapidly causing a creep response of the surface. This mechanosorptive creep relaxes the tension stress enough to prevent checking in many cases. This effect is more pronounced at elevated temperatures. It is shown that high density and low temperature make wood more susceptible to checking, a property closely related to the value of strain at failure. Warping during drying is generally caused by the anisotropic shrinkage of wood. The study of cupping shows that it depends mainly on the differential shrinkage in tangential and radial direction but also that cupping is reduced to some extent by the creep of the surface layer. However during the conditioning treatment cupping increases to values near those reached after stress free drying. Although many methods of drying timber have been tried over the years only a few of these enable drying to be carried out at a reasonable cost and with minimal damage to the timber. The most common method of drying is to extract moisture in the form of water vapour. To do this, heat must be supplied to the wood to provide the latent heat of vaporisation. There are several ways of conveying heat to the wood and removing the evaporated moisture. Nearly all the world's timber is, in fact, dried in air. This can be carried out at ordinary atmospheric temperatures (air drying), or in a kiln at controlled temperatures raised artificially above atmospheric temperature but not usually above 100°C, the boiling point of water. Air drying and kiln drying are fundamentally the same process because, with both, air is the medium which conveys heat to the wood and carries away the evaporated moisture [4-6].

## 2. Factors Influencing the Drying of Wood in Air

The factors which will be described are those which affect wood when dried in air (in the open or in a kiln). There are several other ways in which wood can be dried, in chemicals, in a vacuum etc. Under these conditions different factors come into play.

- Vapour Pressure and Relative Humidity
- Temperature:
- Air Movement:
- Movement of Moisture in the Wood :
- Supply of Heat:

## 3. Methods of Drying Wood in Air

### Air Drying

With air drying there is virtually no control of the temperature, relative humidity or speed of the air passing through the timber stacks. The rate of drying is therefore dependent on all the vagaries of the local climate and can vary between practically zero on a calm, damp day to quite fast enough to cause surface checking during dry, windy weather. With air drying, wood cannot be dried below its equilibrium moisture content and this will vary depending on the atmospheric conditions. So, except in unusually hot and dry weather, the lowest moisture content obtainable is around 16-17%; air drying alone is not sufficient for timber intended for most interior uses in Europe, Japan or N. America where a moisture content of between 8 and 12% is required. In air conditioned buildings moisture contents of about 12% should be anticipated.

### Kiln Drying

In contrast to air drying a modern conventional drying kiln provides temperature control and a steady and adequate flow of air over the timber surface. The air flow rate and direction is controlled by fans and the temperature and relative humidity of the air can be adjusted to suit the species and sizes of timber being dried. It is thus possible to make full use of the increase in drying rate which can be achieved by raising the temperature to the maximum value which a particular timber species can tolerate without excessive degrade. At the same time, the relative humidity can be controlled so that the moisture gradients in the wood are not steep enough to cause surface checking. The same principles apply to the use of heat pump kilns except these recover and re-use a proportion of the energy which in conventional kilns is lost during the drying process when the warm, damp air is vented. In addition to the advantages of more rapid drying and limitation of degrade, the ability to control drying conditions in a kiln means that it is possible to achieve timber moisture contents suitable for specific uses. The direct costs of kiln drying are much higher than those of air drying for they include the capital costs of the equipment and the cost of fuel, electricity and supervision. These costs are partially or wholly offset by the reduction in stock level [7-10].

### Air Drying Followed by Kiln Drying

Kiln drying tends to become uneconomical when the species and size of timber being dried require long kilning times. Therefore, with material taking more than about 4 or 5 weeks to kiln dry from green it will often be more economical to air dry the timber to about 25-30% moisture content before completing the drying in a kiln. The economic advantage of this approach may be lost, however, if the layout or lack of handling facilities necessitates dismantling the air dried stack and repiling for the kiln drying phase. Also with some species, the amount of splitting and checking which occurs during air drying in the dry season can be excessive.

## 4. Moisture Content of Timber

The amount of water in a piece of wood is known as its moisture content. Because this is expressed as a percentage of the dry weight of the piece, not of the total weight, it is possible to have moisture contents of well over 100%. The moisture content of green wood varies greatly from one species to another. Moisture content can vary between apparently similar pieces of the same species and in addition there may be differences, between and within species, in the rates at which moisture is lost from timber during drying. These inherent differences in timber mean that it is important during the drying process to be able to monitor moisture content and check that the drying process is proceeding correctly.

**Moisture Content Determination by the Oven Drying Method**

The oven drying method is the standard way of determining wood moisture content. With this method a piece of wood is initially weighed and then dried in an oven at 103°C ± 2°C. Drying is continued until the piece is completely dry (when no further weight loss occurs) and this oven dry weight recorded. The loss in weight during drying indicates how much water was originally present in the piece and the moisture content can be calculated simply, as follows:

$$\text{Initial moisture content (\%)} = \frac{\text{Initial (wet) weight} - \text{Dry weight}}{\text{Dry weight}} \times 100$$

For example, if the initial weight of the piece was 30.51g and its dry weight 22.60g, then the difference of 7.91g is the weight of moisture initially in the piece and its initial moisture content would be:

(30.51 - 22.60)/22.60 x 100 = 7.91/22.60 x 100 = 35.0% Alternatively the formula can be written:

Moisture content (%) = [(Initial weight/Dry weight) - 1 ] x 100

So that only the division sum needs to be carried out:

[(30.51/22.60) - 1] x 100 = 0.35 x 100 = 35.0 %

## 5. Equipment Required

Oven drying requires a well ventilated oven which can control the temperature to between 101 and 105°C and a balance for weighing the test samples. The balance should have a capacity of about 200g and be capable of detecting differences of 0.005g, an automatic type is recommended as these give an instantaneous reading. Infrared ovens are available for rapid drying. In some of these the heating lamps are directed on to the test section on the pan of a balance (incorporated in the equipment). Drying takes from about 3 to 10 minutes according to species and moisture content. However only one piece can be dried at a time and experience is needed to avoid overheating which can cause inaccurate results [11].

## 6. The Pre-Drying of Timber

A high proportion of the world's timber is either wholly or partly air dried before drying in a kiln. Air drying involves the open piling of fresh-sawn timber out of doors, or in open sheds, so that the wood surfaces are exposed to the surrounding atmosphere. Wind and local convection currents will cause air movement through the stack and this conveys solar heat energy to the wood and carries away evaporated moisture[12]

The employment of correct techniques can reduce air drying times and keep drying degrade to a minimum.

## Sitting and Layout of Yard

Ideally, timber should be stacked well away from trees and buildings on a cleared, level and well drained site which has been concreted, covered with ashes or treated to prevent the regrowth of vegetation [13]. In most situations, the orientation of the stacks has little effect on the drying rate and the most important consideration in planning the yard is to arrange stocks and roadways to facilitate handling. Adjacent stacks should be parallel and oriented with either ends or sides to the roadway depending on the methods used for transport and stacking. Consideration should be given to the possibility of fork-lift trucks or side loaders. Stacks should be erected on good solid foundations and, in order to permit ample ventilation, the bottom layers of timber should be raised well above the ground. The clearance should certainly be no less than 230mm and should preferably be about 460mm. The most convenient form of foundation, and probably the simplest to erect, consists of a series of timber cross-members (bearers) not less than 100 x 100mm in section, preferably preservative-treated and lifted clear of the ground on brick or concrete piers or on treated timber, e.g. railway sleepers (figure 1). The piers should be placed at intervals of 600mm along the whole length of the stack. Stringers or longitudinal timber members can be used to give added strength and rigidity to the foundations, but they are generally only necessary where special stacking arrangements are involved. It is essential that the bearers should all be in one plane, but it is not critical whether these are level or on a slight slope [14]. In either case, any necessary adjustment can be made by varying the height of the brick piers and inserting wooden packing blocks between the bricks and cross-members where required.

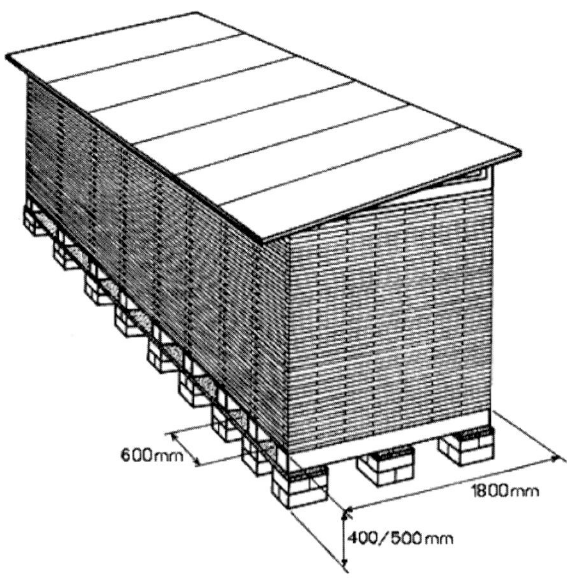


Figure 1. Recommended stack construction for air seasoning of timber.

**The Normal Method of Stacking**

Whenever possible, different species and thicknesses should be stacked separately. It is an advantage if timber can be sorted to length at the outset, and when a variety of lengths has to be stacked it is convenient to place the longest pieces at the bottom and to reduce the length of the stack as the height increases [15]. Alternatively, if sorting beforehand is not practicable, a stack of uniform length may be built by arranging the timber as shown in figure 2. This is sometimes referred to as box-piling.

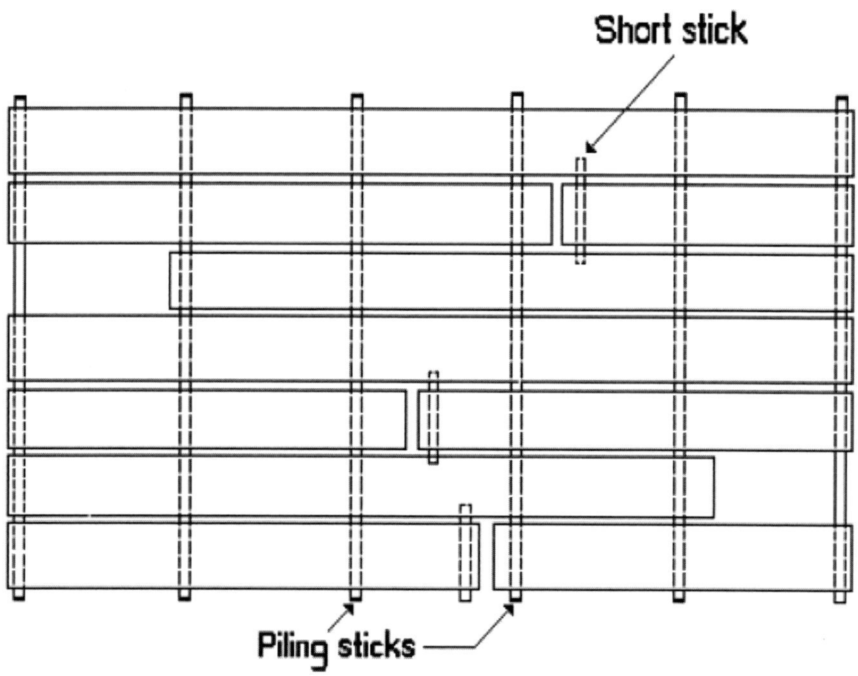


Figure 2. Plan view of layer of boards.

## 7. Kiln Design

In two designs of overhead fan kiln the fans are mounted at regular intervals on a longitudinal drive shaft. The air is diverted by baffle boxes to flow across the top of the kiln above a false ceiling, down the side and through the load as in the designs shown below ( figure 3 ) sometimes called cross circulation kilns. In design A, in spite of correcting plate baffles fixed as indicated, there is a tendency for the circulation to be stronger at the end towards which the fans are blowing. This effect is eliminated in design B in which left and right hand fans are fitted alternately on the shaft, although here a slight loss in efficiency of the fans will occur because opposing pairs will set up back pressure [16-18].

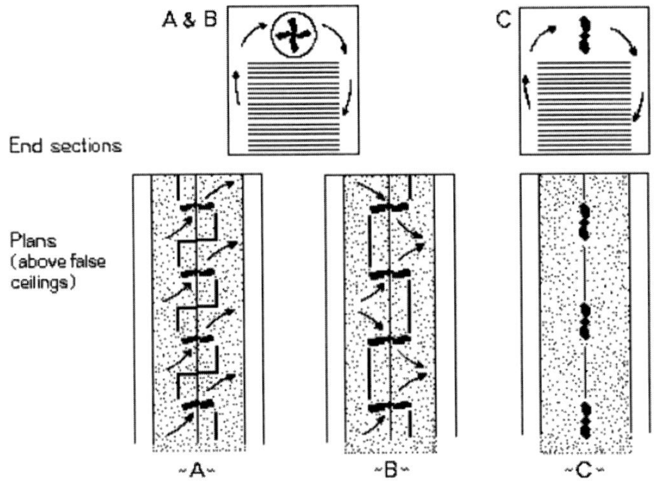


Figure 3. Fan arrangements in three designs of overhead fan kiln.

In these longitudinal shaft kilns, air speeds through the load average only from 0.5 to 1 m/s, unless fans larger than the usual 0.8 to 1.Om diameter are used. In America kilns are built to design B using fans up to 1.8m in diameter. There are two distinctive types of side fan kiln - the vertical flow (figure 4) and the horizontal flow (figure 4, design E). In both, large propeller type fans are placed to one side of the timber load. With side fan designs it is possible to take advantage of the greater energy efficiency of the slower running, larger fans without the need for increased kiln height which would be necessary if these were used in an overhead fan design. In design D1 the air is delivered or returned through a duct which passes air above the timber load and no transverse baffles are required [19-21]. In a variation of this design (D2) smaller fans are mounted to one side of the lower half of the timber load and the air return not through a duct but through the upper half of the load.

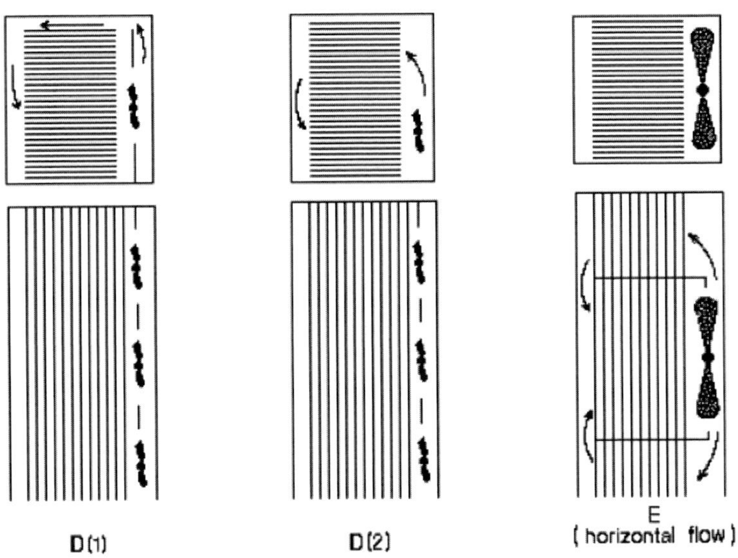


Figure 4. Fan arrangements in three designs of side fan kiln.

In design E, which is a type commonly installed in the UK and in some European countries, the air flow is horizontal throughout. This is achieved by filling the kiln to its full height and, with the wing baffles forming a fan box, air is pulled or pushed (according to the direction of fan rotation) through the portion of the load opposite the fan and pushed or pulled through the two end portions. In designs D2 and E, the air has to pass through the load twice before it is reheated (unless booster heating is placed in a position remote from the fans) and fast air speeds are necessary to minimize the difference in drying rate across the width. Average speeds of the order of 1.5 to 2.4m/s are obtained without excessive power consumption by the use of large fans 1.4-2.4m in diameter. The air speeds through various parts of the timber load are not usually as uniform as in the overhead cross shaft type of kiln but the lowest speed within the load is sufficient for satisfactory drying [18-20]. It is now normal to reverse the direction of air flow frequently and this compensates for local variations in air speed. Side fan kilns can also be built as double load units, the one large fan pulling air through the load on one side of it and pushing the air through that on the other. The traditional and simplest instrument for measuring air conditions is the wet and dry bulb hygrometer (figure 5). The dry bulb thermometer measures the actual temperature in the kiln while the wet bulb reading enables the relative humidity of the air to be estimated. The bulb of the wet bulb thermometer is surrounded by a sleeve which is kept moist with distilled water from a reservoir. Evaporation from this sleeve cools the wet bulb below the temperature of the dry bulb and the magnitude of this wet bulb depression is related to the relative humidity of the air.

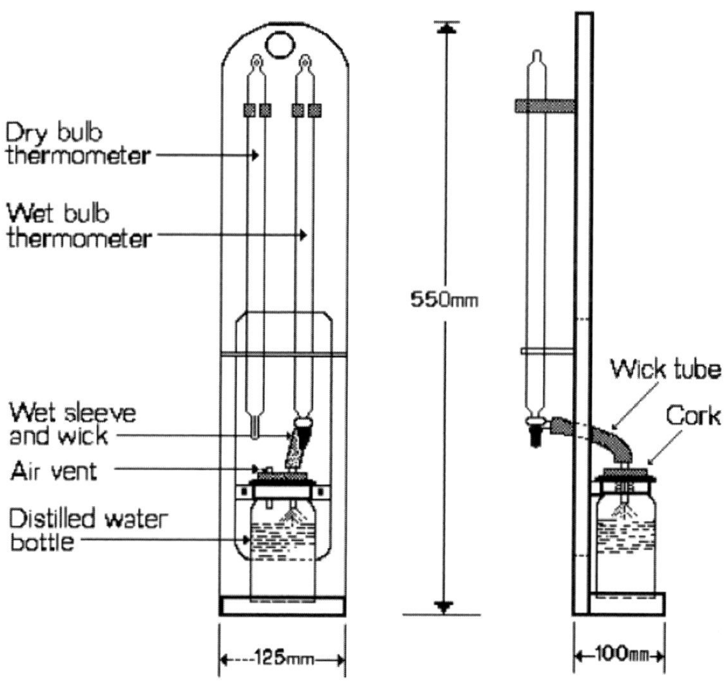


Figure 5. Wet and dry-bulb hygrometer.

Microprocessor and computer technology has already assumed a major role in the development of control systems for kiln drying. Most computer based systems can be programmed to operate the kiln according to a predetermined schedule, while more advanced systems are able to control a sequence of kiln conditions on the basis of the moisture content of the timber, which is monitored by remote sensors in the load. There is still technical difficulty in measuring accurately moisture contents above about 30% and at present fully automatic control systems require a greater margin of safety above this level and consequently drying times may be slightly longer [20-22].

In all types of kiln in which the circulating air returns in a vertical direction to the inlet, the sides of the load should be made as even as possible. Pieces which jut out appreciably will tend to act as deflectors causing an excess of air to pass through one or two spaces at the expense of others (figure 6). An advantage of the side fan kilns with horizontal air flow is that an irregular-faced load has no adverse effect on the circulation.

Figure 6. Effect of irregular pile face on air circulation.

Two other points should be noted in the piling of loads for horizontal flow kilns. The gaps between the face of the load and the end wings of the fan boxes should be as small as possible or should be blanked off to prevent excessive short-circuiting. Secondly, the sticks should be arranged so that there is a vertical tier opposite each end wing to ensure that air entering the load traverses the full width before re-entering and being pulled back into the fan (figure 7). Figures 8B and 8C show how incorrect piling can adversely affect the air circulation.

Figure 7. Piling of timber in side fan horizontal flow kilns.

In any kiln the circulating air will always have a tendency to by-pass the timber through spaces inadvertently formed above, below or along the load. It is recommended that this short-circuiting should be minimized by using canvas curtains or baffles made of wood or other suitable material. Occasionally the volume of timber to be dried may be less than a full load for the kiln available. In such cases the width of the load can be reduced so that it approximately fills the height of the kiln.

Certain timber items cut to standard sizes can conveniently be piled without the use of a large number of sticks. For instance, small furniture parts such as chair leg squares can be self crossed (figure 8). Here the individual pieces are so small that they can be used in place of the normal piling sticks without fear of restricting the air flow[23-25]. In order that the individual stacks which make up a complete load should be stable when drying shrinkage occurs, sticks should be introduced at intervals across the full width of the load.

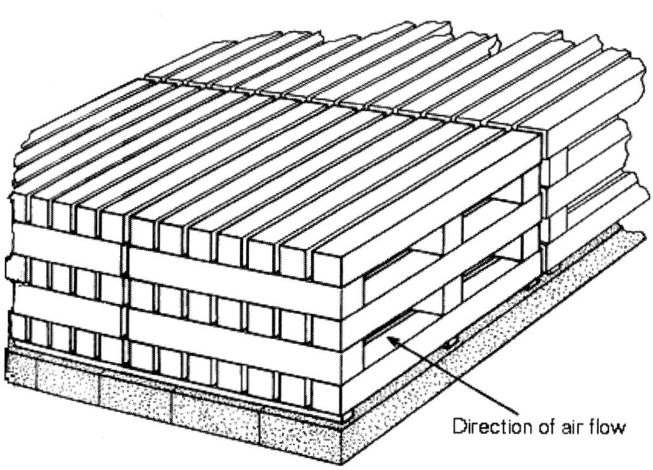


Figure 8. Cross-piling of dimension stock.

For a typical drying operation involving an average size kiln load, and a single species of timber converted through and through to a constant width, it is recommended that the number of sample pieces in the sample series (figure 9) is a minimum of six. This sample series should consist of both plain-sawn and quarter-sawn samples and if possible it should include pieces which are representative of the wettest and driest wood in the kiln load. The number of samples should be increased proportionately in larger kiln loads. However there is a practical and economic limit which may restrict the number that can be monitored [26-28]. For example in large loads containing more than one species of more than one thickness, it may be impractical to have enough samples to provide all the necessary information about the progress of drying. Conversely it may be possible to include less than six samples when kilning timber which has fairly uniform moisture content, and well known and predictable drying characteristics.

Figure 9. Distribution of samples in the load in a kiln (A) with overhead fans (B) with side fans and horizontal flow.

To ensure that the drying response of a sample is typical of the wood in a particular part of the load, samples must be incorporated in a way which does not interfere with local drying conditions. Additionally the method of incorporation must allow easy withdrawal of the sample for weighing. However difficulties of sample withdrawal may sometimes be unavoidable with timber which is prone to distort badly. One method of accommodating withdrawable samples is to place sticks over them which have been notched out to about half the normal thickness. This tends to be time consuming and rather wasteful of sticks unless it is feasible to have samples of a standard width and position, in which case the notched sticks can be used repeatedly. An alternative arrangement, using suitable lengths of sticks of half the normal thickness, is shown in figure 10. Both these methods involve opening the main kiln doors for the removal of the samples but with modern kilns the air conditions soon return to normal [29].

Figure 10. Method of accommodating withdrawable kiln samples.

Samples can be accommodated in the sides of loads by cutting one or more sticks off short (figure 11) leaving the sample free. Access to side samples is usually by the small side doors rather than by the main door and because the operator therefore has to enter the kiln, there may be practical difficulties in retrieving samples under certain kiln conditions.

Samples should always be positioned within the load and not mounted on short projections at the ends of loads, or between two separate loads, where abnormally fast drying may occur.

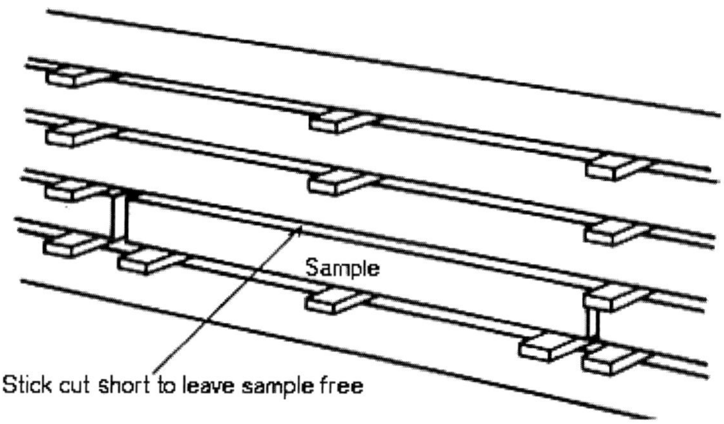


Figure 11. Method of accommodating sample in the side of a kiln stack.

As it is shown, for the duct formed by the boards the aspect ratio, AR, is very low in practice with

$$AR = (\text{duct height})/(\text{duct width}) < 0.1 \quad (1)$$

As duct height is considerably smaller than the duct width, it is assumed that the hydraulic diameter $d_h$ asymptotically approaches twice the duct height (as is in the case of an infinity wide duct), i.e.

$$d_h = (2)\ (\text{duct height}) \quad (2)$$

This is a good approximation in most cases, but should be verified at higher aspect ratio.

The pressure loss of fully developed flow along a channel can be determined by using Moody chart [30] or a friction factor formula as described by White [31]. The dimensionless friction factor formula determined by Haaland [32] represents the turbulent region on the Moody chart and is as follows:

$$f_D^{-1/2} = -1.8\log_{10}\left(\frac{6.9}{\mathrm{Re}_{umd_h}} + \left(\frac{\sigma/d_h}{3.7}\right)^{1.11}\right) \quad (3)$$

The Reynolds number $\mathrm{Re}_{umd_h}$ is based on the average speed $u_m$ and hydraulic diameter $d_h$, i.e.

$$\mathrm{Re}_{umd_h} = \frac{\rho u_m d_h}{\nu} \quad (4)$$

## 8. Estimation of the Average Moisture Content of a Sample

By first estimating and then monitoring the fall in average moisture content of the samples during drying, a given schedule can be followed accurately. The average moisture content for a particular sample is estimated by first measuring, using the oven method the moisture content of one (or more) test sections cut as indicated above. If it is then assumed that this measured moisture content is typical of the remainder of the sample, the dry weight of this remainder can be estimated and changes in its average moisture content can be monitored throughout drying by weighing. Suppose that the initial average moisture content of the sample was estimated to be 35%. If the sample weighed 12.40kg then its dry weight can be estimated:

Dry weight = (wet weight/moisture content)/100 + 1
=12.40/[(35/100) + 1]
=12.40/1.35
=9.18kg

This estimated dry weight remains a constant quantity as long as no further wood is cut from the sample. If after a period of drying the actual weight of the sample has fallen to 11.72kg, its new average moisture content can be estimated as follows:

New average moisture content = [(Current weight/Dry weight) - 1] x 100
=[(11.72/9.18) - 1] x 100
=27.6%

This technique is based on the assumption that the test section(s) do provide an accurate estimate of the average moisture content of the sample from which they were cut. However, even in the ideal situation where a section is removed from each end of the sample (and the values obtained averaged) an appreciable error can occur. This is because moisture content will inevitably vary along the length of a piece and for this reason it is always advisable to predetermine the moisture content of a further batch of test sections towards the end of a kiln run [30]. With certain species, such as teak, there is a marked tendency to retain pockets of moisture along the length. In these cases the average moisture content is difficult to determine in the normal way and, if it is necessary to dry such timbers to uniform moisture content, there will be a need to cut more test sections to obtain a suitably accurate estimate of average moisture content.

## 9. Estimation of the Average Moisture Content of the Load

Having obtained the average moisture contents for the individual samples, it is then quite straight-forward to use this information to estimate the average moisture content of the load. Again, this can be estimated at any time during drying and it helps the kiln operator to judge the progress of drying. It is estimated simply, as follows:

$$\text{Average moisture content of the load} = \frac{\begin{array}{c}\text{Sum of average moisture contents for all}\\ \text{for all samples being used to monitor the load}\end{array}}{\begin{array}{c}\text{Total number of samples}\\ \text{being used to monitor the load}\end{array}}$$

For example, if at a particular time, the samples gave average moisture contents of 16, 14, 13, 12, 10 and 10%: Average moisture content of the load

= (16+ 14+ 13+ 12+ 10+ 10)/6
= 75/6
=12.5%

Once below fiber saturation point, electrical resistance moisture meters can be used to augment the information obtained by oven drying.

## 10. Assessment of Moisture Distribution and Casehardening Stresses

The importance of assessments of moisture distribution and casehardening is in evaluating the risk of drying degrades. Both assessments can only be made by cutting fresh test sections. It is often convenient to cut these at the same time as sections are taken for the average moisture distribution. Casehardening stresses should be assessed at least twice during a kiln run: first before drying is commenced (to ensure that the appropriate drying schedule will be suitable without modification) and again towards the end of the kiln run to ensure that drying is progressing correctly. More assessments may be necessary when drying difficult sizes of timber or species which have unpredictable drying qualities. Moisture distribution is assessed by removal of a test section and by sub-dividing this into strips as shown in figure 12. The strips are cut so that the inner and intermediate strips are representative of increasing depth within the original sample. Moisture distribution is then assessed by measuring by moisture content of each strip separately by the oven method. In parallel (figure 12) a section can be removed for casehardening tests. Although these assessments are time-consuming and labor intensive, careful monitoring in the manner indicated will always be economic if it avoids extensive and unnecessary degrade within a load [32].

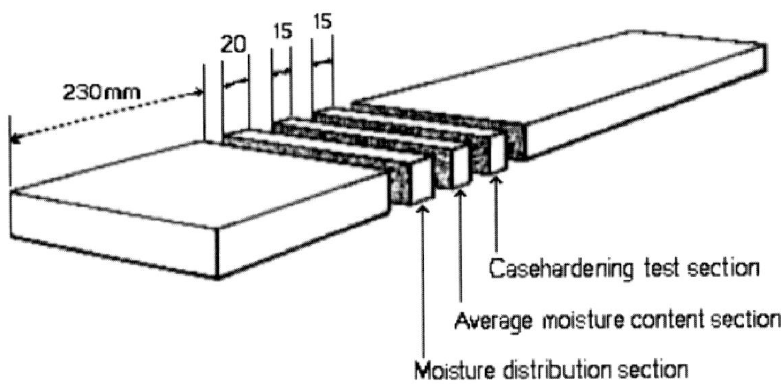


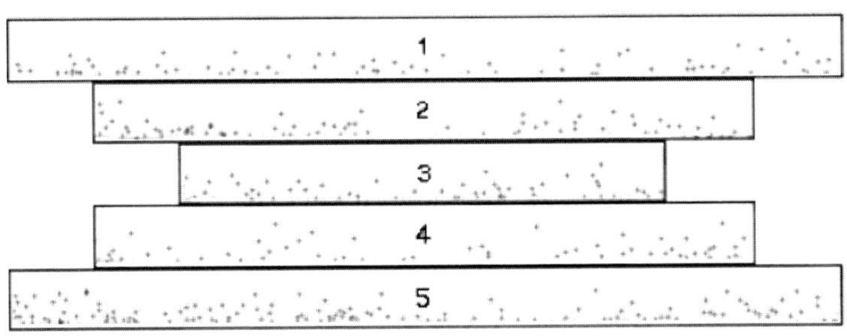


Figure 12. Re-cutting of kiln sample for testing moisture content, average and distribution: also casehardening.

## 11. Moisture Transfer Analysis

During drying transient heat transfer takes place and therefore the Biot number provides a measure of the temperature drop in the solid relative to the temperature difference between the surface and the fluid and hence shows the internal external (surface) resistances to the heat transfer from or to object [33].

$$Bi = kL / D \tag{5}$$

$$F_O = Dt / L^2 \tag{6}$$

The Fourier number is considered as $F_O = (\alpha t)/(L^2)$ where $\alpha$ is the thermal diffusivity, $m^2 s^{-1}$; and t is the process time, s.

The one dimensional transient moisture diffusion equation can be written as:

$$D\left(\partial^2 W / \partial z^2\right) = \left(\partial W / \partial t\right) \text{ and } D\left(\partial^2 \phi / \partial z^2\right) = \left(\partial \phi / \partial t\right) \qquad (7)$$

Equation (7) is subjected to the following initial and boundary conditions:

$$\phi(z,0) = \phi_i = (W_i - W_e)$$

for $0.1 \prec Bi \prec 100$ and $Bi \succ 100$ (8)

$$\left(\partial \phi(0,t) / \partial z\right) = 0$$

for $0.1 \prec Bi \prec 100$ and $Bi \succ 100$ (9)

$$-D\left(\partial \phi(L,t) / \partial z\right) = k\phi(L,t)$$

for $0.1 \prec Bi \prec 100$ ; $\phi(L,t) = 0$

for $Bi \succ 100$ (10)

where $\phi = (W - W_e)$

and the moisture content at a point of the solid object is non-dimensionalized by the following equation:

$$\Phi = (W - W_e)/(W_i - W_e) \qquad (11)$$

Solution to the governing equation (i.e. Eq.(7)) under the corresponding boundry condition with $\Gamma = 0 / z = 0$, yield dimensionless average moisture distribution of the corresponding slab objects in the following form [33] :

$$\Phi = \sum_{n=1}^{\infty} A_n B_n \text{ for } 0.1 \prec Bi \prec 100 \text{ and } Bi \succ 100 \qquad (12)$$

where

$$A_n = (2 \sin \mu_n)/((\mu_n) + (\sin \mu_n \cos \mu_n))$$

for $0.1 \prec Bi \prec 100$ (13)

$$A_n = \left(2(-1)^{n+1}\right)/(\mu_n) \qquad \text{for } B_i \succ 100 \qquad (14)$$

$$B_n = \exp\left(-\mu_n^2 F_O\right) \qquad \text{for } 0.1 \prec Bi \prec 100 \text{ and } B_i \succ 100 \qquad (15)$$

Further simplifications can be made in Eq. (12) by taking $\left(\mu_1^2 F_O\right) \succ 1.2$. Thus, the infinite sum in Eq. (12) is well approximated by the first term only, i.e.[33],

$$\Phi = A_1 B_1 \qquad (16)$$

where $\left(2\sin\mu_1\right)/\left((\mu_1) + (\sin\mu_1 \cos\mu_1)\right)$

for $0.1 \prec Bi \prec 100$ (17)

$$A_1 = \left(2/\mu_1\right) \text{ for } B_i \succ 100 \qquad (18)$$

$$B_1 = \exp\left(-\mu_1^2 F_O\right) \text{ for } 0.1 \prec Bi \prec 100 \text{ and } B_i \succ 100 \qquad (19)$$

## 12. EXPERIMENTAL

Fifty cylindrical green wood samples of Spruce were obtained from Guilan province. The diameter and height of the specimens were approximately 300mm and 21mm respectively. A programmable domestic microwave oven (Deawoo,KOC-1B4K), with a maximum power output of 1000 W at 2450 MHz was used. The oven has the facility to adjust power (Wattage) supply and the time of processing. The hot air drying experiments were performed in a pilot tray dryer consisted a temperature controller. Air was drawn into the duct through a mesh guard by a motor driven axial flow fan impeller whose speed can be controlled in the duct. The infrared dryer was equipped with eight red glass lamps (Philips) with power 175 W, each emitting radiation with peak wavelength 1200 nm. Radiators were arranged in three rows, with three lamps in each row. Dryer was equipped with measuring devices, which made it possible to control air parameters. The amount of water in a piece of wood is known as its moisture content. All the 50 dried samples were tested on a universal Tension Test machine model (Hounsfield HS100KS), with a loading capacity of 100 KN. During the tensile testing, the stress-strain curves as well as the peak load were recorded.

## 13. RESULTS

Conventional hot air drying is one of the most frequently used operations. The drying curves for conventional hot air drying of wood samples are shown in figures 13-18. It can be observed that the drying usually take place in the falling rate period. In essence, air in the oven is saturated, by time, and forms a thick film around the wood sample. That prevents effective separation of the evaporated moisture from the wood. This may be the reason for existence of constant rate period in this study.

Microwave drying is an alternative drying method, which is recently used in different industries. The effect of changing power output in the microwave oven on the moisture content is shown in the figures 18-22. At all power levels, drying curves were tended to end at about the same time. The observed initial acceleration of drying may be caused by allowing rapid evaporation and transport of water.

Infrared radiation is transmitted through water at short wavelength, it is absorbed on the surface. Infrared radiation has some advantages over convective heating. Heat transfer coefficients are high, the process time is short and the cost of energy is low. In this study, the drying time was reduced by nearly 34% compare to hot air drying. The drying curves were plotted in figures 23 and 24. In contrast to the hot air drying curves which had a short constant rate period followed by a falling rate period, figures 23 and 24 indicates that the infrared had only a falling rate period.

All dried samples were tested on a Hounsfield universal tension test machine with a loading capacity of 100 KN (figure 25). . The results of tensile loading of dried samples are presented in figures 26-29. It is clear that the microwave dried spruce specimen with failure strength of 49.6 Mpa has made a significant property improvement (figure 26). The normal stiffness of infrared dried sample is reported as 35.0 Mpa (figure 27) whereas the oven dried sample showed strength of about 44.5 Mpa (figure 28). From figure 29 it is revealed that the natural convection dried specimens are the strongest (50 Mpa). In practice the drying time for this can take up months and years. In figure 30 the strength of dried samples are compared for a better judgment.

## 14. Industrial Application of the Results

In wood drying process we should note that the wood can hold moisture in the cell lumen (cavity) as liquid or "free" water, or as adsorbed or "bound" water attached to the cellulose molecules in the cell wall. Meanwhile, the occurrence of the free water does not affect the properties of wood other than its weight. Bound water, however, does affect many properties of wood, and is more difficult to remove in the drying process. Microscopically, the dimensional change with MC is anisotropic (referring to the fact that wood has very different properties parallel to the fact grain versus the transverse direction). As the MC decreases, wood shrinks; conversely, as the MC increases, wood swells or grows larger. The process of drying focuses on producing wood with an MC about the same as the equilibrium value for the intended service environment.

For the design of dryers it is necessary to carry out drying experiments at various drying conditions. Experimentally determined drying times, transition points, and constant-rate regime temperature can then be used as a base case for the analytical results. Based on the information from the experimental trials, runs with lower amounts of moisture to evaporate, higher dryer temperatures, should be expected to dry faster and reach transition point more rapidly. After an initial increase or decrease of the rate of drying, the drying process enters the constant rate period. This initial change of the rate of drying is caused by a variation of the surface temperature which in turn results in a change of vapour density.

It can be noted that time interval of drying process is solely determined by external conditions. Once the drying process has entered the falling rate period, the external conditions become relatively unimportant compared to the internal parameters.

By comparing runs with the same initial moisture, we see that as oven temperature increases, the transition points are reached more quickly and total drying times are shorter. Sample temperatures are higher because they are exposed to higher heat transfer rates, giving rise to higher mass transfer rates during the constant-rate regime.

The experimental study suggests that the humidity of the free stream should be as low as possible. Partial recirculation, 100% fresh air intake, or dehumidifications are some of the possible ways to accomplish this task, but a cost analysis is imperative before deciding on any option.

Reduction of the drying time in microwave heater seems to be a motivating cost saving factor for industries. In this case a moderate mechanical property is obtained (table 1). To minimize directional variations in use, wood needs to be dry enough to match the service environment.

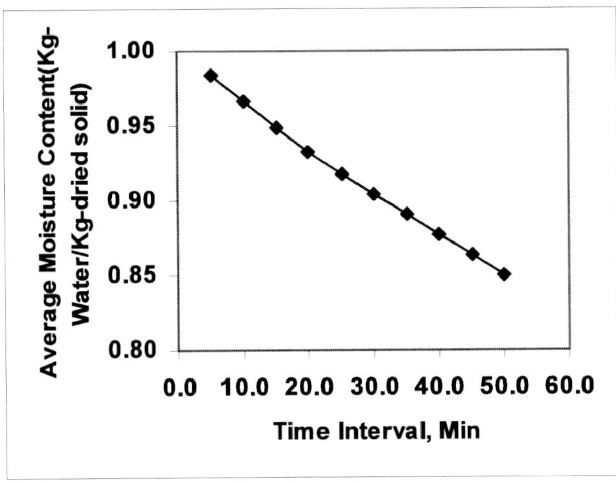


Figure 13. Average moisture content versus time; (Conventional hot air-dried wood at T= 40°C).

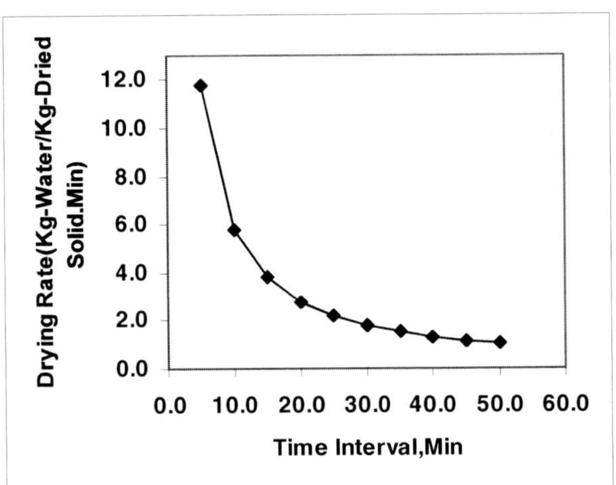


Figure 14. Drying rate curve; (Conventional hot air-dried wood at T=40°C).

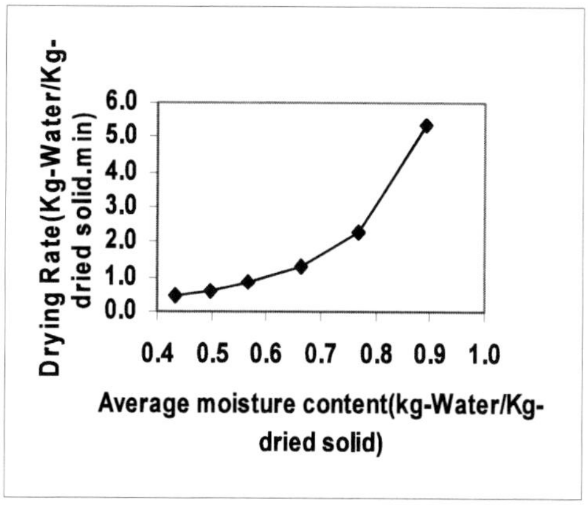


Figure 15. Average moisture content versus time; (Conventional hot air-dried wood at T=100°C).

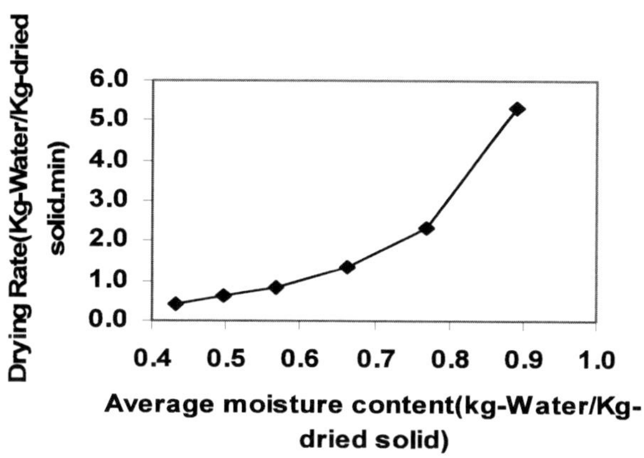


Figure 16. Drying rate curve; (Conventional hot air-dried wood at T=100°C).

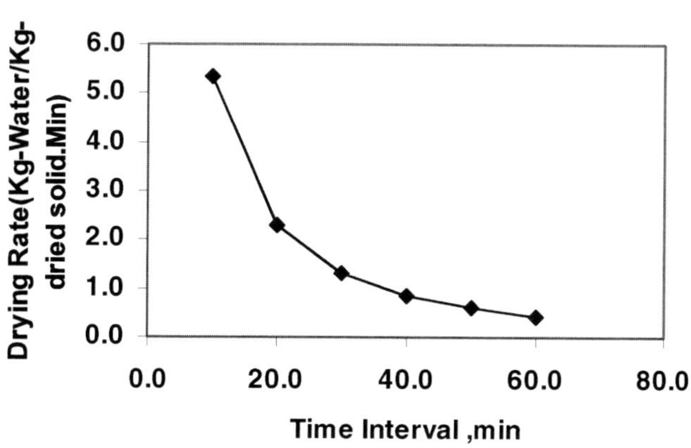


Figure 17. Drying rate curve; (Conventional hot air-dried wood at T=100°C).

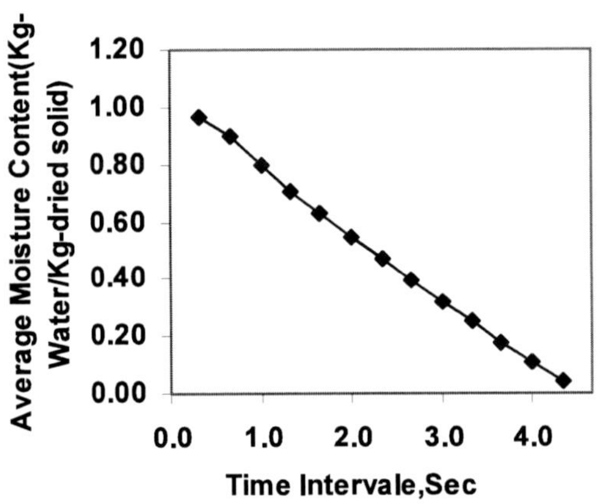


Figure 18. Drying curve for microwave-dried wood at 80powers.

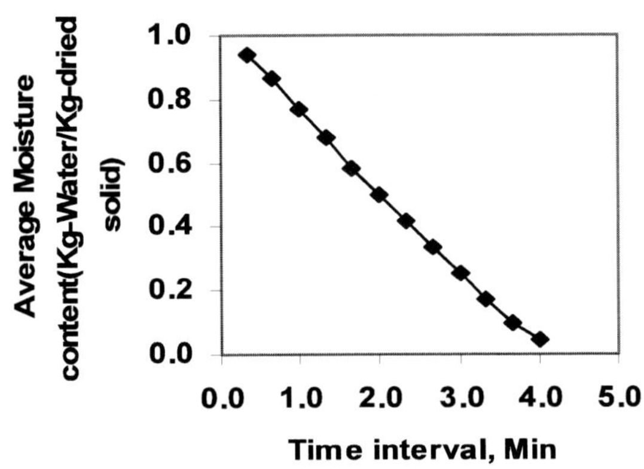


Figure 19. Drying curve for microwave-dried wood at 100% powers.

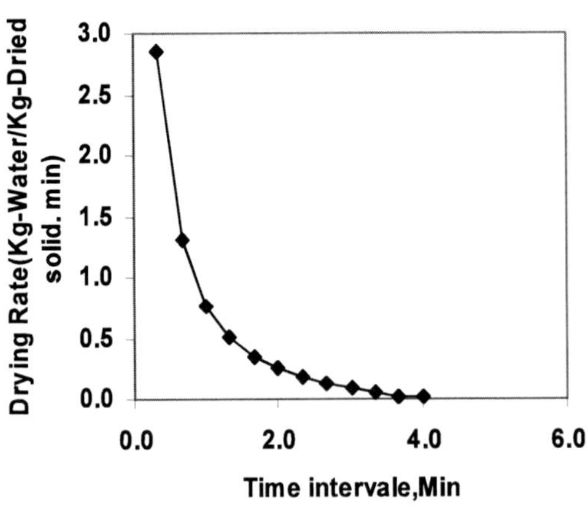


Figure 20. Drying rate curve for microwave-dried wood at 100% power.

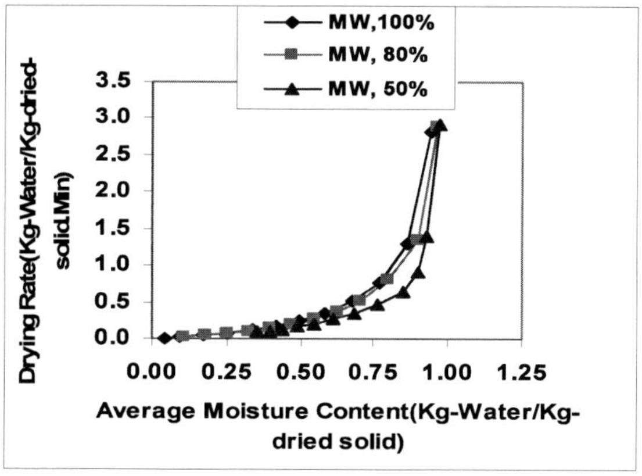


Figure 21. Drying rate curves for microwave-dried wood at three different power.

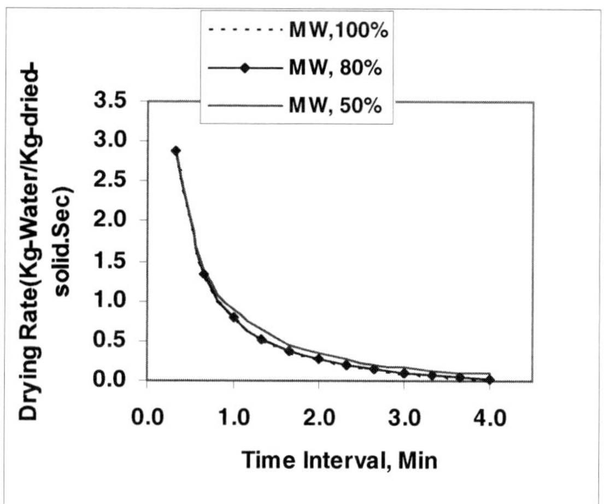


Figure 22. Drying rate curves for microwave-dried wood at three different powers.

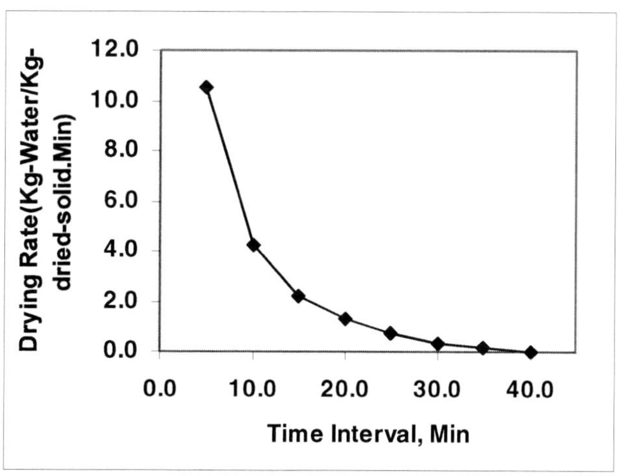


Figure 23. Drying curve for infrared-dried wood at 100% powers.

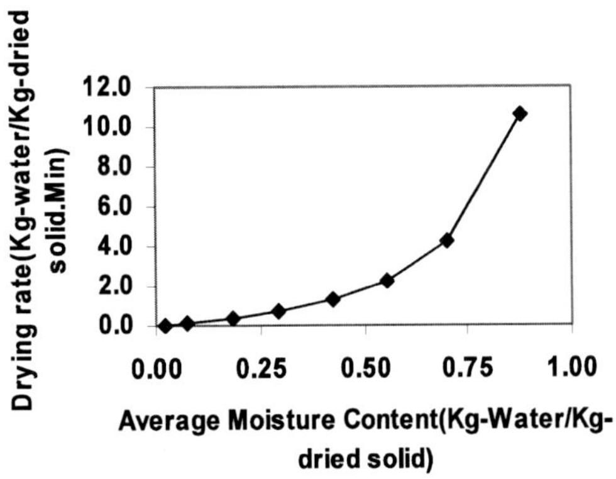


Figure 24. Drying curve for infrared-dried wood at 100% powers.

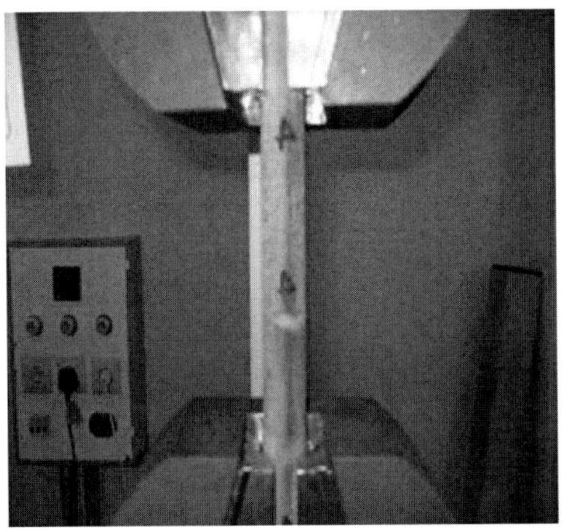

Figure 25. Tension test.

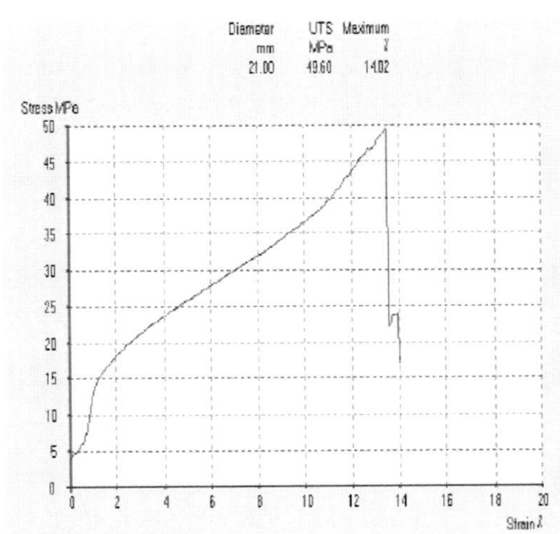


Figure 26. Stress-strain for microwave dried wood.

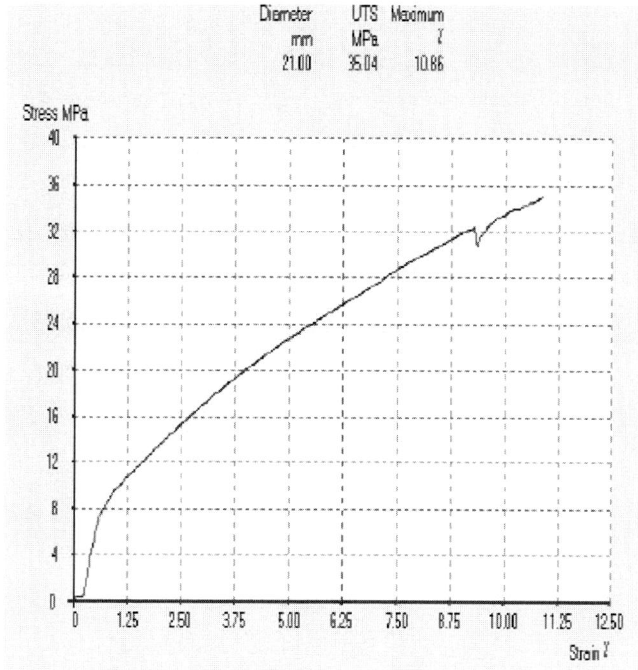


Figure 27. Stress-strain for infrared dried wood.

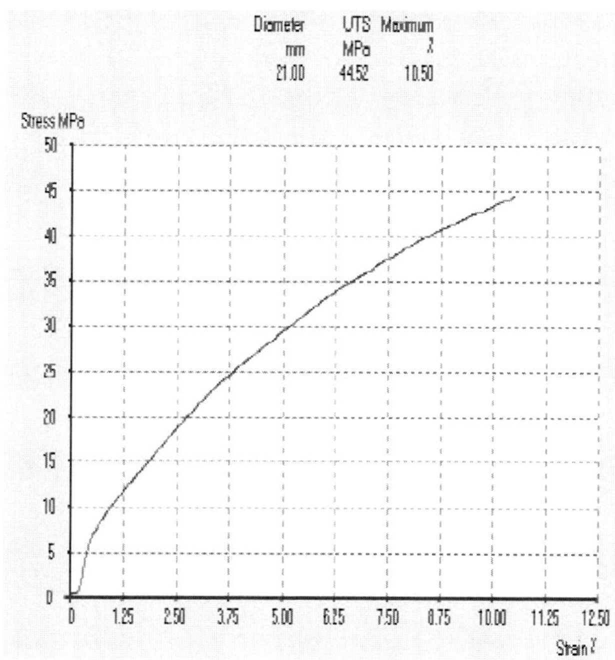


Figure 28. Stress-strain for hot-air dried wood.

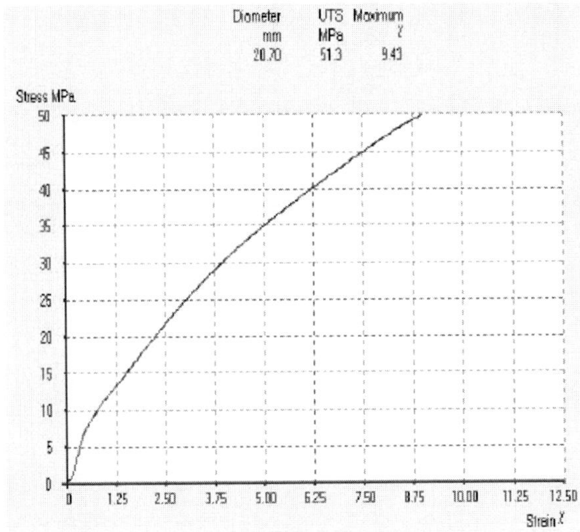


Figure 29. Stress-strain for natural convection dried wood.

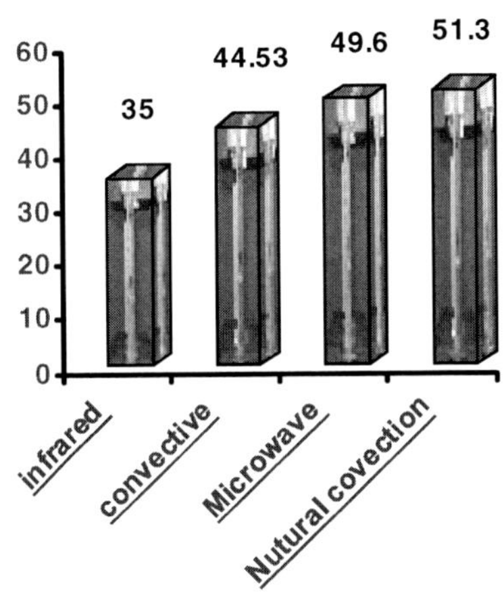


Figure 30. Strength of dried wood samples in three different.

**Table 1. Average strength properties of samples (σ values in brackets refer to standard deviations**

| Drying Method | Failure Strength(Mpa) | Failure Strain % | Yield Strength(Mpa) | Modulus of elasticity(Gpa) |
|---|---|---|---|---|
| Natural convection Dried wood | 51. 3 (σ2.44) | 9. 43 | 12. 8 (σ 0. 615) | 0. 544 (σ 0. 058) |
| Hot air dried wood | 44. 53 (σ 1.72) | 10. 5 | 13. 3 (σ 0. 51) | 0. 424 (σ 0. 05) |
| Infrared dried wood | 35. 04 (σ 1.16) | 10. 86 | 10. 5 σ (0. 35) | 0. 322 (σ 0. 035) |
| Microwave dried wood | 49. 6(σ 4. 51) | 14. 02 | 17. 0 (σ 1. 28) | 0. 354 (σ0. 054) |

## CONCLUDING REMARKS

Convective drying of wood is an old technology, and yet is an immature method. A comprehensive as well as state-of-art review on wood drying process has been made in this paper. Many challenges exist in wood drying process, and a number of fundamental questions remain open. Because of limitations as summarized in this paper, most are considered as perspective in the future. Microwave heating and drying of wood products has not been used to a larger extent by the wood industries and manufactures. This could be explained by the insufficient knowledge of the complex interaction between material and process parameters during heating and drying as well as by the required investment expenses. The knowledge and understanding of the process will be improved as well as applying this technique in the most effective way. Its benefit is the fast drying rate.

Although many methods of drying timber have been tried over the years only a few of these enable drying to be carried out at a reasonable cost and with minimal damage to the timber. The most common method of drying is to extract moisture in the form of water vapor. To do this, heat must be supplied to the wood to provide the latent heat of vaporization. The temperature of a piece of wood and of the air surrounding it will also affect the rate of water evaporation from the wood surface. With kiln drying, warm or hot air is passed over the timber and at the start of the drying process the temperature differential between the air and the wet wood will usually be large. As a result, heat energy will be transferred from the air to the wood surface where it will raise the temperature of both the wood and the water it contains. Water, in the form of vapor, will then be lost from the wood surfaces, provided the surrounding air is not already saturated with moisture. This results in the development of a moisture content gradient from the inside to the outside of the wood. As the temperature is raised this increases not only the steepness of this moisture gradient, but also the rate of moisture movement along the gradient and the rate of loss of water vapor from the surface of the wood. At a given temperature the rate of evaporation is dependent on the vapor pressure difference between the air close to the wood and that of the more mobile air above this zone.

Unfortunately the considerable benefits obtainable by raising the drying temperature cannot always be fully exploited because there are limits to the drying rates which various wood species will tolerate without degrade. In the drying of many species, especially medium density and heavy hardwoods, shrinkage and accompanying distortion may increase as the temperature is raised. So with species which are prone to distort it is normal to use comparatively low kiln temperatures. In contrast to hot air drying a modern radiation drying provides temperature control and a steady and adequate flow of air over the timber surface. The air flow rate and direction is controlled by fans and the temperature and relative humidity of the air can be adjusted to suit the species and sizes of timber being dried. It is thus possible to make full use of the increase in drying rate which can be achieved by raising the temperature to the maximum value which a particular timber species can tolerate without excessive degrade.

# REFERENCES


[1] Scheigger, A. E., "The physics of flow through porous media, University of Toronto Press" 1958.

[2] Plumb, O. A.,and Spolek, G. A., and Olmstead, B. A., "Heat and Mass Transfer in wood drying" *Int. J. Heat Mass Transfer,*Vol. 28, 1985,pp. 1669-1678

[3] Basilico, C., and Martin, M., "Approache expèrimentale des mecanismes de transfert au cours du séchage convective à haute temperature d'un bois"(In French), *Int. J. heat Mass Transfer,Vol.* 27, 1984 ,pp.657-688

[4] Jackson, G. and James, D., The permeability of fibrous porous media, *Can. J. Chem. Eng.* 64, pp. 207-221, 1986.

[5] Hunter, A. , On the basic equation of sorption and isosteric heat, *Wood Sci. Technol.* 25, pp 99-111, 1991.

[6] Hunter, A. On the movement of water through wood-the diffusion coefficient, *Wood Sci. Technol.* 27, pp 401-408, 1993.

[7] Hunter, A., A Complete theoretical isotherm for for wood, based on capillary condensation, *Wood Sci. Technol.* 30, pp. 127-131, 1996.

[8] Hudson, M., Improved method of an apparatus for dehydrating wood and wood products, US Patent No 655062

[9] Kifetew G., Thuvander, F. and Berglund, L., The effect of drying on wood fracture surfaces from specimens loaded in wet condition, *Wood Sci. Tecnol.* 32, pp 83-94, 1998.

[10] Thuvande, F. and Berglund, L., A multiple fracture test for strain to failure distribution in wood, *Wood Sci. Technol.* 32, pp. 227-235, 1998.

[11] Bousquet, D., Drying wood, University of Vermont Extension Brieflet 1326, 1998.

[12] Cech, M. and Pfaff, F., Kiln Operator's Manual for Eastern Canada, Forintek Canada Corp. Eastern Forest Products Laboratory, Ottawa, Canada, 2004.

[13] Culpepper, L., High temperature drying-enhancing kiln operations, Miller Freeman Publications, Inc., San Francisco, 1990.

[14] Hart, C., Kiln overheating when conditioning Lumber, *Forest Prod. J.,* 40, pp. 9-14, 1990.

[15] Kayihan, F., Simultaneous heat and mass transfer with local three-phase equilibra in wood drying, Proc. 3rd International Drying Symposium, 1, pp. 123-134, 1982.

[16] Kelsey, K. and Clarke, L., The heat of sorption of water by wood, Australian *J. Appl. Sci.* 7, pp. 160-175, 1965.

[17] Pang, S., Moisture content gradient in a softwood board during drying, *Wood Sci .Technol.* 30, pp. 165-178, 1996.

[18] Pang, S., External heat and mass transfer coefficients for kiln dring of timber, *Drying Technol.* 14, pp. 859-871, 1996.

[19] Pang, S. Some considerations in simulation of superheated steam drying of softwood lumber, *Drying Technol.* 15, pp. 651-670.

[20] Pang, S., Relative importance of diffusion and convective flow of moiature vapour in simulation of softwood drying, *Drying Technol.* 16, PP. 271-281, 1998.

[21] Pang, S., Langrish, T., Keey, R., The heat of sorption of timber, *Drying Technol.* 11, pp. 1071-1080, 1993.

[22] Pang, S., Langrish, T., Keey, R., Moisture movement in softwood timber at elevated temperatures, *Drying Technol.* 12, pp. 1897-1914.

[23] Simpson, W. and Rosen, H, Equilibrium moisture content of wood at high temperature, *Wood and Fiber*. 13, pp. 150-158, 1981.

[24] Simpson, W., Predicing equilibrium moisture content of wood by mathematical models, *Wood and Fibers*. 5, pp. 41-49, 1973.

[25] Skaar, C., Wood-water relations, Springer-Verlag, Berlin, 1988.

[26] Haghi, A.K., Ghanadzadeh, H., and Rondot, D.," Experimental survey on Microwave Drying of Porous Media" *Iran. J. Chem & Chem. Eng.,* Vol. 24, No. 2, 2005.

[27] Haghi, A. K., "Thermal Analysis of Drying Process- a theoretical approach, *Journal of Thermal Analysis and Calorimetry,"* Vol. 74, pp 827-840, 2003.

[28] Haghi, A. K., and Valizadeh, M., " Experimental Investigation on Microwave Drying" *International Journal of Heat and Technology,* Vol. 22, No. 2, pp167-172, 2004.

[29] Haghi,A.K.,and Ghanadzadeh,H.," A Study of Thermal Drying Process" *Indian Journal of Chemical Technology,* Vol. 12, 2005, pp. 654-663.

[30] Moody, L., Friction factors for pipe flow, *Trans. ASME* 66, pp. 671-684, 1994.

[31] White, F., Fluid Mechanics, third ed., McGraw-Hill, Inc., 1994.

[32] Haaland, S., Simple and explicit formulas for the friction factor in turbulent pipe flows, *J. Fluids Eng,* 1983, pp. 89-90.

[33] Dincer, I., Moisture transfer analysis during drying of slab woods, *Heat and mass Transfer,* 34, pp. 317-320, 1998.

In: Monomers, Oligomers, Polymers, Composites... ISBN: 978-1-60456-877-6
Editors: R. A. Pethrick, G.E. Zaikov et al. 

*Chapter 19*

# DEVELOPMENT OF GREEN ENGINEERED CEMENTITIOUS COMPOSITES

*A. Sadrmomtazi and A. K. Haghi*[*]
University of Guilan, P. O. Box 3756, Rasht, Iran

## ABSTRACT

In this article, a new green composite is developed. The green composite presented in this paper, consisted of renewable raw materials such as soil, cement and natural fibres. Natural fibres used at an appropriate length and amount can develop sufficient bond with the soil-cement to enhance the overall toughness of the green composite.

## 1. INTRODUCTION

Manufacturing high performance engineering materials from renewable resources is one ambitious goal currently being pursued by researchers across the world the ecological benefits of renewable raw materials are clear: they save valuable resources, are environmentally sound and do not cause health problems[1-5].

Natural fibres are getting attention as a reinforcing agent in both thermoplastic and thermoset matrices. These composites possess good mechanical properties Such as relatively high strength and stiffness of natural fibres combined which makes it a covetable substitute for synthetic fibres that are potentially toxic [6-10]. Their low-density values allow producing composites that combine good mechanical properties with a low specific mass [11]. With regard to the environmental aspects it would be very interesting if natural fibres could be used instead of glass fibres as reinforcement in some structural applications. Natural fibres have many advantages compared to glass fibres, for example they have low density, they are recyclable and biodegradable [12, 13-18].

[*] Corresponding author e-mail: Haghi@Guilan.ac.ir

Two major areas of enquiry exist in the field of fibre-matrix adhesion in composite materials. One is the fundamental role that fibre-matrix adhesion plays on composite mechanical properties. The other is what is the best method used to measure fibre-matrix adhesion in composite materials[5, 20-24].. In order to analyze the mechanical behaviour of adhesively bonded joints, several studies have been conducted.

It appears that the most promising method for joining fiber-reinforced composite structures is adhesive bonding. The use of right coupling agent is expected to improve the fiber-matrix adhesion [25]. It is interesting to see that increasing tensile strength with increasing fibre content is only valid for the systems with the coupling agent. . For the systems without coupling agent, the fibre acts as an included filler in the resin matrix, which actually weakens the composite because of poor interfacial adhesion [26].

The mechanical properties of a composite material depend primarily on the strength and modulus of the fibers [27]. An effective use of fibre strength is dependent on both the interfacial adhesion properties and the critical fibre length.

We must consider that the major part of the cellulose consists of a micro-crystalline structure with high order of crystalline regions. Generally, higher cellulose content leads to higher stiffness, in turn the cellulose content have a major influence on the mechanical properties of the resultant composites. Because of the structural features, the high level of moisture absorption and poor wettability of the natural fibre material results in insufficient adhesion between fibres and polymer matrices leading to debonding during use and aging so that the quantity of the fiber in the composite must be optimum [26,28-29].

As aforementioned, the measured average fibre length in the short-fibre composite plate is far below the critical fibre length for all the systems in this study. This allows us to use a modified rule of mixture taking account of interfacial properties for the strength of natural fibre/ polymer composites, $\sigma_c$ , as a function of volume fraction:

$$\sigma c = [\eta L f \tau / (d - \sigma m)] Vf + \sigma m \qquad (1)$$

where η is the fibre orientation factor σm and Vf are tensile strength of matrix and volume fraction of fibre, respectively [30].

Young's modulus reflects the capability of both fibre and matrix material to transfer the elastic deformation in the case of small strains without interface fracture. The Young's modulus of the composite is determined by the equation :

$$Ec = \beta\,(EmVm + EfVf) + (1-\beta)\, EfEm / (EmVf + EfVm) \qquad (2)$$

where β is a factor of efficient stress transfer between fibre and matrix. E is the Young's modulus; c, m, f, refer to the composite, matrix and fibre respectively. For all systems, Ec values increase almost linearly with increasing fibre content.

Micromechanical models that incorporate the properties of the composite and matrix material may be employed to determine the elastic properties of the composite.

## 2. Experimental Program and Results

Soil stabilization is a process to improve certain properties of a soil to make it serve adequately an intended engineering purpose. The improving of the ground properties with various methods is a common case of geotechnical engineering. The aim of the soil stabilization is to decrease the consolidation and permeability capacity and to increase bearing and shear resistance capacity [32].

Natural fibres used at an appropriate length and amount can develop sufficient bond with the soil-cement to enhance the overall toughness of the composite. The slope of stress/strain curve (below yield point) denotes material's stiffness; steeper the curve, the stiffer material, gradient is known as modulus of elasticity or Young's modulus.

The soil used in this experimental program is a common mason sand with a grain size distribution such that 100%, 94%, 56%, 22%, 9%, and 5%, of the material passes the No. 10, No. 20, No. 40, No. 60, No. 120,and No. 200 sieves, respectively. The coefficient of uniformity, $C_u$ and the coefficient of curvature, Cc, for this soil are 2.65 and 1.02, respectively.

Gradation and soil classification of samples is given in figure 1 in terms of particle size distribution and soil classification.

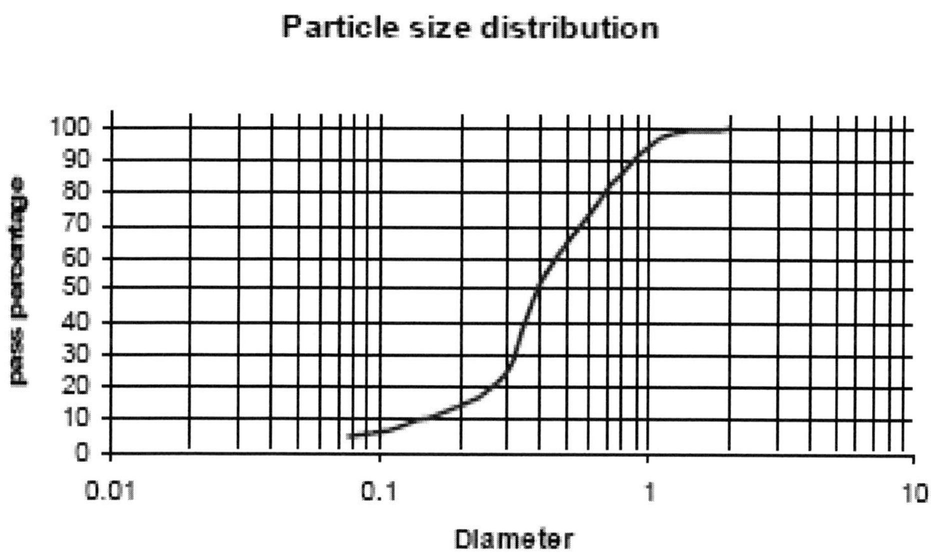


Figure 1. Particle size distribution.

The cement used is ordinary Portland cement. The physical and chemical properties of the cement are given in table 1.

Jute yarn of 1100 tex yarn fineness was obtained from local firms. Jute yarn was dried in an oven at 100 °C for 4 h, then it was tested for moisture absorption by exposing the yarn to 50% and 95% RH atmosphere (in desiccators), at 23° C.

**Table 1. Physical and chemical properties of the cement**

| Physical properties | Cement |
|---|---|
| Fineness | 3.12 |
| Chemical composition | |
| Silica (SiO2) | 20.44% |
| Alumina (Al2O3) | 5.5% |
| Calcium oxide (CaO) | 64.86% |
| Potash (K2O) | 22.31% |
| Magnesia (MgO) | 1.59% |
| Loss on ignition | 1.51% |
| PH | 12.06 |
| 3CaO. SiO2 | 66.48% |
| 2CaO. SiO2 | 10.12% |
| 4CaO. Al2O3. Fe2O3 | 9.43% |
| Free lime | 1.65% |
| 3CaO. Al2O3 | 8.06% |

Brazilian tests were performed to obtain the tensile strength of specimens, based on ASTM C496 for indirect tensile test. The size and curing time of both tensile and compressive samples were similar. The load to failure was recorded and the tensile strength was computed as follows:

$$\sigma_t = \frac{2P}{\pi l d} \qquad (3)$$

Where:

$\sigma_t$ = Indirect tensile strength

P = Applied maximum load

l and d = length and diameter of the specimen, respectively

The experiments were conducted with cement content varying jute percentage (figures 2 and 3).

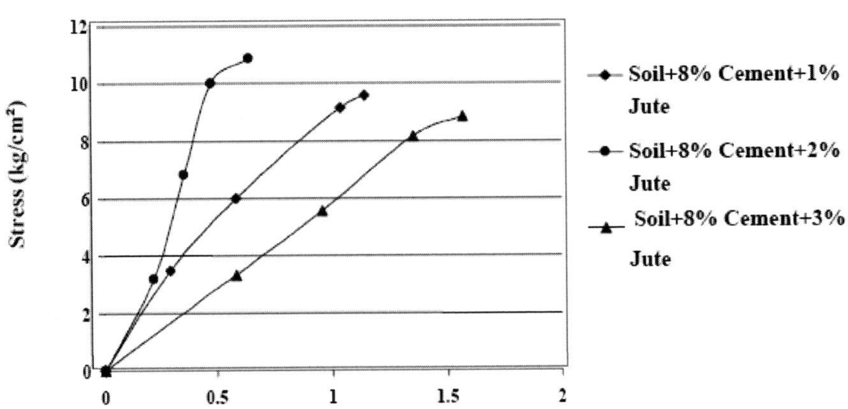

Figure 2. Stress-strain green composite.

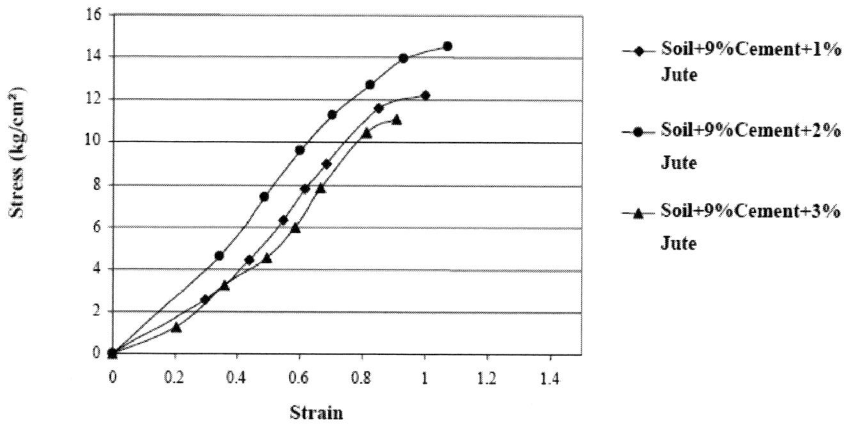


Figure 3. Stress-strain of green composite.

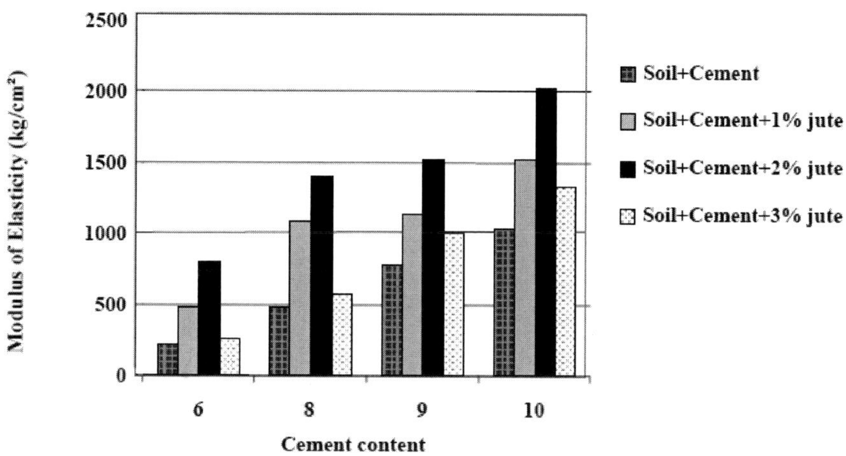


Figure 4. Comparison between modulus of elasticity of reinforced and non-reinforced green composites.

Figure 4 shows comparison of modulus of elasticity between different mixture conditions.

Experimental results show that an increase in the percentage of cement content results in an increase in the modulus of elasticity. Also, adding Fibres by one constant percent of cement results in an increase in the modulus of elasticity and toughness. Figure 4 shows that the maximum modulus of elasticity obtains from mixtures by 2% fibres content and it drops after the improvement of fibres from 2 %.( but it still is higher than similar rates in cases without fibres.)

Indirect tensile strength tests were conducted on stabilized soil specimens containing 8,9,10 percentage of cement, non-reinforced and reinforced with 1, 2, 3 percentage of Jute.

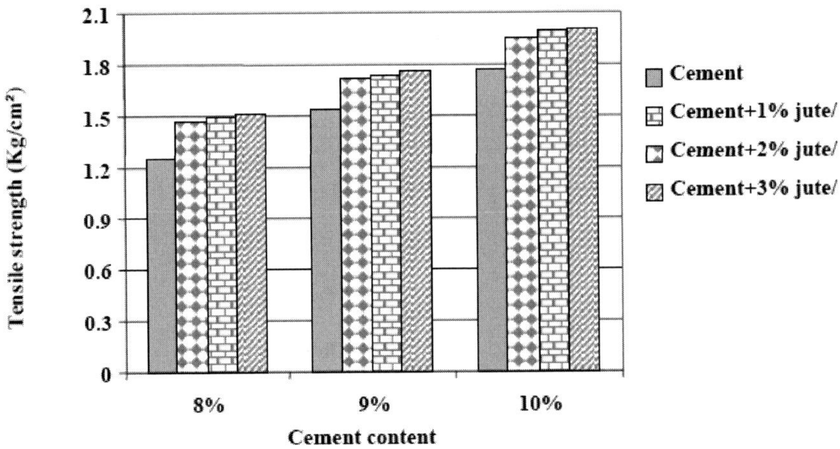


Figure 5. Tensile strength of green composite.

Two important parameters can be considered and discussed, when Jute used as reinforcing fibres in cement stabilized sand samples. First of all, an increase in the tensile strength of specimens can be observed when specimens are reinforced with Jute. Secondly; increase of cement content for a constant Jute percentage increases the tensile strength of mixture.

The results of this experimental investigation in general indicate the importance of modulus of elasticity in evaluation of composite stiffness and interfacial bond between natural fibre and matrix.

## 3. Overview

The elastic properties of short fiber reinforced plastics can be experimentally determined or derived from a variety of mathematical models. The advantage of a comprehensive mathematical model is it reduces costly and time-consuming experiments. Furthermore, a mathematical model may be used to find the best combination of constituent materials to satisfy material design considerations. Lastly, a physical (as opposed to empirical) model can yield insight into the fundamental mechanisms of reinforcement. [33]

The purpose of the micro mechanical models is to predict the properties of a composite based on the properties of each constituent material [48]. Properties such as the elastic modulus Ec, Poisson's ratio ($\nu$) and the relative volume fractions (V) of both fiber and matrix are the fundamental quantities that are used to predict the properties of the composite. In some cases, fiber aspect ratio and fiber orientation are also included [33].

The variation of the amount of fibers in a natural fiber composite can be successfully chosen to correlate with the mechanical properties of composite. The amount of fibers is one of the most important characteristics of any composite material since their mechanical properties are strongly dependent on it [34].

The volume fraction of fibers is commonly used to estimate certain mechanical properties of the composite material. Besides, the type of fibers distribution (aligned or random) and their mechanical properties as well as properties of resin should be available. The volume fraction of a composite is obtained by the following formula:

$$Vf=\rho mWf/(\rho mWf+\rho fWm) \quad (4)$$

Where $Wf$ is the fiber weight fraction, $Wm$ the matrix weight fraction, $\rho f$ the density of the fibers and $\rho m$ the density of the matrix [26,48-49].

The mechanical properties of a composite material depend primarily on the strength and modulus of the fibers, the strength and the chemical stability of the matrix and the effectiveness of the bonding between matrix and fibers in transferring stress across the interface. Generally, the utilization of natural fibers as reinforcing materials in thermoplastics requires strong adhesion between the fiber and the matrix. Cellulose has a strong hydrophilic character due to three hydroxyl groups per monomeric unit, but biopolymers are generally hydrophobic [27]. A large number of research interests were dedicated to theoretical and numerical models with varying degrees of success.

It is necessary to perform comparison of the methods in order to determine the best approach. In this paper the following models are applied to green composites:

1. Rule of mixture (ROM) [35]
2. Inverse rule of mixture (IROM) [36]
3. Cox's model [37]
4. Halpin-Tsai equation [38-39]
5. Our proposed model

### 3.1. Rule of Mixture (ROM)

The simplest available model that can be used to predict the elastic properties of a composite material is the rule of mixtures (ROM). To calculate the elastic modulus of the composite material in the one-direction $E_1$, it is assumed that both the matrix and fiber experience the same strain. This strain is a result of a uniform stress being applied over a uniform cross sectional area. The ROM equation for the apparent Young's modulus in the fiber direction is:

$$E_1 = E_f V_f + E_m V_m \quad (5)$$

Where $E_f$, $E_m$, $V_f$ and $V_m$ are the moduli and volume fractions of the fiber and matrix materials respectively. This model works extremely well for aligned continuous fiber composites where the basic assumption of equal strain in the two components is correct [33, 40-47].

The modified representation of the ROM which was adopted to estimate the modulus of elasticity of a composite material with long randomly distributed fibers is as follows [48-49]:

$$E_1 = \eta E_f V_f + E_m V_m \qquad \text{Or} \qquad E_1 = \eta E_f V_f + E_m (1 - V_f) \quad (6)$$

It takes into account the weakening of the composite due to fibers orientation and fiber length factors through introduced additional coefficient, $\eta < 1$. Some attempts by other researchers have been done in order to estimate $\eta$. For example, in [50] it is suggested to apply $\eta = 0.2$ for a composite reinforced with randomly oriented natural fibers [34].

### 3.2. Inverse Rule of Mixture

The elastic modulus of the composite in the two-direction ($E_2$) is determined by assuming that the applied transverse stress is equal in both the fiber and the matrix (Reuss's assumption) [51]. As result, $E_2$ is determined by an inverse rule of mixtures equation that is given as:

$$E_2 = \frac{1}{\dfrac{V_f}{E_f} + \dfrac{V_m}{E_m}} \quad \text{Or} \quad E_2 = \frac{1}{\dfrac{V_f}{E_f} + \dfrac{1-V_f}{E_m}} \tag{7}$$

### 3.3. Modified Rule of Mixture

The Cox shear lag theory adds to the ROM the shear lag analysis, which includes a fiber length and a stress concentration rate at the fiber's ends. The model is described by equation (2), in which the coefficient $\eta$ is written as follows:

$$\eta = 1 - \frac{\tanh\left(\dfrac{\beta l}{2}\right)}{\left(\dfrac{\beta l}{2}\right)} \tag{8}$$

where l is the length of fibers and $\beta$ is the coefficient of stress concentration rate at the ends of the fibers, which can be described by the following equation. 9:

$$\beta = \frac{l}{r}\sqrt{\frac{E_m}{E_f(1+\nu)\ln\sqrt{\dfrac{\pi}{4V_f}}}} \tag{9}$$

where v is Poisson's ratio of fibers and r is the fiber radius.

### 3.4. Semi-Empirical Equations

The semi-empirical equations developed by Halpin and Tsai are widely used for predicting the elastic properties of SFRT. The following form of the Halpin and Tsai equation is used to predict the tensile modulus of SFRT [33]:

$$E_1 = E_m\left(\frac{1+\xi\eta V_f}{1-\eta V_f}\right) \tag{10}$$

In equation (6) the parameter η is given as:

$$\eta = \frac{(E_f/E_m)-1}{(E_f/E_m)+\xi} \tag{11}$$

Where ζ in equations (6) and (7) is a shape fitting parameter to fit the Halpin–Tsai equation to the experimental data. The significance of the parameter ζ is that it takes into consideration the packing arrangement and the geometry of the reinforcing fibers [33, 48-49]

A variety of empirical equations for ζ are available in the literature, and they depend on the shape of the particle and on the modulus that is being predicted [39]. If the tensile modulus in the principle fibre direction is desired, and the fibers are rectangular or circular in shape, then ζ is given by the following equation [39]:

$$\xi = 2\left(\frac{L}{T}\right) \quad \text{Or} \quad \xi = 2\left(\frac{L}{D}\right) \tag{12}$$

where L refers to the length of a fibre in the one-direction and T or D is the thickness or diameter of the fibre in the three-direction. In equation (8), as L → 0, ζ → 0 and the Halpin–Tsai equation reduces to the IROM equation. In contrast when L → ∞, ζ → ∞ and the Halpin–Tsai equation reduces to the ROM equation [33, 48] .

### 3.5. A Model for Green Composites

A new theoretical model of the modulus estimation for natural fiber composite is necessary, since the existing models (at least those found in literature) can not predict the Young modulus of composites with natural fibers in a reasonable error. A new model should be able to estimate reliably of the modulus of composite with different fiber volume fraction and with different elastic properties of fibers (with constant volume fraction) as well. On the other hand, the development of a new theoretical model from scratch is not reasonable, while combination of the existing models can be used as the basis for a new model.

During the benchmarking of the existing theoretical models it was found that Halpin model and IROM model give the modulus estimation quite close to the experimentally obtained data (figure 6).

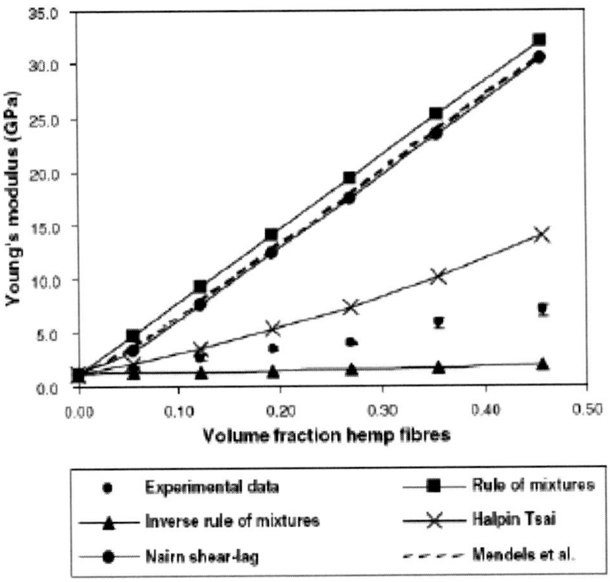


Determination of composite modulus containing hemp fibres.

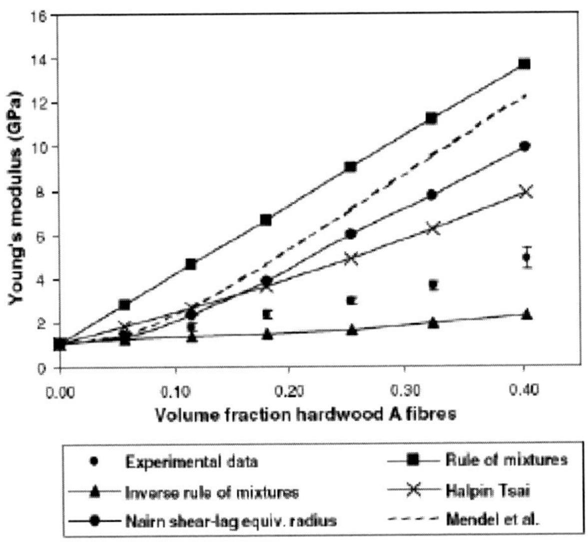


Determination of composite modulus containing hardwood A fibres.

Figure 6. Agreement of IROM and Halpin models with experimental data (graphs from Ref. [1]).

Therefore, the equations of these models can be used as a reference for the development of a new model. In composites with randomly distributed fibers there are fibers which are parallel, series and under an angle orientated to the chosen main direction. Unfortunately ROM model does not take into account the influence of fibers which have orientation perpendicular to the chosen main direction, because in its equation the series part is absent. So the equation of this model has a linear behavior with respect to $V_f$ fiber volume variable which does not give enough flexibility to adjust the model's behavior to interpolate the non-linear trend of experimentally obtained data. But because of great ability of ROM model in main direction, it can be used in conjunction with two other models; say Halpin and IROM.

A try like this approach has been done in Ref [34], combining ROM and IROM models, with two new weight coefficients α and β:

$$E_c = \alpha\left[V_f E_F + (1-V_f)E_m\right] + \beta\left(\frac{1}{V_f/E_f + (1-V_f)/E_m}\right) \tag{13}$$

But a non scientific method has been used to find α and β coefficients.

In this paper, least squares methods with "r" and "s" criteria will be used to find the unique coefficients with the best convergence. The data for PP/Jute composite reinforced with randomly_oriented fibers with different fiber volume fraction will be used, with $E_f = 41000MPa$ and $E_m = 800MPa$ [34].

Following models will be studied and compared:

- Model 1: IROM and modified ROM model (figure 7).
- Model 2: modified ROM and Halpin (figure 8).

**Table 2. Green composite reinforced with randomly oriented fibers with different fiber volume fraction**

| $V_f$ | $E_c$, MPa |
|---|---|
| 0 | 800 |
| 0.06 | 1300 |
| 0.12 | 1650 |
| 0.18 | 1800 |
| 0.24 | 2000 |
| 0.29 | 2100 |
| 0.34 | 2200 |
| 0.45 | 2250 |

***Model 1***

$$E_1 = a\left(\eta E_f V_f + E_m(1-V_f)\right) + b\left(\frac{1}{\frac{V_f}{E_f} + \frac{1-V_f}{E_m}}\right) \tag{14}$$

In this model three weight coefficients, namely a, b and η should be found in order to fit the data of table 2 with best convergence.

With applying standard error and correlation coefficient criteria, a, b and η constants are calculated as follows:

a =5.665136
b = - 4.5332004
$\eta$ =0.05943155
Standard Error: 83.6682820
Correlation Coefficient: 0.9899705

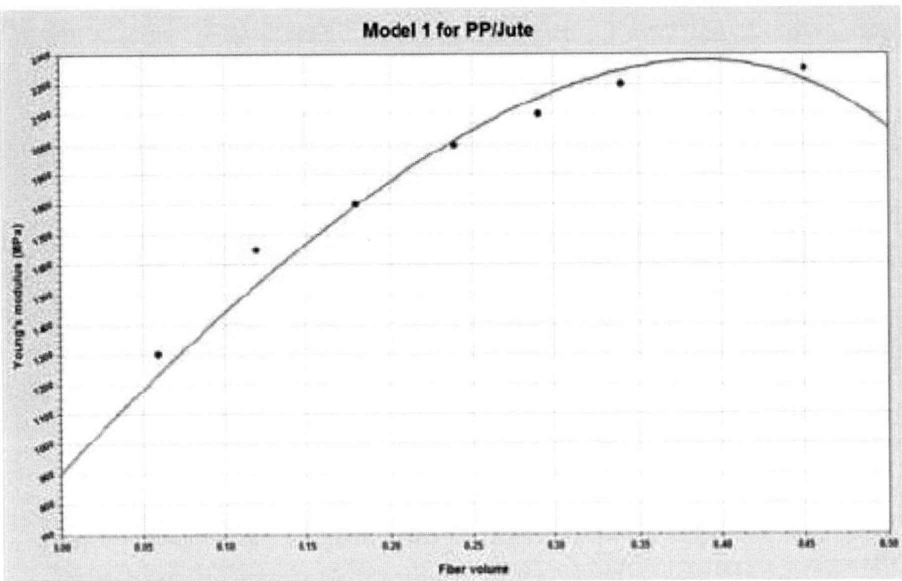


Figure 7. Data fit for IROM and modified ROM model.

***Model 2***

$$E_1 = a\left( E_m \left( \frac{1+\xi V_f \left( \frac{(E_f/E_m)-1}{(E_f/E_m)+\xi} \right)}{1-V_f \left( \frac{(E_f/E_m)-1}{(E_f/E_m)+\xi} \right)} \right) \right) + b\left(\eta E_f V_f + E_m (1-V_f)\right) \tag{15}$$

In this model four weight coefficients, namely a, b, $\zeta$ and $\eta$ should be found in order to fit the data of table 2 with best convergence.

With applying standard error and correlation coefficient criteria, a, b, $\zeta$ and $\eta$ constants are calculated as follows:

$\eta$ =0.65704045
$\zeta$ = - 4.9221785
a =0.82190093
b =0.31977636
Standard Error: 97.3997226
Correlation Coefficient: 0.9891221

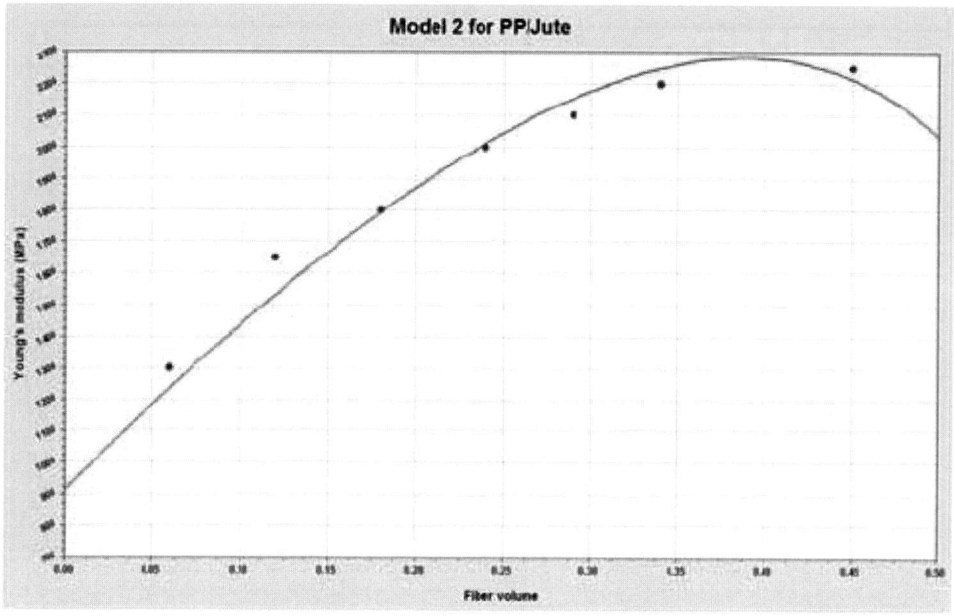


Figure 8. Data fit for modified ROM and Halpin.

# 4. Mathematical Models

When applying these approaches to choose a good model for fitting available data, some interesting properties of them has been considered and some pure mathematical models has been tested. Best series of models was sigmoidal family of models. Processes producing sigmoidal or "S-shaped" growth curves are common in a wide variety of applications such as biology, engineering, agriculture, and economics. These curves start at a fixed point and increase their growth rate monotonically to reach an inflection point. After this, the growth rate approaches a final value asymptotically. This family is actually a subset of the Growth Family, but is separated because of its distinctive behavior:

Gompertz model: $$E = ae^{\left(-e^{(b-cV_f)}\right)} \qquad (16)$$

MMF model: $$E = \frac{ab + cV_f^d}{b + V_f^d} \qquad (17)$$

Logistic model: $$E = \frac{a}{1 + e^{(b-cV_f)}} \qquad (18)$$

Richards model: $$E = \frac{a}{1 + e^{(b-cV_f)^{\frac{1}{d}}}} \qquad (19)$$

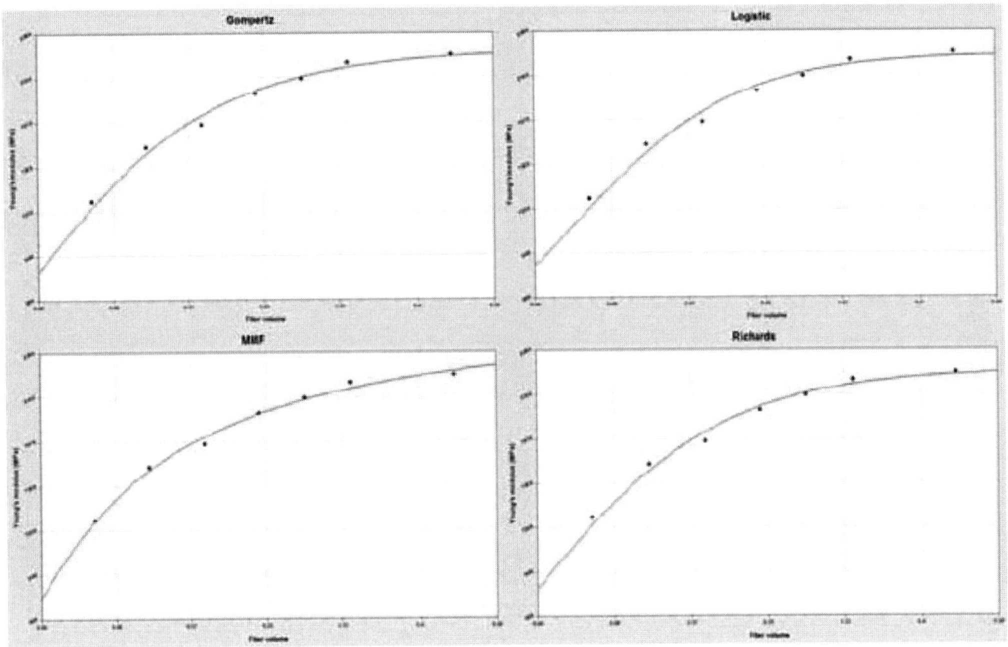


Figure 9. Data fit and comparison of models.

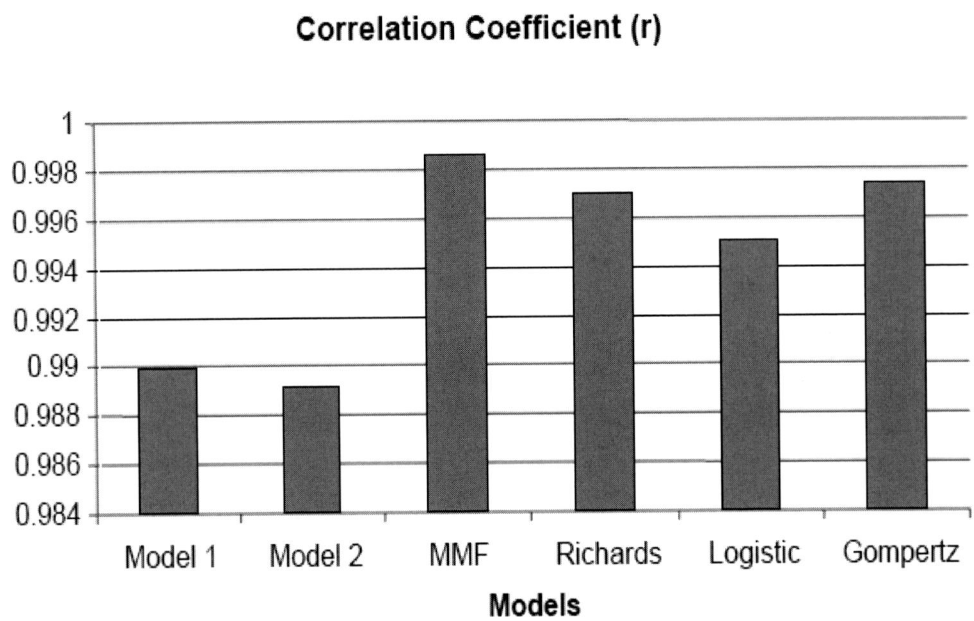


Figure 10. Models correlation coefficients.

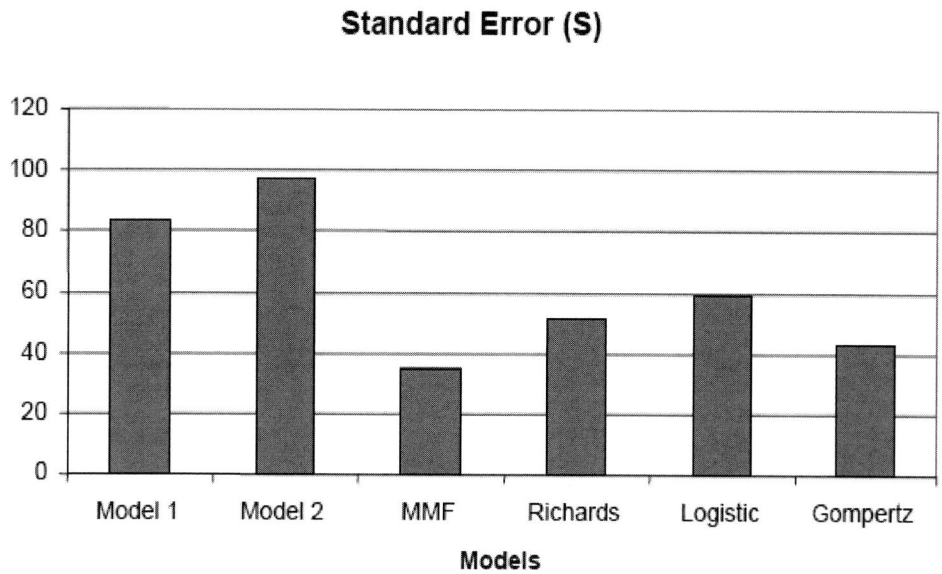


Figure 11. Models standard errors.

## 5. Discussion of Results

The results of this study in general indicated that:

- The alternative base course composite has desirable strength and mechanical characteristics to be considered as a good quality stabilized pavement material.
- The higher the percentage of cement added, the higher the increment in strength and stiffness of treated soil.
- the compressive strength of the mixture increases with the increase of Jute/pp content until it reaches to its optimum value and then the additional fibre more than optimum value has decreasing effect on compressive strength.
- The maximum modulus of elasticity was obtained from mixtures by 2% Jute.
- Reinforcing cement stabilized materials with Jute/pp improved the durability of soil-cement mixtures.
- Reinforcing cement stabilized materials with Jute improved indirect tensile strength of soil-cement mixtures.
- Although our numerical investigation yields results that conform to the expected trends we explained, more research and modifications are necessary to be able to use it confidently. After the period spent studying this particular topic, we believe there is much potential for success in developing a more accurate model. It would be a time-consuming task, but foundations are laid and the benefits are truly worthwhile.

## 6. Conclusion

In this paper a method for predicting the elastic modulus in green composites interfaces was developed. Through theoretical examinations a new model developed to estimate reliably of the modulus of elasticity in green composite interfaces with different fiber volume fraction and elastic properties. This approach allows a simple model for systems without resorting to complicated constitutive equations. The approach presented here, leads to theoretical predictions which can reasonably be explained from the physical point of view. Clearly, the final verification can only reached by systematic experimental investigations which, at present, are being carried out.

## 7. Appendix

Two criteria were adopted to evaluate the goodness of fit of each model, the Correlation Coefficient (r) and the Standard Error (S).

The standard error of the estimate is defined as follows:

$$S = \sqrt{\frac{\sum_{i=i}^{n_{po\,int\,s}} (E_{\exp,i} - E_{\text{pred},i})^2}{n_{po\,int\,s} - n_{param}}} \tag{20}$$

Where $E_{\exp,i}$ is the measured value at point i, and $E_{\text{pred},i}$ is the predicted value at that point, and $n_{param}$ is the number of parameters in the particular model (so that the denominator is the number of degrees of freedom).

To explain the meaning of correlation coefficient, we must define some terms used as follow:

$$S_t = \sum_{i=1}^{n_{po\,int\,s}} (\bar{y} - E_{\exp,i})^2 \tag{21}$$

where, the average of the data points ( $\bar{y}$ ) is simply given by

$$\bar{y} = \frac{1}{n_{po\,int\,s}} \sum_{i=1}^{n_{po\,int\,s}} E_{\exp,i} \tag{22}$$

The quantity $S_t$ considers the spread around a constant line (the mean) as opposed to the spread around the regression model. This is the uncertainty of the dependent variable prior to regression. We also define the deviation from the fitting curve as

$$S_r = \sum_{i=1}^{n_{points}} (E_{\exp,i} - E_{pred,i})^2 \tag{23}$$

Note the similarity of this expression to the standard error of the estimate given above; this quantity likewise measures the spread of the points around the fitting function. In view of the above, the improvement (or error reduction) due to describing the data in terms of a regression model can be quantified by subtracting the two quantities. Because the magnitude of the quantity is dependent on the scale of the data, this difference is normalized to yield

$$r = \sqrt{\frac{S_t - S_r}{S_t}} \tag{24}$$

where, r is defined as the correlation coefficient. As the regression model better describes the data, the correlation coefficient will approach unity. For a perfect fit, the standard error of the estimate will approach S=0 and the correlation coefficient will approach r=1. [52]

The standard error and correlation coefficient values of all models are given in figures 10 and 11.

## REFERENCES


[1] K. Joseph, S. Thomas, C.Pavithran, Effect of chemical treatment on the tensile properties of short sisal fiber-reinforced polyethylene composites, *Polymer.* 37, 5139–45 (1996).

[2] IK. Varma, SR. Ananthakrishnan, S. Krishnamoorthi, Comp of glass/modified jute fabric and unsaturated polyester, *Composites.* 20,383 (1989).

[3] VG. Geethamma, K. Thomas Mathew, R. Lakshminarayanan, S. Thomas, Composite of short coir fibers and natural rubber: effect of chemical modification, loading and orientation of fiber, *Polymer.* 39, 1483 (1998).

[4] MS. Sreekala, MG. Kumaran, S. Thomas, Oil palm fibers: morphology, chemical composition, surface modification and mechanical properties, *J. Appl. Poly. Sci.* 66, 8–821 (1997).

[5] L.A. Pothan , Z. Oommen, S. Thomas, Dynamic mechanical analysis of banana fiber reinforced polyester composites, *Composite Science and Technology.* 63 , 283-293 (2003).

[6] JJ. Aklonis, WJ. MacKnight, Introduction to polymer viscoelasticity, New York , Wiley (1983).

[7] T. Murayama, Dynamic mechanical analysis of polymeric materials, New York , Elsevier (1978)

[8] JD. Ferry JD, Viscoelastic properties of polymers and comp, vol. 2. NY, Wiley (1980).

[9] C. Datta, D. Basu, A. Banerjee , *J. Appl. Poly. Sci.* 85 , 2800–7 (2002).

[10] L. A. Pothan, S. Thomas, G. Groeninckx, The role of fibre/matrix interactions on the dynamic mechanical properties of chemically modified banana fibre/polyester composites, Composite:part A xxx(2005)xxx-xxx.

[11] LA. Pothen, S. Thomas, NR. Neelakandan, J Reinforced Plast Compos 16, 744 (1997).

[12] S. Joseph, M.S. Sreekala, Z. Oommen, P. Koshy, S. Thomas, A comparison of the mechanical properties of phenol formaldehyde composites reinforced with banana fibres and glass fibres,*Composite Science and Technology*. 62 , 1857-1868 (2002).

[13] AR. Sanadi, SV. Prasad, PK. Rohatgi, *Journal of Scientific and Industrial Research*. 44 , 437 (1985).

[14] PJ. Roe, MP. Ansell, Jute reinforced polyester composites, *J. Mater Sci.* 20 , 4015 (1985).

[15] AR. Sanadi, Prasad. SV, Rohatgi. PK, *J. Mater Sci.* 21 , 81 (1986).

[16] Heijenrath. R, T. Peijs, *Advanced Compos Lett.* 5(3) , 81 (1996).

[17] Marchovich. N, Reboredo. M, Aranguren. M, *J. Appl. Poly. Sci.* 61, 119 (1996).

[18] CN. Zarate, MI. Aranguren, MM. Reboredo, *J. Appl. Poly. Sci.* 77, 1832 (2000).

[19] M.S. Sreekala, J. George, M.G. Kumaran, S. Thomas, The mechanical performance of hybrid phenol-formaldehyde-based composites reinforced with glass and oil palm fibres. *Composite Science and Technology*. 62, 339-353 (2002).

[20] S. George, NR. Neelakantan, KT. Varghese, S. Thomas, Dynamic mechanical properties of isotactic polypropylene/nitrile rubber blends: effects of blend ratio, reactive compatibilization, and dynamic vulcanization, *J. Poly Sci.* Part B Polymer Physics 35, 2309–271 (997).

[21] H. Varghese, SS. Bhagawan, Rao. Someswara, S. Thomas, Morphology, mechanical and dynamic mechanical properties of blends of nitrile rubber and ethylene vinyl acetate copolymer, *Eur. Poly. J.* 31(10) , 957–67(1995).

[22] J. George, SS. Bhagawan, S. Thomas, Thermogravimetric and dynamic mechanical thermal analysis of pineapple fiber reinforced polyethylene composites, *J. Ther. Analy.* 47, 1121–40 (1996).

[23] S. Varghese, B. Kuriakose, S. Thomas, Mechanicaland viscoelastic properties of short sisal fiber reinforced natural rubber composites: effects of interfacial adhesion, fiber loading and orientation, *J. Adhes. Sci. Technol.* 8, 234 (1994).

[24] K. Joseph, C. Pavithran, S. Thomas, Dynamic mechanical properties of short sisal fiber reinforced polyethylene composites, *J. Reinf. Plast. Comp.* 12 , 139 (1993).

[25] A.K. Mohanty, A. Wibowo, M. Misra, L.T. Drzal, Effect of process engineering on the performance of natural fiber reinforced cellulose acetate biocomposite , Cpmposite part A 350 ,363-370 (2004).

[26] T.T.Loan. Doan, Sh.L. Gao, E. Ma¨der, Jute/polypropylene composites I. Effect of matrix modification, *Composite Science and Technology*. 66, 952-963 (2006).

[27] M. S. HUDA, A. K. MOHANTY, L.T. DRZAL, E. SCHUT, MISRA. M, “Green” composites from recycled cellulosend poly(lactic acid): Physico-mechanical and morphological properties evaluation, *Journal of Materials A Science*. 40, 4221 – 4229 (2005).

[28] T.T.Loan. Doan, E. Ma¨der, Performance of jute fibre reinforced polypropylene, In: 7th international AVK-TV conference, Baden-Baden (2004).

[29] M. Lucka, AK. Bledzki, J. Michalski, Influence of the hydrophobisation of Flax fibres on the water sensitivity, biological resistance and electrical properties of Flax – polypropylene composites, In: 5th global wood and natural fibre composites symposium, p. A 15.1–7. Kassel, Germany2004.

[30] S-L. Gao, E. Ma¨der, R. Plonka, Surface flaw sensitivity of glass fibres with carbon nanotube/polymer coating, In: International conference on composite materials (ICCM-15), Durban, South Africa(2005). 962 Doan. T.T.Loan, et al, *Composites Science and Technology*. 66, 952–963 (2006).

[31] F.R. Cichocki Jr, J.L. Thomason, Thermoelastic anisotropy of a natural fiber, *Composite Science and Technology*. 66, 669-678 (2002).

[32] A. K. Haghi , M. Arabani , M. Veis Karami, Applications of expanded polystyrene (EPS) beads and polyamide 66 in civil engineering, Part Two: Stabilization of clayey sand by lime/polyamide-66, *Composite Interfaces,* Vol. 13, No. 4-6, pp. 451–459 (2006).

[33] A.G. Facca, M.T. Kortschot, N. Yan, Predicting the elastic modulus of natural fibre reinforced thermoplastics, Available online at www.elsevier.com/locate/compositesa

[34] K.G. Kavelin, Investigation of natural fiber composites heterogeneity with respect to automotive structures, ISBN-10: 90-9020036-3

[35] G.C. Sih, A. Carpinteri, G. Surace, , Advanced Technology for Design and Fabrication of composite Meterials and Structures, Kluwer Academic publishers(1995).

[36] RM. Jones, Mechanics of composite material, Hemisphere Publishing Corporation p. 86, United States(1975).

[37] H.L. Cox, The elasticity and strength of paper and other fibrous materials, *Brit. J. Appl. Phys.* 3 , 72-791 (952).

[38] JC. Halpin, SW. Tai, Effects of environmental factors on composite materials, *AFML-TR.* 7-423 June (1969).

[39] JC. Halpin, JL. Kardos, The Halpin–Tsai equations: a review, *Polym. Eng. Sci.* 16(5), 344–52 (1976).

[40] SW. Tsai, HT. Hahn, Introduction to composite materials, Technomic Publishing Co., Lancaster(PA)( 1980).

[41] RF. Gibson, Principles of composite material mechanics, McGraw-Hill Inc(1994).

[42] MW. Hyer, Stress analysis of fiber reinforced composite materials, McGraw-Hill Inc, (1998).

[43] IM. Daniel, O. Ishai, Engineering mechanics of composite materials, Oxford University Press(1994).

[44] CT. Herakovich, Mechanics of fibrous composites. John Wiley(1998).

[45] AK. Saha, S. Das, D. Bhatta, BC. Mitra, Study of jute fiber reinforced polyester composites by dynamic mechanical analysis, *J. Appl. Polymer. Sci.* 71, 1505–13 (1999).

[46] J. Gassan, AK. Bledski, Possibilities for improving the mechanical properties of jute/epoxy composites by alkali treatment of fibres, *Compos. Sci. Technol.* 59, 1303–9 (1999).

[47] A. Bergander, L. Salmen ,The transverse elastic modulus of the native wood fibre wall, *J. Pulp. Paper Sci.* 26 , 234–8 (2000).

[48] A. Bjurenstedt, F. Lärneklint ,3D Biocomposites for Automotive Interior Parts, MSc Thesis, Department of Applied Physics and Mechanical Engineering, Division of Polymer Engineering,Luleå Tekniska, Universitet ON(2004).

[49] A. André, Fibres for Strengthening of Timber Structures, MSc Thesis,Department of Civil and Environmental Engineering, Division of Structural Engineering ,Luleå Tekniska University of Technology,ON(2006).

[50] C.N. Zárate, M.I. Aranguen, M.M. Reboredo, Resol-vegetable fibers composites, *J. Appl. Polym. Sci.* 77, 1832-1840 (2000).

[51] CL. Tucker, E. Liang, Stiffness predictions for unidirectional shortfibre composites: review and evaluation, *Comp. Sci. Technol.* 59, 655–71 (1999).

[52] A. Daghbandan, A. Hajiloo, A. Farjad, A. K. Haghi, Analytical Determination of the Drying Characteristics, 4th Asia Pacific Drying Conference Proceeding, pp. 802-815(ADC 2005).

[53] JA. Nairn,On the use of shear-lag methods for analysis of stress transfer in unidirectional composites, *Mech. Mater.* 26, 63–80 (1997).

[54] DA. Mendels, Y. Leterrier, JAE. Manson, Stress transfer model for single fibre and platelet composites, *J. Comp. Mater.* 33(16), 1525–43 (1999).

In: Monomers, Oligomers, Polymers, Composites... ISBN: 978-1-60456-877-6
Editors: R. A. Pethrick, G.E. Zaikov et al. 

*Chapter 20*

# PHYSICAL MODIFICATION AND NEW METHODS IN TECHNOLOGY OF POLYMER COMPOSITES, REINFORCED BY FIBERS

***V.N. Stoudentsov****
Technological Institute of Saratov State Technical University

## ABSTRACT

In this work, we report main directions of study, produced by authors in Technological Institute of SSTU. Purposes of these researches are:

- realization economizing technologies of polymer composite materials (PCM),
- improvement technological properties of half-finished products,
- regulation characteristics of PCM by the help of different physical influences.

It is done minimum of theoretical bases of observing phenomena in this paper, main attention is given to practical results. Some new physical influences may be used as supplemental stages in traditional technology.

Main positive effect consists in new property appearance or in increasing of strength characteristics of new materials on several tens percents in comparison with materials, produced by traditional technology. Comparatively high durability of improved reinforced by fibers materials permits to use these materials as in the manufacture of consumer goods as in more responsible application.

## ABBREVIATIONS AND CONDITIONAL MARKS

APFR – aniline-phenol-formaldehyde resin;
bustilate – emulsion of butadienestyrene rubber;
Capron – polycaproamide fibers (nylon-6);

* xt@techn.sstu.ru

CF – carbon fibers;
CLP – components layer putting;
CMC – carboxymethyl cellulose;
CMF – constant magnetic field;
CSP – components separate putting;
DAF – cellulose diacetate fibers;
$E_b$ – modulus of elasticity under static bending;
ER – epoxy resin;
f, Hz – frequency of vibration;
Fenilon – aromatic polyamide fibers;
$H_B$ – solidity;
MF – magnetic field;
MT – magnetic treatment;
Nitron – polyacrylonitrile fibers;
PC – polymer composite;
PCF – polycondensational filling;
PCM – polymer composite material;
PEPA – polyethylenpolyamine;
PF – polymerization filling;
PN-15 –(non-limited) unsaturated polyester resin;
PVA – polyvinylacetate;
$T_{stab}$ – thermostability;
TAF – cellulose triacetate fibers;
TEA – triethanolamine;
TETRA – triethylenetriamine;
VF – viscose fibers;
VT – vibratory treatment;
W – daily water absorption;
$\alpha$, kJ/m$^2$ – impact strength;
$\lambda$, W/ m K – coefficient of heat conduction;
$\rho$, kg/m$^3$ – density;
$\sigma_b$, MPa – stress causing failure under static bending;
$\sigma_{eff}$, C/m$^2$ – surface concentration of electric charge;
$\sigma_t$, MPa – tensile strength.

## INTRODUCTION

In the traditional way of wares production from materials, reinforced by fibers (PC) on the basis of cross-linked polymers (figure 1), thread from bobbin 1 goes to the bath 2, where resin solution is saturating fibers. Thread, saturated by binding, runs to reception mechanism 4 over warming-up pipe 3, which is necessary for removal the solvent and for initial binder hardening appearance. Different versions of produced half-finished products conversion exist. It is possible to produce wares by means of continuous fibers reception by mechanism 4 with following thermal treating. If it is necessary, we cut saturated half-finished product and then process it, for example, by straight pressing.

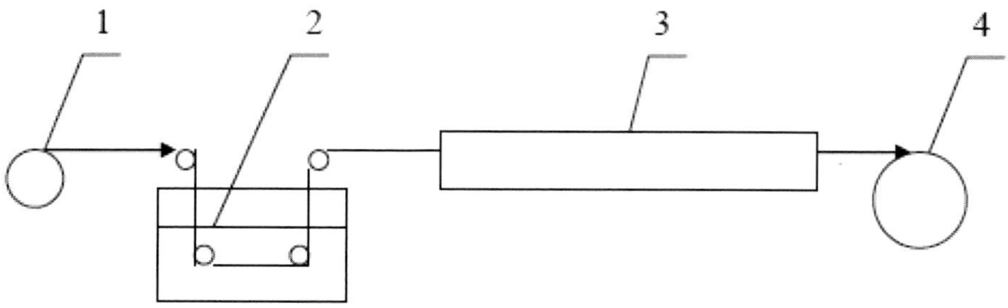


Figure 1. Scheme of traditional method of wares production of PC, reinforced by fibers: 1 – bobbin with thread; 1 – saturating bath; 3 – warming up pipe; 4 – reception mechanism.

## 1. Polycondensation Filling and New Polyesters

Polycondensational filling (PCF) – is logical continuation of polymerization filling (PF) [1] . General line of PCF and PF is realization of polymer synthesis in presence of filler. Differences of these methods: synthesis of polymer has mechanism of polymerization for PF and mechanism of polycondensatoin for PCF respectively.

In general PCF scheme is similar to scheme of traditional method. Distinction is: in the bath 2 thread is saturated by mixture of monomers, not by mixture of resin and hardener, for example in PCF of epoxy resin thread may be saturated by solution of 1,2-epoxy-3-chlorpropane and diphenylolpropane.

Synthesis of resin realizes in the warming-up pipe 3. The point of gel-formation must be obtained (degree of conversion 60-70%) in pipe 3. Unification of resin synthesis stage and stage of filler saturating shortens the common amount of technological process stages. This is dignity of PCF method. Difficulty of hardener penetration to the synthesized resin is deficiency of PCF method (table 1), that may aggravate PC characteristics in comparison with material, produced by traditional method. New technological methods and new methods of physical modification are given further.

**Table 1. Characteristics of materials, reinforced by short fibers and produced by polycondensational filling of epoxy resin**

| Filler | Method of PC production | $\sigma_b$, MPa | $\sigma_t$, MPa | $\alpha$, kJ/m$^2$ | $\rho$, kg/m$^3$ | W,% |
|---|---|---|---|---|---|---|
| Capron | Traditional | 9 | 4 | 46 | 1060 | 26 |
| Capron | PCF | 15 | 4 | 41 | 980 | 23 |
| VF | PCF | 19 | 14 | 8 | 1100 | 16 |
| VF | PCF+MT+CLP | 8 | 28 | 40 | 950 | 31 |

The way of new copolymer of APFR and ER production by PCF method is suggested [2].

PCF method is connected with production of new polyesters. For example, polyester, produced from organic base – glycerine (it has 3 functional groups, f=3)- and organic acid, differing from well-known polyesters by contents of organic base (with f ≥ 2) as supplement – threeethanolamine (TEA) or polyethylenepolyamine (PEPA), and as organic acid it is taken adipine acid (AA) (f=2) either tartar acid (TA) (f=2) or lemon acid (LA) (f=3). In method of this polyester production synthesis and hardening run simultaneously at the same reactive volume [3].

Glycerine, TEA and PEPA contain functional groups of base type. These groups react chemically with acid groups of AA, TA, LA. Maximum degree of initial components conversion into cross-linked polymer achieves, if quantities of acid and base groups are equal, that is at stoichiometric proportion between initial components or at proportions near this proportion.

Produced materials don't contain dangerous volatile matters. Resin synthesis runs at the same time with its hardening. New materials are more cheap in consequence of production, packing, storage and transportation expending absence for semi-product – oligoresin.

New polyester in PCM is suggested. It is produced by soak of fiber filler by non-saturated polyester resin and by new hardening system with following pressing, differing from well-known matters (substances) by application of kapron, nitron or VF as fiber filler and by application of 60-80% mass. solution of RF-342A in acetone as hardening system for resin PE-15 [ 4 ]. Also it is possible to apply following mixture as hardening system, % mass.:

| | |
|---|---|
| Aniline-phenol-formaldehyde resin RP-342A | 9,3 – 9,8 |
| Carboximetyle-cellulose (CMC) | 26,5 – 30,8 |
| Acetone | 3,1 - 3,3 |
| Water | 56,8 - 60,5 |

It is necessary at firstly to saturate fiber filler by non-limited polyester resin PE-15, secondly – by hardening system. Produced material contain, % mass.:

| | |
|---|---|
| Resin PE-15 | 40 - 61 |
| Hardening system | 12 - 42 |
| Fiber filler | 18 – 33. |

In this material toxic and expensive hardeners (organic peroxides) are absent. New materials have following characteristics:

$\rho$ = 1160 - 1510 kg/m$^3$, $\sigma_b$ = 67 - 151 MPa, $\alpha$ = 112 - 208 kJ/m$^2$.

## 2. Methods of Increase Permissible Storage of Prepregs

The problem of increasing permissible storage of prepregs, when we use binder, for which application of hardener is necessary, is solved successfully by selection hardening system now. Two methods of deciding this problem by technological way mainly, when we use hardeners, which may work at the normal temperature, are suggested in this paper:

1. In the method of components separate putting (CSP) some elements of filler is saturated by binder with plenty (surplus) of resin function groups and thus type 1 prepreg is produced, other elements of filler is saturated by binder with surplus of hardener function groups [5] comparatively to stoichiometric proportion of resin and hardener, type 2 prepreg is produced by this way. Prepregs of both types may be preserved separately about two months in the normal conditions. Prepreg, which is produced by traditional method and contains epoxy resin and one among examined hardeners, for example, PEPA, must be treated several hours past production of it. Prepregs of both types in ware formation conditions are mixed and resin in the type 1 prepreg is hardened by plenty of hardener of the type 2 prepreg. Diffusion difficulties increase in this method, it enables to produce porous materials. Such materials may be used, for example, as heat insulation materials (table 2) with more high toughness, than usual heat insulating materials. Low coefficient of heat conduction is index of high heat insulation properties of material, water absorption is in proportion of porosity.
2. In the method of components layer putting (CLP) every element of filler, that is to say, every thread of filler is soaked (saturated) by resin at first, then by hardening system (figure 2). The lay of resin is hardened in the result of hardener molecules diffusion from external (second) lay in the conditions of ware formation [6]. CLP increases permissible storage of prepregs to two weeks and more and improves all the PC durability characteristics comparatively with materials, produced by traditional method (table 3) [6, 7,8].

Main parameters (concentrations of solutions, composition of hardening system, linear speed of thread, temperature of between-baths thermal treatment in warming-up pipe 3) essential influence on the properties of produced PC. Hardening system cans contain protective polymer (bustylate, CMC or PVA). Different types of CLP depend on combination of those parameters.

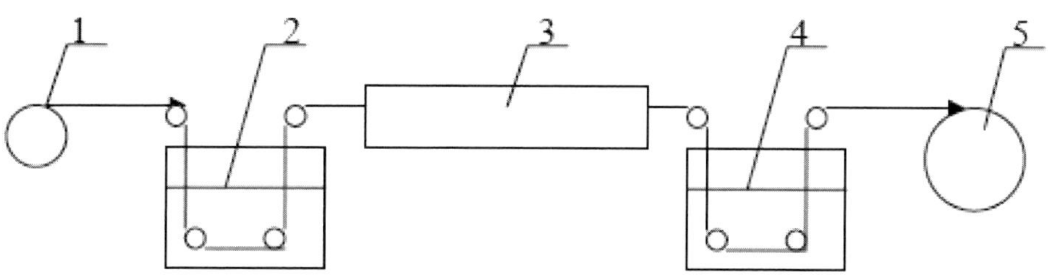


Figure 2. Scheme of method of components layer putting (CLP): 1 – bobbin of thread; 2 – bath of solution of resin; 3 – warming-up pipe; 4 – bath for hardening system; 5 – reception mechanism.

**Table 2. Characteristics of materials on the basis of epoxy resin, filled by short fibers**

| Method of PC production | Hardener | $\lambda$, W/ m K | $\sigma_b$, MPa | $\sigma_t$, Mpa | $\alpha$, kJ/m$^2$ | $\rho$, kg/m$^3$ | W,% |
|---|---|---|---|---|---|---|---|
| Traditional | TETRA | 0,096 | 81 | 28 | 80 | 1120 | 36 |
| CSP | TETRA | 0,057 | 5 | 3 | 10 | 1050 | 68 |
| CSP | TEA | 0,032 | 8 | 2 | 8 | 1000 | 61 |
| Coniferous wood | - | 0,07-0,16 | 72-79 | 108-115 | - | 450-550 | - |

**Table 3. Characteristics of materials on the basis of epoxy resin, filled by continuous threads**

| Method of PC production | Filler, reference | $\sigma_b$, MPa | $\Delta\sigma_b$, % | $\alpha$, kJ/m$^2$ | $\Delta\alpha$, % | $\rho$, kg/m$^3$ | W,% |
|---|---|---|---|---|---|---|---|
| Traditional | Capron, [7] | 31 | - | 53 | - | 930 | 3,7 |
| CLP | Capron, [7] | 64 | +106 | 55 | +4 | 1080 | 4,0 |
| Traditional | Nitron, [7] | 39 | - | 68 | - | 1030 | 4,8 |
| CLP | Nitron, [7] | 42 | +8 | 74 | +9 | 1150 | 5,5 |
| Traditional | VF, [8] | 45 | - | 144 | - | 1200 | 3,4 |
| CLP | VF, [8] | 71 | +68 | 152 | +6 | 1100 | 2,8 |

## 3. MAGNETIC TREATMENT

Magnetic treatment (MT) and other non traditional influences may be employed as the additional stage of traditional technology [9, 10, 11, 12] or as the stage of new technology [8]. In result of interaction between external magnetic field (MF) and magnetic moments of resin's molecules or of segments of macromolecules probability of definite orientation of binding particles increases [13], that leads to formation of anisotropy structure of polymer in the process of hardening. This is able to have positive influence on the PC durability (table 4). External MF increases adhesion between binder and filler, that influences positively on all durability characteristics of PC. Necessary type of MT depends on temperature and duration of treating, on its place in the technological process, on characteristics of MF (orientation, polarity, stress). We used constant and homogeneous MF only.

MT of MN-type is more economical and more effective in accordance with following reasons:

- it is necessary plus short times of MT (about several seconds, but not about several hours);
- it is necessary large values of working volume for placing of all product wholly, since only one thread, saturated by binder, is subjected by MT;
- segments of binder in saturating solution are more agile, than in hardened product.

**Table 4. Influence of MT on the toughness of PC, reinforced by long threads ( MN – magnetic treatment of thread, saturated by binder; NM – hardening in CMF; ⊥ and ‖ - orientation of thread perpendicular or parallel to external CMF; $\Delta\sigma_b$ , $\Delta\alpha$ – change of value $\sigma_b$, $\alpha$ relatively meaning for PC, produced by traditional method)**

| Example | Method of PC production, type of MT | Resin | Filler | $\sigma_b$, Mpa | $\Delta\sigma_b$, % | α, kJ/m² | Δα,% | Reference |
|---|---|---|---|---|---|---|---|---|
| 1 | Tradi-tional | APFR | fenilon | 107 | - | 17 | - | 8 |
| 2 | MT, ⊥MN | APFR | fenilon | 110 | +3 | 39 | +129 | 8 |
| 3 | MT, NM‖ | APFR | fenilon | 204 | +91 | 25 | +47 | 8 |
| 4 | MT, NM⊥ | APFR | fenilon | 169 | +58 | 40 | +135 | 8 |
| 5 | Tradi-tional | ER-20 | capron | 94 | - | 157 | - | 10 |
| 6 | MT, ⊥MN | ER-20 | capron | 221 | +135 | 283 | +80 | 10 |
| 7 | Tradi-tional | ER-16 | CF | 187 | - | 82 | - | 10 |
| 8 | MT, ⊥MN | ER-16 | CF | 364 | +95 | 189 | +130 | 10 |
| 9 | Tradi-tional | ER-16 | DAF | 90 | - | 308 | - | 10 |
| 10 | MT, NM⊥ | ER-16 | DAF* | 151 | +68 | 377 | +22 | 10 |
| 11 | Tradi-tional | ER-16 | TAF | 73 | - | 165 | - | 10 |
| 12 | MT, NM⊥ | ER-16 | TAF* | 148 | +103 | 235 | +42 | 10 |
| 13 | Tradi-tional | ER-100 | CF | 170 | - | 89 | - | 9 |
| 14 | MT, ⊥MN | ER-100 | CF | 188 | +11 | 120 | +35 | 9 |
| 15 | Tradi-tional | ER-100 | CF | 532 | - | 104 | - | 9 |
| 16 | MT, ⊥MN | ER-100 | CF | 591 | +11 | 120 | +15 | 9 |

* - structure of fibers decays in the process of hardening in CMF, PC is homogeneous mixture of two polymers: polymer of binder plus polymer of filler.

Orientation of threads ‖ leads to increasing of concentration of cross-linked chains, orientation of threads ⊥ leads to decreasing of concentration of cross-linked chains, that is why ‖ orientation increases value of $\sigma_b$ mainly ( $\Delta\sigma_b$ = + 91%), but ⊥ orientation increases value of $\alpha$ mainly ($\Delta\alpha$ = + 135%) comparatively with traditional method of production (table 4, examples 3,4). In general MTs strengthen materials in limits from 3 to 135 %.

It is possible to combine magnetic treatment and formation of product by pressing in single device [14]. For this effect prepreg is placed in interpole volume, the effort of pressing is provided by mutual attraction of poles of electromagnet in this device.

## 4. Vibratory Treatment

Used vibratory treatment (VT) consists in following: filler, saturated by solution of binder, is subjected in addition to vibratory treatment through solution of binder by vibration in optimum interval of frequencies [15]. Such treatment may be used in traditional method or in any new method, for example in CLP method. As rule vibration leads to increasing of toughness characteristics of PCM (table 5). Vibratory treatment improves penetration of binder to thread, material becomes more homogeneous, but thread loses compactness simultaneously herewith.

**Table 5. Influence of vibration on the toughness of PCs, filled by different threads ($\Delta\sigma_b$, $\Delta\alpha$ – relative changes in consequence of VT or new method)**

| Binder | Method | f, Hz | $\sigma_b$, MPa | $\Delta\sigma_b$, % | $\alpha$, kJ/m$^2$ | $\Delta\alpha$,% |
|---|---|---|---|---|---|---|
| Capron | Traditional | 0 | 96 | - | 31 | - |
| Capron | Traditional | 66 | 110 | +14 | 50 | +61 |
| Capron | CLP | 0 | 133 | +38 | 54 | +74 |
| Capron | CLP | 66 | 135 | +2 | 83 | +54 |
| Nitron | Traditional | 0 | 23 | - | 93 | - |
| Nitron | Traditional | 66 | 42 | +83 | 95 | +2 |
| Nitron | CLP | 0 | 51 | +128 | 110 | +18 |
| Nitron | CLP | 66 | 50 | - | 114 | +4 |
| VF | Traditional | 0 | 105 | - | 56 | - |
| VF | Traditional | 66 | 110 | +5 | 50 | - |
| VF | CLP | 0 | 137 | +30 | 63 | +12 |
| VF | CLP | 66 | 145 | +6 | 66 | +5 |

## 5. Electric Polarization of Cross-Linked Polymers

Electric polarization of polymers may be considered as reception of physical modification of PCM. Materials, saving stable electric charge during long time (weeks, months), are named by electrets. In our time electrets are produced from thermoplastics by drive of warmed thermoplastic between two tapes of different metals mainly with following cooling of material. Such materials save stable electric charge during several weeks. In new method [16] electrets are produced by hardening of thermosetting (filled or not filled) systems between two tapes of different metals. New method provides following advantages by comparison with known electrets (table 6):

- more long conservation period of value of surface concentration $\sigma_{eff}$ of electric charge in consequence of location and isolation of charged particles into elements of cross-linked structure of polymer;
- more high thermostability is stipulated by temperature of hardening of suggested systems;
- more high toughness characteristics are stipulated by application of reinforcing fibers.

**Table 6. Comparison of characteristics of new and known electrets**

| Physical, mechanical characteristics | New electrets | Known electrets on basis of thermoplastics |
|---|---|---|
| $\tau_e$, days | 200 | 7 - 30 |
| $\sigma_{eff}$, C/ m$^2$ | $10^{-6} - 10^{-7}$ | $10^{-5}$ |
| $\sigma_b$, MPa | 150 | 20 |
| $\alpha$, kJ/m$^2$ | 50 | 2 - 150 |
| $T_{stab}$, K | 453 | 403 |
| $E_b$, MPa | 2000 | 900 |
| $H_B$, MPa | 120 | 14 – 58 |
| $\rho$, kg/m$^3$ | 1500 | 930 |

## Conclusions

1. Technically simple and effective methods of physical modification are offered, these methods permit to increase toughness characteristics of PC, reinforced by fibers:

| Method | Increase of $\sigma_b$, % | Increase of $\alpha$, % |
|---|---|---|
| CLP | 8 –128 | 4-9 |
| MT | 3-135 | 12-135 |
| VT | 6 – 83 | 5 - 61 |

2. Method of components separate putting and method of components layer putting increase permissible storage of prepregs, containing the hardener of hot hardening, to several months (instead of several hours in traditional method).
3. Offered syntheses of new resins and formation of cross-linked polymers are proceeding simultaneously in production of new polyesters.
4. New cross-linked polymer electrets differ by magnified conservation period of value of surface concentration of electric charges.

## REFERENCES


[1] Diachkovski F.S., Novokshenova L.A. Syntheses and properties of polymerization-filled polyolefins // *Successes of chemistry*. – 1984. – V. 53, №2. – P. 200 – 222.

[2] Artemenko S.E., Kardash M.M., Titova T.P., Stoudentsov V.N. The way of polymer press-composition production. Author's certificate № 1616930 USSR// BI. - 1990, № 48.

[3] Poliach E.V., Stoudentsov V.N. Polyester and the way of its production. Patent № 2272047 RU// BI. - 2006, № 8.

[4] Stoudentsov V.N., Levkin A.N., Tcheremouhina I.V. Composite material on basis of unsaturated polyester resin. Patent № 2232175 RU// BI. –2004,№ 2.

[5] Stoudentsov V.N., Ahrameieva E.V., Rozenberg B.A., Smirnov Yu.N. The way of polymer composite materials production. Author's certificate № 1796638 USSR// BI. – 1993, № 7.

[6] Stoudentsov V.N., Rozenberg B.A., Hazizova A.K. The way of production of prepreg. Patent № 2028322 RU// BI. – 1995, № 4.

[7] Stoudentsov V.N., Sergiyenko A.S., Samkov D.V. The way of production of polymer materials, reinforced by fibers. Patent № 2132341 RU// BI. – 1999, № 18.

[8] Stoudentsov V.N., Karpova I.V. The way of production of polymer materials, reinforced by fibers. Patent № 2135530 RU// BI. - 1999, № 24.

[9] Stoudentsov V.N., Panjushkina L.A. The way of production of composite material, reinforced by fibers. Author's certificate № 1785909 USSR// BI. –1993, № 1.

[10] Stoudentsov V.N., Mihaylov M.Ju., Tsarev V.F. The way of production of polymer composite materials, reinforced by fibers. Patent № 2102407 RU// BI. – 1998, № 2.

[11] Stoudentsov V.N., Komarova M. Gh., Poltoretski E.V., Kojevnikova T.N. The way of production of polymer composite material. Patent № 2119502 RU// BI. – 1998, № 27.

[12] Stoudentsov V.N., Mizintsov A.A. The way of production of polymer composite material, reinforced by fibers. Patent № 2182079 RU// BI. – 2002, № 13.

[13] Rodin Yu.P., Molchanov Yu.M. Influence of conformational changes, called by homogeneous constant magnetic field, on process of hardening of epoxy resin // Mechanics of composite materials. – 1988, № 3. – P. 495 – 502.

[14] Stoudentsov V.N. Device for pressure and magnetic treatment. Patent № 2162391 RU// BI. – 2001, № 3.

[15] Tcheremouhina I.V., Stoudentsov V.N. The way of production of polymer composite materials, reinforced by fibers. Patent № 2280655 RU// BI. – 2006, № 21.

[16] Stoudentsov V.N., Levin R.V., Skoudajev E.A., Doroshenko L.M. The way of production of polymer electret. Invention № 2005139922 RU from 20.12.2005.

In: Monomers, Oligomers, Polymers, Composites... ISBN: 978-1-60456-877-6
Editors: R. A. Pethrick, G.E. Zaikov et al. 

*Chapter 21*

# TECHNOLOGICAL AND ECOLOGICAL ASPECTS OF THE PRACTICAL APPLICATION OF QUATERNARY AMMONIUM SALTS IN RUSSIA IN PRODUCTION OF SYNTHETIC EMULSION RUBBERS

***V.M. Misin and S.S. Nikulin***
Emanuel Institute of Biochemical Physics of RAS, Moscow, Russia

Coagulation of the rubber emulsion is one of the main stages during production of synthetic emulsion rubbers. The routine quite efficient way of emulsion rubbers extraction from latexes is the application of inorganic salts (first of all, NaCl) under acidification of coagulating system with mineral acid [1]. It is known that for coagulation of 1 ton of the industrial emulsion of different rubbers (butadiene-styrene, butadiene-nitrile, polybutadiene ones or some others) it is necessary to use from 250 to 1500 kg of NaCl. Mineral salts in the process of wastewater treatment at the waste disposal plants are not decomposed and entrapped but are drained to the natural wells. This results in pollution of the environment, soil and drinking water salinization . For example, if production capacity of emulsion butadiene-styrene rubbers is of 100000 tons per year waste discharge of salts in the form of aqueous solution attains 30000 tons per year just for the extraction workshops. Thus, the annual waste discharge from the extraction workshops at emulsion rubbers production to the natural wells is of hundreds thousand tons of NaCl and other salts making an irreversible ecological damage.

The most efficient way for the perfection of technology of rubbers extraction from latexes is the elaboration of principally new coagulating agents providing a decrease of salt components consumption or their complete elimination from the technological process. Such coagulating agents can be water-soluble ammonium salts, particularly, quaternary ammonium salts (QAS).

However, only in Russia a full-scale investigation work has been made concerning:

- the study of physico-chemical processes of latexes coagulation for industrial rubbers production with the use of cationic polyelectrolytes;

- the study of rubbers properties, their mixtures and vulcanizers;
- applied investigations on the study of a possible application of cationic polyelectrolytes in industry.

Previously, flocculation mechanisms of polystyrene latexes [2-4] and some rubbers [4-7] with the use of QAS were investigated. However, the works aimed at solution of the chemical and technological problems of the industrial production of synthetic emulsion rubbers were made later in Russia.

This article presents the results of investigations of the authors concerned with the application of different QAS's as coagulating and flocculating agents of industrial latexes used in the production of synthetic rubbers. The works that are directly aimed at the elaboration of new approaches to the technology of extraction of synthetic rubbers from butadiene-styrene latexes are considered in most details.

According to the abbreviations accepted in Russia emulsion rubbers are denoted as follows

- SKS-30 – butadiene-styrene rubber (styrene content is 30 %);
- SKMS-25 – butadiene-($\alpha$-methyl) styrene rubber (styrene content is 25 %);
- EPB – emulsion polybutadiene;
- SKN-26 SM – butadiene-nitrile rubber (acrylonitrile content is 26 %);
- Letters A, R, K, P, M, O, C after the numbers denote: A - means the rubber of low-temperature polymerization, R - means regulated, K – colophony emulsifier, P – waxed emulsifier, M – oil-filled, O – oxide-filled, C – carbon-filled (technical carbon), respectively.

## Low-Molecular Ammonium Salts

In [8] a reduction of coagulating activity of ammonium halogenides was observed in a series of $NH_4F > NH_4Cl > NH_4Br > NH_4J$ for the discharge intensity of these salts equal to 20, 25, 50 and 100 kg/ton of rubber (pH = 2.5-3.0; temperature of 60°C). Efficiency of SKS-30 ARK and ARKPN rubber extraction from latexes in the presence of $NH_4Cl$ in a dependence on different parameters was studied in details in [9]. The estimation of the rubber properties as well as compounded rubbers and vulcanizing agents on their basis demonstrated that these properties do not change in fact under the change of the usually applied coagulating agent NaCl by $NH_4Cl$ (see table 1).

By the example of SKS-30 ARKP and SKS-30 ARK latexes it was shown [10,11] that with the use of $Me_4NCl$, $Et_4NCl$, $Et_4NBr$ and n-$Bu_4NI$ a complete coagulation of latex can be achieved for consumption standards of 60 ÷ 150 kg/ton of a rubber. These values are by 2-5 times less than consumption standard for the industrial coagulating agent – NaCl.

As a whole, a coagulating ability of all the low-molecular salts was not very high.

**Table 1. Properties of SKS-30 ARKPN rubbers extracted with the use of NaCl and $NH_4Cl$ as well as rubber compounds and rubber resins**

| Quality performance | Coagulating agents | |
|---|---|---|
| | NaCl | Ammonium chloride |
| Mooney viscosity | 45 | 46 |
| Conditional toughness under stretching, Mpa | 25,0/26,9 | 27,5/26,9 |
| Relative extension under fracture, % | 690/670 | 675/650 |
| Relative residual deformation after fracture, % | 16/18 | 14/16 |
| Rebound elasticity, % | 42/40 | 41/39 |
| Content of antioxidant (Agidol-2), % | 1,0/1,0 | 1,2/1,0 |
| Mass fraction of organic acids, % | 4,92 | 6,06 |
| Mass fraction of saponaceous organic acids, % | 0,16 | 0,05 |
| Loss of mass under drying, % | 0,17 | 0,12 |
| Mass fraction of the bound styrene, % | 22,5 | 22,5 |

## Mixtures of the Routine Coagulating Agents with Poly-N,N-Dimethyl- N,N-Diallylammonium Chloride

By the example of SKS-30 ARKP (ARK, AKO, ARKM) and butadiene-(α-methyl) styrene latexes SKMS-30 ARKP (ARK, ARKM) of industrial rubbers [12-15] it was shown that addition of PDMDAACl to the routine coagulating agents (NaCl, leather glue, protein hydrolyzate) allowed:

- to provide long-term preservation of protein coagulating agents in aqueous solutions at 20-22°C without their decomposition and appearance of unpleasant smells;
- a part of any of coagulating agents in a composition of 2- or 3-component mixture was considerably less than under their individual usage; that allowed to reduce consumption of a leather glue by 1.5-2.0 times;
- application of NaCl- PDMDAACl mixtures allowed to reduce NaCl consumption by 5-10 times up to the values of 20-50 kg/ton of rubber;
- complete coagulation was attained in a wider range of pH values pH = 2.5-4.5 making it possible to reduce consumption of $H_2SO_4$ from 15-18 to 8-12 kg/ton of rubber;
- consumption rate of PDMDAACl was of 0.4-1.2 kg/ton of rubber;
- extracted rubbers satisfied all-Union State Standard (GOST) and technical specifications (TU) requirements for the corresponding grade marks of rubbers;
- properties of vulcanizates did not yield to the properties of check samples.

All of the oil-filled rubbers and their vulcanizates satisfied the requirements of the Russian national standards for the corresponding grade marks of rubbers [13]. The role of PDMDAACl was a decisive one in the experiments and if the dosage of PDMDAACl was more than 2 kg/ton of rubber the possibility of a complete elimination of NaCl application was demonstrated in this case.

## POLY-N,N-DIMETHYL- N,N-DIALLYLAMMONIUM CHLORIDE

Since according to [12-15] an individual PDMDAACl is an efficient flocculating agent for the industrial emulsions of rubbers a large amount of investigations was performed concerning the influence of different parameters (latex and coagulating agent concentrations, polyelectrolyte consumption and its molecular mass, temperature of the process) on the process of extraction and on the properties of different rubber grade marks – SKS and SKMS. The rubbers extracted with the use of NaCl were used as reference samples [16-20].

Mass of the extracted coagulum was shown to increase with an increase of PDMDAACl amount added into the latex [17, 18]. Flocculation completeness was achieved for the consumption rate of PDMDAACl ≈ 4 kg/ton of rubber and rate of application for $H_2SO_4$ ≈ 15 kg/ton of rubber. Here the application rate of cationic polyelectrolyte required for attaining of a complete SKS-30 ARK latex coagulation depends on temperature: optimal coagulation temperature was 60°C. Application of higher temperatures did not result in a considerable increase of a coagulum yield. For the temperatures of 20 and 80°C flocculation curves did not in fact depend on the value of PDMDAACl molecular mass (172000, 62000 and 16000). Concentration of the initial aqueous solution of cationic polyelectrolyte did not have any considerable effect on its consumption rate necessary for a complete extraction of a rubber from latex.

The extracted rubbers very slightly differed from the check rubber samples by their chemical composition and satisfied the requirements of Russian standards. The main quality coefficients of vulcanizates on the basis of the experimental and check samples were equivalent (see table 2). However, rubber compounds on the basis of SKS-20 ARK rubber extracted with the use of PDMDAACl were vulcanized more rapidly. According to [17-19] the role of vulcanization activators could belong to polymer QAS remained in a rubber after its flocculation and/or products of its interaction with the components of emulsion system.

Under investigation of flocculation of the emulsive polybutadiene (EPB) [21] it was found that the consumption rate value for PDMDAACl was 8.0 kg/ton of rubber. At 60-80°C the rate value was reduced up to 5.0 kg/ton of rubber. The change of concentration of the operation PDMDAACl solution from 2.0 to 45.0% did not have effect on its consumption rate value. For $H_2SO_4$ consumption rate value 11-15 kg/ton of rubber a complete latex flocculation takes place. But the amount of coagulum was regularly reduced from 96-98% to 94-96% under decrease of $H_2SO_4$ consumption rate value from 10 to 9 kg/ton of rubber. As a whole, the process of flocculation is less sensitive to the dosage of $H_2SO_4$ than under the use of NaCl as a coagulating agent. Rubbers, rubber compounds and vulcanizates of EPB extracted with the use of PDMDAACl and NaCl were equivalent by the main quality parameters. Just as in case of butadiene-styrene rubbers EPB rubber compounds were vulcanized a little bit more rapidly (table 3) [21].

Similar regularities were found for latex flocculation process of butadiene-nitrile rubber SKN-26 SM with the use of PDMDAACl [22] as well as for its two fractions with mean-viscous molecular masses of $11.5 \cdot 10^4$, $2.3 \cdot 10^4$, $17.4 \cdot 10^4$, respectively.

**Table 2. Properties of SKS-30 ARK rubbers extracted with the use of PDMDAACl and NaCl as well as rubber compounds and vulcanizates based on these rubbers**

| Quality performance | Flocculating agent | |
|---|---|---|
| | PDMDAACl | NaCl |
| Mass content, %<br>free organic acids<br>bound organic acids<br>antioxidant VTS-150<br>ashes<br>bound styrene | <br>6,3-6,8<br>absent<br>1,3<br>0,12<br>22,5 | <br>5,8<br>0,15<br>1,3<br>0,18<br>22,5 |
| Mooney viscosity | 54 | 52 |
| Loss of mass at 105 °C, % | 0,13 | 0,19 |
| Duration of vulcanization, min | 60 | 80 |
| Elastic recovery, mm | 3,0 | 3,0 |
| Strain under 300% extension , MPa | 9,4 | 8,3 |
| Toughness under extension, MPA | 27,8 | 28,8 |
| Relative extension under fracture, % | 580 | 630 |
| Relative residual deformation after fracture, % | 14 | 10 |
| Rebound elasticity, % | 40 | 42 |

**Table 3. Properties of rubbers extracted with the use of NaCl and flocculating agent PDMDAACl as well as rubber compounds and vulcanizates based on oil-filled EPB**

| Quality performance | Flocculating agent | |
|---|---|---|
| | PDMDAACl | NaCl |
| Mass content, %<br>free organic acids<br>bound organic acids<br>antioxidant VTS-150<br>ashes<br>oils of PN-6K | <br>6,0<br>absent<br>0,25<br>0,12<br>15 | <br>5,6<br>0,10<br>0,25<br>0,18<br>15,0 |
| Loss of mass at 105 °C, % | 0,15 | 0,20 |
| Mooney viscosity | 40 | 40 |
| Strain under 300% extension , Mpa | 12,0 | 10,3 |
| Toughness under extension, MPA | 17,7 | 18,6 |
| Relative extension under fracture, % | 430 | 480 |
| Relative residual deformation after fracture, % | 6 | 10 |
| Plasticity | 0,37 | 0,37 |

## PDMDAACl Copolymer with Sulphur Dioxide

Flocculation activity of alternative low-molecular copolymer of N,N-dimethyl- N,N-diallylammonium chloride with sulphur oxide (PDMDAACl-OS) was studied for the industrial latex samples of SKS-30 ARK and EPB [23]. Completeness of the flocculation for SKS-30 ARK latex was attained at the flocculating agent supply rate of 18-20 kg/ton of a rubber, while for EPB latex – 14-15 kg/ton of rubber. Introduction of the increased amount of sulphuric acid as the acidifying agent under the optimal discharge rate of PDMDAACl-OS did not have a considerable effect on the amount of the obtained coagulum. However, the influence of $H_2SO_4$ dosage was more significant under reduced discharge rate of the flocculating agent. For example, discharge rate of PDMDAACl-OS was almost twice reduced up to 9.0 kg/ton of rubber for the discharge rate of sulphuric acid equal to 8.0 kg/ton of rubber. Rubber compounds on the basis of SKS-30 ARK rubber resin extracted from the latex with the flocculating agent of PDMDAACl-OS were vulcanized more rapidly than the check sample due to the presence of the flocculating agent or the products of its interaction with the components of emulsion system. Physico-mechanical quality indexes of vulcanizates on the basis of SKS-30 ARK rubber correspond to the requirements of Russian standards.

Similar regularities were observed under flocculation of EPB latex. For the discharge rate of PDMDAACl-OS 10.8 kg/ton of a rubber completeness of the flocculation for EPB latex attained under the discharge rate of $H_2SO_4$ 6.0 kg/ton of a rubber. Thus, the use of PDMDAACl-OS as a flocculating agent would require its high-precision dosage.

## Poly-(N,N-Dimethyl-2-Oxypropyleneammonium) Chloride

Dependence of the flocculation process for SKS-30 ARK latex on the concentration of dispersed phase in the presence of poly-(N,N-dimethyl-2-oxypropyleneammonium) chloride (PDMOPACl) was investigated in [24]. The formula of this salt is:

$$\begin{array}{l} \quad\;\; CH_3 \\ \quad\;\;\; | \\ [-^{+}N{-}CH_2{-}CH{-}CH_2{-}]_n \\ \quad\;\;\; |\, Cl^- \quad\;\; | \\ \quad\;\; CH_3 \qquad OH \end{array}$$

A maximum (optimal point) of flocculation was found just as in case of latexes extracted with the use of PDMDAACl. According to [24] this maximum was related with two factors (neutralization and bridge ones) that can influence on the flocculation mechanism. The value of the concentration of dispersed phase in the range of values 50, 100, 150 g/l was insignificant in the discharge rate value of the flocculating agent.

The change of temperature did not have a considerable effect on the process of rubber extraction from latex. Nevertheless, a slight increase of mass of the forming coagulum was observed with the increase of temperature from 20 to 80°C at the initial stage of the extraction process (for low discharge rate of PDMOPACl). Decrease of PDMOPACl discharge rate from 4 to 3 kg/ton of rubber allowed to attain a complete flocculation of SKS-30 ARK latex only in the case of the process performance at high temperatures of 80-95°C and discharge rate of acidifying agent up to 30 kg/ton of rubber. If the discharge rate of PDMOPACl was reduced up to 2 kg/ton of rubber, complete extraction of the rubber was not achieved even at these temperatures.

Discharge rate of acidifying agent demonstrated a greater influence on the flocculation process than the temperature and concentration of dispersed phase in the investigated intervals of the process conditions. For example, under discharge rate of PDMOPACl ~ 4 kg/ton of rubber mass of the formed coagulum was regularly increased with an increase of amount of the introduced $H_2SO_4$. Coagulum mass attained ~ 100% for the discharge rate of sulphuric acid 15 kg/ton of rubber that proved to be the optimal technological parameters of the process.

Rubber compounds and vulcanizates did not surrender the check sample (table 4) [24].

**Table 4. Properties of rubber compounds and vulcanizates on the basis of SKS-30 ARK rubber extracted with the use of NaCl and PDMOPACl flocculating agents and rubber resins**

| Quality performance | NaCl | PDMOPACl |
|---|---|---|
| Mooney viscosity | 53 | 42,5 |
| Carrer placticity, arb. un | 0,30 | 0,28 |
| Recovery, mm | 1,86 | 1,84 |
| Optimum of vulcanization at 143 °C, min | 80 | 60 |
| Conditional strain under 300% stretching, Mpa | 8,4 | 14,0 |
| Conditional toughness under stretching, Mpa | 27,0 | 27,6 |
| Relative extension under fracture, % | 600 | 540 |
| Relative residual deformation, % | 16 | 15 |
| Rebound elasticity, % | | |
| at 20 °C | 37 | 32 |
| at 100 °C | 50 | 46 |
| Shore hardness, arb. un. | 59 | 65 |
| Shopper-Schlobach abrasion, $10^{-3}$ $cm^3/m$ | 1,80 | 1,31 |
| Resistance to the growth of cuts up to 12 mm with a puncture, thousands of cycles | 39200 | 115200 |
| Conditional toughness under stretching after ageing (100°C, 72 h) | 18,0 | 20,0 |
| Relative extension after ageing (100°C, 72 h) | 242 | 257 |

## IMPLEMENTATION OF THE NEW TECHNOLOGY IN THE INDUSTRY OF RUSSIA

Performed investigations demonstrated a high efficiency of the application of QAS's as flocculating agents. According to a set of properties the most perspective one proved to be PDMDAACl. As a result of the complete elaboration of the technological conditions for the process of rubber extraction from latex in the laboratory it was demonstrated:

- PDMDAACl is a high-efficient flocculating agent for rubber latexes;
- optimal technological parameters of the flocculating process for latex have been elaborated;
- application of cationic polyelectrolyte PDMDAACl does not have a negative effect on the properties of the obtained rubbers, rubber compounds and vulcanizates on the basis of these compounds.

During 1992-94 period of time realization of the obtained laboratory results was performed at the synthetic rubber plant, namely:

- operating modes of the industrial flocculation for SKS-30 latex were completely elaborated;
- more than 372 tons of SKS-30 rubber was produced;
- application of PDMDAACl did not require considerable changes in the technology of rubber extraction from latexes as well as no any capital investments were needed;
- factory treatment works operated in normal mode;
- the produced rubber satisfied Russian standards.

Thus, in case of transition to a new technology of all the manufacturers of emulsion rubbers of only SKS marks in Russia (production volume is about 240 thousands of tons) the annual ecological damage can be reduced due to:

- the absence of the discharge of 100000-135000 tons/year of sodium chloride;
- decrease of the amount of industrial waste discharge by 8-10% (240000 $m^3$ per year).

As a result of the performed investigations the fundamental of a new ecologically reasonable industrial technology of latex flocculation for the emulsion rubbers were elaborated for the first time in the world with the use of cationic polyelectrolyte poly-N,N-dimethyl- N,N-diallylammonium chloride. This technology was realized in industry in Russia.

## REFERENCES

[1] Kirpichnikov P.A. et al. Chemistry and Technology of Synthetic Rubbers, Russia, Leningrad: *Chemistry,* 1987, 424 P.

[2] Baran A.A. Dokl. AN Ukr.SSR, 1979, ser. B, № 7, p. 529-533

[3] Verezhnikov V.N., Nikulin S.S., Misin V.M., Pojarkova T.N. Russian Polymer News, 1999, v.4, № 4, P. 36-41

[4] Verezhnikov V.N., Kashlinskaya P.E., Pojarkova T.N. *Colloid. Journal* (rus), 1991, v. 53, № 5, P. 822-825

[5] Solomentseva I..M., Teslenko A.Ya., Baran A.A. et al. *Khimia i tekhnologia vody* (rus), 1983, v. 5, № 5, P. 459-462

[6] Patents of the USSR 859377, 1065424

[7] European patent 84837

[8] Nikulin S.S., Verezhnikov V.N., Pojarkova T.N. et al. Production and usage of elastomers (rus), 1997, № 4, P. 10-12

[9] Nikulin S.S., Verezhnikov V.N., Pojarkova T.N. *Chemical technology.* (rus), 2005, № 9, P. 16-19

[10] Nikulin S.S., Verezhnikov V.N., Pojarkova T.N. Procs. of VI regional conf. "Problems of chemistry and chemical technology", Russia, Voronezh, 1998, v. 3, P. 42-46.
[11] Nikulin S.S., Verezhnikov V.N., Pojarkova T.N. et al. *Journal of applied chemistry.* (rus), 1999, Iss. 72, № 7, P. 1188-1191
[12] Garshin A.P., Nikulin S.S., Shapovalova N.N. et al. Production and usage of elastomers (rus), 1994, № 11, P. 2-6
[13] Garshin A.P., Nikulin S.S., Naumova Yu.M. et al. Production and usage of elastomers (rus), 1995, № 6, P. 14-18
[14] Garshin A.P., Nikulin S.S., Shapovalova N.N. et al. Production and usage of elastomers (rus), 1994, № 12, P. 9-14
[15] Nikulin S.S., Verezhnikov V.N., Misin V.M., Pojarkova T.N. *Transactions "Technology. Series. Constructions from composite materials"* (rus), 1998, № 3-4, P. 44-46
[16] Nikulin S.S., Verezhnikov V.N., Pojarkova T.N., Vostrikova G.Yu. *Journal of applied chemistry.* (rus), 2000, Iss. 73, № 10, P. 1720-1724
[17] Nikulin S.S., Verezhnikov V.N., Pojarkova T.N., Dankovtsev V.A. Rubber and vulcanizates (rus), 2000, № 5, P. 2-4
[18] Nikulin S.S., Verezhnikov V.N., Misin V.M., Pojarkova T.N. Russian Polymer News, 2002, V. 7, № 1, P. 1-6
[19] Verezhnikov V.N., Nikulin S.S., Pojarkova T.N., Vostrikova G.Yu. *Journal of applied chemistry.* (rus), 2002, v. 75, Iss. 3, P. 472-475
[20] Verezhnikov V.N., Nikulin S.S., Pojarkova T.N., Misin V.M. Essential Results in Chemical Physics and Physical Chemistry, 2005, Nova Science Publishers Inc., NY, Chapter 10, P. 123-134
[21] Nikulin S.S., Verezhnikov V.N., Pojarkova T.N., Dankovtsev V.A. *Journal of applied chemistry.* (rus), 2000, v. 73, Iss. 5, P. 833-836
[22] Verezhnikov V.N., Vostrikova G.Yu, Pojarkova T.N. *Journal of applied chemistry.* (rus), 2003, v. 76, Iss. 8, P. 1359-1362
[23] Verezhnikov V.N., Nikulin S.S., Pojarkova T.N., Misin V.M. *Journal of applied chemistry.* (rus), 2001, v. 74, Iss. 7, P. 1191-1194
[24] Nikulin S.S., Pojarkova T.N., Misin V.M. *Journal of applied chemistry.* (rus), 2004, v. 77, Iss. 6, P. 996-1000

In: Monomers, Oligomers, Polymers, Composites…
Editors: R. A. Pethrick, G.E. Zaikov et al.
ISBN: 978-1-60456-877-6


*Chapter 22*

# Fibrous Materials - As the Technological Additive in Manufacture of Butadien-Styrene Rubbers and Elastoplastics

***S.S. Nikulin, I.N. Pugacheva, V.M. Misin and V.A. Sedyh***
Emanuel Institute of Biochemical Physics of RAS, Moscow, Russia

Preservation of an environment is one of the major problems of mankind. It is connected to increase of extraction and processing of natural resources. Growth of industrial potential is accompanied by education and accumulation of a significant amount of waste products. For manufacture of a target industrial output it is spent about 1/3 consumed source of raw materials, 2/3 make waste products and by-products. Amplification technological influences on a nature has caused a line of problems of ecological character.

Last years the heightened interest to fibrous a fillers which raw sources are huge is shown. Waste products of fibrous materials in plenties are formed at textile factories and the enterprises of a clothing industry: tangled fibrous, the ends of a yarn and a string, a rag, failures, combings and pile from a hairstyle of artificial fur etc. [1, 2]. Therefore an actual technical task is search of the most perspective directions in application of waste products of fibrous materials [1].

In various industries some part of formed textile waste products is used: strongly littered cotton fibrous waste products staunch oil wells at drilling. Rags and scraps can be used in manufacture of roofing felt and roofing material. Difficultly utilized lining materials are exposed of turning into fibrous and are used as cotton wool as fillers in manufacture of building materials, in quality is warm and soundproofing basis under linoleum [3].

In the published works [4, 5] fibrous fillers entered on rollers during preparation of rubber mixes. Thus rubber mixes got necessary rigidity, and durability their parameters vulcanizates в were improved. However such technology introduction fibrous filler did not allow to achieve uniform distribution filler in volume of a rubber mix that was reflected in physicomechanical parameters vulcanizates. Besides introduction fibrous filler on rollers demanded additional expenses of energy.

Therefore more interesting the following way of reception of the rubbers filled with fibres is represented:

1. preliminary introduction of a fibre in latex of rubber;
2. the subsequent coagulation of the latex filled with a fibre with reception of the rubber filled with a fibre;
3. preparation from rubber of a rubber mix and her vulcanization.

The purpose of the given work - studying of influence of the cotton, viscose and kapron fibres entered into latex butadiene-styrene of rubber of mark SKS-30 ARK, on process of coagulation, and also on properties of rubber, rubber mixes and them vulcanizates.

## EXPERIMENTAL PART

Process of allocation of rubber of latex studied on coagulator, representing the capacity supplied with mixing device. Into him loaded 20 ml of latex SKS-30 ARK (the dry rest ~ 18 %). Coagulator thermostated 15-20 minutes at temperature $60^0$C. After that entered 24 % a water solution coagulant - chloride of sodium. Finished process of coagulation by addition of 1-2 % of a water solution of a sulfuric acid up to size pH≈2.0-2.5. The dropped out deposit (coagulum) separated from a water phase filtering, washed out water, dried in a drying case at temperature 80-$85^0$C and weighed. Calculation of weight formed coagulum carried spent, proceeding from the dry rest of initial latex. After upholding a filtrate (or as it name in the industry of synthetic rubber, - "serum") in him visually defined possible presence small dispersion crumbs coagulum.

Fibrous filler (the cotton, viscose, kapron) entered into latex at different stages of process of coagulation:

- in a dry kind before addition coagulant in latex;
- moistened with water before addition coagulant in latex;
- in a solution tall soaps;
- simultaneously with coagulant as water-salt dispersion;
- simultaneously with a solution of a sulfuric acid as water-acid dispersion.

Length of a fibre (2; 5 and 10 mm) and (from 0.1 up to 1.0 % on weight of rubber) changed his contents during experiment.

Kinetic swelling vulcanizates filled with various fibres, studied in solvents of various polarity by the following technique. Cut out samples vulcanizates as squares in the size 1x1 sm which then weighed. Quantity of samples for each series of measurements - 5. Samples placed in solvents on eight. In each hour of them took out, made gauging their geometrical sizes and weighed. Last point of definition of the size and weights of a sample - in 24 hours. Then carried spent processing the received data:

- for a finding of a degree of swelling a (% of weights.) from weight swell a sample subtracted weight of an initial sample; the received weight of solvent divided into weight of an initial sample and multiplied on 100 %; from five received results for each sample chose the greatest (equilibrium) size $\alpha_{max}$;

- a constant of speed of swelling found under the formula

$k = (1/\tau) \cdot (\ln [\alpha_{max} /(\alpha_{max} - \alpha_{\tau})])$ , where
$\tau$ – time, (h); $\alpha_{\tau}$ - size of the current degree of swelling in time $\tau$.

## Results and Their Discussion

### Influence of a Way of Introduction, Quantity and the Size Filler on Process of Coagulation of Latex

At the first stage of experiment some kinds of processing fibrous filler before its mixture with latex butadiene-styrene rubber SKS-30 ARK were investigated. The size of fibres and their dosage maintained constants, accordingly: 2 mm and 0.5 % of weights of rubber. Fibrous filler entered to rubber five various ways:

1. Dry without preliminary processing. Filler entered directly into latex, mixed within 15-20 minutes. Then coagulated latex on above given technique.
2. Moistened with water. Filler immersed in a small amount of water for 5 minutes at temperature $20^0$C so that it only was moistened. Damp filler entered into latex, mixed within 15-20 minutes. Then latex coagulated.
3. Moistened with a solution tall soaps. In a small amount of 5 % of a water solution tall soaps for 5 minutes at temperature $20^0$C placed fibrous filler so that it was moistened. Moistened in a solution tall soaps filler entered into latex, mixed within 15-20 minutes. Then latex coagulated.
4. With a solution coagulant. For preparation such coagulation mixes fibrous filler entered into a solution coagulant (24% chloride of sodium) which is taken in the volume necessary for coagulation. Mixed within 15-20 minutes. Coagulation of latex carried spent with use as this coagulation mixes: a fibre + a solution of chloride of sodium.
5. With a solution of acidifying agent (2 % a sulfuric acid). Preliminary fibrous filler entered into a solution of a sulfuric acid which is taken in the quantity necessary for coagulation of latex. Mixed within 15-20 minutes. Then the received mix (a fibre + a sulfuric acid) used as acidifying agent, and coagulation carried spent by the standard technique.

Uniformity of distribution of fibres in polymer investigated by the following technique. A sample of latex with fibrous filler placed on glass and dried up in a drying case before formation latex a film having inclusions of a fibre. The samples received thus considered in an optical microscope. Morphology fixed with the help of the camera. Besides an estimation of distribution of a fibre in a matrix carried spent on cuts coagulum, latex formed after coagulation.

In samples in which the fibre entered without any preliminary processing (dry) or only moistened water contained, did not observe a positive effect consisting in uniform distribution of a fibre in a matrix of latex. Dry fibres in latex were confused and formed agglomerates which further at hashing stuck to a mixer. Fibres moistened with water, also were not in regular intervals distributed in a matrix, forming lumps.

Introduction of a dispersion of a fibre with a solution tall soaps rendered more favorable influence on uniformity of distribution of a fibre in volume of latex. Fibres settled down separately from each other in a sample, not forming congestions.

Introduction fibrous filler with coagulation resulted almost in uniform distribution filler though in a sample there were congestions of fibres.

At research of a sample into which entered fibres simultaneously with a sulfuric acid, uniform distribution of fibres in volume is revealed. Fibres settled down separately from each other, and congestions fibrous filler in a matrix did not observe.

Proceeding from above stated, it is possible to draw a conclusion, that introduction of a fibre dry or moistened with water not expediently. Therefore in the further researches these kinds of preparation of composite materials did not use. Thus, for the further experiment the following most rational kinds of processing fibrous filler were chosen: the moistened 5 % a solution tall the soaps entered with coagulant (24 % a solution of chloride of sodium) and entered about 2 % a sulfuric acid. It is necessary to emphasize, that preliminary processing of a fibre by a solution tall soaps is accompanied by additional expenses.

Introduction of a fibre simultaneously with coagulation or with acidifying agent is the most technological and perspective. Thus experiments with addition of a sulfuric acid showed on the one hand increase of an output(exit) коагулюма, and on the other hand uniform distribution of a fibre in volume of a formed crumb, in comparison with experiment with introduction of a fibre simultaneously with coagulant.

In table 1 dependences of completeness of coagulation of latex SKS-30 ARK on a way of input of various fibres are submitted at various charges of chloride of sodium.

The positive effect at introduction of a cotton fibre simultaneously with a sulfuric acid, can be connected by that the basic component of the given fibre / cellulose (Cell) is capable to form sour ethers of the following structure with a sulfuric acid:

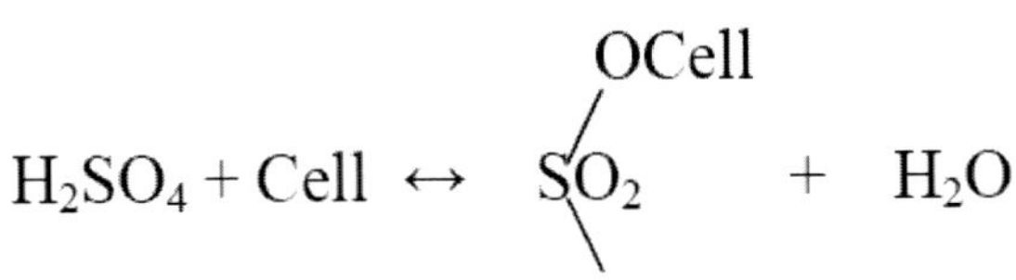


The given reaction is convertible, as ethers of sulfuric acid and cellulose easily are exposed to hydrolysis. Thus, in a water solution between a sulfuric acid and the cotton fibre containing up to 98 % of cellulose, there will be a balance. This phenomenon, most likely, also promotes more uniform distribution given fibrous filler in volume of acidifying agent. On the basis of it it is possible to assume, that process of coagulation at additional introduction in acidifying the agent of a cotton fibre proceeds on more complex mechanism demanding realization additional more of in-depth studies.

The similar explanation of the reason of occurrence of a positive effect is quite applicable also in case of introduction in latex of a fibre of viscose with a sulfuric acid. Really, at reception of a fibre of viscose from xanthate cellulose there is a decomposition xanthate to clearing groups – OH and restoration of significant number of cyclic parts of cellulose up to glucopyranose, characteristic as well for fibres of a cotton.

In view of the received results dependence of process of coagulation on quantity of the cotton, viscose or kapron fibre processed by a sulfuric acid and entered into latex at a finishing stage of coagulation further was investigated. The contents of a fibre in rubber maintained in quantity 0.1; 0.3; 0.5; 0.7; 1.0 % on weight of rubber. The received data are submitted in tables 2-4.

The analysis of experimental data has shown, that application fibrous filler during allocation of rubber from latex results in small increase of an output coagulum. Apparently, it is connected to two reasons:

- reduction of losses **коагулюма**, connected to ablation of a fine crumb of rubber together with serum;
- entry fibrous filler in coagulum.

Researches serum a method of a filtration have shown absence fibrous filler in him, that testifies to full capture of a fibre by a formed crumb of rubber.

The best results were received at a dosage of a fibre in limits from 0.3 up to 0.7 % on weight of rubber.

By results of researches the optimum length of a cotton, viscose and kapron fibre makes 2-5 mm. Change of length of a fibre in this interval renders insignificant influence on change of weight selected coagulum.

It is visually marked, that in separated from serum the contents of a fine crumb of rubber is less than rubber, than in control experiences without application fibrous fillers. It allows:

- to reduce losses of rubber with throw sewage and by that to increase productivity of technological process;
- to reduce an ecological load by clearing constructions.

However the final conclusion about size of losses of a fine crumb can be made after introduction of this technology commercially as in laboratory conditions exact reproduction of existing industrial technological process is not obviously possible.

**Table 1. Influence of a way of input of various fibres (2 mm) on completeness of coagulation (% of weights) latex SKS-30 ARK at various charges of chloride of sodium**

| The charge of chloride of sodium, kg/t of rubber | The contents formed coagulum at various ways of input of a fibre, % of weights. | | | | | | | | | |
|---|---|---|---|---|---|---|---|---|---|---|
| | Without a fibre (control) | Preliminary processed by a solution tall soaps | | | With coagulant | | | With acidifying agent | | |
| | | cotton | viscose | kapron | cotton | viscose | kapron | cotton | viscose | kapron |
| 25 | 8.93 | 13.58 | 11.77 | 9.22 | 8.73 | 9.97 | 9.01 | 11.51 | 10.63 | 10.89 |
| 50 | 21.37 | 19.64 | 19.58 | 22.13 | 15.53 | 22.87 | 18.94 | 19.72 | 21.77 | 22.63 |
| 75 | 32.78 | 28.86 | 34.02 | 31.25 | 25.87 | 22.97 | 24.47 | 29.37 | 31.47 | 32.65 |
| 100 | 62.71 | 40.29 | 47.67 | 41.11 | 40.33 | 43.20 | 38.11 | 45.90 | 46.62 | 46.81 |
| 125 | 80.63 | 79.61 | 77.83 | 78.75 | 68.85 | 82.84 | 75.45 | 79.82 | 79.61 | 84.91 |
| 150 | 93.41 | 94.05 | 94.62 | 90.52 | 92.53 | 90.21 | 88.21 | 97.54 | 95.76 | 99.60 |

The note: a dosage of fibres of 0.5 % of weights on rubber; a solution tall soaps (5 %); coagulant - a solution of chloride of sodium (24 %); acidifying agent - a sulfuric acid (2 %).

**Table 2. Influence of size of the contents (% of weights) and lengths (mm) of the cotton fibre entered with acidifying agent, on completeness of coagulation (% of weights) rubber SKS-30 ARK from latex at various charges of chloride of sodium**

| The charge of chloride of sodium, kg/t of rubber | The contents of a fibre, % of weights on rubber | | | | | | | | | | | | | | | |
|---|---|---|---|---|---|---|---|---|---|---|---|---|---|---|---|---|
| | 0 % | 0.1 % | | | 0.3 % | | | 0.5 % | | | 0.7 % | | | 1.0 % | | |
| | | 2 mm | 5 mm | 10 mm | 2 mm | 5 mm | 10 mm | 2 mm | 5 mm | 10 mm | 2 mm | 5 mm | 10 mm | 2 mm | 5 mm | 10 mm |
| 25 | 8.93 | 11.14 | 12.65 | 12.14 | 12.15 | 12.56 | 12.29 | 11.51 | 13.39 | 11.89 | 13.20 | 12.78 | 13.08 | 13.26 | 12.64 | 11.75 |
| 50 | 21.37 | 20.36 | 21.02 | 19.73 | 20.5 | 20.78 | 19.91 | 19.72 | 21.40 | 20.99 | 20.85 | 23.68 | 21.87 | 20.31 | 23.11 | 21.02 |
| 75 | 32.78 | 31.37 | 30.21 | 29.44 | 29.3 | 31.49 | 28.74 | 29.37 | 31.56 | 28.78 | 29.66 | 32.05 | 29.65 | 29.17 | 31.93 | 30.14 |
| 100 | 62.71 | 42.49 | 40.97 | 40.80 | 45.7 | 43.45 | 40.16 | 45.90 | 41.69 | 42.26 | 45.48 | 41.67 | 45.12 | 43.34 | 41.26 | 39.53 |
| 125 | 80.63 | 88.12 | 80.16 | 75.12 | 87.5 | 76.51 | 75.50 | 79.82 | 73.85 | 76.12 | 88.13 | 75.92 | 78.11 | 80.32 | 78.44 | 76.48 |
| 150 | 93.41 | 96.00 | 95.24 | 93.85 | 95.6 | 97.78 | 94.30 | 97.54 | 97.13 | 95.92 | 95.12 | 97.93 | 95.94 | 95.09 | 95.43 | 93.20 |

**Table 3. Influence of size of the contents (% of weights) and lengths (mm) of the viscose fibre entered with acidifying agent, on completeness of coagulation (% of weights) rubber SKS-30 ARK from latex at various charges of chloride of sodium**

| The charge of chloride of sodium, kg/t of rubber | The contents of a fibre, % of weights on rubber | | | | | | | | | | | | | | | |
|---|---|---|---|---|---|---|---|---|---|---|---|---|---|---|---|---|
| | 0 % | 0.1 % | | | 0.3 % | | | 0.5 % | | | 0.7 % | | | 1.0 % | | |
| | | 2 mm | 5 mm | 10 mm | 2 mm | 5 mm | 10 mm | 2 mm | 5 mm | 10 mm | 2 mm | 5 mm | 10 mm | 2 mm | 5 mm | 10 mm |
| 25 | 8.93 | 12.60 | 11.02 | 9.26 | 12.60 | 9.94 | 7.68 | 10.63 | 10.01 | 8.08 | 9.44 | 10.34 | 9.76 | 11.70 | 11.58 | 10.69 |
| 50 | 21.37 | 23.08 | 22.79 | 20.38 | 29.50 | 20.93 | 20.65 | 21.77 | 20.97 | 19.87 | 21.52 | 23.33 | 20.89 | 20.34 | 22.33 | 17.77 |
| 75 | 32.78 | 31.57 | 28.29 | 32.14 | 29.76 | 30..96 | 28.7 | 31.47 | 31.61 | 30.42 | 33.71 | 30.22 | 34.16 | 29.07 | 28.82 | 26.91 |
| 100 | 62.71 | 44.31 | 43.65 | 42.40 | 46.67 | 46.13 | 39.5 | 46.62 | 39.67 | 37.89 | 45.29 | 46.89 | 42.28 | 44.38 | 41.30 | 37.68 |
| 125 | 80.63 | 83.55 | 84.93 | 89.18 | 92..59 | 92.67 | 81.63 | 79.61 | 84.82 | 83.94 | 86.34 | 85.67 | 82.99 | 70.11 | 75.39 | 88.59 |
| 150 | 93.41 | 97.66 | 98.70 | 96.46 | 95.41 | 99.03 | 96.11 | 95.76 | 97.67 | 95.09 | 94.38 | 96.93 | 94.83 | 93.28 | 93.51 | 94.08 |

**Table 4. Influence of size of the contents (% of weights) and lengths (mm) of the kapron fibre entered with acidifying agent, on completeness of coagulation (% of weights) rubber SKS-30 ARK from latex at various charges of chloride of sodium**

| The charge of chloride of sodium, kg/t of rubber | The contents of a fibre, % of weights on rubber | | | | | | | | | | | | | | | |
|---|---|---|---|---|---|---|---|---|---|---|---|---|---|---|---|---|
| | 0 % | 0.1 % | | | 0.3 % | | | 0.5 % | | | 0.7 % | | | 1.0 % | | |
| | | 2 mm | 5 mm | 10 mm | 2 mm | 5 mm | 10 mm | 2 mm | 5 mm | 10 mm | 2 mm | 5 mm | 10 mm | 2 mm | 5 mm | 10 mm |
| 25 | 8.93 | 10.64 | 9.49 | 11.42 | 10.49 | 10.11 | 12.92 | 10.89 | 9.77 | 10.67 | 11.38 | 9.72 | 9.77 | 10.54 | 9.21 | 11.36 |
| 50 | 21.37 | 21.13 | 19.84 | 22.96 | 21.35 | 22.5 | 20.46 | 22.63 | 20.13 | 20.92 | 22.51 | 20.67 | 22.67 | 24.07 | 21.35 | 22.41 |
| 75 | 32.78 | 30.94 | 28.78 | 32.38 | 31.32 | 29.10 | 33.97 | 32.65 | 32.99 | 28.23 | 28.16 | 29.10 | 35.93 | 31.62 | 29.37 | 31.10 |
| 100 | 62.71 | 40.12 | 38.35 | 42.15 | 39.55 | 41.21 | 41.85 | 46.81 | 35.71 | 41.81 | 43.55 | 39.41 | 40.83 | 44.55 | 42.38 | 39.91 |
| 125 | 80.63 | 82.20 | 85.79 | 85.33 | 82.56 | 88.47 | 87.04 | 84.90 | 90.10 | 89.59 | 85.19 | 86.16 | 90.16 | 84.80 | 83.63 | 88.31 |
| 150 | 93.41 | 94.00 | 93.42 | 93.55 | 94.67 | 96.95 | 94.52 | 99.60 | 99.36 | 99.79 | 95.04 | 96.61 | 98.31 | 94.15 | 94.14 | 95.75 |

**Influences of a Cotton, Viscose and Kapron Fibre on Properties of Rubber Mixes and Vulcanizates**

In all experiments fibrous filler entered into latex simultaneously with acidifying agent at a finishing stage of process of coagulation. Research of properties of the allocated rubbers carried spent according to requirements of GOST 15627-79 on rubber SKS-30 ARK (tables 5-7). The received results compared to properties of standard samples - samples without fibres.

It agrees to the data given in table 5, introduction of a cotton fibre in polymer results in increase of disorder of a parameter of viscosity of rubber and a rubber mix, and also to growth of plasticity of a rubber mix.

Adsorption sewing (structure) rubber of agents (sulfur, accelerators, activators) on a surface of fibres results in reduction of a degree sewing vulcanizate. As a result of it there is a decrease of durability, a strain at 300 % lengthening and relative lengthening at break. At the contents of fibrous 0.7 % of weights increase the size of resistance of tear with 53 up to 56-61 kN/m and reduction of size of resistance to a repeated stretching is observed.

Growth of resistance to ageing ($100^0$C, 72 h) can be explained by end of process of vulcanization in result desorption to any part of sewing agents from a surface of fibres.

From the data given in table 6 it is visible, that presence of a viscose fibre results in insignificant decrease of viscosity as rubber with 55 up to 51 ÷ 53 units, and a rubber mix with 57 up to 55 ÷ 56 units. The minimal twisting moment with 4.8 (at a standard sample) up to 5.5 ÷ 6.5 H×m (table 8) is simultaneously increased. In a matrix of a rubber mix orientation of badly moistened fibres of viscose in a direction, perpendicular to a direction of the appendix of loading (calander effect) is not shown. It the increase rehabilitation samples with 1.4 up to 1 5÷1 8 mm speaks mainly at length of fibres of 5 mm and the contents of 0.7 % of weights. (table 6) and the reduction of creep of rubber mixes connected to it. Presence of additives of a fibre increases disorder of size of plasticity of a rubber mix within the limits of 0.31 ÷ 0.38 units (table 6).

For vulcanizates with a viscose fibre adsorption of a part of sewing agents on fibrous filler also takes place. Owing to pauperization of rubber by sulfur and accelerators the degree of lace vulcanizates decreases. In comparison with vulcanizates without fibrous filler it results to

- to increase of optimum time of vulcanization of composites with 27.5 (at a standard sample) up to 28 ÷ 30 minutes,
- to decrease of durability and a strain at 300 % lengthening,
- to increase relative lengthenings and residual deformation at break.

However the essential increase of sizes of resistance tear with 53 up to 71 ÷ 96 kN/m and resistance to a repeated stretching with 70 до72÷96 the thousand cycles (table 6), proportional to size of the contents and length of viscose fibres in composites is observed. It speaks display reinforcing effect of fibres, before revealed for vulcanizates, filled with fibres a stage mixing rubber [8].

**Table 5. Properties of rubbers, rubber mixes and rubbers on basis SKS-30 ARK, filled with a cotton fibre of various length (mm) and with the various contents (% of weights)**

| The name of a parameter | Without a fibre | 2 mm | | 5 mm | | 10 mm | |
|---|---|---|---|---|---|---|---|
| | | 0.3% | 0.7% | 0.3% | 0.7% | 0.3% | 0.7% |
| Muni's viscosity MV 1+4 ($100^0$C): | | | | | | | |
| - rubber | 55 | 57 | 56 | 52 | 56 | 52 | 55 |
| - a rubber mix | 57 | 58 | 57 | 56 | 57 | 55 | 58 |
| Karrer's plasticity , r/m, conditional units | 0.34 | 0.34 | 0.36 | 0.36 | 0.35 | 0.39 | 0.37 |
| Capacity for restoration, mm | 1.4 | 1.4 | 1.4 | 1.4 | 1.4 | 1.2 | 1.2 |
| Conditional strain at 300 % lengthening, MPa | 9.4 | 9.0 | 8.6 | 8.6 | 8.1 | 8.3 | 7.8 |
| Conditional durability at a stretching, MPa | 26.3 | 25.3 | 24.1 | 23.4 | 21.6 | 25.5 | 20.7 |
| Relative lengthening at break, % | 618 | 605 | 615 | 570 | 486 | 616 | 582 |
| Relative residual deformation, % | 12 | 14 | 12 | 10 | 10 | 14 | 12 |
| Elasticity on recoil, %: | | | | | | | |
| - $20^0$C | 40 | 39 | 39 | 39 | 34 | 34 | 34 |
| - $100^0$C | 53 | 52 | 55 | 54 | 57 | 52 | 60 |
| Shore's hardness, conditional units | 57 | 56 | 57 | 56 | 61 | 58 | 57 |
| Resistance tear, kN/m | 53 | 52 | 61 | 56 | 59 | 52 | 55 |
| Resistance to a repeated stretching, thousand cycles | 70 | 69 | 77 | 57 | 62 | 49 | 53 |
| Factor of ageing ($100^0$C, 72 h): | | | | | | | |
| - on durability | 0.44 | 0.51 | 0.48 | 0.53 | 0.51 | 0.46 | 0.60 |
| - on relative lengthening | 0.33 | 0.37 | 0.33 | 0.39 | 0.41 | 0.33 | 0.41 |

The note: temperature of vulcanization $143^0$C; duration of 60 minutes.

**Table 6. Properties of rubbers, rubber mixes and vulcanizates on basis SKS-30 ARK, filled with a viscose fibre of various length (mm) and with the various contents (% of weights)**

| The name of a parameter | Without a fibre | 2 mm | | 5 mm | | 10 mm |
|---|---|---|---|---|---|---|
| | | 0.3% | 0.7% | 0.3% | 0.7% | 0.3% |
| Muni's viscosity MV 1+4 ($100^0$C): | | | | | | |
| - rubber | 55 | 53 | 53 | 53 | 52 | 51 |
| - a rubber mix | 57 | 55 | 56 | 55 | 56 | 55 |
| Karrer's plasticity , r/m, conditional units | 0.34 | 0.38 | 0.31 | 0.32 | 0.35 | 0.36 |
| Capacity for restoration, mm | 1.4 | 1.4 | 1.7 | 1.8 | 1.5 | 1.4 |
| Conditional strain at 300 % lengthening, MPa | 9.4 | 3.9 | 4.9 | 6.3 | 5.8 | 6.0 |
| Conditional durability at a stretching, MPa | 26.3 | 24.3 | 20.0 | 23.0 | 22.0 | 23.0 |
| Relative lengthening at break, % | 618 | 698 | 690 | 680 | 675 | 688 |
| Relative residual deformation, % | 12 | 18 | 18 | 16 | 14 | 14 |
| Elasticity on recoil, %: | | | | | | |
| - $20^0$C | 40 | 43 | 43 | 46 | 44 | 44 |
| - $100^0$C | 53 | 49 | 50 | 53 | 50 | 51 |
| Shore's hardness, conditional units | 57 | 55 | 56 | 55 | 59 | 57 |
| Resistance tear, kN/m | 53 | 71 | 75 | 81 | 96 | 95 |
| Resistance to a repeated stretching, thousand cycles | 70 | 70 | 72 | 76 | 96 | 85 |
| Factor of ageing ($100^0$C, 72 h): | | | | | | |
| - on durability | 0.44 | 0.63 | 0.77 | 0.53 | 0.56 | 0.60 |
| - on relative lengthening | 0.33 | 0.45 | 0.43 | 0.33 | 0.33 | 0.41 |

The note: temperature of vulcanization $143^0$C; duration of 60 minutes.

**Table 7. Properties of rubbers, rubber mixes and rubbers on basis SKS-30 ARK, filled with a kapron fibre of various length (mm) and with the various contents (% of weights)**

| The name of a parameter | Without a fibre | 2 mm | | 5 mm | | 10 mm | |
|---|---|---|---|---|---|---|---|
| | | 0.3% | 0.7% | 0.3% | 0.7% | 0.3% | 0.7% |
| Muni's viscosity MV 1+4 ($100^0$C): | | | | | | | |
| - rubber | 55 | 57 | 56 | 55 | 57 | 57 | 58 |
| - a rubber mix | 57 | 59 | 58 | 59 | 58 | 58 | 60 |
| Karrer's plasticity , r/m, conditional units | 0.34 | 0.33 | 0.34 | 0.33 | 0.36 | 0.34 | 0.34 |
| Capacity for restoration, mm | 1.4 | 1.3 | 1.2 | 1.4 | 1.2 | 1.3 | 1.3 |
| Conditional strain at 300 % lengthening, MPa | 9.4 | 5.6 | 7.1 | 6.1 | 6.6 | 5.9 | 6.0 |
| Conditional durability at a stretching, MPa | 26.3 | 29.0 | 23.0 | 23.1 | 23.0 | 23.0 | 20.5 |
| Relative lengthening at break, % | 618 | 680 | 610 | 670 | 640 | 670 | 650 |
| Relative residual deformation, % | 12 | 16 | 14 | 14 | 14 | 14 | 14 |
| Elasticity on recoil, %: | | | | | | | |
| - $20^0$C | 40 | 38 | 42 | 40 | 42 | 40 | 43 |
| - $100^0$C | 53 | 50 | 52 | 52 | 50 | 50 | 50 |
| Shore's hardness, conditional units | 57 | 57 | 57 | 54 | 55 | 56 | 57 |
| Resistance tear, kN/m | 53 | 89 | 90 | 81 | 85 | 66 | 73 |
| Resistance to a repeated stretching, thousand cycles | 70 | 78 | 82 | 93 | 73 | 76 | 78 |
| Factor of ageing ($100^0$C, 72 h): | | | | | | | |
| - on durability | 0.44 | 0.65 | 0.67 | 0.72 | 0.69 | 0.80 | 0.96 |
| - on relative lengthening | 0.33 | 0.46 | 0.40 | 0.40 | 0.41 | 0.45 | 0.54 |

The note: temperature of vulcanization $143^0$C; duration of 60 minutes.

**Table 8. Dependence vulcanizations of rubber mixes on the basis of rubber SKS-30 ARK filled with a viscose and kapron fibre**

| Parameter | Standard sample (without additives) | The size (mm) and quantity (% of weights on rubber) fibres | | | | | | | | | | |
|---|---|---|---|---|---|---|---|---|---|---|---|---|
| | | Viscose fibre | | | | | Kapron fibre | | | | | |
| | | 2 mm | | 5 mm | | 10 mm | 2 mm | | 5 mm | | 10 mm | |
| | | 0.3 % | 0.7 % | 0.3 % | 0.7 % | 0.3 % | 0.3 % | 0.7 % | 0.3 % | 0.7 % | 0.3 % | 0.7 % |
| The minimal twisting moment, N × m | 4.8 | 6.3 | 6.0 | 5.5 | 5,5 | 6.5 | 8.5 | 7.5 | 7.0 | 7.5 | 7.5 | 7.0 |
| The maximal twisting moment, N × m | 36.5 | 30 | 32 | 34 | 33 | 36 | 31 | 34 | 32 | 34 | 32 | 32 |
| Time of the beginning of vulcanization, minutes. | 3.5 | 5 | 4 | 4 | 5 | 4 | 5 | 5 | 4 | 5 | 4 | 5 |
| Optimum time of vulcanization, minutes | 27.5 | 30 | 29 | 28 | 29 | 29 | 21 | 20 | 21 | 19 | 15 | 21 |

Growth of resistance to ageing (100$^0$C, 72 h) can be explained with end of process of vulcanization as a result of gradual desorption to any part of sewing agents (sulfur, accelerators, activators) from a surface of these fibres.

Anisotropy elastic - durability properties vulcanizates with fibres results in growth of disorder of hardness, and also to reduction of elasticity on recoil at 100$^0$C with 53 (a sample of comparison) up to 49÷51 % for viscose and up to 50÷52 % for kapron (tables 6, 7).

From table 7 it is visible, that presence of a kapron fibre at quantity from 0.3 up to 0.7 % of weights. And length 2÷10 mm results in insignificant increase of Muny viscosity: - rubber with 55 up to 56 ÷ 58 units; and a rubber mix with 57 up to 58 ÷ 60 units. At introduction of a kapron fibre in comparison with a viscose fibre there is a significant growth of the minimal twisting moment with 4.8 (at a standard sample) up to 7.0 ÷ 8.5 (kapron) against 5.5-6.5 N·m (viscose) (table 8). It indirectly specifies the big compatibility of a matrix butadiene-styrene rubber with a surface of the kapron entered into rubber with acidifying agent at a finishing stage of process of coagulation in comparison with viscose.

Orientation in a matrix butadiene-styrene rubber (SKS-30 ARK) more compatible fibres of kapron in a direction, perpendicular to a direction of the appendix of loading (calander effect), speaks decrease rehabilitation a rubber mix with 1.4 (without a fibre) up to 1.2 ÷ 1.3 mm (with a fibre) (table 7). Presence of additives of a kapron fibre limits disorder of size of plasticity of a rubber mix within the limits of 0.33 ÷ 0.36 units.

Adsorption on a surface of a kapron fibre of a part of sewing agents, results in reduction of a degree сшивки vulcanizates, to reduction of durability and a pressure(voltage) at 300 % lengthening in comparison with vulcanizates without a fibre, to growth of relative residual deformation and lengthening at break. It is marked decrease(reductions) of optimum time of vulcanization with 27.5 (at a standard sample) up to 15 ÷ 21 mines because of presence fibrous filler the basic character (polymer with amide groups) with properties of the accelerator of vulcanization (table 8 [ 10 ]).

The essential increase (table 7) of resistance tear with 53 up to 66 ÷ 90 kN/m and resistance to a repeated stretching with 70 up to 73-93 thousand cycles speaks display reinforcing effect of kapron fibres long 5-10 mm, before revealed for vulcanizates, filled with fibres a stage mixing rubber [9].

Growth of resistance to ageing (100$^0$C, 72 h) speaks end of process of vulcanization in result desorption parts of sewing agents from a surface of fibres.

## Studying Kinetics Swelling Vulcanizates, Filled with Fibres

The important parameter for an estimation of properties of received composites is presence or absence of interphase interaction of a matrix of rubber (vulcanizate) with a surface fibrous filler (viscose or kapron), allocation of rubber entered at a stage from latex.

With the purpose of an estimation of display of interphase interaction of additives of a fibre with a matrix of rubber investigated kinetic swelling filled vulcanizates on a basis butadiene-styrene rubber in solvents of various polarity. Interphase interactions estimated on size of an equilibrium degree of swelling ($\alpha_{max}$) and a constant of speed of swelling (k, $h^{-1}$) of samples vulcanizates, containing viscose or kapron filler. Used solvents n-oktan, toluene,

**Table 9. Influence solvents, sizes and quantity of addition viscose fibre and kapron fibre of an equilibrium degree of swelling ($\alpha_{max}$ , % of weights ) of vulcanizates and a constant of speed of swelling ($k, h^{-1}$) of samples vulcanizates on basis SKS-30 ARK**

| Size / quantity fibres in rubber, mm / % of weights | Viscose fibre | | | | | | Kapron fibre | | | | | |
|---|---|---|---|---|---|---|---|---|---|---|---|---|
| | n-octane | | toluene | | chloroform | | n-octane | | toluene | | chloroform | |
| | $\alpha_{max}$ | k | $\alpha_{max}$ | k | $\alpha_{max}$ | k | $\alpha_{max}$ | k | $\alpha_{max}$ | k | $\alpha_{max}$ | k |
| 2 / 0.3 | 168 | -0.93 | 360 | -1.43 | 675 | -1.07 | 140 | -1.38 | 272 | -1.73 | 550 | -1.38 |
| 5 / 0.3 | 162 | -0.93 | 316 | -1.43 | 600 | -1.07 | 142 | -0.97 | 276 | -1.73 | 525 | -1.38 |
| 10 / 0.3 | 145 | -0.93 | 294 | -1.43 | 582 | -1.43 | 134 | -1.19 | 252 | -1.20 | 520 | -1.66 |
| 2 / 0.7 | 150 | -1.19 | 302 | -1.33 | 590 | -1.33 | 128 | -1.73 | 270 | -1.73 | 540 | -2.14 |
| 5 / 0.7 | 154 | -1.19 | 312 | -1.33 | 614 | -1.43 | 120 | -1.43 | 243 | -0.84 | 514 | -2.25 |
| 10 / 0.7 | - | - | - | - | - | - | 122 | -1.54 | 262 | -0.93 | 524 | -1.96 |

chloroform (table 9) have parameter of solubility (ρ) accordingly 15.4 ÷18.2 and 18.8 (mJ) 0.5 × ($m^{-1.5}$) [9].

***Viscose Fibre***

Irrespective of a nature of solvent of unequivocal dependence of influence of the sizes and size of the contents of a viscose fibre in an interval from 0.3 up to 0.7 % of weights. On size $\alpha_{max}$ for vulcanizates it is not revealed.

With increase of polarity of solvents in a number(line) oktan < toluene < chloroform irrespective of the contents of a viscose fibre is established growth of sizes ($\alpha_{max}$) for these solvents accordingly in intervals 145 ÷ 168, 294 ÷ 360 and 582 ÷ 675 % of weights. The greatest sizes ($\alpha_{max}$), received in chloroform for vulcanizate with viscose filler, specify increase of polarity vulcanizate, containing viscose, and also on approach (approximation) of size of its average parameter of solubility $\rho_{vulc/viscose}$ to parameter of solubility of chloroform $\rho_{ch}$ = 18.8, in comparison with size $\rho_{\kappa rubber}$ = 17.4 $(mJ)^{0.5} \cdot (m^{-1.5})$ for initial unfilled and nonvulcanized rubber SKS-30 ARK [10]. The reason of it is presence of more polar viscose fibre ($\rho_{viscose}$ = 31.9 $(mJ)^{0.5} \cdot (m^{-1.5})$ in a matrix vulcanizate and occurrences additional vulcanization grids [11].

Apparent driving force - size $\alpha_{max}$ did not render influence on speed of swelling filled vulcanizate.

Distinction in thermodynamic compatibility of various solvents and filled with a viscose fibre vulcanizate was defined with speed of swelling of the last. For system « vulcanizate - octan » with the greater size of a square of a difference of parameters of solubility $\beta = (\rho_{rubber} - \rho_{octan})^2 = 4$ (mJ) · ($m^{-3}$) [9,10] smaller speed of swelling (k = 0.93 – 1.19 $h^{-1}$) is revealed. For system « vulcanizate - toluene » (or « vulcanizate - chloroform ») with smaller sizes β = 0.64 (или 1.96) (mJ) · ($m^{-3}$) speeds of swelling k were more: k = 1.33-1.43 (or 1.07 – 1.43 $h^{-1}$) accordingly.

The size β defined kinetic swelling vulcanizates in various solvents. For system « vulcanizate - chloroform » influence of the contents of a viscose fibre and the size of a fibre for speed of swelling is revealed. So doubling of the contents of a fibre with 0.3 up to 0.7 % of weights. In the length 2 and 5 mm has resulted in increase of size k with 1.07 up to 1.33-1.43 $h^{-1}$.

Thus, for viscose fibres of essential influence of their size and the contents in vulcanizate for speed of swelling in n-octane and toluene it is not established. Apparently, the reason was weak interphase interaction of a matrix of rubber with fibres, presence of boundary layers because of the big distinctions in $\rho_{rubber}$ and $\rho_{viscose}$, for which $\beta = (\rho_{rubber} - \rho_{viscose})^2 = 210$ (mJ) · ($m^{-3}$).

***Kapron Fibre***

For vulcanizates with a kapron fibre, irrespective of the contents in them fibres, with increase of polarity of solvents in a line octan < toluene < chloroform is established growth of size $\alpha_{max}$ accordingly, in intervals 120 ÷ 140, 243 ÷ 276 and 514 ÷ 550 % of weights. The greatest size $\alpha_{max}$, received in chloroform, specifies increase of polarity vulcanizate and on approach of his parameter of solubility $\rho_{vulc/kapron}$ to size $\rho_{ch}$ = 18.8 $(mJ)^{0.5} \cdot (m^{-1.5})$ in comparison with size r for initial unfilled and nonvulcanized rubber. The reason of it is presence of additives of a polar fibre at a matrix vulcanizate, for which size $\rho_{\Pi A} = 27.8\ (mJ)^{0.5} \cdot (m^{-1.5})$ [11].

Apparent driving force size $\alpha_{max}$ did not render influence on speed of swelling filled with a kapron fibre vulcanizate.

Distinction in thermodynamic compatibility of solvents and filled with a kapron fibre vulcanizate was defined with its speed of swelling. For system « vulcanizate -octan » with the greater size $\beta = (\rho_{rubber} - \rho_{octan})^2 = 4\ (mJ) \cdot (m^{-3})$ the interval of speed of swelling k = 0.97 – 1.54 $h^{-1}$ is characteristic. For system « vulcanizate - toluene » with smaller size $\beta = 0.64\ (mJ) \cdot (m^{-3})$ the interval of sizes of speed of swelling k has extended (k = 0.84 – 1.73 $h^{-1}$). For system «vulcanizate - chloroform » with $\beta = 1.96\ (mJ)\cdot(m^{-3})$ depending on size of the contents of a kapron fibre the interval of speed of swelling was displaced up to sizes k = 1.38 – 2.25 $h^{-1}$. So doubling of the contents of a fibre with 0.3 up to 0.7 % of weights, size 2 and 5 mm, has resulted in increase of size k with 1.38-1.66 up to 2.14-2.25 $h^{-1}$. Such increase of speed of swelling in chloroform with growth of the contents of a kapron fibre, apparently, speaks the big interphase interaction of a matrix of rubber with filler because of smaller distinctions $\rho_{rubber} = 17.4$ и $\rho_{kapron} = 27.8\ (mJ)^{0.5} \cdot (m^{-1.5})$, for which $\beta = (\rho_{rubber} - \rho_{kapron})^2 = 108\ (mJ) \cdot (m^{-3})$, owing to the high contribution of hydrogen connections at the presence of chloroform [9].

On the contrary, for the toluene which is not forming hydrogen connections, decrease of a constant of speed of swelling k is revealed at increase of the contents of fibres of kapron as a result of display of barrier properties by the boundary layers having high polarity.

## CONCLUSIONS

1. The new technological direction for the successful decision of an environmental problem - problems of recycling of waste products of the fibres, consisting in preliminary mixture of a fibre with acidifying agent before submission of latex on a finishing stage of his coagulation is offered.
2. Scientific bases of the offered approach providing reception of rubbers filled with fibres at a stage of manufacture emulsion of rubber are incorporated.
3. The expediency of the offered approach is experimentally shown:

- the offered approach allows to achieve uniform distribution of a fibre in volume of rubber that is positively reflected in separate properties received vulcanizates;
- the optimum length of a cotton, viscose and kapron fibre makes 2-5 mm, at its contents in rubber in limits from 0.3 up to 0.7 % of weights.;
- introduction of a cotton fibre does not worsen property vulcanizates, and introduction of a viscose and kapron fibre allows to increase such parameters vulcanizates, as stability to thermal ageing, repeated deformations, resistance tear without deterioration of other operational characteristics;
- influence of additives of a kapron fibre differs from influence of additives of cotton and viscose fibres on properties of rubber SKS-30 ARK and rubber mixes because of distinction of a nature of fibres;

- distinction of interphase interaction between a surface of a viscose and kapron fibre and a matrix vulcanizates butadiene-styrene the rubber, reflected for speed of swelling vulcanizate in solvents of a different nature.

4. Deviations реологических, vulcanization properties and physicomechanical parameters vulcanizates are explained at the presence of additives of a viscose and kapron fibre.

## References


[1] Ed. Pakshver Propertys and features of processing of chemical fibres. // Russia , Moscow: *Chemistry,* 1969. 400 P.

[2] Nemchenko E.A., Novikov N.A., Novikova S.A. etc. Properties of chemical fibres and methods of their definition. // Russia , Moscow: *Chemistry,* 1973. 216 P.

[3] Ozerova N.V. Recycling of textile waste products. The collection of materials V International scientific - practical conf. «Economy of wildlife management and preservation of surroundings» // Russia, Penza, 2002. 210 P.

[4] Jagnjatinskaja E.A., Goldberg B.B., Leonov V.V. etc. The manufacturing techniques, properties and features of application of rubbers with fibrous fillerми in elastoplastics // Russia , Moscow: *CNIITENeftehim,* 1979. 54 P.

[5] Hutareva G.V. Textile materials from chemical fibres for manufacture of the basic kinds elastoplastics In: G.V. Hutareva, V.L. Zhul'kov, I.I. Leonov // Russia, Moscow: *CNIITENeftehim,* 1983. 60 P. - (Industry elastoplastics: Subject review).

[6] Verezhnikov V.N., Nikulin S.S., Krutikov M.J., Poyarkova T.N. // *Kolloid. J (rus)* 1999. V. 61. N 1. P. 37-40.

[7] Nikulin S.S., Verezhnikov V.N., Poyarkova T.N., Dankovtcev V.A. // *Rubber and elastoplastic (rus)* 2000. N 5. P. 2-4.

[8] Reznichenko S.V. // *Rubber and elastoplastic* (rus) 2002. № 2. P. 38-43.

[9] Drinberg S.A., Icko E.F. Solvent for paint and varnish materials: Handbook // Russia, Leningrad: *Chemistry,* 1986. 208 P.

[10] Koshelev F.F., Kornev A.F., Bukanov A.M. General technology of rubber // Russia, Moscow: *Chemistry,* 1978. 528 P.

[11] Tugov I.I., Kostrykina G.I. Chemistry and physics of polymers // Russia, Moscow: *Chemistry,* 1989. 432 P.

[12] Ed. Kirpichnikov A. Technology of rubber products // Russia, Leningrad: *Chemistry,* 1991. 352 P.



Perspectivity of recycling of waste products of fibrous materials as reception of elastic composite materials is shown by creation of a composition a fibre - latex with the subsequent coagulation and vulcanization of the filled rubber.

Process of allocation butadiene-styrene rubber from latex SKS-30 ARK with use in quality filler a cotton, viscose and kapron fibre is considered. Influence of the contents and lengths of a fibre is established at various charges coagulant on completeness of allocation of rubber from latex. The optimum contents of a fibre and his length is determined.

Influence of a cotton, viscose and kapron fibre on properties of received rubbers, rubber mixes and vulcanizates is investigated.

Perspectivity of the offered approach for recycling fibres is shown.

In: Monomers, Oligomers, Polymers, Composites... ISBN: 978-1-60456-877-6
Editors: R. A. Pethrick, G.E. Zaikov et al. 

*Chapter 23*

# INTENSIFICATION OF PROCESS OF GAS CLEANING IN THE DEVICE WITH COMBINED SEPARATION STEPS

***R.R. Usmanova*, G.E. Zaikov* and V.G. Zaikov****
Ufa State Technical University of Aviation,
12 Karl Marks str., Ufa 450000, Bashkortostan, Russia
*N.M. Emanuel Institute of Biochemical Physics,
Russian Academy of Sciences,
4 Kosygin str., Moscow 119334, Russia

## ABSTRACT

The design of the new device for wet clearing gas is offered. The device contains combined separation the steps executed on a basis vane swirler, set in motion by a rotating rotor. The combination of such elements of a design allows to intensify processes of an interphase exchange and to raise efficiency of clearing of gas.

Results of industrial tests of the device have shown his high efficiency and reliability in operation.

**Keywords: gas cleaning, combined separation, vane swirler, rotor, steps.**

In the chemical industry devices on a basis vane swirler are one of perspective in engineering of separation water gas systems [1] and are widely used in manufacture at realization of processes wet dust separation [2].

Separators with axial, cylindrical and conic swirler [3] are most distributed.

One of perspective directions of perfection of separators on a basis vane swirler is development combined vorticitys formed various swirler [4]. The basic advantage of such devices in comparison with vane swirler is, that in them various mechanisms of separation

* UsmanovaRR@mail.ru
* Chembio@sky.chph.ras.ru
* Chembio@sky.chph.ras.ru

that allows to combine in one device of advantage, for example, centrifugal and inertial separators are simultaneously realized.

Devices with combined separation steps can work in a counterflow mode. It interferes забиванию blocking channels slime at work in conditions of sticking of a dust, that considerably expands their scope [5].

For clearing smoke gases of furnaces of roasting of limestone was applied gas washer with combined separation steps. The technological diagram of clearing of gases is submitted in figure 1.

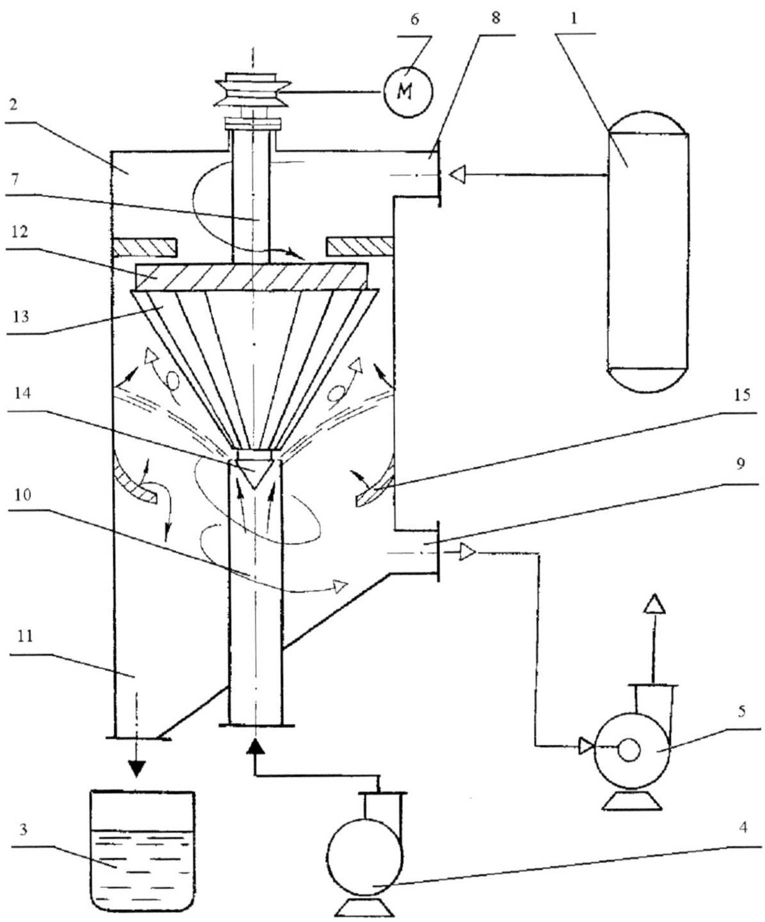


1 - a furnace of roasting; 2 – gas washer; 3 - the filter a slime pond; 4 - the pump; 5 - the fan; 6 - the electric motor; 7 - a rotor; 8,9 - branch pipes of input-output of gas; 10,11 - branch pipes input -output of irrigating liquid; 12 - cylindrical swirler; 13 - conic swirler; 14 - dissector; 15 – entrainment separator.

Figure 1. The Technological diagram of clearing of smoke gases of the furnace of roasting.

Departing from the furnace of roasting 1 gases at temperature 550°C feed in gas washer 2 on a tangential branch pipe 8. On an axial branch pipe 10 in the device the solution of limy milk (PH=11,5-12,5) moves on an irrigation.

Installation conic dissector 14 promotes formation of flat radial jets of irrigating liquid which, being reflected from the case of the device, create a high-intensity surface of contact of phases. In gas washer 2 as the first step it is used cylindrical swirler 12, as the second step - conic swirler 13. Swirler are fixed on a rotor 7 which is resulted in rotation by the electric motor 6.

Gas twists in space between swirler 12 and 13 and in a direction twist acts on the blade conic swirler 13 where there is his irrigation and final clearing.

The centrifugal forces arising at rotation of a rotor 7, provide intensive crushing water gas a stream that results in updating a surface of contact of phases and an intensification of processes of an interphase exchange.

Separated slime from the bottom zone of the device flows down on a branch pipe 11 in the filter - a slim pond 3 where there is a cooling and clarification of limy milk then it again moves the pump 4 on washing of acting gas in a branch pipe 10, forming a circulating contour.

The cooled and cleared gas on a branch pipe 9 with the help of the fan 5 is thrown out in an atmosphere.

Industrial tests were carried spent with the following initial data: a degree of a dust content of a gas stream $(5 \div 50) \cdot 10^{-3}$ kg / m$^3$, speed of a gas stream in the device 15-20 м/с, density of particles 1000÷2000 kg / m$^3$, temperature of gas on an input in the device 550 ° With.

As have shown results of researches, gas was cooled up to 62°C, ablation of caught particles did not exceed 0,5 %, the dust content of a gas stream was reduced up to (0,05÷0,2) $10^{-3}$ kg / m$^3$. The degree of clearing of gas made 98±1 %

## Conclusions

1. In the device with combined separation steps due to action of centrifugal forces, intensive hashing of gas and a liquid and presence of the big interphase surface of contact, occur effective clearing of gas in a foamy layer.
2. Industrial tests gas washer have shown his high technological efficiency and reliability in operation. The developed device can be used at designing new and reconstruction of the operative equipment for gas cleaning.

## References

[1] Rovin L.E.Perspective methods of clearing of gas emissions in foundry manufacture. - Minsk: *Data centre,* 1975,63 p.

[2] Uzhov V.N., Valdberg A.J. Clearing of gases wet filters. - Moscow: *Chemistry,* 1972, 240 p.

[3] Centrifugal dedusters with vane swirler // *Clearing of gases. The survey information.* - Moscow: data centere, 1979, 50 p.
[4] Lakomkin A.A., Ershov A.I. Application of devices with swirler steps. - *the Chemical industry,* 1993, p.p. 50-53.
[5] Rusanov A.A. Handbook on dust separation. - Moscow*: Energy,* 1975, 296 p.

Offered approach for recycling fibres is shown.

In: Monomers, Oligomers, Polymers, Composites… ISBN: 978-1-60456-877-6
Editors: R. A. Pethrick, G.E. Zaikov et al. 

*Chapter 24*

# Research of Critical Modes of Operation of a Separator with Swirler Various Construction

***R.R. Usmanova**[*1]**, G.E. Zaikov**[*2] **and A.K. Panov**[*3]*
[1] Ufa State Technical University of Aviation,
12 Karl Marks str., Ufa 450000, Bashkortostan, Russia
[2] N.M. Emanuel Institute of Biochemical Physics,
Russian Academy of Sciences,
4 Kosygin str., Moscow 119334, Russia
[3] Sterlitomak Branch of Bashkortostan Academy of Sciences,
68 Odesskaya str., Sterlitomak 453120, Bashkortostan, Russia

## Abstract

Experimental researches of separators with swirler various construction are carried spent. Critical modes of their work are investigated, the comparative analysis of size of ablation of a firm phase from researched devices is given.

The major factors influencing for effective work of separators are revealed.

**Keywords: separators, swirler, construction, analysis, size of ablation, critical modes.**

Now an actual problem, as is known, is clearing of gases of the weighed firm particles and a drop liquid. The variety of the requirements showed to quality of clearing, and also conditions of realization of process, has caused creation of set of designs of separators which principle of work is based on use of various forces (centrifugal, inertial etc.) due to what there is a branch of suspensions from a gas phase.

* UsmanovaRR@mail.ru
* Chembio@sky.chph.ras.ru
* Illnur@mail.ru

Practice of operation shows, that the most effective and, hence, designs are perspective, the branch of a suspension in which is carried out in a field of centrifugal forces. To such designs it is possible to attribute direct-flow, and counterflow cyclones, vortical devices, rotor separators.[1]

The expediency of use of this or that type of a separator is determined both a required degree of clearing, and technical and economic calculation. So, cyclones with turn of a gas stream have the high hydraulic resistance and a small range of change of gas loadings at which high efficiency of division of phases [2] is observed. In rotor separators the drive is necessary for rotation of a rotor that complicates a design and narrows area of their possible application [3].

On the skilled installation developed by us comparative experimental researches of some designs of the separators were carried spent, allowing to provide highly effective branch of gas from firm particles in a wide range of change of charges of a gas mix at rather small hydraulic resistance.

For clearing a gas mix of firm particles horizontal devices were developed and investigated: the bubbler-vortical device with an axial sprinkler [4]. *(figure 1, a);* the bubbler-vortical device with screw swirler *[5] (figure 1, в);* bubbler-vortical gas washer [6] *(figure 1, c).*

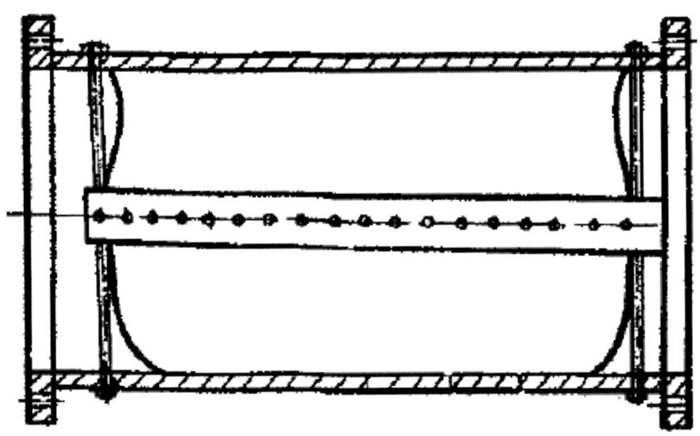

a

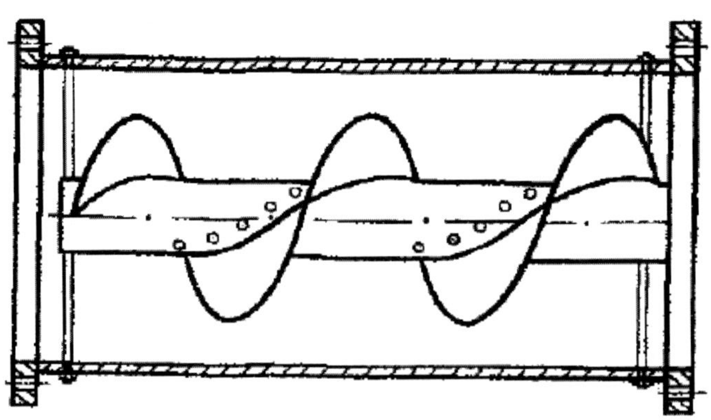

b

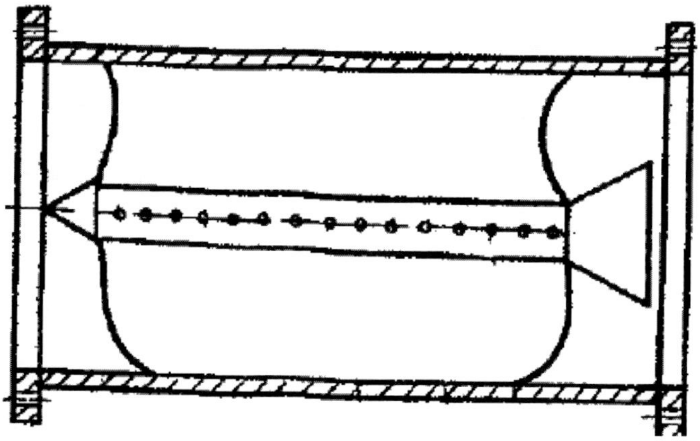

c

a-vane swirler with an axial sprinkler;
в- screw-swirler;
c-vane swirler with conic fairing.

Figure 1. Cylindrical chamber with swirler various construction.

The principle of work of devices is based on use of the centrifugal forces arising about installation static vane or screw swirler. Devices are executed as the cylindrical chamber with swirler variousconstruction. The circuit of skilled installation is submitted in *figure 2.*

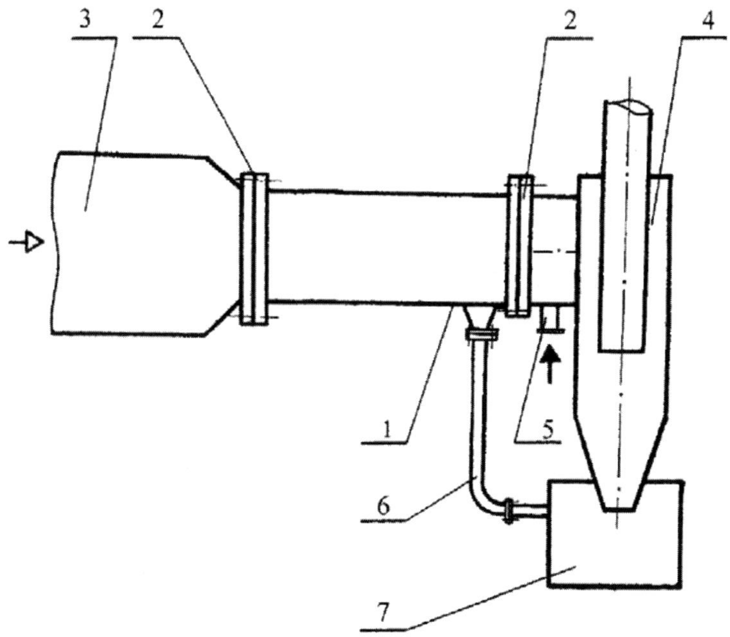


The 1-cylindrical chamber; 2-flanges; 3-pipe of input of gas; 4-cyclone; 5-input of irrigating liquid; 6-pipe of a drain slime; 7-slimecollector.

Figure 2. Circuit of experimental installation.

The cylindrical chamber 1 is mounted with the help of flanges 2 in gas duct, connecting a pipe of input of gas 3 with a wet cyclone 4. The liquid on an irrigation moves in the device on a pipe 5.

Separated slime it is washed off by a liquid and by means of an inclination of the cylindrical chamber 1 it is transported on a pipe discharge slime 6 in slimecollector 7. The subsequent division of suspension occurs in a cyclone 4.

Experiences carried spent at atmospheric pressure upon system an air - firm body. As firm impurity used a powder of talc, sand, table salt. Thus fractions of a dust of various dispersiveness prepared a method of sifting on **ситах** with cells in the size 80-120, 120-160, 200-300 microns. The sizes of particles determined with the help of a microscope. The firm phase moved in a pipe on an input in the device. The degree of a dust content changed within the limits of $(5-50) \bullet 10^{-3}$ kg / m$^3$, speed of a gas stream in the device $\vartheta$=1,2÷4,5 m/s. For catching firm particles carried away from the device the fabric filter was used.

Ablation expected under the formula

$$e=100-\eta,$$

Where $\eta$ - a degree of catching, %

$$\eta = \frac{G_2}{G_1} \cdot 100\%,$$

Where $G^{-1}$ quantity of a dust included in the device, kg;
$G^{-2}$ Quantity of the caught dust, kg.

As have shown results of researches, ablation of a dust for the device (figure 1, a) is higher, than at others. Especially appreciable increase of ablation is observed for devices with vane swirler at speeds of gas, big 4 m/s.

The sharp increase of ablation also occurs at diameter of particles d < 80 microns. At d > 80 microns ablation practically remains a constant which is not exceeding 2 % at speeds of gas in the device within the limits of 1,5÷3,0 m/s. It is established also, that the more hardly a particle, the there is their branch more effectively.

For the device (figure 1, в) the size of ablation is insignificant, however occurs blocking screw channels screw products of clearing that limits use such swirler at work in conditions sediment environments. The diagram of influence of speed of gas in the device on ablation of a dust is submitted on *figure 3.*

In parallel with measurement of quantity of a firm phase carried away from the device determined hydraulic resistance of devices and expected factor hydraulic resistance $\xi$ on speed of gas in the device.

The following results for devices with various swirler are received:

figure 1, a $\xi$=31,12; figure 1, b $\xi$=83,71; figure 1, e $\xi$=22,25.

It was established, that besides influence of speed of gas and the size of particles, occurrence of ablation is influenced significally with a field of centrifugal forces. So, with growth of a tangent of a corner of an inclination of a screw line шнека tg $\alpha$ influence of centrifugal forces decreases, and the increasing influence on size of ablation renders $Re_g$ and, at tg $\alpha$ > 0,74 account characteristics of a liquid phase, and also centrifugal forces already have a little an effect for occurrence of ablation. And, on the contrary, at tg $\alpha$ < 0,3 prevailing influence on process rendered $Re_e$ and centrifugal forces.

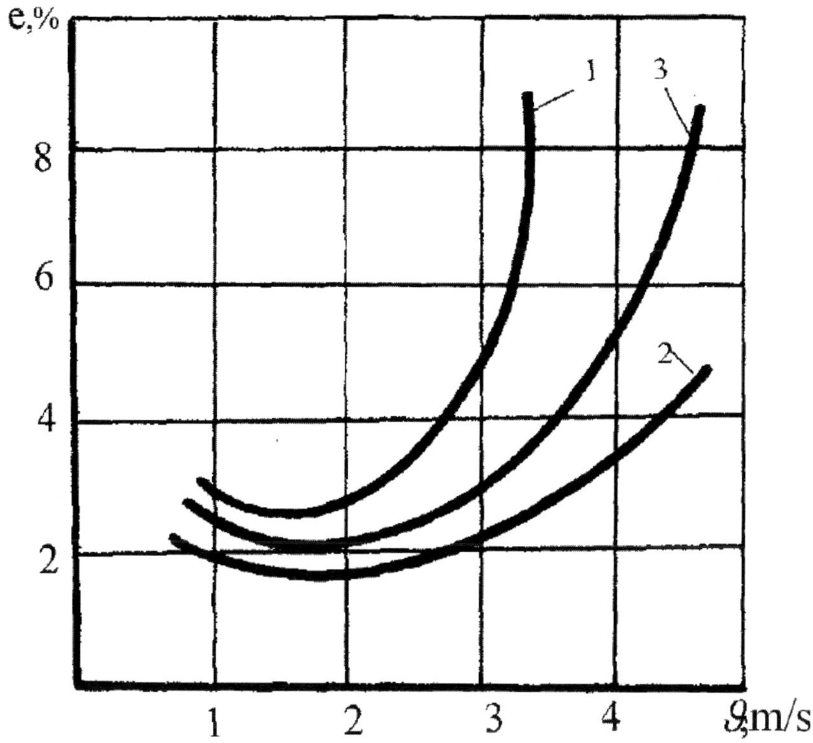


1- for swirler fig. 1, a;
2- for swirler fig. 1, в;
3- for swirler fig. 1, c.

Figure 3. Influence of speed of gas in the device on ablation of a dust.

## Conclusions

Thus, the carried spent complex of experimental researches has allowed to reveal the major factors influencing for effective work of separators.

Results of researches were used at calculation and designing of the devices found practical application for branch of weighed impurity from a gas stream.

## References


[1] Uzhov V.N., Valdberg A.J .Clearing of industrial gases from dust.-Moscow: *Chemistry,* 1981, 392 p.

[2] Bogatich S.A. Cyclone-foamy device.-Leningrad: Mechanical engineering, 1978, 223p.

[3] Usmanova R.R. Rotory the bubbler-vortical device // *the Application for the invention* №2007117109 from 7.05.07.

[4] Usmanova R.R. Bubbler-vortical the device with an axial sprinkler // *the Application for the invention* №2006113869 from 24.04.06

[5] Usmanova R.R. Bubbler-vortical the device with screw swirler // *the Application for the invention* №2006113870 from 24.04.06.

[6] Usmanova R.R. Bubbler-vortical gas washer // *the Application for the invention* №2007117108 from 7.05.07.

In: Monomers, Oligomers, Polymers, Composites… ISBN: 978-1-60456-877-6
Editors: R. A. Pethrick, G.E. Zaikov et al. 

*Chapter 25*

# METHOD OF CALCULATION OF EFFICIENCY DUST SEPARATION IN NEW DESIGNS DYNAMIC GAS WASHER

***R.R. Usmanova*, G.E. Zaikov* and V.G. Zaikov****
Ufa State Technical University of Aviation,
12 Karl Marks str., Ufa 450000, Bashkortostan, Russia
*N.M. Emanuel Institute of Biochemical Physics,
Russian Academy of Sciences,
4 Kosygin str., Moscow 119334, Russia

## ABSTRACT

In article the description of a new design dynamic gas washer, directed on increase of efficiency of clearing of gas emissions and decrease secondary carry –over of dust is given. Experimental researches which analysis has shown essential influence on process of separation of aerodynamics of a stream are carried out and has allowed to develop a new method of calculation of efficiency dust separation.

**Keywords:** dust, separation, new design, gas washer, analysis, stream, method of calculation.

## SYMBOLS

$\eta$ - efficiency dust separation;
t-Time of a relaxation, s;
$\rho$ - density of a dust, kg / m$^3$;

* UsmanovaRR@mail.ru
* Chembio@sky.chph.ras.ru
* Chembio@sky.chph.ras.ru

μ - dynamic viscosity $Pa \cdot s$;
m, n-parameters of distribution;
v-Speed of gas, m/s.

## INTRODUCTION

Now in connection with intensive development chemical, oil refining and other industries all rises a problem of preservation of the environment more sharply.

Millions tons of harmful gaseous substances are thrown out annually in an atmosphere. Creation of high-power plants results in necessity of application of clearing constructions of high efficiency [1].

One of perspective directions of gas purification are separation the devices using effect of action of a field of centrifugal forces, allowing most full to realize advantages new power - intensive technologies.

Centrifugal devices are characterized by high efficiency, simplicity of a design and low metal intensive. Application of such equipment allows also essentially to intensify process mass transfer due to increase of speed of movement of phases [2].

As efficiency of separation is defined by requirements of concrete process, the big value gets correct calculation of speeds of movement of phases at which the high degree of clearing without occurrence of secondary ablation is achieved.

Without the decision of these, and also of some other questions, the further optimization processes of gas purification becomes difficult a problem.

## 1. DEVELOPMENT OF A NEW DESIGN DYNAMIC GAS WASHER

The new design highly effective dynamic gas washer, directed on decrease secondary ablation and a radial drain of particles [3] is developed.

The circuit dynamic gas washer is submitted in *figure 1.*

The gas stream containing mechanical or gaseous impurity, acts in the device on a tangential branch pipe 3. The liquid acts by means of an axial branch pipe 10. The bottom basis swirler 8 is placed at a level top cut an axial branch pipe 10, forming with it a backlash.

Due to this the liquid acts in the device as flat radial jets which, being reflected from the case 1, form a liquid veil that causes high-intensity contact of phases. Swirler contains blank off bottom 8 and the cylindrical top 6 bases connected with each other by means of unidirectional blades 9. Swirler it is fixed on a rotor 12 which is resulted in rotation by means of a belt drive 13. The centrifugal forces arising at rotation of a rotor 12, provide intensive crushing water gas a stream that results in updating a surface of contact of phases and an intensification of processes of an interphase exchange.

The basic advantage of the offered device is that rotation in a field of centrifugal forces, and also the offered organization of liquid jets, allow essentially to change a principle of work of the device and on the basis of it to improve the basic technical and economic parameters (a degree of clearing of gas, productivity, metal consumption).

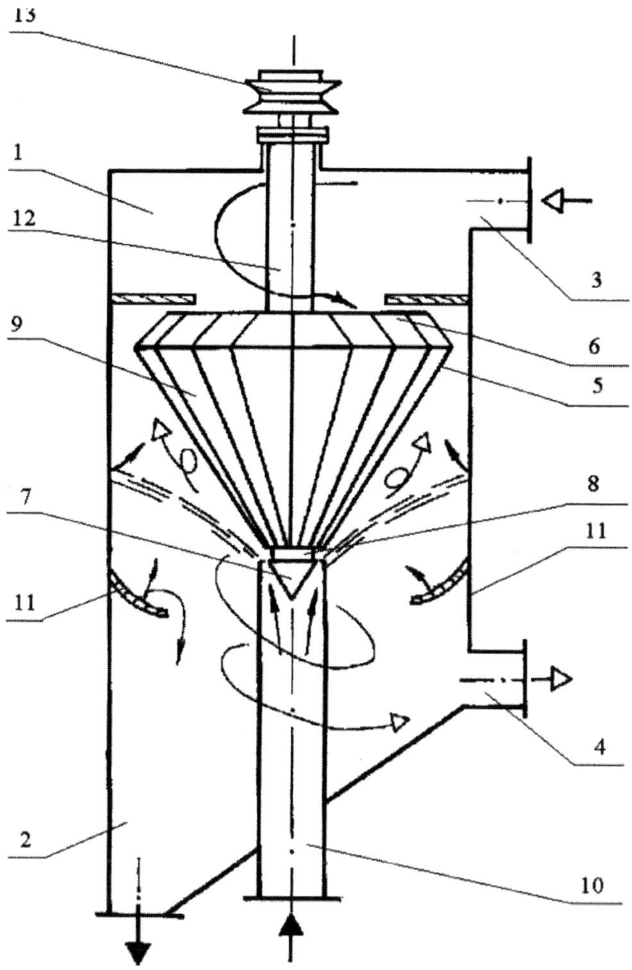


The 1-case; 2-tap(removal) slime; 3,4-branch pipes of an input / conclusion of a gas stream; 5-conic swirler; 6-cylindrical swirler; 7-dissectior; 8-basis swirler; 9-blades; a 10-axial branch pipe 11-entrainment separator; a 13-pulley.

Figure 1. Dynamic gas washer.

## 2. An Experimental Research of Aerodynamics of a Stream

Research of aerodynamic laws of a biphase stream was carried out(spent) on the device in diameter of 0,25 m, height 0,6 m, on system air - water. During researches speed changed within the limits of 15-20 м/c. The charge of a liquid 0-3 $m^3$ / hour, the charge of gas 200-600 $m^3$ / hour.

Measurement of a field of speeds and static pressure was carried out with the help of a focused single-channel cylindrical probe in diameter of 4 mm. Techniques calibration a probe and realization of measurements were standard [4]. The chosen measurement technique does not allow to determine radial making speeds of the gas, however known experimental data specify that radial making speeds is scornfully small, in comparison with axial and tangential [5].

Results of measurements were represented in relative sizes.

$$\overline{\vartheta_t} = \frac{\vartheta_t}{\vartheta},$$

$$\overline{\vartheta}_Z = \frac{\vartheta_Z}{\vartheta},$$

$$\overline{\rho} = \frac{2\rho}{\rho \cdot \vartheta^2},$$

Where $\vartheta_z, \vartheta_t$ - axial and tangential making speeds of gas, m/s;

$p$ - Static pressure, Pa.

The analysis of experimental data has shown, that distinction in structure of the stream, caused by design features of the device, is shown only at a level swirler. For the device repeatability is characteristic Character of structures of speeds at return on a new coil of rotation, breakdown axiall symmetric, and also insignificant decrease of size tangential speed of a stream in process of his promotion.

## 3. Calculation of Efficiency Dust Separation

The carried out analytic review of methods of definition of efficiency dust separation in centrifugal devices has shown, that the new design procedure which is taking into account his design features is necessary for designing the developed device.

The equation for definition of general efficiency looks like:

$$\eta = \frac{\int_0^\infty F(t)\eta(t)dt}{\int_0^\infty F(t)dt},$$

Where t - time of a relaxation of the particle, determined under law Stoks:

$$t = \frac{d^2 \cdot \rho}{18\mu}$$

F (t) - differential function of distribution of particles of a dust on time of their relaxation;

$\eta$ (t) - function of fractional efficiency dust separation.

Use t allows to take into account at realization of the disperse analysis of a dust form particles, their orientation, and also forces of interaction of particles.

Function of distribution of particles is described by equation Rosin-Rammler, and, entering concept of boundary time of a relaxation $t_{lim}$ and relative time of a relaxation $\bar{t}$, we shall receive:

$$\bar{t} = t / t_{\lim},$$

$$F(\bar{t}) = p \cdot n \cdot \bar{t}^{n-1} \cdot \exp(-p \cdot \bar{t}^{n}),$$

$$p = m \cdot t_{\lim}^{n},$$

Where a p-complex which is taking into account dispersiveness and properties of a caught dust (parameters m and n).

The equation for definition of general efficiency will look like:

$$\eta = \exp(-p) + p \cdot n \cdot \int_0^1 \bar{t}^{\,n-1} \cdot \exp(-p \cdot \bar{t}^{n}) \cdot \eta(\bar{t}) d\bar{t}$$

Function $\eta$ ($\bar{t}$) is described by the equation of a kind:

$$\eta(\bar{t}) = \bar{t}^{a} \cdot \exp[b \cdot (1 - \bar{t})]$$

Parameters a=2,47; b=3,08 are determined as a result of experimental researches then function of general efficiency will become:

$$\eta = \exp(-p) + p \cdot n \int_0^1 \bar{t}^{\,n-1,15} \cdot \exp[(-p \cdot \bar{t}^{n}) \cdot 3{,}08 \cdot (1 - \bar{t})] d\bar{t}$$

The received equation allows to define efficiency of gas purification depending on geometrical features of the device and his aerohydrodynamical characteristics.

## Conclusions

1. the new design highly effective dynamic gas washer, directed on improvement of the basic technical and economic parameters (a degree of clearing of gas, productivity, metal consumption) Is developed.

2. the method of calculation of efficiency of gas purification with use of distribution Rosin-Rammler which takes into account influence on process of separation of aerohydrodynamical characteristics of a stream is offered.

## REFERENCES

[1] Kuznetsov I.E., Troizkaia T.M. Protection of air pool from pollution in harmful substance. - Moscow: *Chemistry,* 1979, 344p.

[2] Uzhov V.N., Valdberg A.J. Clearing of gases wet filter.-Moscow: *Chemistry,* 1972, 245p.

[3] Usmanova R.R. Dynamic gas washer. *The application for the invention* №2007 120001 from 29.05.07.

[4] Petunin A.N. Technique and engineering of measurements of parameters of gas stream.-Moscow: *Mechanical engineering,* 1974, 250p.

[5] Stim A.N. Aerodynamics vortex chamber. - Vladivostoc, 1984, 200p.

In: Monomers, Oligomers, Polymers, Composites... ISBN: 978-1-60456-877-6
Editors: R. A. Pethrick, G.E. Zaikov et al. 

*Chapter 26*

# The Bases of the Technological Maintenance of Polymeric Implants' Biocompatibility

***N.I. Bazanova, L.S. Shibryaeva and G.E. Zaikov***
*Emanuel Institute of Biochemical Physics, Russian Academy of Sciences
119334 Russia, Moscow, Kosygina st. 4

## Abstract

The new complex approach to the problem of implants' biocompatibility is proposed. The oriented phototechnology is developed for the modification of the silicon intraocular lens that enables to increase the biocompatibility.

**Keywords:** polymeric implants, biocompatibility, silicon intraocular lens, phototechnology, technological adaptation, IR- spectroscopy, UV-spectroscopy, chromatography.

The wide use of artificial materials for manufacturing of the medical implants determined that the study of laws of their interaction with biological structures is necessary to be provided [1,2].

The accumulation of the experimental data, the expansion of methodical base for research of the biocompatibility of "alive" and "lifeless" substance and the search of the most informative physical and chemical characteristics of polymers have observed during several decades. Finally, it assumes the development of the special technologies for reception of polymeric products for medical purposes.

The retrospective analysis of the problem of bearableness of polymeric implants has shown that positions of the theory of biotechnical systems were not applied in the development the approaches to its decision. That reflected on treatment of the results of toxicological techniques to research from the point of final result ("a black box").

The absence of the complex techniques to research of the interaction of elements of the system "implants - biological environment" on border of division of phases and ignoring of some fundamental laws established within the limits of the physical chemistry of polymers are marked.

By virtue of the specified features there was a certain break between theoretical consideration of questions of the polymeric implants' bearableness and the experimental data complicated the further promotion in the decision of problems of reception the biocompatible polymeric products of various functions and on the basis of the directed technologies.

Thus use of the system positions with reference to an artificial crystalline lens (intraocular lens (IOL)) enabled to determine that despite of the extensiveness and duration of researches of bearableness of lenses, these researchers had the isolated character. Thus, insufficient attention was paid to the research of own physical and chemical characteristics of IOL and their transformation during the manufacturing and at contact with biological environments. Apparently, such unilateral approach has led to what even at full passage tested under special program IOL with a positive estimation « "toxic- syndrome" has not been eliminated. However these results were considered only as the certificate insufficient information both reliability of the developed tests and absence of knowledge etiopathogenesis complications.

The correlation between thermodynamic balance of superficial layer with the contiguous volumetric phases (which are defined by the general conditions of heterogeneous balance) was assumed as a basis of the proposed approach to the problem of the increase of polymeric implants' biocompatibility.

This approach enabled to consider the power and electric characteristics of surface as unified display of the processes proceeding in a superficial layer of a polymeric material in contact with the biological substance. The theoretical positions describing behavior of the superficial layer of polymer enabled to prove the necessity of inclusion in standard technological process of polymeric implant (IOL) production at the special stage of "technological adaptation".

The "technological adaptation of a polymeric implant" is a number of technological operations which introduction in a standard technological process enables to create already at production phase such a physical and chemical structure of a superficial layer of polymer that will be adequate to biological structures.

In other words, already during the manufacturing process it is possible to lower considerably the efforts of the body to bring the system "implant - biological environment" in stability.

It should be noted that the additional technological stage is not limited to carrying out of various kinds of modifications (physical, chemical, etc.), but also includes the creation of focused supramolecular structures of a superficial layer through maintaining the optimum conditions for its relaxation.

The methodical complex included physical, chemical and toxicological methods of the research. The fundamental methods (IR-, UV-spectroscopy, chromatography, physicomechanical tests etc.) and the original techniques, enable to reach optimization of the technological modes of silicon IOL adaptation. The complex biotechnical technique was formed stage by stage.

At the first stage of researches on the basis of photometric a method (copyright certificate SU 1529112 A1) the tool technique of the hygienic control silicon for medical purposes has been developed. The research of the definition of absorption in a range of waves' length (350-

190 nanometers) was carried out. The characteristic strip of absorption of microimpurity was revealed.

A degree of toxicity of the microimpurity contained in extracts, and their influence on the energy metabolism of cells judged by results of: estimations of extracts' toxicity from polymeric materials by studying of the activity of bull sexual cells suspension, caused own mobility; the test for culture fibroblasts a cornea of the rabbit. The quantitative communication between the optical density of an extract on the established length of wave and the results of the toxicological test "alive cells" was estimated. The quantitative assessment of correlation ($r = -0{,}673$) was used as a quantitative estimation of the narrowness of communication of both tests. Based on the statistical processing or empirical results, the confidence bounds were defined for the general correlation factor $-0{,}788 < r < -0{,}520$. The empirical equation of theoretical line of regression is $y=100e^{-4{,}221x}$.

The second stage of researches was directed on the elimination of such lacks of silicon IOL as water repellency and discrepancy of the spectral characteristics of material to spectrum transmission of a natural crystalline lens to UV-areas. The phototechnology has been developed for that purpose.

As the result of silicon lens modification the chemical compound of the superficial layer has changed in particular the hydrophilous groups are formed; not reacted trailer groups are finished; the absorption in UV-area of a spectrum increase.

However the modification of superficial characteristics of silicone appeared insufficient for the adequate coupling of elements of the biotechnical system "implant - biological environment". It is caused by the fact that the structures of IOL surface are non-equilibrium.

The analysis of behavior of the processed silicon material surface in contact with biological environment enabled to determine the stages of the mechanism of its relaxation which final result is formation of the focused secondary structure. Apparently, it will define reaction of an eye to an implant.

The established laws contributed to develop the technique for production of silicon intraocular lenses of the higher biocompatibility.

## REFERENCES

[1] Handbook of Polymer Research Monomers, Oligomers, Polymers and Composites, Richard A. Pethrick, Antonio Ballada, G.E. Zaikov, Nova Science Publishers, Inc., 2007, New York

[2] New trends in Biochemical Physics research, Sergei D. Varfolomeev, Elena B. Burlakova, Anatolii A. Popov, Gennady E. Zaikov, Nova Science Publishers, Inc., 2007, New York

In: Monomers, Oligomers, Polymers, Composites… ISBN: 978-1-60456-877-6
Editors: R. A. Pethrick, G.E. Zaikov et al. 

*Chapter 27*

# STIMULI-RESPONSIVE DRUG DELIVERY SYSTEM

***Raluca Dumitriu*[*1], Cornelia Vasile[1], Geoffrey Mitchell[2] and Ana-Maria Oprea[1]***

[1]"Petru Poni" Institute of Macromolecular Chemistry, Department of Physical Chemistry, 41A Gr. Ghica Voda Alley, 700487, Iasi, Romania

[2]University of Reading, Department of Physics and Polymer Science Centre, Whiteknights, Reading RG6 6AF UK

## ABSTRACT

The interpenetrated network systems (IPN) or hydrogels which are *biodegradable and also stimuli-responsive* are special materials explored lately and a subject of great interest in the last years.

This paper deals with synthesis, characterization and drug release behavior of a new biodegradable stimuli-sensitive hydrogel containing alginic acid (ALG) and N-isopropylacryl amide (NIPAM).

The properties and morphology of the mixed alginic acid (ALG) /N-isopropylacryl amide (NIPAM) hydrogel were investigated through swelling kinetic studies performed at different temperatures and pHs, rheology and environmental scanning electron microscopy (ESEM). The controlled release of ketoprofen from this hydrogel was also followed. The studies performed allowed us to ascertain the sol-gel transition temperature and the gel point for the investigated hydrogel, to observe the morpgology of the gel as the water content changes and also thermal behavior.

**Keywords:** hydrogel, polysaccharide, biodegradable, stimuli-responsive

[*] E-mail: rdumi@icmpp.ro

## INTRODUCTION

The development of new polymeric systems, which are *biodegradable and also stimuli-responsive* is a subject of great interest in the last years. The hydrogels capable to undergo drastic volume changes (volume phase transition) in response to small variations in the external conditions, such as temperature, pH, solvent, ionic concentration, electric field or light irradiation were named "stimuli-responsive" or "intelligent" gels, having a great potential for applications in high-performance areas such as biosensors, bioseparators, bioreactors, in tissue engineering, or in controlled drug delivery systems through responding to environmental stimuli by swelling / deswelling.

In an attempt to obtain biodegradable materials with sensitivity to external stimuli, like pH and/or temperature, biopolymers from renewably resources were associated with thermo-sensitive macromolecules. [1,2]

Various stimuli-responsive hydrogels have been developed for drug delivery, based especially on poly(*N*-isopropylacrylamide) (PNIPAM), which is one of the most widely studied thermosensitive polymers, exhibiting a volume phase transition at a lower critical solution temperature (LCST) in water of around 32$^0$C [3], close to physiological temperature.

Alginate, a natural polysaccharide derived from brown seaweeds composed of D-mannuronic acid and D-guluronic acid is one of the biodegradable polymers used to obtain pH-sensitive hydrogels Alginate hydrogels are pH-sensitive and biocompatible, with a relatively low cost. [4]

Ketoprofenul is a derivate of the propionic acid with the following structure

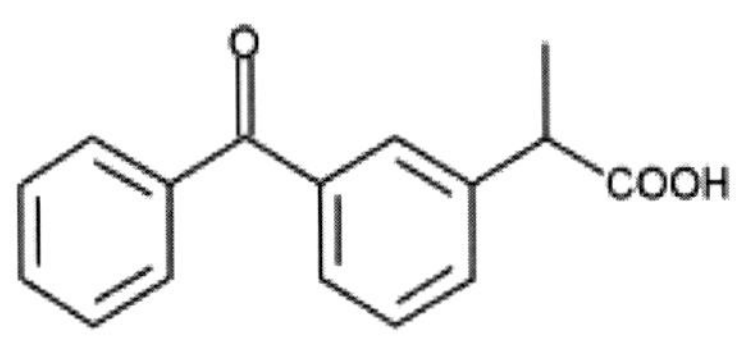


Ketoprofen is a white crystalline powder insoluble in water but soluble in acetone, ethanol methylene chloride, chloroform, ether and benzene. Ketoprofen is an anti-inflammatory, analgezic and antipiretic drug efficient in the treatment of the rheumatoid arthritis and osteoarthritis, spondylitis, as well as soft tissue injuries, such as tendinitis and orthopedic pain surgery, inflammatory desease-gout, to relieve minor aches and pain from headaches, toothaches, the common cold, muscle aches, and backaches, Reiter's syndrome, dysmenorrhoea, metastatic bone pain, pyrexia and to reduce fever. It works by stopping the body's production of a substance that causes pain, fever, and inflammation.[5] Its anti-inflammatory action is 20 times higher than that of ibuprofen, of 80 times higher than that of phenylbutazone and of 160 times higher than that of aspirin. [6-9]

Because is water-insoluble several techniques have been tested to improve the solubility, as solid dispersion, complexation, etc. Frequent administration is necessary because of its rapid elimination from organism (half-life time of 2 – 2.5 hours). The high quantities of ketoprofen in stomach determine gastric troubles as ulcers, bleeding, or holes in the stomach or intestine. Controlled release of this drug should permit a less frequent administration and reduction of the gastrointestinal pains. It is known the use of the microspheres of ethylcellulose as matrices for controlled release or coating with chitosan to increase accesibility by bioadehsion.[10]

# EXPERIMENTAL

Alginic acid (ALG) from brown algae (Fluka), Mw 48 000 - 186 000, $\eta_{red}^{25^\circ C} = 2.41$ ml·g$^{-1}$, c = 0.2 g/dL; N-isopropylacryl amide (Aldrich) 97% (NIPAM); crosslinking agent: N,N'-methylenebisacrylamide (Fluka) (bis). The hydrogel with 75 % NIPAM / 25% ALG composition was obtained by simultaneously polymerization/crosslinking reaction using 4 wt% crosslinking agent amount in respect with NIPAM. It was purified by five repeated washings with twice-distilled water.

*Swelling kinetic studies* – performed by weight measurements, at different temperatures and pHs. Swelling ratio (SR) determined according to the equation:

$$SR(\%) = \frac{W_S - W_D}{W_D} \times 100 \qquad (1)$$

Temperature range was 20 – 40 °C while pH values studied were: 2.2, 5 and 7.2.

*Thermo-gravimetry analysis (TG/DTG)* - performed by a thermo-gravimetric analyser Mettler STARe SW 8.10; samples weight ~3 mg, heat rate 10 °C/min.

*Rheology measurements* – performed with a Bohlin rheometer C-VOR, equipped with a Pelltier stage for temperature variation; frequency range 0.01-1 Hz, constant strain 10%.

*ESEM studies* – performed with a FEI Quanta 600 FEG Environmental Scanning Electron Microscope; measurements at different degrees of relative humidity (RH) and magnifications.

*Drug release studies* - Loading of the Ketoprofen in 75% NIPAM/25%ALG hydrogel matrix was done by swelling in ethanol solution followed by freeze-drying.

The release profile of Ketoprofen from 75% NIPAM/25%ALG hydrogel matrix having various degrees of swelling at the same drug concentration was determined based on UV-VIS spectroscopy measurements. With this aim an UV-VIS spectrofotometer HP 8540A type was used. The calibration was done with solutions of ketoprofen in ethanol of concentrations varying $10^{-5}$ to $10^{-2}$ g/L. The characteristic wavelength of ketoprofen is 254 nm. Prevealing of the samples was made at constant time intervals.

This paper deals with synthesis, characterization and drug release behavior of a new biodegradable stimuli-sensitive hydrogel containing alginic acid (ALG) and N-isopropylacryl amide (NIPAM).

# RESULTS AND DISCUSSION

Polymerization/chemical crosslinkig reaction of NIPAM in presence of alginic acid and N,N'-methylene bisacrylamide leads to a semi-transparent hydrogel stable in water, and acidic solutions a very long time keeping its shape and characteristics. It is hydrophilic absorbing from atmosphere approximately 10 wt% of water, as it resulted from thermogravimetric analysis and repeated weighing of sample kept in room conditions. It is also thermally stable as it appears from DTG curves presented in figure 1.

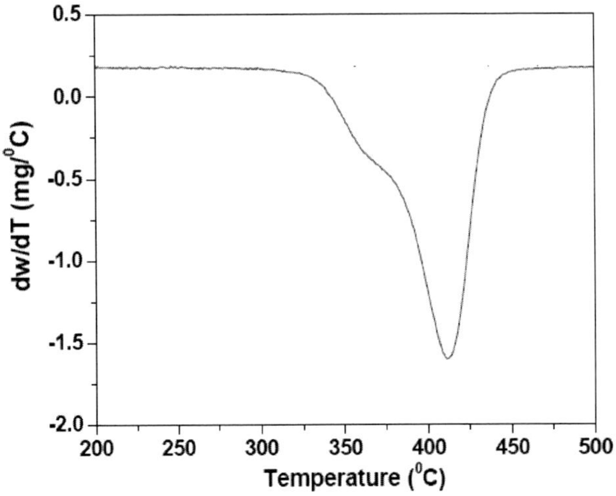


Figure 1. DTG curve of the 75/25 NIPAM/ALG dry hydrogel.

It decomposes in nitrogen atmosphere starting with 290 °C reaches a maximum of decomposition at 409 °C, the final weight loss in this main decomposition process being of 86 wt% - table 1.

**Table 1. Thermogravimetric data for 75/25 NIPAM/ALG dried hydrogel**

| Sample | Δw (%) | Characteristic Temperatures (°C) | | |
|---|---|---|---|---|
| | | Onset temperature of degradation ($T_i$) | Temperature corresponding to maximum weight loss ($T_m$) | Final temperature ($T_f$) |
| 75 %NIPAM/25% ALG | 86 | 290 | 408 | 477 |

The swelling behavior of the sample in response to temperature and pH of the external media is manifested by the decrease of the swelling capacity with increasing temperature (table 2). It was observed a similar behavior at pH 2.2 and 7.2, with a slightly increased capacity of swelling at acid pH – table 3.

**Table 2. Maximum swelling degree values in ethanol for the 75%NIPAM / 25 %ALG hydrogel at different temperatures**

| Temperature (°C) | $Q_{max}$ (wt %) |
|---|---|
| 20°C | 3734 |
| 30°C | 2679 |
| 33°C | 1817 |
| 35°C | 844 |
| 40°C | 485 |

**Table 3. Maximum swelling degree values in ethanol for the 75%NIPAM / 25 %ALG hydrogel at different pH values**

| pH | $Q_{max}$ (wt %) |
|---|---|
| 2.2 | 486 |
| 5 | 911 |
| 7.2 | 339 |

By analysing the graphical representation of the maximum swelling degree versus temperature -figure 2 – it can be seen a sudden variation which can be associated with a transition temperature.

For accurate determination of the transition temperature the curve was fitted by a Boltzmann function (using Origin 6.1 program) given by equation:

$$y = \frac{A_1 - A_2}{1 + e^{\frac{(x - x_0)}{dx}}} + A_2$$

where:

$A_1$- minimum value of the function

$A_2$ – maximum value of the function

$x_0$ - value on the x axis corresponding to the inflexion of the curve, equivalent to the transition temperature;

dx - domain in which this value is found

The reduced chi-squared, for all the fitted curves was of $\chi^2 \leq 0.1$, therefore the use of this function is a good approach.

A transition at 31.9 °C was found which can be associated with gel point of the system.

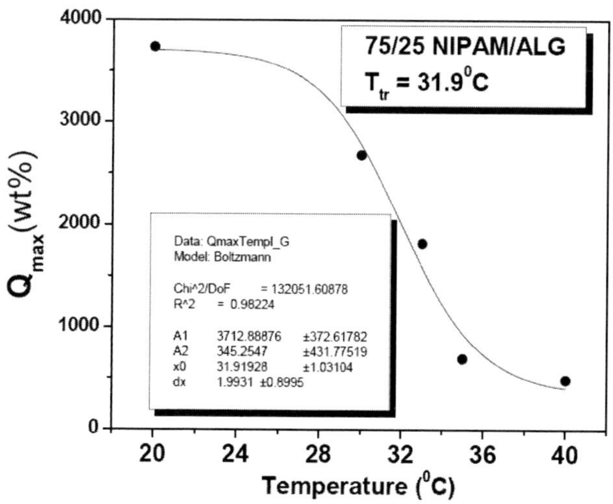


Figure 2. Qmax versus temperature for 75%NIPAM / 25 %ALG hydrogel.

Rheological behavior is influenced by temperature changes proofing the thermo-responsive properties of the 75%NIPAM / 25 %ALG hydrogel.

It can easily remarked that the gel point shifted toward lower frequency with increasing temperature – figure 3.

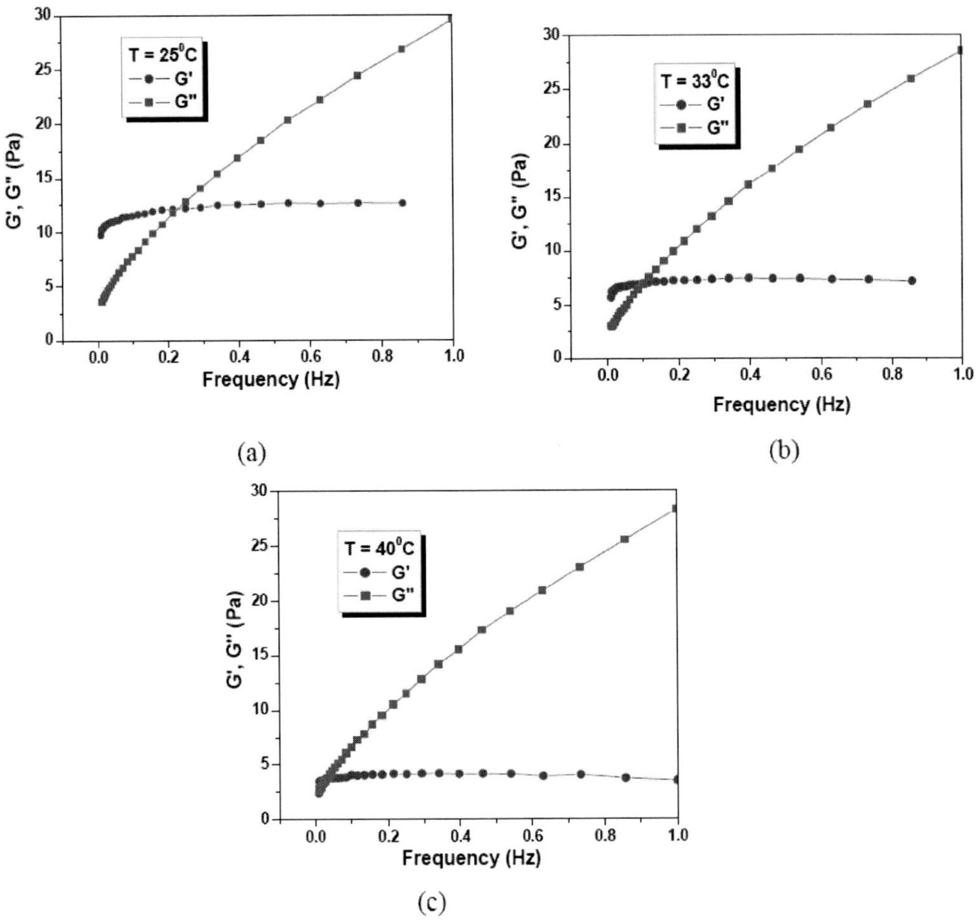


Figure 3. G' and G" of the 75%NIPAM / 25 %ALG hydrogel versus frequence at different temperatures.

ESEM study reveals a porous morphology with a honeycomb-like structure – figure 4.

Morphological aspect and cavities dimensions (increase) modify as the water content changes.

The release profile of the ketoprofen from 75%NIPAM / 25 %ALG hydrogel matrix is different depending on solvent content in the gel, the higher solvent content the lower is the quqntity of the released drug which is ~ 100% from loaded quantity for un-swollen matrix that means physically mixing of matrix with drug and decreases at about 60 % from loaded quantity when the matrix was swollen at maximum SR = 3300%. The matrix swollen in drug solution up to a SR = = 1650 % show and intermediary behaviour between physically loading and the case when maximum SR was used. – figure 5.

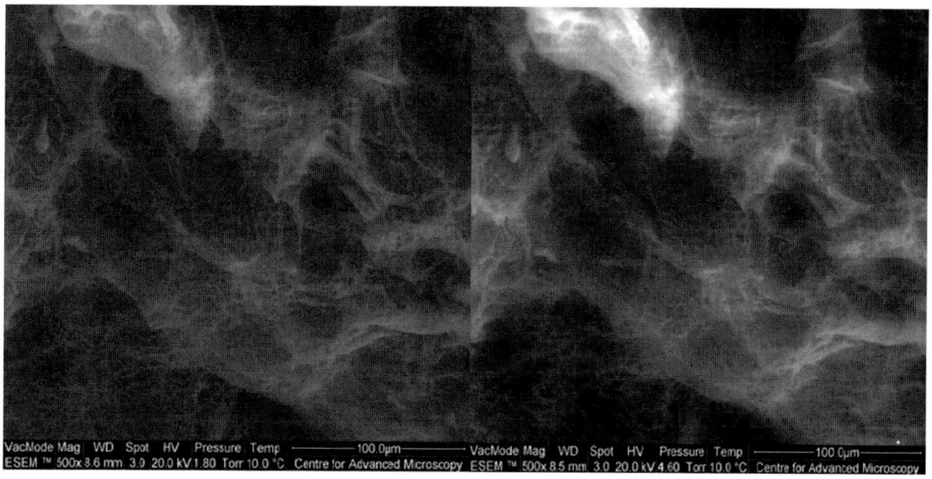

a b

Figure 4. ESEM micrographs of the 75NIPAM/25 ALG hydrogel at RH 20% (a) and 50% (b).

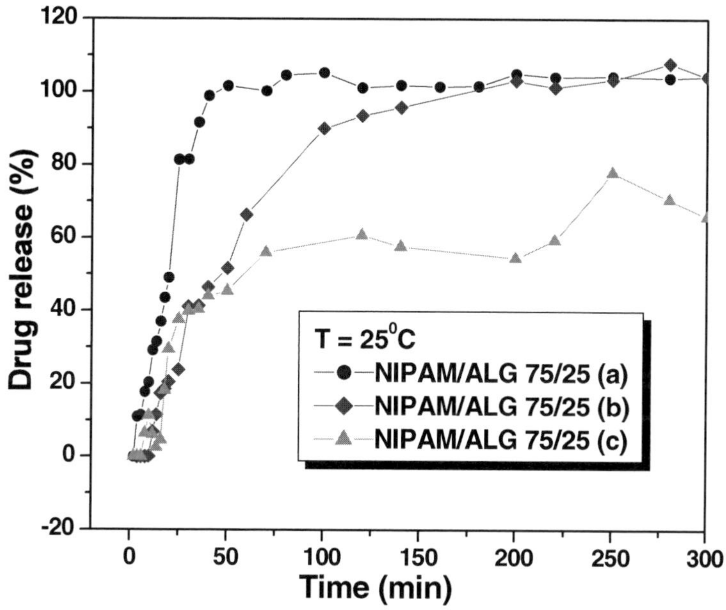


Figure 5. Drug release profile of ketoprofen at $25^0$C from 75%NIPAM/25% ALG hydrogel with different swelling ratios in ethanol a) unswollen; b) SR = 1650 %; c) SR = 3300 %.

A difference appear also in the release rate of the three kinds of matrix loading with ketoprofen – figure 6. The higher is SR of hydrogel matrix the slower release rate of the drug from the matrix

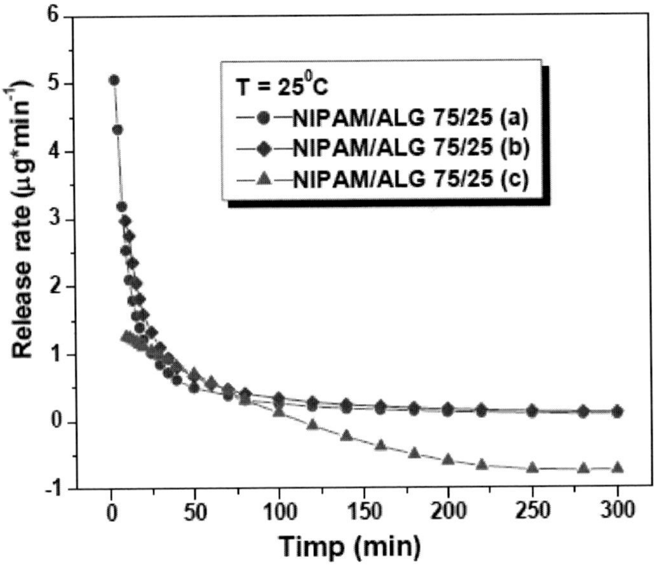


Figure 6. Drug release rate of ketoprofen at 25$^0$C from 75%NIPAM/25% ALG hydrogel with different swelling ratios in ethanol a) unswollen; b) SR = 1650 %; c) SR = 3300 %.

## CONCLUSIONS

The swelling and rheology studies performed allowed us to ascertain that the 75 wt% NIPAM/25% ALG semi-interpenetrating network (hydrogel) obtained possess thermo- and pH-responsive properties. Morphological examination by ESEM microscopy showed a porous morphology which modifies when the water content changes. The drug release profiles depend on solvent quantity used for drug loading.

## ACKNOWLEDGMENTS

The authors thank for financial support of COST P12 Action "Structuring of Polymers" and Romanian ANCS and CNCSIS through national projects.

## REFERENCES

[1] Kim S.Y., Cho S.M., Lee Y.M., Kim S.J., *J. Appl. Polym. Sci., 78,* 1381-1391 (2000).
[2] Marsano E., Bianchi E., Viscardi A. *Polymer* , *45*, 157-163 (2004).
[3] Schild H.G., *Prog. Polym Sci., 17*, 163-249, (1992).
[4] Shi J., Alves N.M., Mano J.F., *Macromol. Biosci., 6*, 358-363, (2006).
[5] http://www.aapspharmscitech.org/view.asp?art=pt060104
[6] http://www.google.ro/search?hl=ro&q=ketoprofen&btnG=C%C4%83utare+Google&meta

[7] http://www.pdrhealth.com/drug_info/rxdrugprofiles/drugs/oru1313.shtml
[8] http://www.rxlist.com/cgi/generic/ketoprof.htm
[9] http://www.pharmgkb.org/views/index.jsp?objId=PA450149&objCls=Drug
[10] http://www.google.ro/search?hl=ro&q=ketoprofen&btnG=C%C4%83utare+Google&meta

In: Monomers, Oligomers, Polymers, Composites… ISBN: 978-1-60456-877-6
Editors: R. A. Pethrick, G.E. Zaikov et al. 

*Chapter 28*

# NOVEL POLYMERIC CARRIER FOR CONTROLLED DRUG DELIVERY SYSTEMS FROM RENEWABLE SOURCES

***Catalina Duncianu, Ana Maria Oprea and Cornelia Vasile***
"Petru Poni" Institute of Macromolecular Chemistry,
41 A, Gr.Ghica Voda Alley, 700487, Iasi, Romania

## ABSTRACT

The hydrogen-bonded interpolymeric complexes (IPCs) have gained a great interest in last decades showing a similar behavior with natural systems and also because of their distinct physical and chemical properties in comparison with pure components; IPCs are considered promising materials in the development of different drug formulations.

The interpolymeric complex based on a natural polymer like alginic acid (AgA) and poly (ethylene glycol) (PEG) was tested by UV-VIS spectroscopy to investigate the possibility of using it as a new matrix in the active principles delivery. The kinetic profile of the procaine hydrochloride release from 16 %AgA/ 84% PEG complex at various pHs of 1.14; 2.16 and 3.09 and temperatures was studied.

The interpolymeric complex between AgA and PEG showed a good behavior in acidic medium and it can be considered a promising material for the release of active substances in stomach.

**Keywords:** alginic acid, polyethylene glycol, procaine, interpolymeric complex, controlled delivery.

## INTRODUCTION

The hydrogen-bonded interpolymeric complexes (IPCs)[1] have attracted a great interest from the pharmaceutical scientists due to the similar behavior with natural systems and their unique physical and chemical properties in comparison with pure components and also their potential applications in the development of different drug formulations[2-3]. The design and control of the release mechanism of the active principles have gained an increased interest in last years. The main goal of the controlled release of the active agents is to prolong their action time, to minimize the undesired reactions, and also to increase the release efficiency of the active principle. A controlled release system should have a stable chemical structure which satisfies the conditions of biodegradability and biocompatibility and also to have also a suitable release rate of the active substance at targeting site in a definite time[3-4].

The selection of an adequate support for the active principles delivery/transport is important in order to obtain an efficient sustained release system. An ideal vehicle for an active principle should satisfy also some specific requirements like: to posses a high loading ability according to the therapeutic dose, to be able to penetrate or to localize at a targeting site and to release in a controlled way of an active principle. It should not be toxic and it has to be biocompatible and biodegradable, especially in the case of intraocular administrations.

The use of a natural polymer such as a weak polyacid e.g. alginic acid to obtain interpolymeric complexes represents an attractive target. Alginic acid is a natural non-toxic, biodegradable hydrophilic polymer which can be extracted from different brown seaweeds e.g. *Macrocystis pyrifera, Ascophyllum nodosum.[5]*

PEG is widely used in pharmaceutical industry and cosmetics; it is non-volatile and inert from physiologically point of view and it can be used for the different ointments, emulsions, pastes, lotions and suppositories production [6].

In the present work the interpolymeric associations between alginic acid (AgA) with poly (ethylene glycol) (PEG) was tested by UV-VIS spectroscopy in order to investigate the possibility of using it as a new matrix in the controlled delivery of certain drugs.

For testing the formed IPC as drug carrier the procaine hydrochloride was used. The structure of procaine hydrochloride is shown in figure 1.

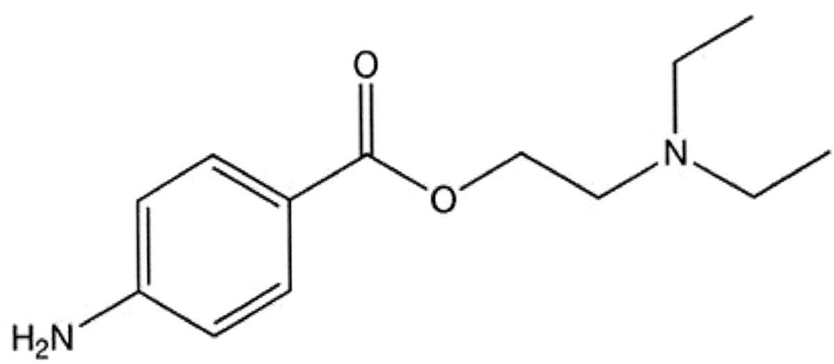


Figure 1. The structure of procaine hydrochloride [8]

Procaine hydrochloride is a well-known active substance used as local anaesthetic, which blocks the generation and conduction of nerve impulses by decreasing the permeability of the nerve membrane to ions, thereby inhibiting depolarisation, loss of pain sensation, other sensory functions, and finally motor activity [7]. Its prolonged action should be interesting.

# MATERIALS AND METHODS

## Materials

It was used alginic acid, a Fluka product, with an average molecular weight of 48,000-186,000; the reduced viscosity in water at 25 °C for an aqueous solution of c= 0.2 wt % is $\eta_{red}$= 2.41 ml $g^{-1}$, with a drying loss ≤ 10 wt % and ash ≤ 3 wt%.

PEG has a molecular weight of 35,000 and a melting temperature of 60-65°C.

## The Preparation of the IPC from Alginic Acid, AgA and Poly (Ethylene Glycol), PEG

The composition of the interpolymeric complex between AgA and PEG was identified by studying many mixing ratios between components which were carefully characterized by viscometric and potentiometric methods [9]. It was established that the most stable IPC has the composition 16% AgA / 84 % PEG. The solutions of components in twice distilled water, adjusted at pH = 4 and c = 0.2% were mixed and the obtained complex was isolated from solution by drying during 48 hours in a freezing-drying apparatus.

Viscometry have been performed by means of an Ubbelhode type viscometer with dilution and suspended level, at 25 °C ± 0.02 ° C and flow times were measured with an accuracy of ± 0.1s.

pH measurements were performed at 25 °C ± 0.02 ° C, in a thermostated bath, with a Consort C835 multimeter equipped with a separate pH glass electrode suitable for diluted solutions domain.

## The Swelling Process of the Support Based on Alginic Acid and PEG

The swelling profile (figure 2) of the IPC support shows its ability to absorb the acidic solution (pH = 1 – 4). In this way a temporary enlargement of the intermacromolecular spaces occurs. Also, it can be observed that the maximum absorption capacity of the modified polymer was reached in about two hours. After that, a swelling equilibrium is installed. There is no absolute dissolution even after a long period of time, in this pH interval the IPC being stable. The loading of the support with procaine was done by swelling it with acidic solution of procaine followed by freeze-drying.

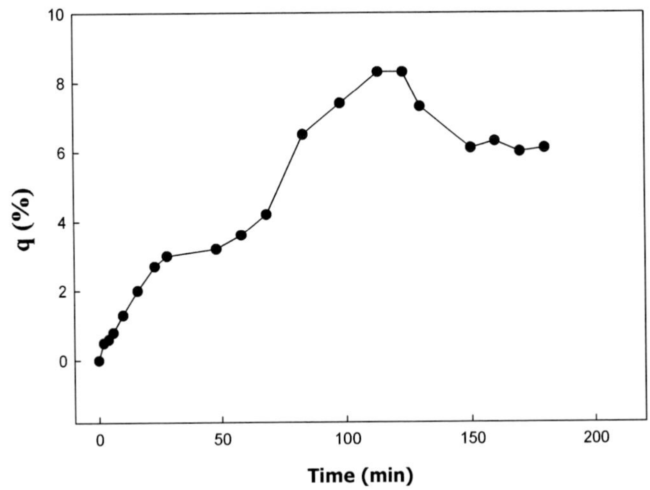


Figure 2. Swelling curve in acidic twice-distilled water of the IPC based on AgA/PEG with procaine hydrochloride entrapped.

## UV VIS Method

The controlled release mechanism was evidenced by using photocolorimetric method. A UV-VIS HP 8540A spectrophotometer was used; the calibration curve was achieved with acidic solutions of various pH (pH=1-4) of different concentrations $10^{-5}$-$10^{-2}$ g/L range. The released procaine has been distinguished at a $\lambda_{max}$=194, 221, and 291 nm [10]. The most suitable wavelength for determination was at 221 nm. The ability of the complex of AgA/PEG to release procaine entrapped into carrier was tested at different flow rates of the solvent of 0.2; 0.3 and 0.6 ml $min^{-1}$).

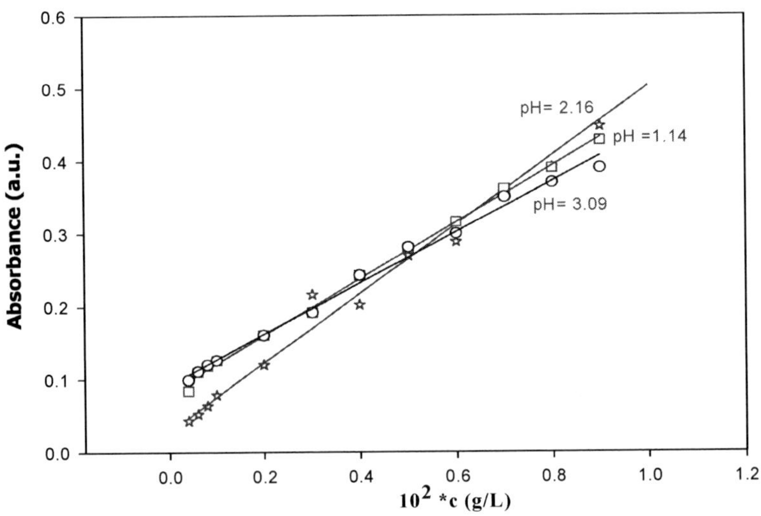


Figure 3. Calibration curves of the procaine hydrochloride at different pHs (1.14; 2.16 and 3.06), $\lambda$ =221 nm.

## RESULTS AND DISCUSSIONS

The interpolymeric associations between AgA and PEG were selected after study of the mixed solutions of polymers, adjusted at pH= 4 and c=0.2 %, in different compositions ranging 0-100 wt% AgA and then characterized by viscometric and potentiometric methods. The obtained results are summarized in figure 1 (for details see ref.9).

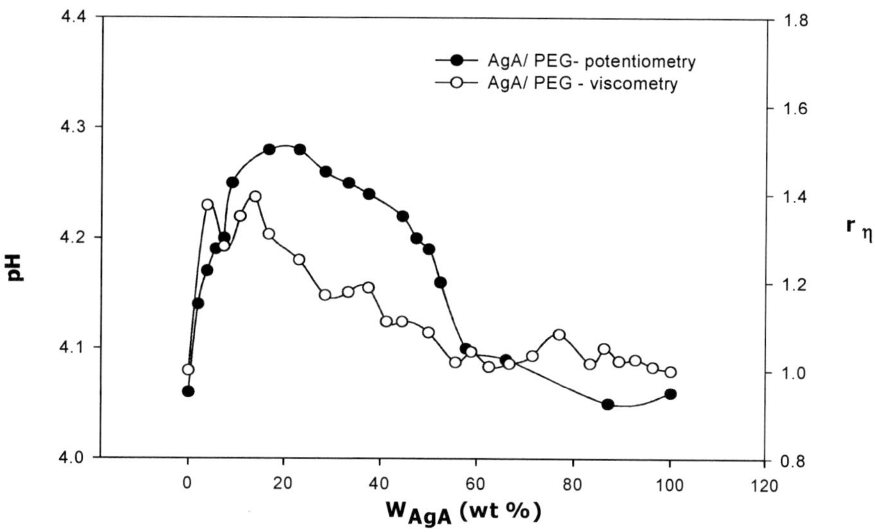


Figure 4. The dependence of the viscosity ratio on the composition of the system AgA/ PEG, in aqueous solution with pH=4 at 25 °C (o-right axis); pH- values (•- left axis) obtained at AgA titration with PEG in twice-distilled water, at 25 °C.

Correlating data of viscosimetry and potentiometry, maximum value of both viscosity ratio and pH was found in the 2 -20 wt% AgA/ 98-80 wt% PEG weight composition range - figure 4. It was established that a more stable interpolymeric association occurs at 16 % AgA / 84 % PEG which can be considered as an interpolymeric complex. After the stoichiometry of the IPC was established; the IPC was separated. The stability of the obtained 16 % AgA / 84 % PEG interpolymeric complex was also evaluated and it was found that the stability constant is about 21 $(l*mol^{-1})$[9].

The possibility of using this IPC as a new matrix in the active principles delivery was tested by UV-VIS spectroscopy. The kinetic profile of the procaine hydrochloride release from 16% AgA/ 84% PEG complex at various pHs of 1.14; 2.16 and 3.09 (figure 5) and temperatures was studied because it was established that pH = 4 represents the limit of stability of the complex [9].

The stability of the AgA/ PEG support without the procaine entrapped in the acidic solutions of pH= 1.14; 2.16 and 3.09 was tested by keeping it in such solutions long time. No dilution was observed and no bands in UV-VIS method were found.

Within the study it were evaluated the dependences of the released percent of the procaine hydrochloride from the IPC matrix in time (figure 5a) at pH = 1.14; 2.16 and 3.09 and of the release rate of the procaine hydrochloride from the same solutions in time (figure 5b).

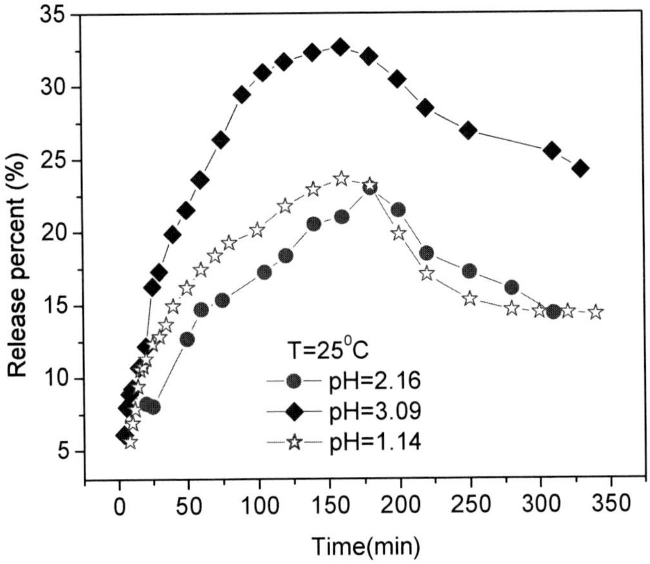


a

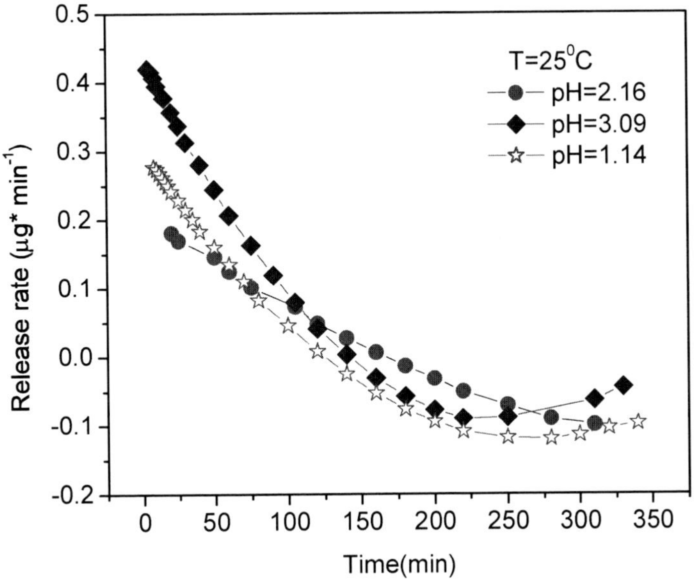


b

Figure 5. Release profile for the procaine hydrochloride from 16 % AgA / 84 % PEG interpolymeric complex at different pH values (a); Rate release of procaine hydrochloride from AgA/PEG interpolymeric complex at different pH values (b).

The release profile of the procaine hydrochloride from IPC matrix with 16 % AgA / 84 % PEG composition shows in the first two hours a higher released amount of procaine at pH= 3.09 (figure 5a), of about 24% from the overall quantity of the procaine entrapped in the IPC.

After 3 hours the amount of procaine released is decreasing showing a constant level up to 5 hours.

In the first 3 hours, the release profile of procaine at pH= 1.14 shows that about 22% of procaine was released from the polymeric matrix. The profile from pH = 2.16 indicates a faster release of procaine from the matrix and after 3 hours the amount of procaine released is approximately the same as in the case of the profile from pH= 1.14. During 3- 5 hours, the shape of both profiles are similar reaching a constant concentration of the delivery. The rate release curves showed in figure 5b indicate different shapes between the all three dependencies in the first 100 minutes, having higher values of release rate for the profile at pH= 3.09 about 0.4 μg/min, a sharp decrease in the case of profile at pH= 2.16, recording almost a linear decrease of the rate release in time. The dependence from pH= 1.14 shows that the procaine hydrochloride was released with a maximum rate about 0.3 μg/min while at pH= 2.16, the rate release was 0.18 μg/min.

The table 1 shows a decrease of the half time release, $t_{1/2}$ with the increase of the pH of the medium. For an efficient controlled delivery it is necessary that the time of the delivery to be long so that the half time of release should reach low values. Within the study it is observed that the half time of release of procaine at pH= 3.09 is approximately two times longer (40 min) and the shortest value of 20 min is recorded at pH= 1.14. The lower values of the half time of the release of procaine at pH 3.09 indicates that at this pH, closer to pH=4, the AgA/PEG matrix can suffer some modifications in its structure due to the increased solubility of alginic acid at pH closer to pH=4 which is the limit of solubility in water. Therefore, the matrix became swollen and its structure allows to the active principle entrapped inside to be slower released.

**Table 1. The values of the half time of the controlled release process**

| Sample | $t_{1/2}$ (min) |
|---|---|
| AgA/PEG- procaine hydrochloride, pH= 1.14 | 40 |
| AgA/PEG- procaine hydrochloride, pH= 2.16 | 25 |
| AgA/PEG- procaine hydrochloride, pH= 3.09 | 20 |

The influence of flow rate of the acidic solutions through the complex of 16%AgA/84%PEG with entrapped procaine hydrochloride is described by the curves in figure 6. Three different flow rates about 0.2; 0.3 and 0.6 ml $min^{-1}$ were used. Figure 6a illustrates a different profile of release at a low flow rate. It can be observed that using a lower flow rate of solvent (0.2 ml/ min) through the matrix of IPC- active principle, it can be obtained a higher percent of released procaine, about 24% from total amount of procaine entrapped into the polymeric matrix after 3 hours. In the second region of the profile, after 3 hours, the amount released is decreasing reaching a constant level after 5 hours.

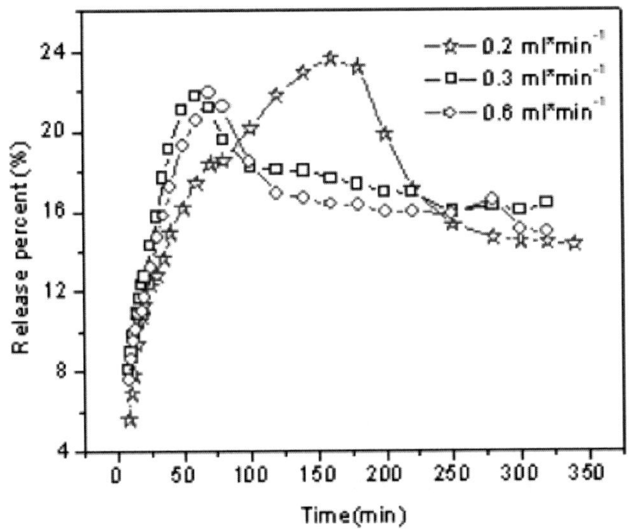


(a)

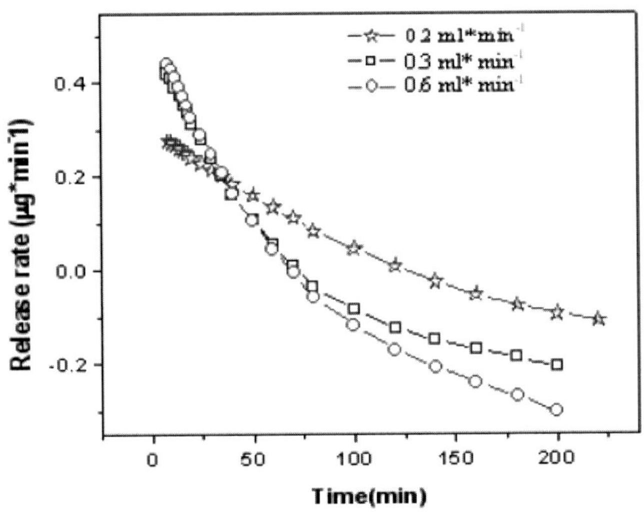


(b)

Figure 6. Release profile for the procaine hydrochloride from AgA/PEG interpolymeric complex at different flow rates of the twice- distilled water through the sample at pH= 1.14 (a); Rate release of procaine hydrochloride from AgA/PEG interpolymeric complex at different flow rates of the twice-distilled water through the sample at pH= 1.14 (b).

In the case of a higher flow rate of 0.3 and 0.6 ml min$^{-1}$ the amount of released procaine was only 22 % in the first 1.5 hours reaching the constant concentration after 2 hours. So that a more efficient release mechanism at low flow rate of the solvent and a better controlled release of the procaine from the IPC complex between alginic acid and poly (ethylene glycol) was found.

The release rates in the case of a high flow rate through the sample (0.3 respectively 0.6 ml/ min) show a sharp decrease in the first 50 min., while at a low flow rate (0.2 ml /min) the decrease of the rate release is monotonically.

## CONCLUSIONS

An interpolymeric complex based on a natural polymer like alginic acid and poly (ethylene glycol) with composition 16 %AgA/84 %PEG was tested for controlled delivery of procaine.

The obtained profile shows that the optimal pH for the release of procaine hydrochloride is 1.14 - 2, similar with the pH of the physiological medium from stomach. The interpolymeric complex between AgA and PEG showed a good behavior in acidic medium. Therefore the support based on AgA/PEG can be a promising material for the release of active substances in stomach (at acidic pHs).

## REFERENCES

[1] Jiang M., Li M., Xiang M., Zhou H., *Adv. in Polym. Sci.*, 146:121-128 (1999)
[2] Ozeki T., Yuasa H., Kanaya Y., *J. Controlled Release*, 63, 287-293 (2000)
[3] Campbell L. K., White J. R., Campbell R. K., *Ann. Pharmacother*, 30:1255-1262, (1996)
[4] Kang S., Brange J., Burch A., *Diabetes Care,* 14:942-947 (1991)
[5] Lele B. S., Hoffman A. S., *J. Controlled Release*, 69:237-244 (2000)
[6] http://www.arpc-ir.net/PDF/catalogue/ChemicalSpec/Ethoxylates/PEG-Chemical%20 Grade.pdf
[7] Whistler R. L., BeMiller J.N., Third edition, Academic Press, San Diego, 105-120 (1993)
[8] http://en.wikipedia.org/wiki/Novocaine
[9] Duncianu C., Vasile C., *Nova Science*, submitted
[10] Merino C., Junquera E., Jimenez-Barbero J., Aicart E., *Langmuir*, 16:1557-1565, (2000)

In: Monomers, Oligomers, Polymers, Composites... ISBN: 978-1-60456-877-6
Editors: R. A. Pethrick, G.E. Zaikov et al. 

*Chapter 29*

# DISSOCIATIVE ATTACHMENT OF LOW-ENERGY ELECTRONS (BELOW IONIZATION OR ELECTRONIC EXCITATION THRESHOLDS) IN FROZEN AQUEOUS PHOSPHATE SOLUTIONS

***O. S. Nedelina*, O. N. Brzhevskaya, E.N. Degtyarev and A.V. Zubkov***

Emanuel Institute of Biochemical Physics, Russian Academy of Sciences, ul. Kosygina 4, Moscow, 117977 Russia, 939 74 69

## ABSTRACT

We investigated the ESR spectra of the products of interaction of $H_2PO_4^-$ with low-energy electrons emitted by irradiated fluorophores (Flu). Our selection of this particular source of electrons enabled an efficient electron-injection process upon near UV light excitation (4,0-4,5eB). Our experiments were conducted in frozen aqueous dilute solutions of $H_2PO_4^-$ following the scheme: Flu+ hν →(Flu)*→Flu $^+$ $e_{aq}$,, $e_{aq}$+ $H_2PO_4^-$ → H. + $HPO_4^{2-}$, (Flu -ferrocyanide ions, acetate, tryptophan (λ>240nm), NADH, phenothiazine, 1,3,6,8-pyrenetetrasulfonic acid (λ>340nm). The photoinduced ESR spectra of solvated electrons (g=1,9987, ΔHpp 0,15Gauss), of the hydrogen atom (g=2,0043, ΔHpp ~508Гс), and of secondary acceptors were the basic indicators provided evidence for the electron attachment to $H_2PO_4^-$ and for the subsequent interaction of an electron-adduct $[H_2PO_4^-]^\cdot$ with secondary acceptors (specifically with vanadate). In our experiment we observed (i) the reverse relation between the ESR intensity of the hydrogen atom and the free electron with acetate in SDS-micelle as fluorophore, (ii) an interdependent ESR signal relation of the hydrogen atom and the donor-acceptor system, and (iii) disappearance of the hydrogen-atom spectra after addition of the electron scavenger $KNO_3$. We used the ESR method to visualize the discharge channel of photoejected electron(or some of the form of its relaxation to $e_{aq}$) to the dissociative attachment $e^-$ + $H_2PO_4^-$ → $H^\cdot$ + $HPO_4^{2-}$ . Our experimental results suggest that the

* olga.nedelina@gmail.com

possible range of the functional activity of phosphates can be extended to direct involvement in redox reactions.

Electron transfer is an elementary and very important step in many chemical and biochemical reactions, where certain molecules act as electron donors and others as electron acceptors. Molecular anions formed in the electron-target interaction are considered to be the driving force for the respective enzymatic reaction. The chemical behavior of enzymes can largely be controlled by electron transfer from the substrate. The cation and anion produced in this way are better oxidants and reductants, respectively, than either of the neutral ground state molecule. Determining redox intermediates is necessary to establish the mechanism of elementary chemical events of enzymatic reactions.

In recent years, increasing importance has been associated with electron-induced chemical processes in biological environments. Identifying the underlying mechanisms involves several research methodologies, including studies of the electron interaction with the bio-molecules such as DNA, its component subunits, and amino acids [1,2]. Using ionizing radiation as a method for electron extraction and electron attachment is very powerful for preparing and studying the electron-gain and electron-loss species; the utility and simplicity of the method is stressed in [3]. Electrons are removed or added without the need of oxidizing or reducing agents whose presence often leads to complications.

In studying reaction mechanisms, model photochemical systems with photon energies up to 4,5eV($\lambda$>250nm) are preferable to radiolytic systems, because they are conducive to direct stabilization of the functional reactive intermediates, while avoiding concurrent generation of highly active products of water radiolysis, which initiate the formation of nonspecific half-products. This region is not absorbed by water, and some solutes so photochemical reactions are only possible in presence of some recyclable absorbing sensitizers or fluorophores. Photochemical methods (namely fluorophores photoionization associated with monochromatized electron release) are advantageous for studying redox process intermediates and, in particular, are used frequently to determine potential biological electron donors or acceptors. Many of these products may be intermediates of dark chemical reactions. [4].

Photoinduced electron detachment from fluorophores is a rapid, efficient charge-transfer reaction. In this reaction, charge separation between a photoexcited sensitizer and an electron carrier is one of the most important steps in production of long-lived photoinduced charge separation for energy conversion and storage. To harvest the light energy put in the system, the oxidizing and reducing power of the photoinduced species must be used before the electrons are transferred back. In presence of appropriate acceptors, the net yield of charge separation grows, and the back electron transfer slows down significantly, compared with the values of the same parameters in their absence [5].

It is advantageous to detect the products of electron impact in frozen solutions, in which the products are matrix isolated and chemical transformations of primary photoproducts are stopped. The advantage of using low-temperature rigid matrices is that highly reactive species are rendered impotent by immobility, while the lifespan of the unimolecularly unstable species gets extended by the low temperatures and sometimes by the inhibitory effects of the matrix on their tendency to fragment. The results are very similar in aqueous and frozen solutions; this strongly suggests that both phases generate the same radiation-produced intermediates, which, in turn, react in similar ways [6].

In competitive reactions, electron capture can be remarkably specific in the solid state, suggesting that, at least in certain media, electrons migrate over large distances prior to being captured. The reactions are strongly favored by the environment because of the charge, which reflects the role of the solvent. Thus, if protic solvents are used, solvation is rapid and strong for anions, including electrons. As a result, the ejected electrons ($e^-_D$) are very rapidly solvated ($e^-_s$) and thereby stabilized. In contrast, electrons in rigid media may not be solvated at all, even in such media as glassy alcohols, or may be solvated very slowly, and are far more likely than in fluid solutions to be captured by reactive solutes prior to being deeply trapped. Photoejected electrons $e^-_D$ are likely to be more reactive than $e^-_s$ , the margin being close to the solvation energy. This is in good agreement with the functional shape of the observed electron decays and with the dependence of the decay lifetimes on the scavenger concentrations and on the initial electron yield [7].

Photochemical methods combined with low-temperature ESR spectroscopy make it possible to distinguish between (1) direct electron capture by molecule ABC to give thermodynamically stable anion $[ABC]^{\cdot-}$ , and (2) dissociative electron capture to give $A^-$ + $BC^{\cdot}$ [3]. The same is true for resonant capture of free electrons at subexcitation energies by molecule M $e+M\rightarrow(M^{-(*)})$, which forms either unstable species ($M^{-(*)}$) or a transient molecular anion, referred to as "temporary negative ion" (TNI) [8, 9]. TNI typically involves multi-electron resonance, where extra electrons are bound temporarily to electronically excited molecules. A TNI can decay either via electron autodetachment or via dissociative electron attachment (DEA) $M^{-(*)}\rightarrow M^{\cdot}R+X^-$. It has been shown that, at energies well below the ionization potential of M, DEA is the only mechanism that efficiently controls molecular fragmentation. Furthermore, recent studies on DEA in low-energy electron attachment to gas phase molecules in the biology context have shown that hydrogen loss is the predominant reaction channel [10].

In an ESR study of biologically significant frozen aqueous matrices modeling the medium for UV-light-induced ATP synthesis (ADP + Pi), we were the first to demonstrate [11] the presence of atomic hydrogen in the multicomponent mixture of free radicals. In such a system, it is possible to observe both stable anions and products of disintegration by the DEA mechanism. The doublet with the 508G separation has been assigned to trapped hydrogen atoms produced from the reaction of the monophosphate ions present in the matrix with electrons at subexcitation energies (below 4,5 eV) resulting from photoionization of the adenine base with $\lambda$ >260nm.

The assumption of electron attachment to phosphate was confirmed by our experiments with nanosecond laser photolysis, in which we showed that hydrated electrons are quenched by phosphate [12]. Pulse photoexcitation of an aqueous solution of NADH (0.2 mM) or pyrene with 337-nm light of $N_2$-laser produced two intermediate products in time less than the resolution time of the recording system (10 ns). These products were (1) hydrated electrons, with a typical absorption spectrum with the maximum at 715 nm and lifespan about 120 ns (quenched by $O_2$ and NADH), and (2) cation-radical $NADH^+$ , with absorption maximum at approximately 550 nm. Introducing $NaH_2PO_4$ into the solution did not influence the kinetics of degradation of the NADH+ radical, but decreased the lifespan of eaq and its release. The rate constant of the dynamic quenching was about $1x10^7$l/(mol s).

The process of electron interaction with phosphate does not produce stable phosphate anion. In fact, in our experiments we did not obtain ESR spectra with extremely large $P^{31}$ hyperfine interaction, which would correspond to stable phosphate anion. Indeed, it is known that attempts to add electrons to monophosphate anions or their salts in various solvents have failed even in radiolysis, as determined by ESR spectroscopy [13].

Hydrogen atoms found in these systems seem to indicate another important channel of low-energy photoejected electron consumption, namely electron attachment to phosphate $e_{aq} + H_2PO_4^- \rightarrow [H_2PO_4^{\cdot}] \rightarrow H^{\cdot} + HPO_4^{2-}$, which leads to the formation of an important intermediate of photochemical conversions.

Thus, our photochemical experiment combined with ESR spectroscopy confirmed the early results reported in [14, 15], which indicate that photochemically and radiolytically solvated electrons can be converted into hydrogen atoms via interaction with protons or Brønsted acids: $e_{aq} + HX \rightarrow H + X^-$, where HX is any proton-containing acid. It is reported in [16] that this role may be implemented by proton donors (oxyacids). Lately, electrochemical experiments have demonstrated direct cathode release of hydrogen from phosphate and from some acids, rather than from water [17, 18].

This postulate of electron attachment is based primarily on the suppression of the yields of trapped hydrogen atoms by added electron scavengers. The yields of hydrogen atoms are decreased by added electron scavengers such as nitrate ion in ices containing mononegative ions; the addition was not effective in polynegative oxyanion solutes. The relative rates of decrease by nitrate and scavengers forming stable anion indicate that electron is a mobile precursor of the trapped hydrogen atoms [16].

In case of pulse radiolysis and photolysis of aqueous solutions under exposure to light with $\lambda < 200$ nm, when the medium is filled with extra electrons of various origins, there are several channels through which the hydrogen atom is formed and utilized, including the products both of water radiolysis and of photodissociation of dissolved molecules. Hence, the proportion of H atoms formed as a result of electron attachment to monophosphate ions is small (0.36% of its concentration) [15].

In view of this, it was reasonable to restrict the scope of our experiment, limiting it to a collision study of only two components: extra low-energy electron and phosphate. In this study we attempted to directly detect the products of interaction between orthophosphate and monochromatized low-energy electrons in a medium containing predominantly these components.

In this work, we investigated the interaction of $H_2PO_4^-$ with low-energy photoejected electrons, whose sources were photoexcited fluorophores Flu* ; this particular choice allowed for producing an efficient process of monochromatized electron injection upon near-UV irradiation 340nm or 260nm with quantum energy 4,0-4,5eV. The investigation was conducted in frozen aqueous dilute solutions following the scheme $Flu + h\nu \rightarrow Flu^* \rightarrow Flu^+ + e^-$, $e^- + H_2PO_4^- \rightarrow [H_2PO_4^{\cdot}] \rightarrow H^{\cdot} + HPO_4^{2-}$ [Flu -ferrocyanide ions, acetate, tryptophan ($\lambda$>240nm), phenothiazine, 1,3,6,8-pyrenetetrasulfonic acid ($\lambda$>340 nm)]. Modern experimental approaches, such as ESR, fluorescence, and nanosecond laser photolysis, permit both detection of all intermediates in these main processes and selective determination of specific properties of the high-energy products generated in these processes (such as excited states, free radicals including atomic hydrogen, or solvated electrons).

A DRSh-1000 high-pressure mercury lamp, equipped with light filters UFS-1 (240 nm $<\lambda<$ 300 nm) and BS-6 ($\lambda>$ 320 nm), was used as an illumination source. The excitation light was passed through an UFS-1 light filter (240 nm $<\lambda<$ 300 nm) for tryptophan and acetate, and through a BS–6 filter ($\lambda>$ 320 nm) for NADH and 1,3,6,8-pyrenetetrasulfonic acid. The irradiation time was 8 min. The quantum yields of these fluorophores are different, the lowest (10%) being characteristic of NADH. In this study, the following reagents were used: $NaH_2PO_4$ (extra pure), tryptophan and sodium acetate of reagent grade, and NADH from Acros Organics. The EPR absorption of frozen samples irradiated at 77 K was measured using a Bruker EMX-8 EPR spectrometer (frequency 9.6 GHz) at 77 K.

In continuous photolysis of frozen solutions of fluorophores in presence of phosphate, ESR spectra, including lines from counter radical and hydrogen atoms (doublets with splitting of ~508 Gs), were found in broad scanning (650 Gs) of EPR spectra (77 K) in all systems containing different photosensitizers (ferrocyanide ions, acetate, tryptophan - $\lambda>$240nm, NADH, phenothiazine, 1,3,6,8-pyrenetetrasulfonic acid $\lambda>$340nm).

Among photochemical events in fluorophores, photoionization is paramount in producing both electrons and fluorophores cation $Flu^* \rightarrow Flu^{\cdot+} + e^-$. Electrons in the excited state can either revert to the ground state or may be stabilized either by physical trapping or by electron capture with the electron acceptors present in the matrix. In absence of these events, electrons may return to their cations. These experiments did not detect directly the ESR line of photoejected electron. Instead, we used the effect of phosphate reactants to demonstrate its presence as recorded in the EPR spectra of the fluorophore cation radical and of free radicals of acceptors Ac ascribed to stable radical $Ac^{\cdot-}$ formed in the fast bimolecular reaction $e^- + Ac > Ac^{\cdot-}$ or products of DEA $e^- + H_2PO_4^- \rightarrow [H_2PO_4^{\cdot}] \rightarrow H^{\cdot} + HPO_4^{2-}$.

Figure 1 shows photoelectron capture by $H_2PO_4^-$ or $D_2PO_4^-$, as monitored by ESR spectra of hydrogen or deuterium atoms in the tryptophan ($5x10^{-4}$M solutions in presence of 0,5M $NaH_2PO_4$ (pH 4,9) irradiated with $\lambda>$240nm at 77K.

We established that the signal from atomic hydrogen was recorded only in presence of both fluorophore and monophosphate anion $H_2PO_4^-$ in a weak acid medium, the signal intensity being dependent on the phosphate concentration. Our investigation of the pH dependence of hydrogen atom showed that the intensity of the signal from hydrogen was maximal in the range of maximum concentration of the monoanion (figure 2,3).

We observed increases in the total yield of hydrogen and cation radical with increasing concentration of the phosphate acceptor. One can interpret the low hydrogen yield and the low cation radical yield in alkaline media, as opposed to the respective yields in weak acid media, as evidence of absence of another competing process, caused by the presence of an electron acceptor, namely of monoanion phosphate. The absence of monophosphate as electron scavenger is evidently responsible for the rapid decrease in yields of both hydrogen and cation radical spectra (figure 3).

Note that, under these conditions, the structure of the signal from atomic hydrogen was similar in all the systems that we studied. This also indicates that hydrogen atoms had the same fixation locus, and that the environment of this locus was homogeneous. By using partially deuterated $NaD_2PO_4$, with deuterium connected to all oxygen atoms, it is possible to show that all these resonances originate from abstraction of hydrogen from oxygen sites but not from water (figure 1).

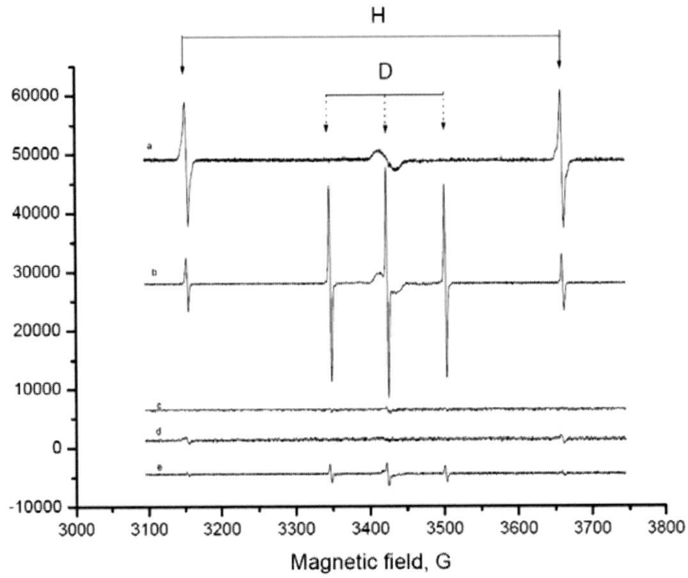


Figure 1. The X-band ESR spectra of 0.5 M solutions of $NaH_2PO_4$ (pH= 4.5) in $H_2O$ or $D_2O$ after 8 min UV radiation at 77 K: a – in presence of Trp ($5 \cdot 10^{-4}$ M) in $H_2O$; b – the same in $D_2O$; c, d – in $H_2O$; e – in $D_2O$. (Microwave power 2 μW, modulation amplitude 3 G. Conditions of UV irradiation (1 kW mercury lamp) as follows: a, b, c - λ>240nm; d,e – without filter).

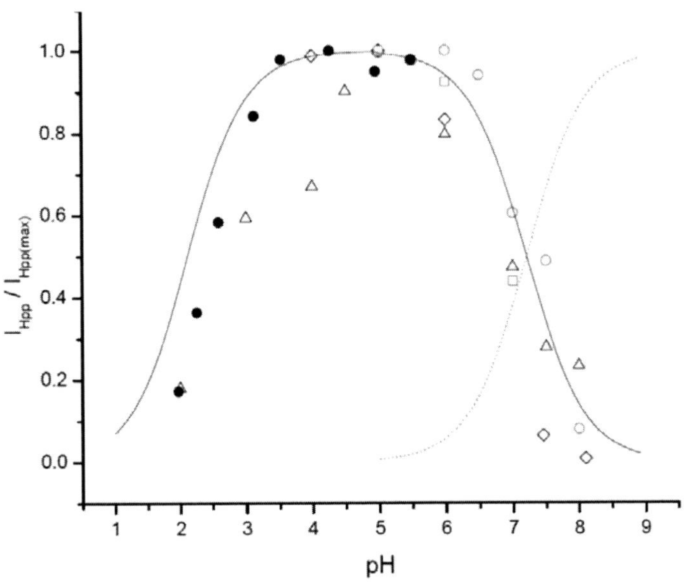


Figure 2. The dependence of the normalized intensities of the H-atom ESR spectrum high-field component on pH for various photosensitizers after UV irradiation at 77 K in 0.5 M solution of $NaH_2PO_4$ in water (●,○ - Ade ($5 \cdot 10^{-4}$ M); ◇ - Trp ($5 \cdot 10^{-4}$ M); △ - NADH ($5 \cdot 10^{-4}$ M); □ - $K_4(CN)_6$ ($5 \cdot 10^{-3}$ M)). Solid and dotted lines show the mole fraction $f$ of $H_2PO_4^-$ and $HPO_4^{2-}$ ions in total phosphate concentration ($f(H_2PO_4^-)+f(HPO_4^{2-}) = 1$). (Microwave power 2 μW, modulation amplitude 1 G. Conditions of UV irradiation (1 kW mercury lamp) as follows: Ade, Trp - λ>240nm; NADH - λ>320nm; $K_4(CN)_6$ – without filter).

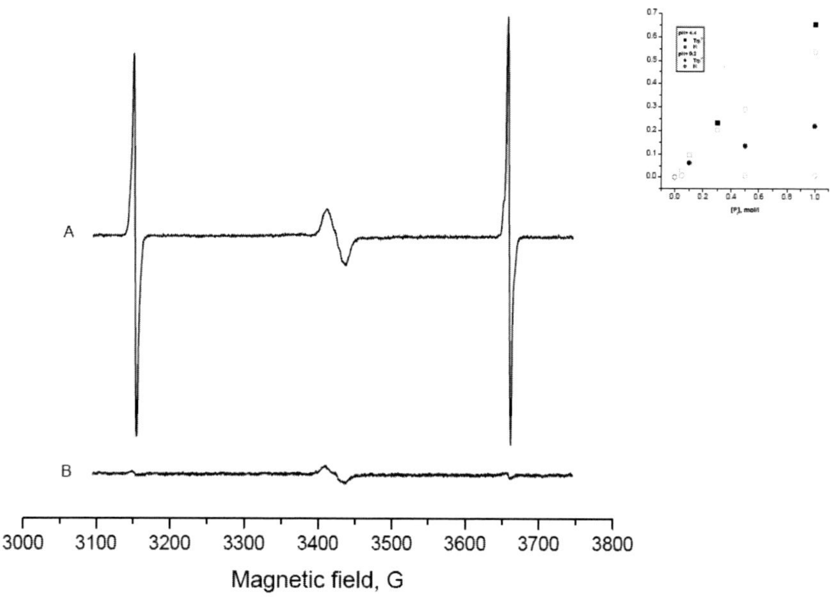


Figure 3. The X-band ESR spectra of 0.3 M phosphate buffer solutions in presence of Trp ($5 \cdot 10^{-4}$ M) after 8 min UV radiation at 77 K for different pH: A – pH = 4.4; B - pH = 9.2. The inset shows the dependence of the double integral of Trp radical and H-atom ESR spectrum components on the phosphate buffer concentration (■, □ - pH = 4,4; ●, ○ - pH = 9,2 accordingly). The conditions of irradiation and recording are the same as in figure 1.

The acceptor properties of phosphate with respect to a photoejected electron were additionally corroborated by the results obtained by the method of competing electron acceptors. These results demonstrated competitive relations between well-known electron scavenger $KNO_3$ forming stable anion and $H_2PO_4^-$, revealed on signal intensity of atomic hydrogen.

The ratio $k_{(e^- + NO_3^-)} / k_{(e^- + PO_4^-)}$ = 340±20 obtained in our experiments agrees well with the value reported by Kevan [16] for ϒ-irradiated phosphates (figure 4) and characterizes different degrees of affinity of $KNO_3$ and phosphate for the electron.

The dependence of the pattern of the ESR spectrum of $e_{aq}$ on the presence of orthophosphate or the electron scavenger $KNO_3$ was also studied in a medium containing the anionic detergent sodium dodecyl sulfate (SDS). The reason is, a more reliable line corresponding to $e_{aq}$ is detected in a micellar structure, rather than in an aqueous solution. Since photoionization in micelles is a sufficiently effective way of generation of $e_{aq}$, this is a good method for a spectroscopic study of chemical reactions, in an aqueous solution, of organic compounds triggered by electron trapping. Photoejected electrons are released into the ambient aqueous medium, and their return into anionic micelles is hampered by the repulsing electrostatic potential. Thus, in microheterogeneous micellar structures, photoproducts are separated via hydrophilic–hydrophobic interactions [19]. Photoejected electrons may be scavenged by dissolved acceptors located on the periphery of micelles. In this case, the line corresponding to $e_{aq}$ disappears, and the spectrum of the products of interaction of the electron with the acceptor appears instead.

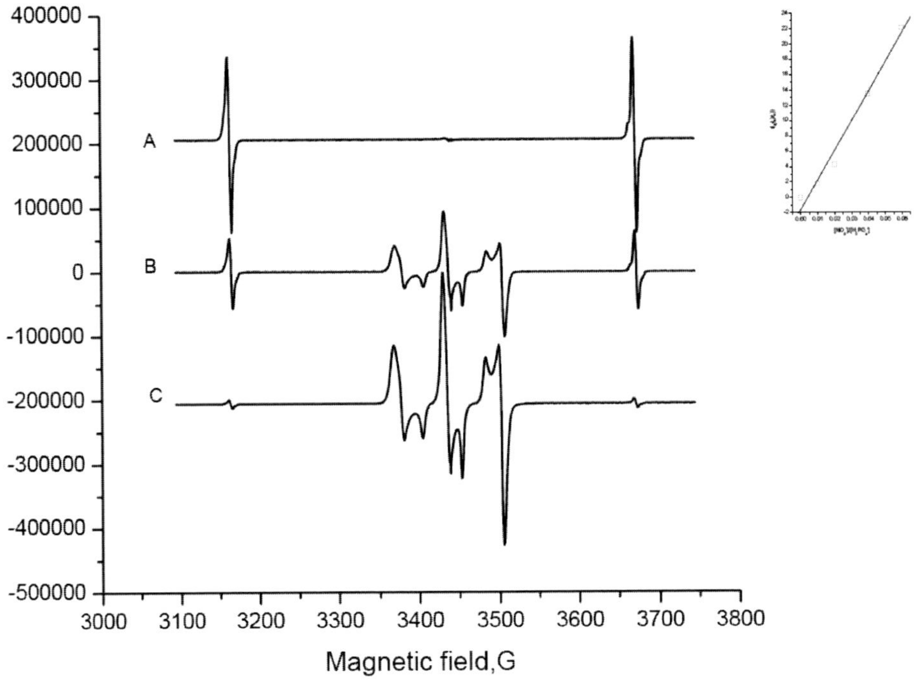


Figure 4. The ESR spectra of 0.5 M solutions of $NaH_2PO_4$ (pH= 4.4) in $H_2O$, containing $5\cdot10^{-3}$ M of $K_4(CN)_6$ and different amounts of $KNO_3$ after 8 min UV radiation at 77 K.(A – 0 M, B – 0.005 M, C – 0.03 M $KNO_3$ accordingly). The insert shows kinetic plot for competitions of electrons between $H_2PO_4^-$ and $NO_3^-$. From this slope the ratio $k_{(e^-+NO_3^-)}/k_{(e^-+PO_4^-)}$ = 340±20 was determined.

Figure 5 shows the EPR spectra obtained as a result of photolysis of frozen samples both in the absence and in presence of phosphate. In our experiments, we recorded the $e_{aq}$ spectrum ($g$= 1.9987,$H$pp= 0.15 G) in a phosphate-free medium. The $g$-factor value obtained in our experiments was somewhat smaller than the $g$-factor value characteristic of a free electron (2.0023); this can be explained by the fact that, similarly to F-centers, electrons in frozen systems are trapped in the field of molecules and ions. Depending on the structure of the system, traps have different characteristics. Using EPR spectroscopy it was shown that the $g$-factor of the electron in the F-center may be 1.995 [20]. This fact indicates that bound electrons in ions contribute markedly to the magnetic moment of the electron contained in the F-center. Therefore, a considerable part of the lifespan of the electron in the F-center passes in the vicinity of ions surrounding the site at which it was trapped.

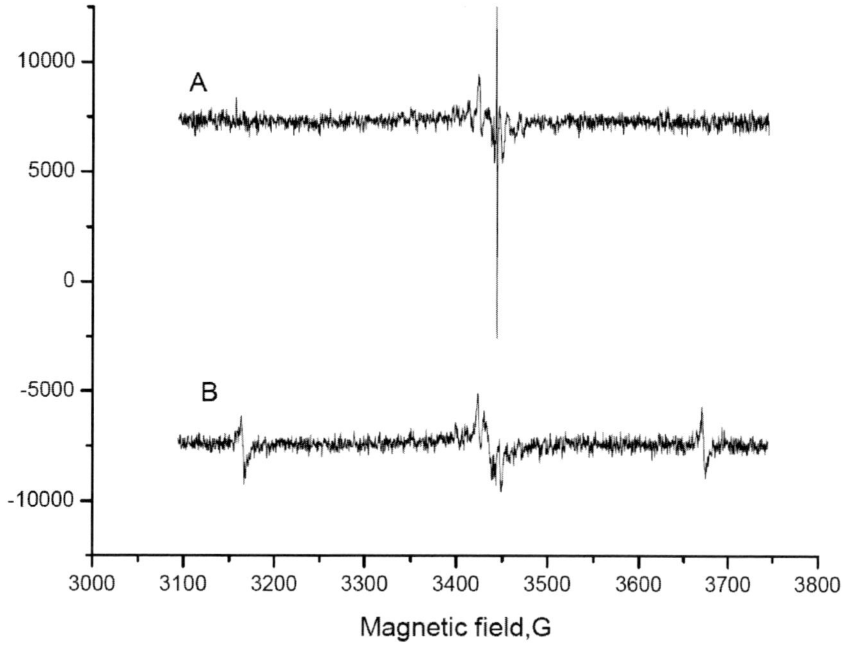


Figure 5. The ESR spectra of sodium acetate ($5\cdot10^{-2}$ M) and SDS ($2\cdot10^{-2}$ M) solutions (pH= 4.9) in water after 8 min UV radiation at 77 K in absence (A) and presence (B) of 0.5 M $NaH_2PO_4$.

The signal of $e_{aq}$ ($g$= 1.9987, $H$pp= 0.15 G) was quenched when both phosphate and $KNO_3$ were added. The disappearance of the line corresponding to $e_{aq}$ was accompanied by the appearance of an EPR spectrum of the product of interaction of $e_{aq}$ with the aforementioned acceptor — either the hydrogen signal or a characteristic triplet, respectively.

In a micellar solution we observed the same dependencies as in our previous study [12], when investigating the intensity of signal from atomic hydrogen in dependence on pH, orthophosphate concentration, and the effect of deuteration led us to assume that the signal from the hydrogen atom characterizes the acceptor interaction of $H_2PO_4^-$ with a photoejected electron.

In this study, in contrast to radiolytic conditions (mobile electron energy up to 20 eV), hydrogen atoms were stabilized under relatively mild conditions (~ 4-4,5eV) when there is no energy transfer from the excited states of a sensitizer to the phosphate molecule. In this case, direct photodissociation of phosphate $H_2PO_4^- \rightarrow H + HPO_4^{2-}$ is excluded (we never saw the ESR signal of $HPO_4^{\cdot -}$, figure 1), and their precursors seem to be a mobile electron trapped by phosphate $e^- + H_2PO_4^- \rightarrow [H_2PO_4^-]^{\cdot} \rightarrow H^{\bullet} + HPO_4^{2-}$ in a dissociative electron attachment DEA.

Note that redox processes involving phosphate may be initiated not only by light or radiation but also by electron transmission in dark processes. For example, dark one-electron reduction of vanadate by ascorbate and NADH and related free-radical generation in a phosphate buffer were investigated by ESR and ESR spin trapping [21, 22]. The vanadate reduction was stimulated by phosphate, the vanadium (IV) yield increasing with increasing phosphate concentration. Addition of formate to the incubation mixture generated the carboxylate radical ($^{\cdot}COO^-$), which indicated the formation of reactive species in the vanadium reduction mechanism. We posit that a phosphate electron adduct revealed by hydrogen loss may be mediating this process. In this system of vanadate reduction by photoejected electrons we observed a decrease in appearance of atomic hydrogen and vanadium in presence of phosphate (figure 6).

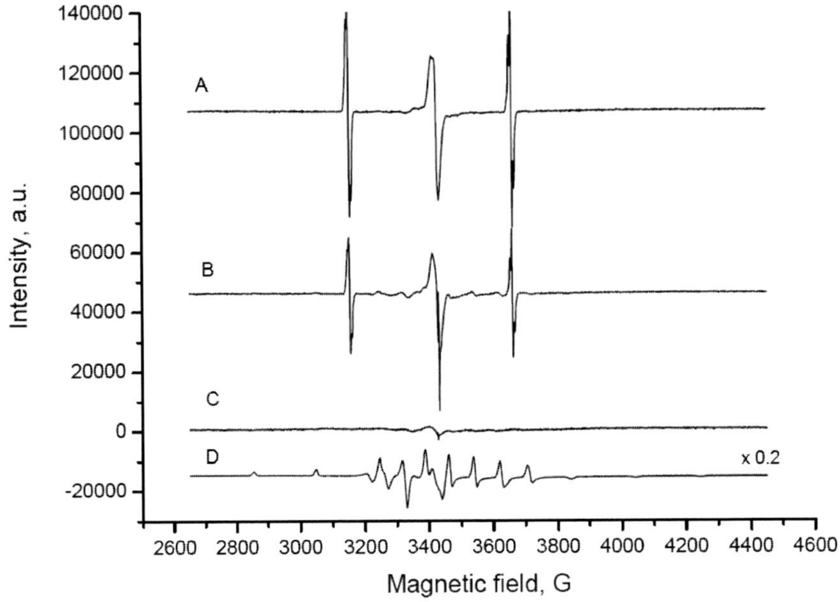


Figure 6. The X-band ESR spectra of $5\cdot10^{-4}$ M solutions of Trp (pH= 4.4) in water, containing: (A - 0.5 M $NaH_2PO_4$ ; B - $5\cdot10^{-4}$ M $VO_3^-$ +0.5 M $NaH_2PO_4$; C - $5\cdot10^{-4}$ M $VO_3^-$) after 8 min UV radiation at 77 K. Fig.6D shows ESR spectrum of the paramagnetic $5\cdot10^{-4}$ M solution of vanadyl $VO^{2+}$ in 0.5 M solution of $NaH_2PO_4$ in water at 77 K.

The interrelations of photoinduced ESR spectra of photoejected electron, the hydrogen atom, and the fluorophore cation radical were suggested to be the basic indicators of the process of electron attachment to $H_2PO_4^-$ and, in some cases, of subsequent interaction of an electron-adduct $[H_2PO_4^-]^{\cdot}$ with secondary electron acceptors (vanadate).

This confims the role of phosphate monoanion and its electron adduct as acceptor-donor intermediates in models of photochemical and dark-electron transport. The ESR method is assumed to visualize the discharge channel of the photoejected electron (or some form of its relaxation to $e_{aq}$) to the dissociative electron attachment $e^- + H_2PO_4^- \rightarrow [H_2PO_4^-]^{\cdot} \rightarrow H^{\cdot} + HPO_4^{2-}$. In the study by Atkins [23], it was assumed that electrons apparently attack phosphate at the electrophilic hydrogen atom, which then leads to its cleavage, the acceptor level being the localized σ* O-H bond. Acceptor interaction of phosphate with electrons in the physiological pH range is of interest for studying the mechanism of many biochemical reactions involving orthophosphate, including the synthesis of ATP, because the highly active intermediates obtained in the interaction are included in the reaction. According to these studies, the possible range of the functional activity of phosphates can be extended to direct involvement in redox reactions.

This process may be important for studying the radiolytic and photolytic chemistry of biological systems, because solvated electrons are the main reagent in these processes. In presence of acceptors converting them into hydrogen atoms, reactions mediated by the latter may be decisive for the observed results. This aspect may also be of interest in studying the effect of low-energy electrons on the DNA damage that results in free-radical dissociation of the C-O-P bond. Our results coincide with the outcomes of the computational work [24], which assumes that the most direct mechanism of single strand breaks occurring in DNA at subexcitation energies (<4eV) is due to resonance electron capture DEA directly to the phosphate group.

## REFERENCES


[1] J.Berdys, I.Anusiewicz, P. Skurski, and J. Simons, *J. Am. Chem. Soc.* 128, 6445 (2004).

[2] Y.V.Vasil'ev, B.J.Figard, V.G.Voinov, D.F. Barofsky, M.L.Deinzer, *J. Am. Chem. Soc.,* 128, 5506 (2004).

[3] M.C.R. Symons *Pure & Appl. Chem.,* 53, 223 (1981).

[4] L.I.Grossweiner, J.F.Baugher, *J.Phys.Chem.* 81, 93(1983).

[5] J.R.Bolton, *Solar Energy* 57, 37 (1996).

[6] L.J.Kevan, *Phys.Chem.*, 70, 2529 (1966).

[7] S.P. Mishra, M.C.R. Symons, *Faraday Discuss.Chem.Soc.*, 63, 175(1977).

[8] O. Ingolfsson, F. Weik, E. Illenberger, *International Journal of Mass Spectrometry and Ion Processes*, 155, 1(1996).

[9] Y.Zheng, P.Cloutier, D.J. Hunting, L.Sanche, J.R.Wagner, *J. Am. Chem. Soc.* 127, 16592 (2003).

[10] B..Liu, P. Hvelplund, S.B.Nielsen, S.Tomita, *J.Chem.Phys.* 121, 4175 (2004).

[11] S.N.Dobryakov, O.N.Brzhevskaya, I.S.Solov'ev, E.M.Sheksheev, O.S.Nedelina, *Dokl. Akad. Nauk*, 384, 119 (2002) [*Dokl. Biochem. Biophys. (Engl. Transl.)*, 384, 136 (2002).]

[12] O.N.Brzhevskaya, E.N Degtyarev, P.P.Levin, T.A.Lozinova, O.S.Nedelina, *Dokl. Biochem. Biophys. (Engl. Transl.)*, 405, 395 (2005) [*Dokl. Akad. Nauk,* 405, 259(2005)].

[13] S.P Mishra, D.J.Nelson, M.C.R.Symons, *Int.J.Radiat.Phys.Chem.*, 7, 581 (1975)

[14] J. Jortner, M. Ottolenghi, J. Rabani, G. Stein, *J. Chem. Phys.*, 37, 2488 (1962).

[15] L. Kevan, P.N. Moorthy, J.J. Weiss, *J.Am.Chem.Soc.*, 86, 771 (1964).

[16] L.Kevan, C.Fine, *J.Am.Chem.Soc.*, 88, 869 (1966).

[17] V.Marinovich, A.Despic, *Electrochimiya,* 33, 1044 (1997) [*Russ.J.Electrochem. (Engl.Trans.)*, 33, 965 (1997)].

[18] V.Marinovich, A.Despic, *Electrochimica Acta,* 44, 4073 (1999).

[19] M.Gratzel, J.K.Thomas, *J.Phys.Chem.* 78, 2248 (1983).

[20] E.J.Hart, M.Anbar. *The hydrated electron,* Wiley-Interscience, New York (1970).

[21] M.Ding, P.M.Ganett, Y.Rojanasakul, K.Lui, X.Shi, J.Inorg.Biochem.55, 101 (1994).

[22] S.Yoshino, G. Sullivan, A.Stern, Arch Biochem Biophys., 272, 76 (1989)

[23] P.W. Atkins, N. Keen, M.C.R. Symons, H.W. Wardale, *J.Chem.Soc.,* 5594 (1963).

[24] C.König, J. Kopyra, I.Bald, E.Illenberger, *Phys. Rev. Lett.* 97, 018105 (2006).

In: Monomers, Oligomers, Polymers, Composites… ISBN: 978-1-60456-877-6
Editors: R. A. Pethrick, G.E. Zaikov et al. 

*Chapter 30*

# BIODEGRADATION OF COMPOSITE MATERIALS ON POLYMER BASE IN SOILS

***O.A. Legonkova****
Moscow State University of Applied Biotechnology, Russia

## ABSTRACT

During incubation of polymer composite materials in soils it was revealed that the structure of composite materials, unsoundness, physical and mechanical properties have changed. The replacement of microorganisms groups with each other in time in the layer bordering to the composite materials was displayed. Durability of composite materials decreases with increasing of surface and volumetric unsoundness of the samples, occurring after incubation in soils. The selectivity of microorganisms' impact on polymer composites was disclosed. The mechanism of fracture of composite materials is suggested.

Polymer materials essentially improve our everyday life, as they are being used in transport, food, agriculture industries. So, the problem of utilization of the great amount of synthetic plastics arises. Creation of composite materials on polymer base with admittedly biodegradable filler could be one of the ways of solving the problem of utilization of synthetic polymers. That's why the investigation of behavior of composite materials and polymer base in different soils was the aim of the present work.

The following polymers were taken as a base for composite materials: co-polymer of acrylic acid and styrene (Lentex), co-polymer of ethylene and vinyl acetate (sevilene) , co-polymer of hexamethylenhydrazine and adipinic acid and sebacic acid (PA), polyurethane (PU), polyvinyl alcohol (PVA). Waste of seed processing and mineral fertilizing (which is a mix of salts - $(NH_4)_2SO_4$, $NH_4H_2PO_4$ , $KNO_3$, $MgSO_4*7H_2O$) were chosen as organic and inorganic fillers, correspondingly. Composite materials contained up to 50% of organic filler and 30% inorganic filler, that depends on the technology of sample getting and, finally, of

* OALegonkovaPB@mail.ru

getting the factory-made foods [1]. Two samples of soils, differing from each other with agrochemical characteristics, were used in the investigation (table 1).

**Table 1. Agrochemical characteristic of soils**

| Samples | humis, % | pH | $H_r$ | $P_2O_5$ | $K_2O$ | $N_2$, % |
|---|---|---|---|---|---|---|
| | | | mg-equivalent/ 100 g of soil | | | |
| Sample 1 | 5,45 | 6,63 | 1,26 | 52,56 | 30,64 | 0,37 |
| Sample 2 | 14,25 | 3,50 | 17,3 | 26,22 | 6,74 | 0,51 |

Durability is considered to be an index, reflecting the total impact of a great amount of different factors on behavior of material in various conditions of exploitation.

Changes, having taken place while incubation of the filled composite materials in soils #1 and #2 during 8 months, are presented in the tables 2 and 3.

**Table 2. Dynamics of changes of durability of polymers while incubation in different soils (speed of tension 10 mm/min)**

| Soils | Polymers | Durability, MPa | | | | |
|---|---|---|---|---|---|---|
| | | Time of incubation, months | | | | |
| | | 0 | 2 | 3,5 | 5 | 8 |
| Soil #1 | PA | 18,8 | 9,3 | 8,5 | 8,2 | 6,7 |
| | PU | 51,3 | 41,3 | 44,3 | 32,2 | 37,8 |
| | PVA | 120,0 | 58,9 | 64,4 | 54,2 | 56,8 |
| | Sevilene | 6,4 | 6,5 | 6,6 | 7,0 | 5,8 |
| | Lentex | 1,5 | 2,2 | 5,2 | 5,7 | 7,8 |
| Soil #2 | PA | 18,8 | 9,3 | 7,3 | 7,1 | 6,5 |
| | PU | 51,3 | 49,0 | 51,4 | 57,2 | 48,6 |
| | PVA | 120,0 | 100 | 92,2 | 80 | 71,4 |
| | Sevilene | 6,4 | 7,4 | 7,0 | 6,8 | 7,2 |
| | Lentex | 1,5 | 2,5 | 5,0 | 5,8 | 6,8 |

As it is shown, the durability of individual polymers (not filled) such as PA, PU and PVA diminishes in time. Durability doesn't change in sevilene and in case of Lentex samples it increases (within the experimental mistake). One of the reason of changes in durability is the state of the surface, its unsoundness. That's why, electron-microscopical pictures of the surfaces of polymers after incubation in soils are presented on the picture 1. On these pictures we can see unsoundness (cracks, deepenings) of the surface of all investigated polymers. It should be noticed that these cracks settle on the surface irregularly, have chaotic character.

There are morphological changes on the Lentex surface, but there are no obvious defects. So, the increase of durability at decreasing of deformation at break can be explained with displaying of relaxation processes that take place under influence of sorbed water [2].

Durability of composite materials based on PA, PU, sevilene decreases. Durability of composite materials based on PVA increase, that can be explained not only with structuring of macromolecules but also with the possibility of arising of ion-coordinating bonds between macromolecules of polymer and metal ions of inorganic filler in the presence of water [3-5].

It should be paid attention to the fact that composite materials based on Lentex fragmented. The sizes of fragments were from 2 mm till 20 mm. And samples for durability investigation were prepared from the generated residues, and it was noticed that the durability of these samples increases in 2 times (from 0,7 till 1,5 MPa, table 2).

**Table 3. Dynamics of changes of durability of composite materials while incubation in different soils (speed of tension 10 mm/min)**

| Soils | Composite materials based on | Durability, MPa | | | | |
|---|---|---|---|---|---|---|
| | | Time of incubation, months | | | | |
| | | 0 | 2 | 3,5 | 5 | 8 |
| Soil #1 | PA | 9,7 | 6,7 | 8,7 | 6,0 | 5,7 |
| | PU | 8,5 | 1,8 | 2,7 | 2,0 | 1,9 |
| | PVA | 4,5 | 12,3 | 6,8 | 17,7 | 10,6 |
| | Sevilene | 1,8 | 0,54 | 0,6 | 0,55 | 0,6 |
| | Lentex | Fragmentation | | | | |
| Soil #2 | PA | 9,7 | 5,6 | 6,3 | 5,2 | 4,5 |
| | PU | 8,5 | 2,1 | 3,0 | 3,1 | 2,9 |
| | PVA | 4,5 | 16,4 | 10,7 | 13,7 | 20,2 |
| | Sevilene | 1,8 | 0,5 | 0,6 | 0,3 | 0,3 |
| | Lentex | Fragmentation | | | | |

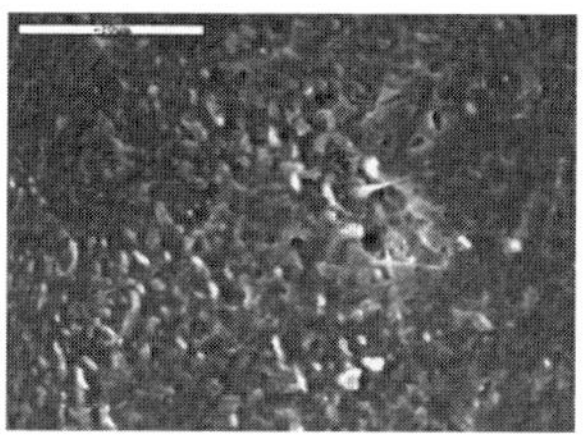

sevilen

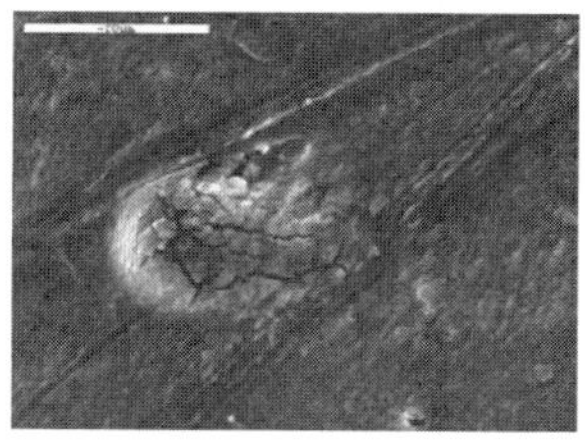

PVA

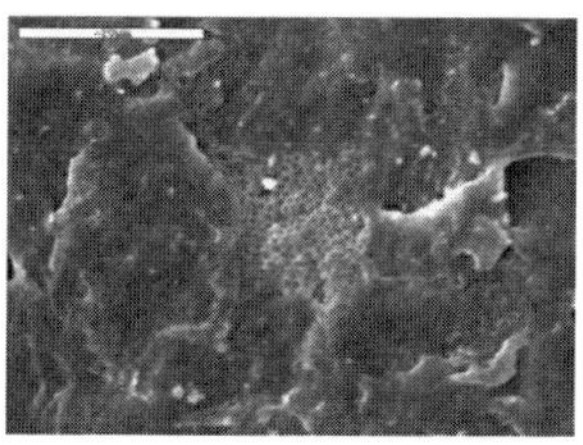

Lentex

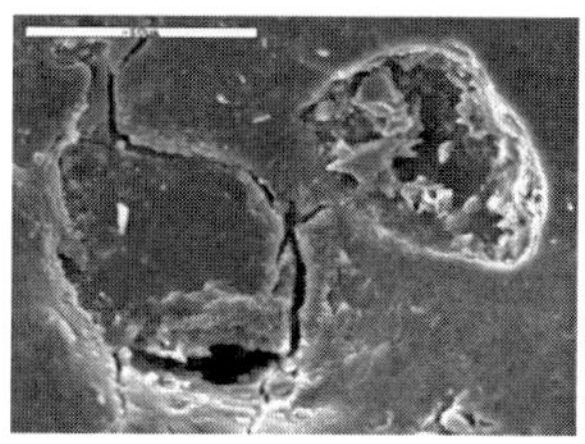

PA

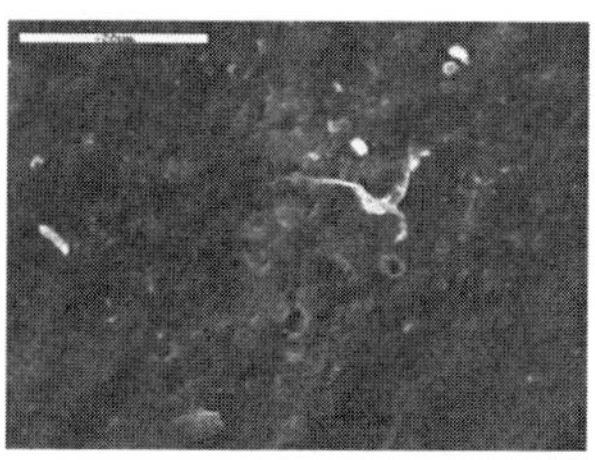

PU

Picture 1. Electron-microscopic photos of the surface of not filled polymers, incubated in soil during 8 months (enlargement x2000).

In order to explain the decreasing of durability changes of composite materials, electron-microscopic photos of the chips of composite materials based on different polymers were got, picture 2. While incubation of composite materials the unsoundness of samples in bulk increases. At the same time the influence of different soils is not so evident.

A. B.

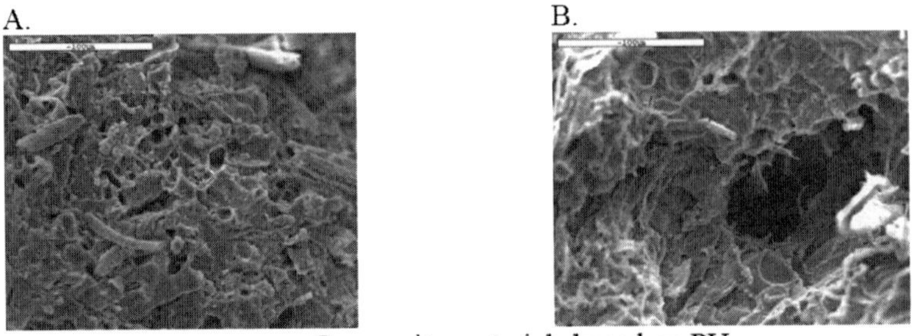

Composite materials based on PU

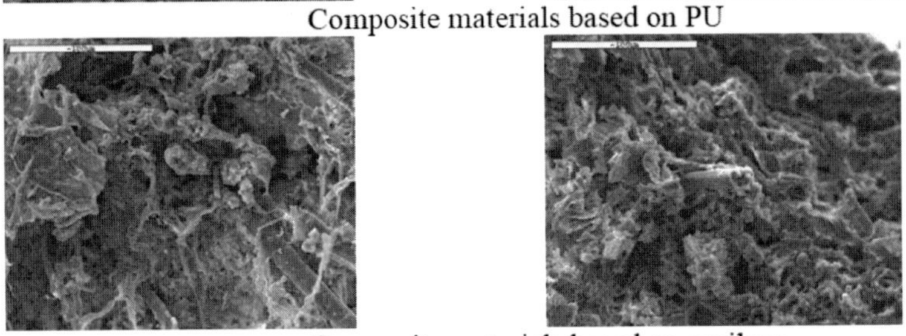

Composite materials based on sevilene

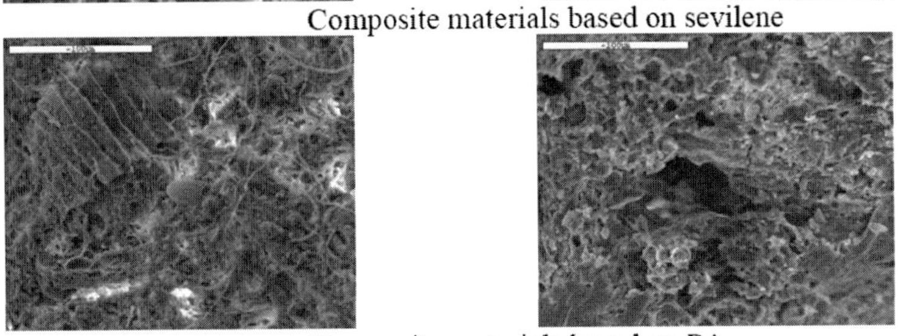

Composite materials based on PA

Pictue 2. Electron-microscopic photos of the chips of composite materials based on different polymers: A - soil #1, B - soil #2 (enlargement x500); during 8 months of incubation.

The results of investigation on permeability changes can be the evidence of increasing of defects in the bulk of composite materials. Thus, the coefficient of permeability of nitrogen gas ($P_{N2}$) through the initial PU samples is $1{,}51*10^{-8}$ $cm^3/(cm^2*c*atm)$. After incubation during 8 months in the soil #1 $P_{N2}$ was $2{,}12*10^{-8}$. In case of the filled sample with PU base, coefficient of permeability was 3,2 and after incubation - 9,69 $cm^3/(cm^2*c*atm)$.

The coefficient of permeability of sevilene samples was $1{,}14*10^{-8}$ $cm^3/(cm^2*c*atm)$. After incubation of these samples in soils the meaning of the coefficient of permeability remained nearly the same ($1{,}20*10^{-8}$). The meaning of $P_{N2}$ for composite material was 29,52 and after incubation this figure came practically to 79,45 $cm^3/(cm^2*c*atm)$.

The received data witness that permeability of individual samples as well as composite materials changes in time during their incubation in soils. However, the permeability of individual samples increases slightly (in 1,1-1,5 times), while the permeability of composite materials increases in 3-4 times. While the share of free volume also increases in the same amount. The analogous data were received when samples based on the other polymer bases were measured before and after incubation. The increase of permeability is the sequence of the extension of porosity of samples and is confirmed with electron-microscopic data, picture.2.

It has to be noticed that the temperature transitions of individual samples after incubation don't change.

As it was shown in the previous works [6], the decrease of durability of polymers is connected with fungi impact. In order to reveal the soil polymer destructor, the surfaces of polymers were covered with fungi, singled out from soil layer, contacted with polymer [7]. The total results are presented in table 4, where the rate of fungi growth is presented with figures: 0 - the investigated material is not a nourishing medium for fungi; 1,2,3 - material contains nourishing substances that promotes negligible growth of fungi; 4,5, - material doesn't resistant to fungi impact and contains nutritious substances promoting fungi growth.

As it is shown, fungi impact on the polymers selectively: the surface of PU became cluttered with Thrihoderma viride, Pen.cyclopium, Pen.chrisogenum, Thrihoderma harsianum, Clonostayis solani; surface of PVA accumulates Fusariium solani, Thrihoderma harsianum, Clonostahys rosea, Ulocladium botrytis, Pen.chrysogenum, Asp.nidulans, Mucor circinelloides, Umbellopsis romanianys; surface of Lentex accumulates Thrihoderma harsianum, Clonostahys solani, Acremonium strictum, Mucor hiemalis; surface of PA accumulates Aspergillius ohraceus, Acremonium strictum, Fusarium solani; Pen. cyclopium, Ulocladium botrytis, Thrihoderma harsianum; surface of sevilene became cluttered with Fusarium solani, Clonostayis rosea, Thrihoderma harsianum, Fusarium sambuciunm, Aspergillius flavous, Mucor hiemalis, Asp. ohraceus .

From biodegradation point of view, the most complex component is synthetic polymer.

According to classical mechanics one of the great amount of reasons of durability decrease is aggressive medium impact, for example, water. But during incubation of samples in soils the durability decreases greater then after enduring them in water: the decrease of durability of PU, sevilene, PA samples is 1,5- 2 times while after incubation in soils durability decreases in 2-2,8 times. So, we can say that durability decrease of incubated samples is reinforced with fungi impact.

Thus, during the carried out investigations it was revealed that polymer surface is exposed to biocorrosion, picture 1. The fungi impact on composite materials is not restricted only with defects on the surface of polymers. The volume changes has taken place during incubation of samples in soils: coefficient of permeability increases in 3-4 times that is connected with biodegradation of organic filler and consumption of inorganic filler. As it was revealed in the work the organic filler (being the organic waste) has fungi that able to evoke biocorrosion of polymer on the inside. So, biodegradation of polymer filler can weaken polymer matrix.

As water doesn't change the mechanism of polymer destruction, its main role in biodamaging of composite materials is found in being a nutritious medium for fungi growth.

As fungi accumulate on the polymer surface irregularly, under the law of chaos, the porosity increase of composite materials promotes the fungi adhesion on the inner side, their adaptation and growth in volume.

In order to force plastic to biodegradation it's necessary to fracture it on small parts capable to assimilate in the environment. Creation of composite materials with biodegradable filler helps to solve the task of fracture of material entirety and, finally, fragmentation.

Biodegradation of composite materials on polymer base with biodegradable fillers under impact of soil fungi consists of the following stages: surface biocorrosion, increase of porosity, biodegradation of filler and inner biocorrosion (due to fungi adhesion on inner roughnesses), spreading of biocorrosion, fragmentation.

**Table 4. Estimation of fungi impact on polymer materials (GOST 9.049-91)**

| Fungi | Days | PU | PVA | Lentex | PA | Sevilene |
|---|---|---|---|---|---|---|
| Pen. | 7 | 2 | | | 4 | |
| cyclopium | 14 | 3 | | | 4 | |
| | 21 | 4 | | | 4 | |
| Pen. | 7 | 3 | 3 | 1 | | |
| chrysogenum | 14 | 3 | 4 | 2 | | |
| | 21 | 4 | 4 | 2 | | |
| Thrihoderma viride | 7 | 4 | | | | |
| | 14 | 4 | | | | |
| | 21 | 5 | | | | |
| Thrihoderma | 7 | 4 | 5 | 2 | | 4 |
| harsianum | 14 | 4 | 5 | 2 | | 5 |
| | 21 | 5 | 5 | 2 | | 5 |
| Clonostahis | 7 | 3 | | 2 | | |
| solani | 14 | 3 | | 3 | | |
| | 21 | 4 | | 3 | | |
| Fusarium | 7 | | 5 | | 5 | 5 |
| solani | 14 | | 5 | | 5 | 5 |
| | 21 | | 5 | | 5 | 5 |
| Clonostahis | 7 | | 4 | | | 3 |
| rosea | 14 | | 5 | | | 3 |
| | 21 | | 5 | | | 4 |
| Ulocladium | 7 | | 4 | 2 | 3 | |
| botritis | 14 | | 4 | 2 | 4 | |
| | 21 | | 4 | 2 | 3 | |
| Aspergillius | 7 | | 3 | | | |
| nidulans | 14 | | 4 | | | |
| | 21 | | 4 | | | |
| Mucor circinelloides | 7 | | 4 | | | |
| | 14 | | 4 | | | |
| | 21 | | 5 | | | |
| Umbellopsis | 7 | | 3 | | | |
| romanianys | 14 | | 4 | | | |
| | 21 | | 4 | | | |
| Aspergillius ohraceus | 7 | | | 2 | 4 | 3 |
| | 14 | | | 2 | 4 | 4 |
| | 21 | | | 2 | 5 | 3 |
| Mucor hiemalis | 7 | | | 2 | 4 | 3 |
| | 14 | | | 3 | 3 | 3 |
| | 21 | | | 3 | 3 | 2 |
| Acrmonium | 7 | | | | 3 | |
| strictum | 14 | | | | 3 | |
| | 21 | | | | 3 | |
| Fusarium | 7 | | | | | 5 |
| sambucinum | 14 | | | | | 5 |
| | 21 | | | | | 5 |
| Aspergillius | 7 | | | | | 4 |
| flavous | 14 | | | | | 4 |
| | 21 | | | | | 4 |

## References


[1] Patent #2257045, RF. Nutritive composition for growing of seedings.

[2] O.A.Legonkova, A.A.Bokarev, V.S.Ivolgin. Swelling of Filled Polymer Compositions. *J. of Balkan Tribological Association,* 2007, v.13, #1, pp.67-72.

[3] Lipatov J.S. Polymer composite materials. Kiev, *"Znanie",* 1979, 60 s.

[4] Lipatov J.S. Colloid chemistry of polymers. Kiev, *"Naykova Dumka",* 1984,340s.

[5] Manson J., Sperling L. Polymer mixtures and composites, Moscow, *Chemistry,* 1979, 430s.

[6] Дж. Мэнсон, Л.Сперлинг. Полимерные смеси и композиты. М.: *Химия.* 1979. 430 с.

[7] V.Torsvi, J. Goksoryl, F.L. Daae, R.Sorheim, J.Michalsen and R. Salte in Beyond the Biomass: Compositional and Functional Analysis of Soil Microbial Communities, Eds., R.Ritz, J.Dighton and K.E.Gille, Wiley, London, UK, 1994, 39 p.

[8] O.A.Legonkova, O.V.Selitskaya. Behaviour of composite materials under microorganism of soil. *J. of Appl. Polym. Sci.,* 2007, v.105, #6.

In: Monomers, Oligomers, Polymers, Composites...
Editors: R. A. Pethrick, G.E. Zaikov et al.
ISBN: 978-1-60456-877-6


*Chapter 31*

# POLYMER-COLLOID COMPLEXES BASED ON CHITOSAN AND THEIR COMPUTER MODELING

***Y.P. Ioshchenko*, V.F. Kablov* and G.E. Zaikov*****

* Volzhsky Politechnical Instititute (branch) of Volgograd State Technical University; 42a Engels St., Volzhsky 404121, Volgograd Region

**N.M. Emanuel Institute of Biochemical Physics, Russian Academy of Science; 4 Kosygin St., Moscow 119991

## ABSTRACT

Polymer-colloid complexes based on biopolymers and synthetic polymers were studied. The computer modeling of the conformational and geometrical properties of these complexes and their macromolecules was developed. Thermodynamic properties of the polymer-colloid complexes were defined; the most advantageous, energetical their positions were shown. Using the multifactor modeling the behavioral patters of overcoats based on the complexes and their fire-resisting and heat-protection properties were determined in different operational conditions. It is possible to receive different sorbents, coatings and films with increased efficacy basing on the polymer-colloid complexes.

**Keywords:** polymers, polymer-colloid complexes, chitosan, computer modeling, thermodynamics, multifactor modeling.

## INTRODUCTION

Methods of computer characteristics' modeling are rather informative and effective because they allow to fasten the process of the determination of the row physical characteristics of the complicated chemical construction polymers and to define such characteristics which are rather difficult to determine during the experiments [1-3]. Chitin, chitosan, lactoserum proteins, gelatin and etc. can be referred to such polymers [4-8]. Properties' research by the means of computer modeling of the polymer-colloid complexes based on chitosan with various synthetic and biopolymers are also rather interesting. The computer modeling allows to build not only the volume model representation of these complexes, but also to understand their structurization peculiarities and to define by what means happens their creation.

## Materials and Methods

In this work were studied biopolymers - chitosan, gelatin, lactoserum protein, gelatin; synthetic polymers - polyvinyl alcohol, methylcellulose [9], and their polymer-colloid complexes such as:

- chitosan-polyvinyl alcohol (ChS-PVA);
- chitosan-methylcellulose (ChS-MC);
- chitosan-lactoserum protein (ChS-LP);
- chitosan-gelatin (ChS-G);

The following volume characteristics of studied macromolecules [10] were calculated: increments of different atoms and their main groups, cohesive energy, density of cohesive energy, considering overmolecular polymer structure; surface tension was defined basing on chemical structure of substances and considering the contribution of every molecular group and specific molecular interaction and etc. [1].

Thermodynamic characteristics of the polymer-colloid complexes: enthalpy, entropy, Gibbs energy and equilibrium constant were defined by given thermodynamic functions of initial components. It was necessary to estimate the possibility of the substances' participation in chemical process. Thermodynamic calculation was made for one link of a macromolecule using semiexperimental methods of calculation [11].

Thermodynamic substances functions: $\Delta H^0_{298}$, $\Delta G^0_{298}$, $C^0_p$ were determined by the given chemical structure. Equilibrium constant was defined by Gibbs equation: $\ln(K_{eq}) = \frac{-\Delta G}{RT}$ [12].

Physical-chemical characteristic quantities of the studied polymer molecules that took part in complex construction were given to create their volume models, later these characteristic quantities were optimized, and it led to the complex structurization.

Using the multifactor modeling and «Teplo» [13] software the evaluation of thermophysical and heat protection properties of overcoats based on the polymer-colloid complexes were carried out, considering the thermophysical characteristics of the complexes and heating conditions, and accounting for the physical-chemical transformations in the coating layer (pyrolysis, bursting expansion, and cavitation).

## Results and Discussion

The following volume characteristics of the macromolecules were calculated: the average distance between molecule ends (*h*) which characterizes the reactivity of the macromolecule during flocculation and sorption, and the hydrodynamic volume ($V_M^{\Gamma}$) taken up by the macromolecule mass unit and determining the total size of the macromolecule [10]. The calculated values of the main macromolecule characteristics are given in table 1.

**Table 1. Main macromolecule characteristics**

| Macromolecules | $V_M^{\Gamma}$, $nm^3$ | $h$, $10^6$, $cm$ |
|---|---|---|
| Chitosan | 6,5 | 4,5 |
| Methylcellulose | 11,5 | 4,2 |
| Lactoserum protein | 0,2 | 2,0 |
| Gelatin | 2,3 | 2,1 |
| PVA | 1,3 | 3,5 |

During the analysis of the volume characteristics it was defined that the distance between the end groups in the macromolecules of chitosan, methylcellulose and PVA was 2 to 4 times greater than that in the molecules of gelatin and lactoserum protein, which indicates a higher reactivity and activity of their end groups in the course of sorption and flocculation.

The hydrodynamic volume in the chitosan and methylcellulose molecules is considerably higher than the volume of the other macromolecules under study, which exhibits their more unfolded and volume structure with the functional groups being more readily available for intermolecular interaction.

The properties of the fragments of polymer molecular structures were calculated by means of computer modeling (table 2).

**Table 2. The calculated properties of the fragments of polymer molecular structures**

| Properties | Chitosan | PVA | MC | Gelatin | LP |
|---|---|---|---|---|---|
| Van der Vaals volume, $\sum_i \Delta V_i$, $\overset{0}{A}{}^3$ | 132,4 | 41,5 | 152,0 | 44,5 | 57,5 |
| Density, $\rho$, $g/cm^3$ | 1,40 | 1,35 | 1,41 | 1,60 | 1,06 |
| Cohesive energy, $\sum_i \Delta E_i^*$, J/mole | 56272 | 22421 | 89870 | 28259 | 37737 |
| Density of cohesive energy $\delta$, $(J/cm^3)^{1/2}$ | 26,6 | 30,0 | 52,9 | 32,5 | 33,0 |
| Surface tension, $\gamma_n$, $din/cm$ | 36,6 | 46,2 | 46,4 | 49,5 | 51,0 |

All these characteristics allow estimating the polymers' combination with each other and solution of polymers in different organic dissolvents. Solubility criterion for system polymer 1: polymer 2 is the maintenance of correlation $\gamma_{p1} < \gamma_{p2}$. As it shown in the table 2, when you compare different systems like chitosan:polymer (PVA, LP and etc.), the solubility criterion fulfills and it proves the possibility of the complex structurization between these polymer molecules.

Thermodynamic characteristics of the complex structurization between chitosan and proteins (table 3) showed that total heat (ΔH) decreased during their synthesis. This fact can be explained by the more advantageous conformations of initial substances in the polymer-colloid complex. These conformations are caused by the creation of ion-coordinative centers of interaction between positive charged protein amino groups of chitosan $\sim NH_3^+$ and carboxyl anion - which is an amino acid remainder of protein $\sim COO^-$. Exothermic reactions that give 21,9 и 23,4 *kJ/mole* take place during of chitosan-gelatin and chitosan-lactoserum protein complex structurization. This proves a higher stability of chitosan-lactoserum complex.

Chitosan-methylcellulose complex characterize by increased thermodynamic stability, caused by the considerable amount of coordinative interactions.

During chitosan-PVA complex structurization intermolecular interaction happens intermolecular interaction, caused by the hydrogen connection between chitosan ($\sim NH_2$ and ~OH) and PVA polar groups (~OH). The structurization of this complex can be characterized by lower Gibbs energy comparing with chitosan-methylcellulose complex. So this complex is the less stable one.

**Table 3. Thermodynamic characteristics data of the polymers and their polymer-colloid complexes**

| Polymers and complexes | $\Delta H^0_{298}$, kJ/mole | $\Delta S^0_{298}$, J/mole ·K | $\Delta G^0_{298}$, kJ/mole | $C^0_p$, J/mole ·K | $K_{eq}$ |
|---|---|---|---|---|---|
| Chitosan | -823 | 242 | -896 | 54 | - |
| Methylcellulose | -1049 | 257 | -1126 | 73 | - |
| Gelatin | -601 | 72 | -629 | 65 | - |
| Lactoserum protein | -452 | 72 | -474 | 65 | - |
| PVA | -835 | 24 | -842 | 59 | - |
| ChS- MC | -1889 | 506 | -2039 | 145 | 1389 |
| ChS-G | -1447 | 329 | -1545 | 127 | 40459 |
| ChS-LP | -1299 | 330 | -1398 | 133 | 87026 |
| ChS-PVA | -1680 | 283 | -1765 | 123 | 52847 |

Representation of polymer-colloid complexes received by the methods of computer modeling allows direct viewing their construction with the metal ions that they sorb.

Figure 1 shows us that the metal ions sorption takes in different cavities of the polymer-colloid complex. The size of the cavities in the complexes is much larger than the size of the ions of the sorbed metals (table 4) [14]; the chemical metal ions sorption in the macromolecular cavities makes them more resistant to the retention of metal ions, whereas the considerable mass of the metal complex can lead to particle sedimentation.

**Table 4. Geometric characteristics of the complexes and the metal ions that they absorb**

| | |
|---|---|
| The size of particle complexes | 80-130 nm |
| The size of the metal ion capture cavity | 5-9 nm |
| The size of metal ions: | $Cu^{2+}$=0,071 nm; $Fe^{3+}$=0,063 nm; $Zn^{2+}$=0,083 nm; $Cd^{2+}$=0,092 nm; $Ni^{2+}$=0,069 nm. |

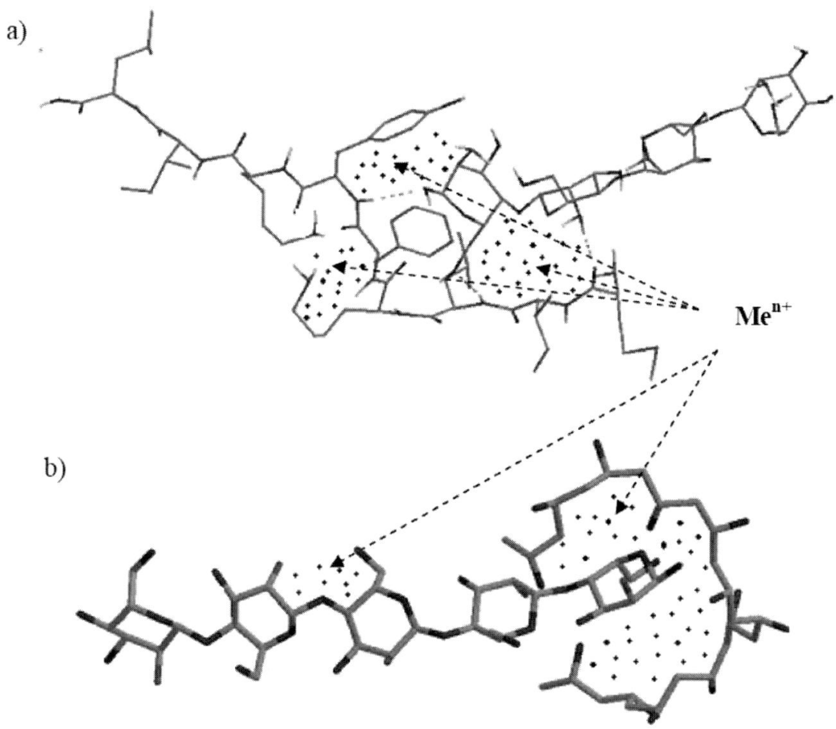


Figure 1. The structures of the fragments of chitosan-lactoserum protein (a) and chitosan-PVA (b) complexes.

The scheme of ion metals sorption on functional groups of chitosan-lactoserum protein complex is shown on the figure 2:

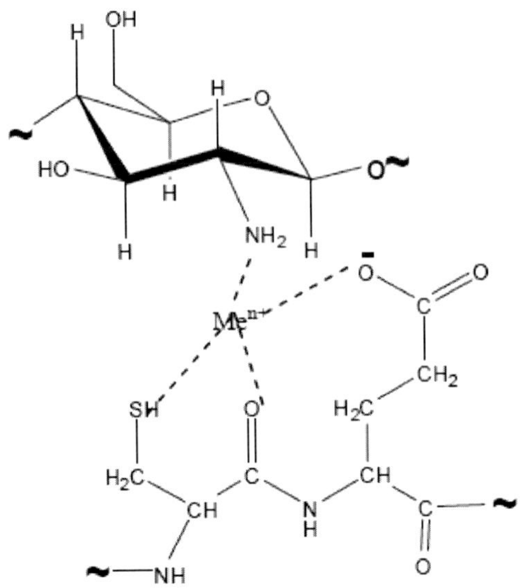


Figure 2. The schematic representation of the fragments of chitosan-lactoserum protein complex.

Using a multifactor model for the heating of fire- and heat-protection materials and software «Teplo» [13], the modeling of the fire-resisting and heat-protection properties of the polymer complex-based coatings was carried out, considering the thermophysical characteristics of the complexes and heating conditions, and accounting for the physical-chemical transformations in the coating layer (pyrolysis, bursting expansion, and cavitation). The software is based on the automodel mode of heating. The coatings of this model consist of swelling coke and elastic material. As a result of bursting expansion, the working surface moves to the source of heat, and the pyrolysis' border spreads deeply into the material. So, the partitioning of temperatures is determined by two mutual-connected, variable magnitudes: thickness and time. By these means the automodel mode is realized.

It was shown that the most significant influence on the operating characteristics was produced by the bursting expansion effect, coke formation and the amount of fixed water. Thus, at the water retention level in the coating equal to 20 % the aggregate endothermic effect can rise by over 30 %. Changing the thermal effect of pyrolysis makes it possible to change the coating's thickness which is necessary to heat the material. So, the thermal effect of pyrolysis can be increased more than 1,3 times due to the rising of water's content [15].

Upon data processing, the following characteristics of water-free polymer complex-based coatings and coatings with fixed water were determined (table 5):

- structure breaking layer ($\delta_D$) and heated layer ($\delta_h$) thicknesses;
- necessary coating thickness for heat ($\delta$), where $\delta = \delta_D + \delta_h$;
- efficiency parameter ($L$) considering the destruction rate ($V_g$) and material density ($\rho$), where $L = \dfrac{1}{V_g \cdot \rho}$, $V_g = \xi_z / \sqrt{t}$; $\xi_z$ is a parameter connected with thermal conductivity, bursting expansion strain, and pyrolysis-driven thermal effect; $t$ – time of exposure, s.

Data analysis shows that the most effective coating is chitosan-gelatin polymer-colloid complex. The relation between the coating thickness $\delta$ and the bursting expansion strain $\varepsilon_в$ in chitosan-gelatin complex is given in figure 3 for different thermal effects of pyrolysis $Q$ and material densities $\rho$.

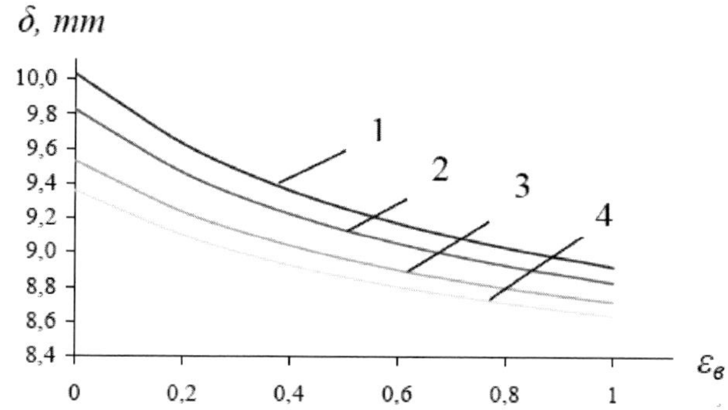


Figure 3. Relation between the coating thickness $\delta$ and the bursting expansion strain $\varepsilon_в$ in ChS-G complex: with water-free - 1, 2; with fixed water – 3, 4: 1. $Q$=980 kJ/kg, $\rho$=1,45 kg/m$^3$; 2. $Q$=1 261kJ/kg, $\rho$=1,45 kg/m$^3$; 3. $Q$=980 kJ/kg, $\rho$=1,39 kg/m$^3$; 4. $Q$=1 261 kJ/kg, $\rho$=1,39 kg/m$^3$.

**Table 5. Thermophysical characteristics of water-free polymer complex-based coatings and coatings with fixed water**

| | Magnitudes | Polymer-colloid complexes | | | |
|---|---|---|---|---|---|
| | | ChS- MC | ChS-LP | ChS- G | ChS-PVA |
| Water-free coatings | Q, kJ/kg | 950 | 980 | 970 | 1000 |
| | ρ, g/sm$^3$ | 1,35 | 1,22 | 1,45 | 1,32 |
| | $\delta_д$, mm | 3,4 | 3,3 | 3,0 | 3,0 |
| | δ, mm | 9,4 | 9,7 | 8,9 | 9,6 |
| | L, sm$^2$·s/g | 27,2 | 30,1 | 33 | 30,2 |
| Coatings with fixed water | Q, kJ/kg | 1235 | 1274 | 1261 | 1300 |
| | ρ, g/sm$^3$ | 1,23 | 1,02 | 1,39 | 1,22 |
| | $\delta_д$, mm | 2,8 | 2,3 | 2,3 | 2,4 |
| | δ, mm | 9,2 | 8,6 | 8,6 | 9,1 |
| | L, sm$^2$·s/g | 34,5 | 36,2 | 36,2 | 41,7 |

## CONCLUSIONS

1. The computer modeling of the conformational and geometrical properties of the macromolecules and polymer-colloid complexes based on biopolymers and synthetic polymers was developed. It was shown that the process of metal ions and organic contextures sorption takes place in macromolecular cavities of these complexes.
2. Thermodynamic properties of the polymer-colloid complexes were defined; the most advantageous, energetical positions were shown.
3. Using the multifactor modeling the behavioral patters of overcoats based on the complexes and their fire-resisting and heat-protection properties were determined in

different operational conditions. Thanks to this, it is possible to receive different sorbents, coatings and films based on the polymer-colloid complexes with increased efficacy.

4. Materials based on the polymer-colloid complexes can be used as fire-resisting and heat- protection materials, especially in conditions of increased requests to the toxic components.

## REFERENCES


[1] A.A. Askadsky, V.I. Kondrashchenko. Computer-based Polymer Material Study. Vol. 1. Atomic and Molecular Level. – Moscow: *World of Science,* 1999. 544 p.

[2] O.V. Solovyov, M.M. Solovyov. Computer Chemistry. – Moscow: Solon-Press, 2005. 536 p.

[3] V.I. Vershinin, B.G. Derendyayev, K.S. Lebedev. Computer Identification of Organic Compounds. – Moscow: *Akademkniga,* 2002. 197 p.

[4] Chitin and Chitosan: Production, Properties and Application / Edited by K.G. Skryabin, G.A. Vikhoreva, V.P. Varlamova. – Moscow: *Nauka,* 2002. 368 p.

[5] B.E. Geller, A.A. Geller, V.G. Chergilov. Manual of the physic-chemical fiber-forming polymers: Textbook for Chem. Universities. - Moscow: *Khimiya,* 1996. 432 p.

[6] A.G. Khramtcov. Lactoserum protein. - Moscow: *Agropromizdat,* 1990. 240 p.

[7] O.V. Mikhaiylov. Gelatin and immobilized metal complexes. - Moscow: *World of Science,* 2004. 236 p.

[8] A.V. Finkelshtein, O.B. Ptitcin. Protein physics: lecture course with stereoscopic illustrations. - Moscow: Publishing house *"Universitet",* 2002. 376 p.

[9] Encyclopedia of polymers // Edited by V.A. Kabanov, et al. 3 volumes. Moscow: *«Soviet Encylopedia»,* 1974. 1032 p.

[10] T.V. Shevchenko, V.L. Osadchy, M.A. Yakovchenko, E.V. Ulrikh. // *Chemical Industry Today.* 2004. – No. 11. p. 38-41.

[11] A.G. Stromberg, D.P. Semchenko. Physical chemistry: Textbook for Chem. Universities. - Moscow: *Vysshaya shkola,* 2001. 527 p.

[12] V.F. Kablov, Y.P. Ioshchenko, D.A. Kondrutsky. // *Vestnik MITHT,* 2006. – Vol. 1 – No. 5. p. 49-53.

[13] V.F. Kablov. //*Resin and Rubber.* – 1997. – No. 1. p. 8-10.

[14] Chemical encyclopedia // Edited by I.L. Knunyants, et al. 5 volumes. Moscow: *«Soviet Encylopedia»,* 1988. 623 p.

[15] Y.P. Ioshchenko. Preparation and study of chitosan polymer complexes with proteins and hydroxylous polymers // *Thesis for a Degree of Candidate of Technical Science.* – Volgograd, 2006. 119 pgs.

# INDEX

## A

## B

## C

## D

## E

## F

## I

## J



## K

## P

## Q

## R

## S

## T